Mechanism Design

This updated edition of *Mechanism Design: Visual and Programmable Approaches using MATLAB® and Simscape* Multibody™ offers a comprehensive introduction to kinematic synthesis, covering motion, path, and function generation techniques for a wide range of planar and spatial single- and multi-loop linkage systems.

This book presents foundational concepts alongside practical methodologies, making it an accessible resource for both students and practitioners in the field. In this revised edition, real-world application of the presented methods is supported through the integration of MATLAB® and its powerful simulation and visualization toolbox, Simscape Multibody™. These tools help bridge theory and practice, allowing readers to implement kinematic synthesis techniques and observe system behavior through dynamic visualizations. New content expands this book's scope, including topics such as geared five-bar kinematic synthesis and both forward and inverse kinematics for robotic systems.

Designed as a complete introduction to kinematic synthesis, this book is an essential resource for students in mechanical engineering and related disciplines seeking to master the principles and practicalities of mechanism design.

The new edition also includes a solution manual and MATLAB as an online resource for instructors to support the topics discussed in this book.

Mechanism Design

Visual and Programmable Approaches using MATLAB® and Simscape Multibody™

Second Edition

Kevin Russell, Qiong "John" Shen,
and Raj S. Sodhi

CRC Press
Taylor & Francis Group
Boca Raton London New York

CRC Press is an imprint of the
Taylor & Francis Group, an **informa** business

Designed cover image: Qiong "John" Shen

Second edition published 2026
by CRC Press
2385 NW Executive Center Drive, Suite 320, Boca Raton FL 33431

and by CRC Press
4 Park Square, Milton Park, Abingdon, Oxon, OX14 4RN

CRC Press is an imprint of Taylor & Francis Group, LLC

First edition published by CRC Press 2014

ISBN: 9781032956800 (hbk)
ISBN: 9781032969404 (pbk)
ISBN: 9781003591375 (ebk)

DOI: 10.1201/9781003591375

Typeset in Times
by codeMantra

Access the Instructor and Student Resources: https://www.routledge.com/9781032956800

In memory of Willie J. Russell and Ella Russell.

Kevin Russell

To my wife Lili Qiu, my daughters Catherine Shen and Christine Shen who have been supporting me in all aspects so that I can bring forth the exciting new contents using future-proof technologies.

Qiong "John" Shen

This book is affectionately dedicated to my wife Rani, my daughters Preeti and Shelly and my son-in-law, Andrew.

Raj S. Sodhi

Contents

Preface

In *dimensional synthesis*, mechanism dimensions are determined to achieve the prescribed output motion. Current kinematics books devote only a few sections or a chapter to dimensional synthesis, which is not enough to provide a comprehensive introduction. Many books present *qualitative* methods for dimensional synthesis, which are based on hand drafting techniques. Although qualitative methods are easy to teach and implement, they are limited to two-dimensional, small-scale problems. In recent times, personal computers and mathematical modeling software have facilitated the design of complex systems. Computation-based or *quantitative* dimensional synthesis methods are far more robust in terms of the variety and scale of problems that can be considered. Also, the computing technology available today, both hardware and software, makes quantitative methods very practical.

This book, which is dedicated entirely to dimensional synthesis, is based on our teaching, research, and industry experience in mechanism design. Its purpose is to give the reader an introduction to dimensional synthesis that is both broad and practical. This book presents applications of *planar* and *spatial* motion, path, and function generation methodologies (both classical and modern). Additional topics include *number synthesis*, synthesis using *finite and multiply separated positions*, the synthesis of *adjustable mechanisms*, and the *forward* and *inverse kinematics* of robotic manipulators.

To improve both the practicality of the concepts covered as well as the clarity in their presentation, the second edition of *Mechanism Design: Visual and Programmable Approaches using MATLAB® and Simscape Multibody*™ includes the following updates:

- **the integration of MATLAB and its latest mechanical simulation toolbox: Simscape Multibody**

  In the prior edition of the authors' book, MATLAB's former mechanical simulation toolbox *Simmechanics* was integrated throughout its chapters and appendices. Considering the release of MATLAB's most recent mechanical simulation toolbox, *Simscape Multibody*™, it has been integrated throughout the chapters and appendices in the second edition. In addition, a new library of MATLAB, and now Simscape Multibody, files have been produced and are available for download with this book. These files are available for download at the publisher's web page https://www.routledge.com/9781032956800 and click the *Downloads* tab. These files can also be downloaded from the authors' server http://www.softalink.com/kruss/mech_des_vis_prog_app_2nd_ed/MATLAB_SIMMULT_FILES.zip. This new content improves clarity for the reader regarding the preparation of book problems for use in MATLAB and Simscape Multibody as well as maintain book compatibility with the latest MATLAB releases.
- **the inclusion of the Planar Geared Five-bar Mechanism**

  This book now includes the kinematic analysis and synthesis of the *Geared Five-bar Mechanism*. Models for the kinematic analysis and synthesis of this mechanism are formulated here. Also, MATLAB and Simscape Multibody files are included in this textbook for geared five-bar mechanism kinematic analysis, motion generation, and path generation.
- **the inclusion of Robotic Manipulator Forward and Inverse Kinematics**

  This book now includes robotic manipulator forward and inverse kinematic analyses. The industrial robotic manipulators included here are the *Cartesian P-P-P*, *Cylindrical R-P-P*, *Spherical R-R-P*, *Articulated R-R-R*, and *SCARA R-R-C*. Models for their kinematic analysis and inverse kinematics are formulated here. Also, MATLAB and Simscape Multibody files are included in this book for the forward and inverse kinematic analyses of these robotic systems.

Because this work is dedicated to dimensional synthesis—a subdiscipline in kinematics—it is hoped that the reader would have been exposed to the content covered in a typical undergraduate course in mechanism kinematics. Also, this book was written to accommodate readers with no working knowledge of MATLAB. In terms of MATLAB knowledge, the ideal user should know how to launch MATLAB and have access to the MATLAB software package itself. Any version of MATLAB post-2020 is suitable to run the MATLAB and Simscape Multibody files associated with this book (provided that all the required toolkits listed in Appendix A.1 are installed).

The reader should note that the library of MATLAB nonlinear optimization models that accompany this textbook can produce different results (given the same input) depending on which version of MATLAB is used. MATLAB version 2021b for Windows was used for all in-chapter examples and post-chapter problems in this book.

Dimensional synthesis is a *creative* means to produce mechanism solutions. There is an endless variety of possible mechanism solutions that the user is free to explore, realize, and evaluate for any given problem. Even though this work provides a variety of synthesis examples and end-of-chapter synthesis problems, the user can invent their own specialized synthesis problems according to their interests. The resources in this book help make this possible. The book can be used

- as a text for an undergraduate or graduate course in advanced (computer-based) kinematics
- as a reference text for mechanical engineering research
- as a reference text for the application of MATLAB and Simscape Multibody in mechanical engineering

We encourage you to these uses and look forward to any feedback you may have. For e-mail correspondence, we can be reached at *kevin.russell@njit.edu*. We hope you enjoy utilizing this work as much as we have enjoyed preparing it. Thank you.

Kevin Russell
Qiong "John" Shen
Raj S. Sodhi

About the Authors

Kevin Russell, Ph.D., P.E., is a member of the teaching faculty in the Department of Mechanical and Industrial Engineering at New Jersey Institute of Technology (NJIT). At NJIT, Dr. Russell teaches courses in computer-aided design, computer-aided simulation, kinematics, machine design, and mechanical design. Formerly, Dr. Russell was a Senior Mechanical Engineer at the U.S. Army Research, Development and Engineering Center (ARDEC) at Picatinny, New Jersey. His responsibilities at ARDEC included the utilization of computer-aided design and modeling and simulation tools for small- and medium-caliber weapon system improvement, concept development, and failure investigations. A Fellow of the American Society of Mechanical Engineers (ASME) and a registered Professional Engineer in New Jersey, Dr. Russell holds several patents for his design contributions relating to small- and medium-caliber weapon systems, linkage-based inspection systems, and human prosthetics. He has published extensively among mechanical engineering journals in areas such as kinematic synthesis, theoretical kinematics, and machine design.

Qiong "John" Shen, Ph.D., is the founder of Softalink LLC, New Jersey, a consulting company that applies cloud-computing and big data technologies to help automate and optimize business processes, and transform traditional marketing and strategic planning into a data-driven manner. Besides business activities, Dr. Shen is also active, as an adjunct professor at New Jersey Institute of Technology (NJIT), in preparing college students for ever-growing challenges in engineering and management. Dr. Shen received a Ph.D. degree from a joint program between Mechanical Engineering and Electrical Engineering Departments at NJIT. He has made substantial contributions to research in Robotics and Mechanism Synthesis by applying technologies from distributed parallel computing, machine learning, visualization, and simulation.

Raj S. Sodhi, Ph.D., P.E. is a professor in the Department of Mechanical and Industrial Engineering at NJIT. He has over 30 years of experience in research and education related to mechanical design, mechanisms synthesis, and manufacturing engineering. Dr. Sodhi is the author or co-author of over 100 refereed papers in scientific journals and conference proceedings. He was awarded the Society of Manufacturing Engineering's *University Lead Award* in recognition of leadership and excellence in the application and development of computer integrated manufacturing. He also received the *N. Watrous Procter & Gamble Award* from the Society of Applied Mechanisms and Robotics for significant contributions to the science of mechanisms and robotics and the *Ralph R. Teetor New Engineering Educator Award* from the Society of Automotive Engineers. Dr. Sodhi is a registered Professional Engineer in Texas.

1 Introduction to Kinematics

CONCEPT OVERVIEW

In this chapter, the reader will gain a central understanding regarding the following:

1. *Kinematic*s and its use in engineering design
2. Distinctions between *kinematic chains* and *mechanisms*
3. Planar and spatial mechanism *mobility*
4. Types of mechanism motion
5. Distinctions between *kinematic analysis* and *kinematic synthesis*

1.1 KINEMATICS

Kinematics is the study of motion without considering forces driving it. In kinematic analysis, positions, displacements, velocities, and accelerations are calculated for mechanical system components without considering the loads that actually govern them. In comparison to other engineering design studies such as *statics*, where motion and governing loads are considered according to Newton's first law and *dynamics*, where motion and governing loads are considered according to Newton's second law, kinematics is the most fundamental engineering design study. It is often necessary in the design of a mechanical system to not only consider the motion of its components but also the following:

- static and/or dynamic loads acting on the components (considered in statics and dynamics)
- component material stress and strain responses to the loads (considered in stress analysis)
- required component dimensions for the working stresses (considered in machine design)

Because of this, static, dynamic, stress, and machine design analyses often follow a kinematic analysis.

Figure 1.1 includes kinematics, statics and dynamics, stress analysis, and machine design in increasing order of progression. This order follows the intended order of use of these studies in

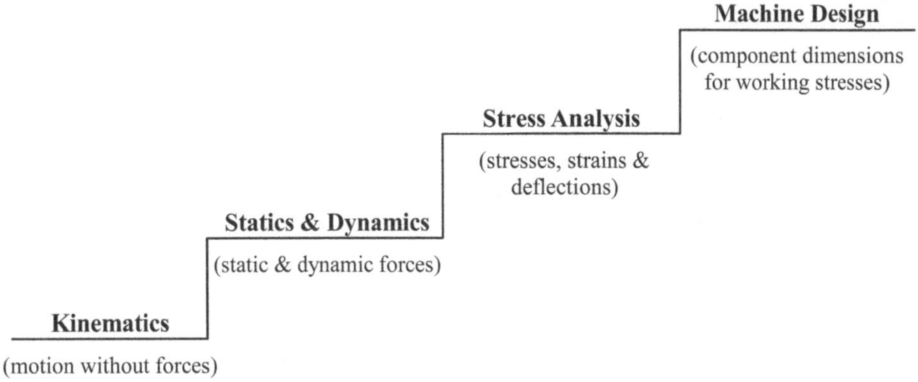

FIGURE 1.1 Kinematics in relation to other associated engineering design studies.

DOI: 10.1201/9781003591375-1

mechanical design. After a mechanical system has first been determined to be kinematically feasible, the static and/or dynamic loads acting on the system components are then considered. After static and/or dynamic feasibility has been achieved, the stresses and strains produced in the mechanical system components are then considered. Lastly, machine design principles and methodologies are employed to ensure that the material and dimensions of the mechanical system components (and subsequently the entire mechanical system) are satisfactory for the known working stresses.[1]

As illustrated in Figure 1.1, kinematics is the most fundamental of the engineering design studies listed. When a design is not kinematically sound, this will often appear as an issue in the other engineering design studies. For example, a discontinuous displacement profile calculated in a kinematic analysis would be revealed as excessive acceleration in a dynamic analysis, which in turn could produce excessive dynamic forces, which would likely produce high stresses. These high stresses may require a material type and/or component dimensions that can lead to impracticalities in the overall component design for the intended design application. Kinematic feasibility, therefore, must be established first before considering the follow-on engineering design studies in Figure 1.1.

1.2 KINEMATIC CHAINS AND MECHANISMS

This book focuses primarily on the kinematic analysis and kinematic synthesis of mechanical systems or *mechanisms* as they are commonly called.[2] A *kinematic chain*, an overarching classification that includes mechanisms, is an assembly of links interconnected by joints where the motion of one link compels the motion of another link (which compels the motion of another link and so on depending on the number of mechanism links).[3,4] Complex mechanical systems such as an automobile engine, for example, comprises multiple kinematic chains, while a single kinematic chain can constitute an entire mechanical system in the case of a simple tool. Figure 1.2 illustrates a commonly used kinematic chain: locking pliers.

Moving the lower handle (link L_3) toward the upper handle (link L_1) or vice-versa compels the motion of the remaining links, including the lower grip (link L_4), which produces a gripping and locking action. Having links compel the motion of each other link in a controlled manner is important because the fundamental objective in the design of a mechanical system is to provide a controlled output motion in response to a supplied input motion.

One characteristic that distinguishes *mechanisms* from other kinematic chains is that the former has a *ground* link [1]. A ground link is attached to a reference coordinate frame (making it immobile). Some mechanisms have a permanent ground link through friction, gravity, or fastening members (e.g., bolts, screws, and welds), whereas with the example of locking pliers, the ground link can be established according to one's preferences.

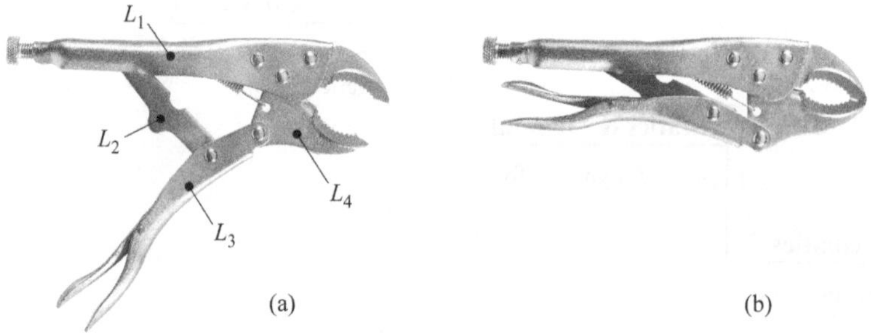

FIGURE 1.2 Locking pliers in (a) open and (b) closed positions.

1.3 MOBILITY, PLANAR, AND SPATIAL MECHANISMS

The *mobility* or the number of *degrees of freedom* of a mechanism is the number of independent parameters required to uniquely define its position in space. Knowing the mobility of a mechanism is particularly important when formulating equation systems for the kinematic analysis or synthesis of the mechanism. This is because the equation systems must include sufficient parameters to fully define the motion of each mechanism component. Three independent parameters are required to fully define the position of a body in two-dimensional or *planar* space at an instant in time. Consider the example of the parking automobile in Figure 1.3a, where the planar Cartesian coordinate frame is affixed to the parking space. At any time, the position of the automobile can be measured with respect to the coordinate frame given three independent parameters. The *X* and *Y* coordinates of a point on the automobile are two of the three parameters required to define the planar position of a body. Because the parking automobile also rotates in the coordinate frame's Z-axis, its angular position is also required to fully define its position in the coordinate frame. Therefore, the three parameters required to define a body's planar position are the planar Cartesian coordinates of a location on the body and its orientation angle about the Z-axis. *Because three independent parameters are required to define the position of the body in 2D space, an individual mechanism link restricted to planar motion can have a mobility of up to three or up to three degrees of freedom.*

Six independent parameters are required to fully define the position of a body in three-dimensional space at an instant in time. Consider the flying aircraft example in Figure 1.3b where the spatial Cartesian coordinate frame is affixed in space. At any time, the position of the aircraft can be measured with respect to the coordinate frame given six independent parameters. The *X*, *Y*, and *Z* coordinates of a point on the aircraft are three of the six parameters required to define the spatial position of a body. Because the aircraft can also rotate about each coordinate frame axis, these three angular positions are also required to fully define its position in the coordinate frame.[5] Therefore, the six parameters required to define a body's spatial position are the spatial Cartesian coordinates of a location on the body and its orientation angles about each coordinate frame axis. *Because six independent parameters are required to define the* position of the body in 3D space, an individual mechanism link restricted to spatial motion can *have a mobility of up to six or up to six degrees of freedom.*

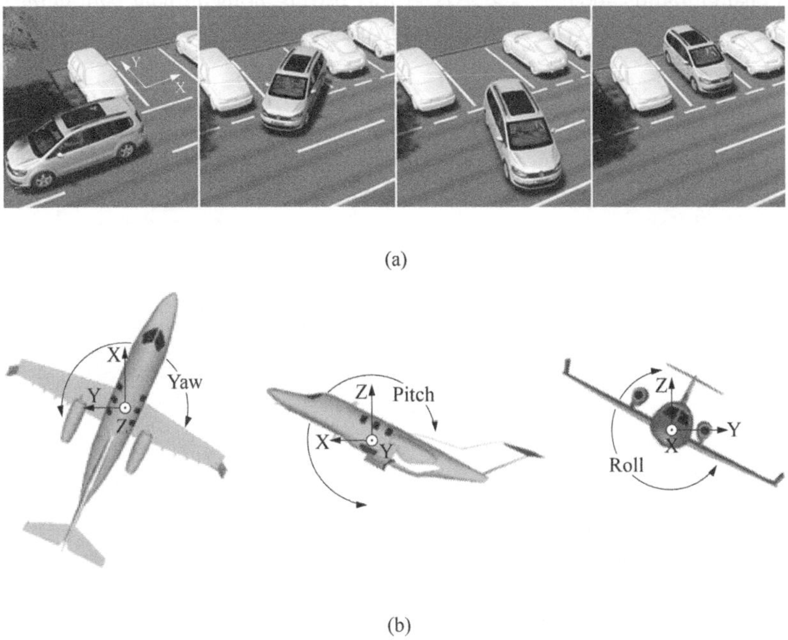

FIGURE 1.3 (a) Parking automobile and (b) aircraft in flight.

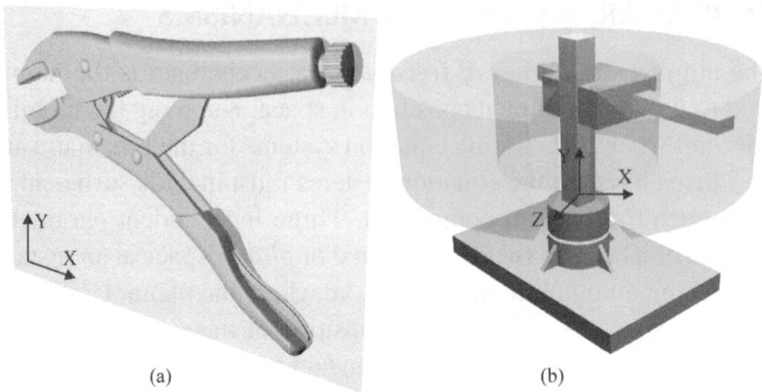

FIGURE 1.4 (a) Pliers in plane and (b) robotic manipulator in spatial workspace.

Figure 1.4a illustrates a pair of pliers. As indicated by the overlapping plane, the motion exhibited by this mechanism is restricted to 2D space. A pair of pliers is an example of a commonly used *planar mechanism*. Each link in this mechanism has a single degree of freedom—which is consistent with the condition that a maximum mobility of 3 is possible with a body in planar motion.

In comparison to the pair of pliers, Figure 1.4b illustrates a particular type of robotic manipulator—the Revolute-Prismatic-Prismatic *(RPP) robotic manipulator* (presented in Chapter 10). As indicated by the overlapping cylindrical workspace, this mechanism can exhibit motion in 3D space.[6] This robotic manipulator is an example of a *spatial mechanism*. Each link in this particular robotic manipulator has a single degree of freedom—which is also consistent with the previously stated condition that a maximum mobility of 6 is possible with a body in spatial motion.

1.4 TYPES OF MECHANISM MOTION

The motion exhibited by any mechanism link in 2D or 3D space can be classified as one of four types of motion. These four types are illustrated in Figure 1.5. In *pure rotation* (Figure 1.5a), a link rotates at a constant radius about a fixed axis. A link travels along a linear path in *pure translation* (Figure 1.5b).[7] *Complex motion* is a type of planar motion that includes both link rotations and

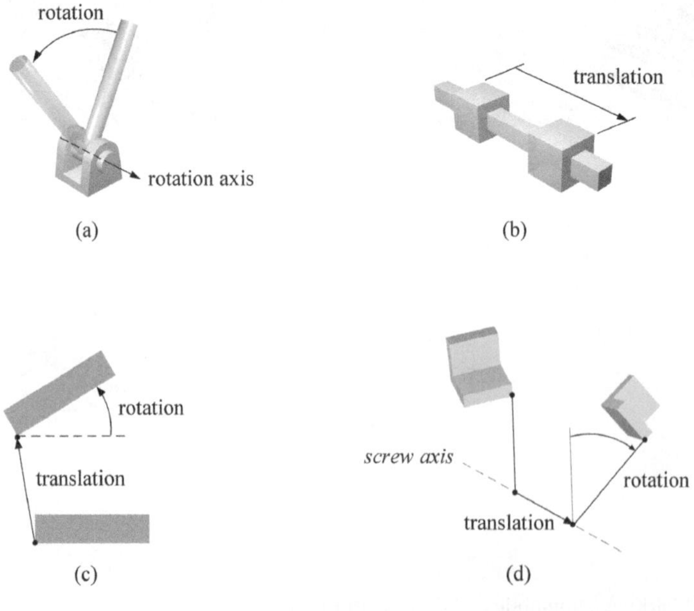

FIGURE 1.5 Links in (a) pure rotation, (b) pure translation, (c) complex, and (d) screw motion.

translations simultaneously (Figure 1.5c). Any two spatial link positions can be expressed as a rotation about and a translation along a spatial axis (called a *screw axis*). This type of spatial motion is called *screw motion* (Figure 1.5d). Pure rotation and translation can be exhibited by mechanisms in both 2D and 3D spaces, whereas only mechanisms in 2D space can exhibit complex motion, and only mechanisms in 3D space can exhibit screw motion.

1.5 KINEMATIC SYNTHESIS

In kinematic analyses, the dimensions of mechanism links are known, and link motion specifications such as positions, displacements, velocities, and accelerations are calculated. Understanding the kinematic behavior of a given mechanism is the scope of a kinematic analysis.

In comparison to kinematic analysis, in *kinematic synthesis*, the problem is considered in reverse. Here, the mechanism required to fulfill specific predetermined conditions is produced.

There are two categories in kinematic synthesis. One category involves the determination of the *type* of mechanism needed to accomplish a given purpose. This category is called *type synthesis* [2]. *Number synthesis* fits this category. In number synthesis (presented in Section 3.3), mechanisms that match a given mechanism mobility are produced.

The other category in kinematic synthesis involves the determination of the mechanism *dimensions* needed to achieve a given motion sequence. This category is referred to as *dimensional synthesis* since it involves determining specific mechanism dimensions [3]. Chapters 5–9 and 11 cover planar and spatial four-link mechanism dimensional synthesis.

1.6 SOFTWARE RESOURCES

The commercial mathematical software package MATLAB® (version 2021b) and its mechanical modeling and simulation toolbox Simscape Multibody™ are thoroughly integrated in this book for applied kinematic synthesis and analysis [4]. Both software resources are well established (and often the *de facto* standard for mathematical analysis and simulation) in many colleges and universities worldwide. A library of MATLAB and Simscape Multibody files developed for this book can be downloaded from https://www.routledge.com/9781032956800. With these files, the user can develop solutions for the equation systems presented throughout this book and conduct mechanical simulations to independently verify or evaluate equation system results.

1.7 SUMMARY

Kinematics—the study of motion without considering governing forces—is the most fundamental engineering study in mechanical system design. In mechanical system design, kinematic feasibility should be determined before considering other engineering design studies such as *statics*, *dynamics*, *stress analysis*, and *machine design*.

Mechanical systems are comprised of *kinematic chains*—an assembly of interconnected *links* where the motion of one link compels the motion of another link in a controlled manner. Achieving controlled output motion in response to a supplied input motion is the fundamental objective in mechanical system design. Kinematic chain is an overarching classification that includes *mechanisms* (also called *linkages*). The presence of an established ground link is a primary characteristic that distinguishes mechanisms from kinematic chains.

The *mobility* or the number of *degrees of freedom* of a mechanism is the number of independent parameters required to uniquely define its position in space. Knowing the mobility of a mechanism is important when formulating equation systems for mechanism kinematic or synthesis. This is because the equation systems must include a sufficient number of variables to fully define the motion of each mechanism component. An individual link confined to planar motion can have up to three degrees of freedom, and an individual link confined to spatial motion can have up to six degrees of freedom.

A mechanism link in 2D or 3D space can exhibit *pure rotation*, *pure translation*, *complex motion*, or *screw motion*. In pure rotation, a link rotates at a constant radius about a fixed axis. A link travels along a linear path in pure translation. Complex motion is a type of planar motion that includes simultaneous link rotations and translations. Screw motion is a type of spatial motion that includes simultaneous link rotations about and translations along a spatial axis called a *screw axis*.

In *kinematic analyses*, the dimensions of mechanism links are known, and link motion specifications such as positions, displacements, velocities, and accelerations are calculated. In comparison to kinematic analysis, the problem is considered in reverse in *kinematic synthesis*. Here, the mechanism dimensions required to fulfill specific predetermined conditions are calculated. One category in kinematic synthesis, called *type synthesis*, involves the determination of the type of mechanism needed to accomplish a given purpose. *Number synthesis* fits this category. In number synthesis, mechanisms that match a given mechanism mobility are produced.

The other category in kinematic synthesis involves the determination of the mechanism *dimensions* needed to achieve a given motion sequence. This category is called *dimensional synthesis* since it involves determining specific mechanism dimensions. Chapters 5 through 9 and 11 cover planar and spatial four-link mechanism dimensional synthesis.

The mathematical software package MATLAB® and Simscape Multibody™ are fully integrated throughout the remaining book chapters for applied kinematic synthesis and analysis. The library of MATLAB and Simscape Multibody files for this book is available for download at https://www.routledge.com/9781032956800.

NOTES

1 In addition to engineering design factors pertaining to kinematics, statics, dynamics, and machine design—also called *traditional* engineering factors, *non-traditional* or *modern* engineering factors (including *producibility*, *cost*, *environmental impact*, *disposal*, *aesthetics*, *ergonomics*, and *human factors*)—are often equally important.
2 The distinctions between *kinematic analysis* and *kinematic synthesis* are first presented in Section 1.5.
3 Because a mechanism is an assembly of links, it is also called a *linkage*.
4 Links are generally assumed to be non-deforming or *rigid* in kinematics.
5 As shown in Figure 1.3b, the three aircraft rotation angles are called the *roll*, *pitch*, and *yaw* angles and are about the *X*, *Y*, and *Z* axes, respectively.
6 The space (2D or 3D) that encompasses all possible positions achieved by a mechanism is called its *workspace*.
7 Pure rotation at a radius of infinity becomes pure translation.

REFERENCES

1. Norton, R. L. 2008. *Design of Machinery*, 4th Edition. pages 30–40. New York: McGraw-Hill.
2. Sandor, G. N. and Erdman, A. G. 1984. *Advanced Mechanism Design: Analysis and Synthesis*, Volume 2. pages 49–50. Englewood Cliffs: Prentice-Hall.
3. Russell, K., Shen, Q. and Sodhi, R. 2023. *Kinematics and Dynamics of Mechanical Systems: Implementation in MATLAB® and Simscape Multibody™*. chapter 5. Boca Raton: CRC Press.
4. "Products and Services," Available at: https://www.mathworks.com. (Accessed on: July 18, 2024).

ADDITIONAL READING

Myszka, D. H. 2005. *Machines and Mechanisms: Applied Kinematic Analysis*, 3rd Edition. chapter 1. Saddle River: Prentice Hall.

Waldron, K. J. and Kinzel, G. L. 2004. *Kinematics, Dynamics and Design of Machinery*, 2nd Edition. chapter 1. Saddle River: Prentice-Hall.

Wilson, C. E. and Sadler, J. P. 2003. *Kinematics and Dynamics of Machinery*, 3rd Edition. chapter 1. Saddle River: Prentice Hall.

2 Mathematical Concepts in Kinematics

CONCEPT OVERVIEW

In this chapter, the reader will gain a central understanding regarding the following:

1. Characteristics of vectors, complex vectors, and complex vector forms
2. The formulation of vector loop displacement, velocity, and acceleration equations using complex vectors
3. Characteristics of point-based vectors and their application in mechanism motion equations
4. Characteristics of linear simultaneous equations and their representation in matrix form
5. Fundamental matrix operations and the identity matrix
6. Matrix inversion and its application in solving linear simultaneous equations
7. Intermediate and total spatial motion and their application in mechanism kinematics
8. The general transformation matrix and its application in the kinematic analysis of robotic manipulators

2.1 INTRODUCTION

This chapter introduces the mathematical concepts and methodologies the reader should become familiar with to gain a solid understanding of the equation systems presented throughout this textbook. These mathematical concepts relate to *complex numbers, linear simultaneous equations, matrices, spatial motion,* the *general transformation matrix and nonlinear simultaneous equations* for constrained and unconstrained optimization problems.

2.2 COMPLEX NUMBERS AND OPERATIONS

2.2.1 Complex Number Forms

Vectors are commonly used in the formulation of mechanism equation systems because, being quantities that have both *magnitude* and *direction*, vectors can appropriately define mechanism motion (specifically the motion of linkage-based mechanisms). One way to represent a two-dimensional vector (a vector that lies on a plane) is with a *complex number.* A complex number is comprised of both a *real component* and an *imaginary component.*

Figure 2.1 illustrates a vector $\mathbf{V}_1$ in two-dimensional complex space. The real and imaginary components of this vector are V_{1x} and V_{1y}, respectively. Vector $\mathbf{V}_1$ has the magnitude V_1, where $V_1 = \sqrt{V_{1x}^2 + V_{1y}^2}$, and its direction angle is θ_1, where $\theta_1 = \tan^{-1}\left(V_{1y}/V_{1x}\right)$. The real and imaginary components of $\mathbf{V}_1$ can also be expressed as $V_1\cos\theta_1$ and $V_1\sin\theta_1$, respectively, due to the vector direction angle.[1] Vector $\mathbf{V}_1$ can be expressed in the following four forms:

$$\mathbf{V}_1 = V_{1x} + iV_{1y} = V_1\left(\cos\theta_1 + i\sin\theta_1\right) = V_1 e^{i\theta_1} \tag{2.1}$$

where $i = \sqrt{-1}$ and is called the *imaginary unit.*[2] The second form in Equation (2.1) is the *rectangular form* of the vector $\mathbf{V}_1$—the complex number. The third and last forms in Equation (2.1) are the *polar forms* of $\mathbf{V}_1$ (the last form being the *polar exponential form*).[3]

DOI: 10.1201/9781003591375-2

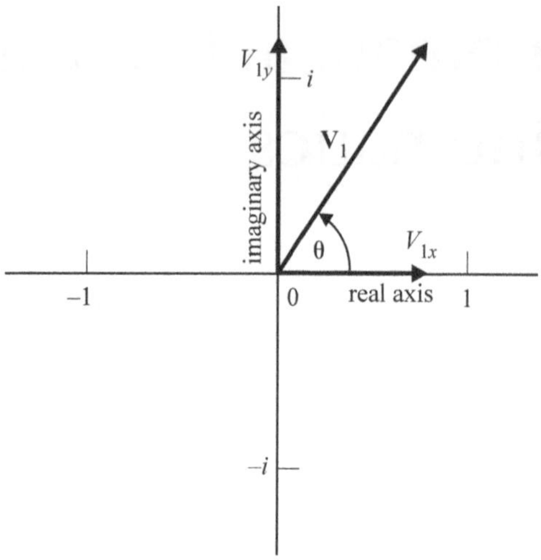

FIGURE 2.1 A vector in two-dimensional complex space.

MATLAB includes a library of functions covering a wide range of conventional calculation methods. A table of the MATLAB functions used in this textbook is included in Section A.2. From here on, this book will primarily utilize MATLAB functions. There will be occasional examples, however, where manual calculation methods will be utilized to give the reader greater insights into the calculation method or to convey specific techniques in MATLAB.

Example 2.1

Problem Statement: Calculate the magnitude and direction angle of vector $\mathbf{V}_1$, where $\mathbf{V}_1 = 1.5 + i2$.

Known Information: Vector $\mathbf{V}_1$ and Section 2.2.1

Solution Approach 1 (Using MATLAB Functions): Figure E.2.1 includes the calculation procedure in MATLAB's command window. After specifying the given values for the real and imaginary components of $\mathbf{V}_1$, this vector is then defined. The magnitude and direction angle of $\mathbf{V}_1$ were produced using MATLAB functions.

```
>> V1x = 1.5;
>> V1y = 2;
>> V1 = V1x + i*V1y

V1 =

   1.5000 + 2.0000i

>> Magnitude = abs(V1)

Magnitude =

    2.5000

>> Direction = angle(V1)*180/pi

Direction =

   53.1301

>>
```

FIGURE E.2.1 Example 2.1 calculation procedure (for Solution Approach 1) in MATLAB.

Solution Approach 2 (Using Manual Calculations in MATLAB): Figure E.2.2 includes the calculation procedure in MATLAB's command window where the magnitude and direction angle of $\mathbf{V}_1$ were produced using manual calculations.

```
>> V1x = 1.5;
>> V1y = 2;
>> V1 = V1x + i*V1y

V1 =

   1.5000 + 2.0000i

>> Magnitude = sqrt(V1x^2 + V1y^2)

Magnitude =

    2.5000

>> Direction = atan2(V1y, V1x)*180/pi

Direction =

   53.1301

>>
```

FIGURE E.2.2 Example 2.1 calculation procedure (for Solution Approach 2) in MATLAB.

2.2.2 Complex Number Addition

Equation systems for mechanisms can be formulated by producing vector loops for the mechanisms and taking the sum of the individual vectors in the loops [1]. Figure 2.2 illustrates a loop comprised of vectors $\mathbf{V}_1$, $\mathbf{V}_2$, $\mathbf{V}_3$, and $\mathbf{V}_4$ in two-dimensional complex space. Taking the sum of the vectors in a clockwise loop produces

$$\mathbf{V}_1 + \mathbf{V}_2 + \mathbf{V}_3 - \mathbf{V}_4 = 0 \tag{2.2}$$

If we conclude from Figure 2.2 that $\mathbf{V}_1 = V_1 e^{i\theta_1}$, $\mathbf{V}_2 = V_2 e^{i\theta_2}$, $\mathbf{V}_3 = V_3 e^{i\theta_3}$, and $\mathbf{V}_4 = V_4 e^{i\theta_4}$, each vector in the loop can be expressed in polar form as

$$\begin{aligned} &V_1(\cos\theta_1 + i\sin\theta_1) + V_2(\cos\theta_2 + i\sin\theta_2) + V_3(\cos\theta_3 + i\sin\theta_3) \\ &-V_4(\cos\theta_4 + i\sin\theta_4) = 0 \end{aligned} \tag{2.3}$$

and in rectangular form as

$$(V_{1x} + iV_{1y}) + (V_{2x} + iV_{2y}) + (V_{3x} + iV_{3y}) - (V_{4x} + iV_{4y}) = 0 \tag{2.4}$$ [4]

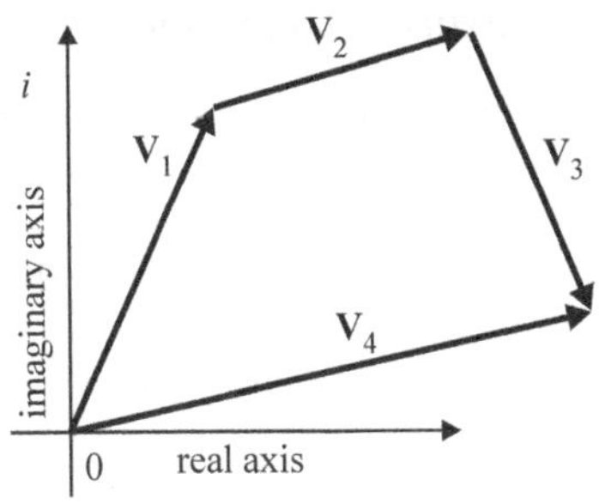

FIGURE 2.2 A vector loop in two-dimensional complex space.

After grouping and separating the real and imaginary terms in Equation (2.4), the two equations in Equation (2.5) are produced. The imaginary unit in the second equation (being common among all equation terms) can be removed if preferred.

$$\begin{aligned} V_{1x} + V_{2x} + V_{3x} - V_{4x} &= 0 \\ i\left(V_{1y} + V_{2y} + V_{3y} - V_{4y}\right) &= 0 \end{aligned} \tag{2.5}$$

Example 2.2

Problem Statement: Determine if the sum $\mathbf{V}_1 + \mathbf{V}_2 + \mathbf{V}_3 + \mathbf{V}_4$ of the following vectors forms a closed loop: $\mathbf{V}_1 = 1.5 + i2$, $\mathbf{V}_2 = -i0.5$, $\mathbf{V}_3 = -1.25 - i2.25$, and $\mathbf{V}_4 = 0.25 - i0.25$.

Known Information: Vectors $\mathbf{V}_1$, $\mathbf{V}_2$, $\mathbf{V}_3$, and $\mathbf{V}_4$ and Section 2.2.2

Solution Approach: Figure E.2.3 includes the calculation procedure in MATLAB's command window. After specifying the given vectors and calculating the vector sum, the vectors are not found to form a closed loop since the vector sum (called *Sum* in Figure E.2.3) is not 0.

```
>> V1 = 1.5 + i*2

V1 =

   1.5000 + 2.0000i

>> V2 = -0.5

V2 =

   -0.5000i

>> V3 = -1.25 - i*2.25

V3 =

   -1.2500 - 2.2500i

>> V4 = 0.25 - i*0.25

V4 =

   0.2500 - 0.2500i

>> Sum = V1 + V2 + V3 + V4

Sum =

   0.5000 - 1.0000i

>>
```

FIGURE E.2.3 Example 2.2 solution calculation procedure in MATLAB.

2.2.3 Complex Number Multiplication and Differentiation

The product of two complex numbers (vectors $\mathbf{V}_1 = V_1 e^{i\theta_1}$ and $\mathbf{V}_2 = V_2 e^{i\theta_2}$, for example) in polar exponential form is

$$\mathbf{V}_1\mathbf{V}_2 = V_1 e^{i\theta_1} V_2 e^{i\theta_2} = V_1 V_2 e^{i(\theta_1+\theta_2)} = V_1 V_2\left[\cos(\theta_1+\theta_2) + i\sin(\theta_1+\theta_2)\right]^5 \tag{2.6}$$

The first derivative with respect to time of a complex number (vector $\mathbf{V}_1 = V_1 e^{i\theta_1}$, for example) is

$$\frac{d\mathbf{V}_1}{dt} = i\frac{d\theta_1}{dt} V_1 e^{i\theta_1} = i\dot{\theta}_1 \mathbf{V}_1 \tag{2.7}$$

where vector $\mathbf{V}_1$ represents a *rigid* link (a link having a fixed length).[6] The second derivative of Equation (2.7) is

$$\frac{d^2\mathbf{V}_1}{dt^2} = -\left(\frac{d\theta_1}{dt}\right)^2 V_1 e^{i\theta_1} + i\frac{d^2\theta_1}{dt^2} V_1 e^{i\theta_1} = \mathbf{V}_1\left(i\ddot{\theta}_1 - \dot{\theta}_1^2\right) \tag{2.8}$$

The multiplication and differentiation of complex numbers are used in the formulation of vector loop equations for higher-order mechanism motion quantities such as velocity and acceleration [2].

Example 2.3

Problem Statement: Formulate an equation system for vector $\mathbf{V}$ for the vector loop illustrated in Figure E.2.4. In this vector loop, $\mathbf{W} = We^{i\alpha}$ and $\mathbf{X} = Xe^{i\delta}$. Also, formulate equation systems for the first and second derivatives of $\mathbf{V}$ ($\dot{\mathbf{V}}$ and $\ddot{\mathbf{V}}$, respectively) manually and symbolically in MATLAB. Assume all vectors represent rigid links.

Known Information: Figure E.2.4 and Sections 2.2.2 and 2.2.3

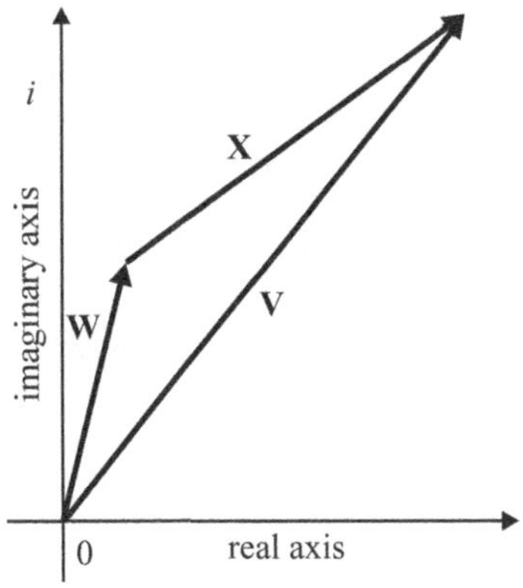

FIGURE E.2.4 Vector loop in two-dimensional complex space.

SOLUTION APPROACH

Initial Formulation

Taking a clockwise vector loop sum for the vector loop in Figure E.2.4 and solving for vector $\mathbf{V}$ produces

$$\mathbf{V} = \mathbf{W} + \mathbf{X} = We^{i\alpha} + Xe^{i\delta} \tag{2.9}$$

After expanding the polar exponential form of $\mathbf{V}$, Equation (2.10) is produced and Equation (2.11) is produced after grouping and separating the real and imaginary terms.

$$\mathbf{V} = W(\cos\alpha + i\sin\alpha) + X(\cos\delta + i\sin\delta) = W_x + iW_y + X_x + iX_y \tag{2.10}$$

$$\begin{aligned} V_x &= W\cos\alpha + X\cos\delta = W_x + X_x \\ V_y &= W\sin\alpha + X\sin\delta = W_y + X_y \end{aligned} \tag{2.11}$$

First Derivative Formulation

Taking the first derivative of Equation (2.9) produces

$$\dot{\mathbf{V}} = i\dot{\alpha}We^{i\alpha} + i\dot{\delta}Xe^{i\delta} \tag{2.12}$$

Figure E.2.5 includes this calculation procedure in MATLAB's command window. Upon observation, Equation (2.12) agrees perfectly with the first derivative produced symbolically in MATLAB. In Figure E.2.5, the terms *alpha(t)* and *diff(alpha(t),t)* represent the terms α and $\dot{\alpha}$, respectively, in Equation (2.12), and the terms *delta(t)* and *diff(delta(t),t)* represent the terms δ and $\dot{\delta}$, respectively.

```
>> syms W X t alpha(t) delta(t) real;
>> V = W*exp(i*alpha(t)) + X*exp(i*delta(t));
>> dV = diff(V)

dV =

X*exp(delta(t)*i)*diff(delta(t), t)*i +
W*exp(alpha(t)*i)*diff(alpha(t), t)*i

>>
```

FIGURE E.2.5 Example 2.3 vector first-order differentiation procedure in MATLAB.

After expanding the polar exponential form of $\dot{\mathbf{V}}$, Equation (2.13) is produced and Equation (2.14) is produced after grouping and separating the real and imaginary terms.

$$\dot{\mathbf{V}} = \dot{\alpha}W\left(i\cos\alpha - \sin\alpha\right) + \dot{\delta}X\left(i\cos\delta - \sin\delta\right) = i\dot{\alpha}W_x - \dot{\alpha}W_y + i\dot{\delta}X_x - \dot{\delta}X_y \quad (2.13)$$

$$\dot{V}_x = -\dot{\alpha}W\sin\alpha - \dot{\delta}X\sin\delta = -\dot{\alpha}W_y - \dot{\delta}X_y$$
$$\dot{V}_y = \dot{\alpha}W\cos\alpha + \dot{\delta}X\cos\delta = \dot{\alpha}W_x + \dot{\delta}X_x \quad (2.14)$$

Second Derivative Formulation

Taking the first derivative of Equation (2.12) produces

$$\ddot{\mathbf{V}} = -\dot{\alpha}^2We^{i\alpha} + i\ddot{\alpha}We^{i\alpha} - \dot{\delta}^2Xe^{i\delta} + i\ddot{\delta}Xe^{i\delta} \quad (2.15)$$

Figure E.2.6 includes this calculation procedure in MATLAB's command window. Upon observation, it can be seen that Equation (2.15) agrees perfectly with the second derivative produced symbolically in MATLAB. In Figure E.2.6, the terms *diff(alpha(t),t,t)* and *diff(delta(t),t,t)* represent the terms $\ddot{\alpha}$ and $\ddot{\delta}$, respectively, in Equation (2.15).

```
>> syms W X t alpha(t) delta(t) real;
>> V = W*exp(i*alpha(t)) + X*exp(i*delta(t));
>> ddV = diff(V,2)

ddV =

- W*exp(alpha(t)*i)*diff(alpha(t), t)^2 -
X*exp(delta(t)*i)*diff(delta(t), t)^2 +
W*exp(alpha(t)*i)*diff(alpha(t), t, t)*i +
X*exp(delta(t)*i)*diff(delta(t), t, t)*i

>>
```

FIGURE E.2.6 Example 2.3 vector second-order differentiation procedure in MATLAB.

After expanding the polar exponential form of $\ddot{\mathbf{V}}$, Equation (2.16) is produced and Equation (2.17) is produced after grouping and separating the real and imaginary terms.

$$\begin{aligned}\ddot{\mathbf{V}} &= -\dot{\alpha}^2W(\cos\alpha + i\sin\alpha) + \ddot{\alpha}W(i\cos\alpha - \sin\alpha) \\ &\quad -\dot{\delta}^2X\left(\cos\delta + i\sin\delta\right) + \ddot{\delta}X\left(i\cos\delta - \sin\delta\right) \\ &= -\dot{\alpha}^2W_x - i\dot{\alpha}^2W_y + i\ddot{\alpha}W_x - \ddot{\alpha}W_y - \dot{\delta}^2X_x - i\dot{\delta}^2X_y + i\ddot{\delta}X_x - \ddot{\delta}X_y\end{aligned} \quad (2.16)$$

$$\begin{aligned}\ddot{V}_x &= -\dot{\alpha}^2 W\cos\alpha - \ddot{\alpha} W\sin\alpha - \dot{\delta}^2 X\cos\delta - \ddot{\delta} X\sin\delta = -\dot{\alpha}^2 W_x - \ddot{\alpha} W_y - \dot{\delta}^2 X_x - \ddot{\delta} X_y \\ \ddot{V}_y &= -\dot{\alpha}^2 W\sin\alpha + \ddot{\alpha} W\cos\alpha - \dot{\delta}^2 X\sin\delta + \ddot{\delta} X\cos\delta = -\dot{\alpha}^2 W_y + \ddot{\alpha} W_x - \dot{\delta}^2 X_y + \ddot{\delta} X_x\end{aligned} \tag{2.17}$$

Example 2.4

Problem Statement: Calculate the vector product **VW**, where $\mathbf{V} = V_1(\cos\theta_1 + i\sin\theta_1)$ and $\mathbf{W} = W_1(\cos\alpha_1 - i\sin\alpha_1)$. Also, formulate **VW** symbolically in MATLAB.

Known Information: Vectors **V**, **W**, and Section 2.2.3

Solution Approach: Equation (2.18) includes the basic form of the product **VW**. Equation (2.19) includes the vector product with the trigonometric identities for $\cos(\alpha_1 - \theta_1)$ and $\sin(\alpha_1 - \theta_1)$ included.

$$\mathbf{VW} = V_1W_1(\cos\theta_1\cos\alpha_1 - i\cos\theta_1\sin\alpha_1 + i\sin\theta_1\cos\alpha_1 + \sin\theta_1\sin\alpha_1) \tag{2.18}$$

$$\mathbf{VW} = V_1W_1\left(\cos(\alpha_1 - \theta_1) - i\sin(\alpha_1 - \theta_1)\right) \tag{2.19}$$

Figure E.2.7 includes the calculation procedure for **VW** in MATLAB's command window.

```
>> syms theta1 alpha1 V1 W1 real;
>> V = V1*(cos(theta1) + i*sin(theta1));
>> W = W1*(cos(alpha1) - i*sin(alpha1));
>> Product = simplify(expand(V*W))

Product =

-V1*W1*(sin(alpha1 -theta1)*i - cos(alpha1 - theta1))

>>
```

FIGURE E.2.7 Example 2.4 solution calculation procedure in MATLAB.

2.3 VECTOR AND POINT REPRESENTATION

Not only are vectors used to formulate equation systems for mechanisms, but *points* can also be used to define vectors. Both the magnitude and direction can be calculated for a vector, given the coordinates of is endpoints. Vector **V** and its endpoints $\mathbf{p}_1$ and $\mathbf{p}_2$ in two-dimensional and three-dimensional space are illustrated in Figure 2.3. In terms of point coordinates, the vector magnitude V can be expressed in 2D space as

$$V = \|\mathbf{p}_2 - \mathbf{p}_1\| = \sqrt{(p_{2x} - p_{1x})^2 + (p_{2y} - p_{1y})^2} \tag{2.20}$$

where $\mathbf{p}_1 = [p_{1x}, p_{1y}]^T$ and $\mathbf{p}_2 = [p_{2x}, p_{2y}]^T$.[7] The direction of this vector is expressed in the x- and y-components $p_{2x} - p_{1x}$ and $p_{2y} - p_{1y}$, respectively.

In 3D space, the vector magnitude can be expressed as

$$V = \|\mathbf{p}_2 - \mathbf{p}_1\| = \sqrt{(p_{2x} - p_{1x})^2 + (p_{2y} - p_{1y})^2 + (p_{2z} - p_{1z})^2} \tag{2.21}$$

where $\mathbf{p}_1 = [p_{1x}, p_{1y}, p_{1z}]^T$ and $\mathbf{p}_2 = [p_{2x}, p_{2y}, p_{2z}]^T$. The direction of this vector is expressed in the x, y, and z-components $p_{2x} - p_{1x}$, $p_{2y} - p_{1y}$, and $p_{2z} - p_{1z}$, respectively. Point-based vectors are used

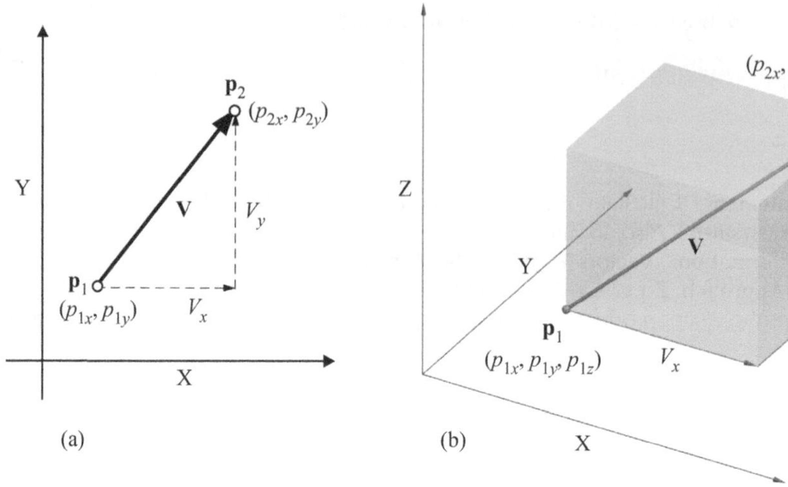

FIGURE 2.3 Point and vector representations in (a) 2D and (b) 3D spaces.

in planar mechanism equation systems in Sections 6.5–6.7 and in the spatial mechanism equation systems in Chapter 8 [3].

Example 2.5

Problem Statement: Calculate the components and magnitude of vector **V**, where this 2D vector is comprised of points $\mathbf{p}_1 = (1.25, -5)$ and $\mathbf{p}_2 = (-2, 9.65)$. Also, calculate the components and magnitude of vector **V** where this 3D vector is comprised of points $\mathbf{p}_1 = (1.25, -5, 0)$ and $\mathbf{p}_2 = (-2, 9.65, 6)$.

```
>> p1 = [1.25, -5]';
>> p2 = [-2, 9.65]';
>> V = p2 - p1

V =

   -3.2500
   14.6500

>> Magnitude = norm(V)

Magnitude =

   15.0062

>> p1 = [1.25, -5, 0]';
>> p2 = [-2, 9.65, 6]';
>> V = p2 - p1

V =

   -3.2500
   14.6500
    6.0000

>> Magnitude = norm(V)

Magnitude =

   16.1612

>>
```

FIGURE E.2.8 Example 2.5 solution calculation procedure in MATLAB.

Known Information: Planar and spatial points $\mathbf{p}_1$ and $\mathbf{p}_2$ and Section 2.3

Solution Approach: Figure E.2.8 includes the calculation procedure in MATLAB's command window.

2.4 LINEAR SIMULTANEOUS EQUATIONS, MATRICES, AND MATRIX OPERATIONS

2.4.1 Linear Simultaneous Equation Systems and Matrices

A *linear equation* includes linear or *first-order* variables.[8] A system of linear equations (or a *linear system*) is collection of linear equations including the same variables. If a common solution is sought among a system of linear equations, it is called a set of *simultaneous equations*. An arbitrary set of linear simultaneous equations is given in Equation (2.22). Variables x_1, x_2, x_3, and/or x_4 are included among the four linear equations, and true x_1, x_2, x_3, and x_4 solutions for these simultaneous equations must satisfy each equation in the set.

$$\begin{aligned} x_1 - 2x_2 + x_3 - 3x_4 &= 1 \\ 2x_1 + x_2 + 2x_3 - 2x_4 &= -2 \\ -x_1 + 2x_2 - 4x_3 + x_4 &= 1 \\ 3x_1 \qquad\qquad\qquad - 3x_4 &= 3 \end{aligned} \tag{2.22}$$

A set of linear simultaneous equations can be expressed in matrix form. A *matrix* is a rectangular array of numbers, symbols, or expressions arranged in rows and columns. This form can be particularly convenient when solving for the variables in a simultaneous equation set. When expressed in matrix form, Equation (2.20) becomes

$$\begin{bmatrix} 1 & -2 & 1 & -3 \\ 2 & 1 & 2 & -2 \\ -1 & 2 & -4 & 1 \\ 3 & 0 & 0 & -3 \end{bmatrix} \begin{Bmatrix} x_1 \\ x_2 \\ x_3 \\ x_4 \end{Bmatrix} = \begin{bmatrix} 1 \\ -2 \\ 1 \\ 3 \end{bmatrix} \quad \text{or} \quad [A]\mathbf{x} = \mathbf{b} \tag{2.23}$$

$$\text{where } [A] = [A]_{4\times4} = \begin{bmatrix} 1 & -2 & 1 & -3 \\ 2 & 1 & 2 & -2 \\ -1 & 2 & -4 & 1 \\ 3 & 0 & 0 & -3 \end{bmatrix}, \ \mathbf{x} = \mathbf{x}_{4\times1} = \begin{Bmatrix} x_1 \\ x_2 \\ x_3 \\ x_4 \end{Bmatrix}, \text{ and } \mathbf{b} = \mathbf{b}_{4\times1} = \begin{bmatrix} 1 \\ -2 \\ 1 \\ 3 \end{bmatrix}.$$

Matrix $[A]$ includes the variable coefficients and is subsequently called the *coefficient matrix*. For simultaneous equation sets having n equations and n variables, the coefficient matrix will always be *square*-having n rows and n columns ($[A]_{n\times n}$).

Matrices $\mathbf{x}$ and $\mathbf{b}$, which include the equation variables and remaining quantities, respectively, are called *column matrices* or *column vectors*. For simultaneous equations sets having n equations and n variables, the column matrices will always have n rows and 1 column ($\mathbf{x}_{n\times1}$ and $\mathbf{b}_{n\times1}$).

Therefore, considering Equation (2.24)—a general set of linear simultaneous equations having n equations and n variables,

$$\begin{aligned} a_{11}x_1 + a_{12}x_2 + a_{13}x_3 + \cdots + a_{1n}x_n &= b_1 \\ a_{21}x_1 + a_{22}x_2 + a_{23}x_3 + \cdots + a_{2n}x_n &= b_2 \\ a_{31}x_1 + a_{32}x_2 + a_{33}x_3 + \cdots + a_{3n}x_n &= b_3 \\ &\vdots \\ a_{n1}x_1 + a_{n2}x_2 + a_{n3}x_3 + \cdots + a_{nn}x_n &= b_n \end{aligned} \tag{2.24}$$

the matrix form expression of this equation set becomes

$$\begin{bmatrix} a_{11} & a_{12} & a_{13} & \cdots & a_{1n} \\ a_{21} & a_{22} & a_{23} & \cdots & a_{2n} \\ a_{31} & a_{32} & a_{33} & \cdots & a_{3n} \\ \vdots & \vdots & \vdots & & \vdots \\ a_{n1} & a_{n2} & a_{n3} & \cdots & a_{nn} \end{bmatrix} \begin{Bmatrix} x_1 \\ x_2 \\ x_3 \\ \vdots \\ x_n \end{Bmatrix} = \begin{bmatrix} b_1 \\ b_2 \\ b_3 \\ \vdots \\ b_n \end{bmatrix} \tag{2.25}$$

$$\text{where } [A] = [A]_{n\times n} = \begin{bmatrix} a_{11} & a_{12} & a_{13} & \cdots & a_{1n} \\ a_{21} & a_{22} & a_{23} & \cdots & a_{2n} \\ a_{31} & a_{32} & a_{33} & \cdots & a_{3n} \\ \vdots & \vdots & \vdots & & \vdots \\ a_{n1} & a_{n2} & a_{n3} & \cdots & a_{nn} \end{bmatrix},$$

$$\mathbf{x} = \mathbf{x}_{n\times 1} = \begin{Bmatrix} x_1 \\ x_2 \\ x_3 \\ \vdots \\ x_n \end{Bmatrix} \text{ and } \mathbf{b} = \mathbf{b}_{n\times 1} = \begin{bmatrix} b_1 \\ b_2 \\ b_3 \\ \vdots \\ b_n \end{bmatrix}.$$

2.4.2 Matrix Transpose, Addition, Subtraction, and Multiplication

This section presents the most common matrix operations used in this book. One operation, which is used in this book for column matrices, is called the *transpose*. In this matrix operation, the rows and columns of a matrix are interchanged. Considering the column matrices in Equations (2.25), the transpose of **x** and **b** becomes

$$\mathbf{x}^T = \mathbf{x}_{1\times n} = \begin{bmatrix} x_1 & x_2 & x_3 & \cdots & x_n \end{bmatrix} \text{ and } \mathbf{b}^T = \mathbf{b}_{1\times n} = \begin{bmatrix} b_1 & b_2 & b_3 & \cdots & b_n \end{bmatrix}$$

where the superscript T in $\mathbf{x}^T$ and $\mathbf{b}^T$ denotes the transpose operation. The transpose of a column matrix produces a *row matrix* or *row vector.*

When adding or subtracting column matrices, the cells in each matching row are added or subtracted. Considering the column matrices in Equations (2.25), the sum $\mathbf{x}+\mathbf{b}$ and difference $\mathbf{x}-\mathbf{b}$ become

$$\mathbf{x}+\mathbf{b} = \mathbf{x}_{n\times 1} + \mathbf{b}_{n\times 1} = [\mathbf{x}+\mathbf{b}]_{n+1} = \begin{bmatrix} x_1+b_1 \\ x_2+b_2 \\ x_3+b_3 \\ \vdots \\ x_n+b_n \end{bmatrix} \text{ and } \mathbf{x}-\mathbf{b} = \mathbf{x}_{n\times 1} - \mathbf{b}_{n\times 1} = [\mathbf{x}-\mathbf{b}]_{n+1} = \begin{bmatrix} x_1-b_1 \\ x_2-b_2 \\ x_3-b_3 \\ \vdots \\ x_n-b_n \end{bmatrix}$$

When calculating the product of a matrix and a scalar quantity (e.g., $k[A]$, where k is a scalar quantity), the product of each cell in the matrix and the scalar quantity is taken (distributing the constant throughout the matrix). For the product of a square matrix and a column matrix, each row in the square matrix is multiplied by the column matrix. Considering the square and column matrices in Equations (2.25), the product $[A]\mathbf{x}$ becomes

$$[A]\mathbf{x} = [A]_{n\times n}\,\mathbf{x}_{n\times 1} = \left([A]\mathbf{x}\right)_{n\times 1} = \begin{bmatrix} a_{11}x_1 & + & a_{12}x_2 & + & a_{13}x_3 & + & \cdots & + & a_{1n}x_n \\ a_{21}x_1 & + & a_{22}x_2 & + & a_{23}x_3 & + & \cdots & + & a_{2n}x_n \\ a_{31}x_1 & + & a_{32}x_2 & + & a_{33}x_3 & + & \cdots & + & a_{3n}x_n \\ & & & & \vdots & & & & \\ a_{n1}x_1 & + & a_{n2}x_2 & + & a_{n3}x_3 & + & \cdots & + & a_{nn}x_n \end{bmatrix}$$

For the product of two column matrices, the first matrix is transposed so that the product of a row matrix and a column matrix is what is actually taken.[9] Considering the column matrices in Equations (2.25), the product of **x** and **b** becomes

$$(\mathbf{x})^T(\mathbf{b}) = (\mathbf{x}_{1\times n})(\mathbf{b}_{n\times 1}) = \{(\mathbf{x})(\mathbf{b})\}_{1+1} = x_1b_1 + x_2b_2 + x_3b_3 + \cdots + x_nb_n$$

It can be observed that the product of a row vector and a column vector is a single scalar quantity.

Example 2.6

Problem Statement: For the given matrices:

$$\mathbf{v} = \begin{bmatrix} 2.5 & -2 & 1 & 0.5 \end{bmatrix},\ \mathbf{w} = \begin{bmatrix} -5 & -1 & 0.75 & -3 \end{bmatrix},\ \text{and}\ [A] = \begin{bmatrix} 3 & 2 & -1 & 1 \\ 5 & 3 & 2 & 1 \\ 3 & 1 & 3 & 2 \\ -6 & -4 & 2 & -2 \end{bmatrix},$$

calculate $\mathbf{v}^T$, $\mathbf{w}^T$, $\mathbf{v}^T + \mathbf{w}^T$, $[A]\mathbf{v}^T$, and $(\mathbf{v})(\mathbf{w})^T$.

Known Information: Given matrices and Section 2.4.2 equations

Solution Approach: Figure E.2.9 includes the calculation procedure in MATLAB's command window.

Considering the product of two general matrices $[A]$ and $[B]$ (as shown in Matrix (2.27)), where $[A] = [A]_{m\times n}$ and $[B] = [B]_{n\times o}$, the entry in each cell in the product $[A][B]$ (which is labeled AB_{ij} in Matrix (2.27)) is given by

$$AB_{ij} = a_{i1}b_{1j} + a_{i2}b_{2j} + a_{i3}b_{3j} + \cdots + a_{in}b_{nj} \quad (2.26)$$

[10]

$$[A][B] = \begin{bmatrix} a_{11} & a_{12} & \cdots & a_{1n} \\ a_{21} & a_{22} & \cdots & a_{2n} \\ \vdots & \vdots & & \vdots \\ a_{i1} & a_{i2} & \cdots & a_{in} \\ \vdots & \vdots & & \vdots \\ a_{m1} & a_{m2} & \cdots & a_{mn} \end{bmatrix} \begin{bmatrix} b_{11} & b_{12} & \cdots & b_{1j} & \cdots & b_{1o} \\ b_{21} & b_{22} & \cdots & b_{2j} & \cdots & b_{2o} \\ \vdots & \vdots & & \vdots & & \vdots \\ b_{n1} & b_{n2} & \cdots & b_{nj} & \cdots & b_{no} \end{bmatrix}$$
$$= \begin{bmatrix} AB_{11} & AB_{12} & \cdots & AB_{1j} & \cdots & AB_{1o} \\ AB_{21} & AB_{22} & \cdots & AB_{2j} & \cdots & AB_{2o} \\ \vdots & \vdots & & \vdots & & \vdots \\ AB_{i1} & AB_{i2} & \cdots & AB_{ij} & \cdots & AB_{io} \\ \vdots & \vdots & & \vdots & & \vdots \\ AB_{m1} & AB_{m2} & \cdots & AB_{mj} & \cdots & AB_{mo} \end{bmatrix} \quad (2.27)$$

```
>> v = [2.5, -2, 1, 0.5];
>> w = [-5, -1, 0.75, -3];
>> A = [3, 2, -1, 1
5, 3, 2, 1
3, 1, 3, 2
-6,-4, 2, -2];
>> v_T = v'

v_T =

    2.5000
   -2.0000
    1.0000
    0.5000

>> w_T = w'

w_T =

   -5.0000
   -1.0000
    0.7500
   -3.0000

>> v_T + w_T

ans =

   -2.5000
   -3.0000
    1.7500
   -2.5000

>> A*v_T

ans =

    3.0000
    9.0000
    9.5000
   -6.0000

>> v*w_T

ans =

  -11.2500

>>
```

FIGURE E.2.9 Example 2.6 solution calculation procedure in MATLAB.

The subscript i in Equation (2.26) corresponds to the rows in matrix $[A]$, and the subscript j in the same equation corresponds to the columns in matrix $[B]$. The subscript pair ij in Equation (2.26) corresponds to the cells in the matrix product $[A][B]$. Matrix (2.27) includes matrices $[A]$, $[B]$, and $[A][B]$ with an arbitrary row i, column j, and cell ij shaded.

When calculating the product of three or more matrices, the product is calculated from right to left. Therefore, to calculate the product of matrices $[A]$, $[B]$, and $[C]$, for example, the product $[B][C]$ is first calculated, and the result is multiplied by $[A]$ (or $[A][B][C]=[A]([B][C])$).

Example 2.7

Problem Statement: Calculate the matrix products $[A][B]$ and $[A][B][C]$, where

$$[A]=\begin{bmatrix} a_{11} & a_{12} \\ a_{21} & a_{22} \end{bmatrix}, [B]=\begin{bmatrix} b_{11} & b_{12} \\ b_{21} & b_{22} \end{bmatrix}, \text{ and } [C]=\begin{bmatrix} c_{11} & c_{12} \\ c_{21} & c_{22} \end{bmatrix}.$$

Known Information: Given matrices

Solution Approach: Figure E.2.10 includes the calculation procedure in MATLAB's command window.

```
>> syms a11 a12 a21 a22 real;
>> syms b11 b12 b21 b22 real;
>> syms c11 c12 c21 c22 real;
>> A = [a11, a12; a21, a22];
>> B = [b11, b12; b21, b22];
>> C = [c11, c12; c21, c22];
>> A*B

ans =

[ a11*b11 + a12*b21, a11*b12 + a12*b22]
[ a21*b11 + a22*b21, a21*b12 + a22*b22]

>> A*B*C

ans =

[ c11*(a11*b11 + a12*b21) + c21*(a11*b12 + a12*b22), c12*(a11*b11 +
a12*b21) + c22*(a11*b12 + a12*b22)]
[ c11*(a21*b11 + a22*b21) + c21*(a21*b12 + a22*b22), c12*(a21*b11 +
a22*b21) + c22*(a21*b12 + a22*b22)]

>>
```

FIGURE E.2.10 Example 2.7 solution calculation procedure in MATLAB.

2.4.3 The Identity Matrix and Matrix Inversion

The *unit matrix* or *identity matrix* (denoted by I) is a square matrix having the number 1 along its main diagonal (with all other cells having the number 0).[11] The identity matrix is the matrix equivalent of the number 1. The general form of the identity matrix can be expressed as

$$I = \begin{bmatrix} 1 & 0 & 0 & \cdots & 0 \\ 0 & 1 & 0 & \cdots & 0 \\ 0 & 0 & 1 & \cdots & 0 \\ \vdots & \vdots & \vdots & \ddots & \vdots \\ 0 & 0 & 0 & \cdots & 1 \end{bmatrix} \tag{2.28}$$

As noted in Section 2.4.1, expressing a set of linear simultaneous equations in matrix form can be particularly convenient when solving for the variables in the equation set. Considering the matrix form $[A]\mathbf{x} = \mathbf{b}$, the column vector of variables $\mathbf{x}$ is the result of the product $[A]^{-1}\,\mathbf{b}$ or $\mathbf{x} = [A]^{-1}\,\mathbf{b}$, where the superscript -1 in $[A]^{-1}$ represents the *inverse* of $[A]$. Assuming $[A]$ is invertible, its inverse can be defined as

$$[A]^{-1} = \frac{1}{\det[A]}\text{adj}[A] \tag{2.29}$$

where *det* and *adj* are the *determinant* and *adjoint* (two matrix functions) of $[A]$, respectively. Equation (2.29) is used in *Cramer's Rule*—a formula for the solution of linear simultaneous equations having n equations and n unknown variables.

This book does not include descriptions of determinant and adjoint functions. This is in part because the procedures for manually calculating the determinant and adjoint of a matrix become increasingly involved for matrices having dimensions beyond 2×2. Another reason is because matrix inversion is a simple procedure in the mathematical analysis software MATLAB.

For those interested in becoming more familiar with the determinant and adjoint functions (as well as Cramer's Rule), we recommended that you refer to either online resources or books that include the fundamentals of *linear algebra*.[12]

The *order of operations* used in computer programming and arithmetic operations for scalar quantities also applies to matrix operations. This order is as follows: (1) parentheses, (2) exponents, (3) multiplication, (4) division, (5) addition, and (6) subtraction.

Example 2.8

Problem Statement: Calculate $[A]^{-1}$ where

$$[A]=\begin{bmatrix} a_{11} & a_{12} & a_{13} \\ a_{21} & a_{22} & a_{23} \\ a_{31} & a_{32} & a_{33} \end{bmatrix}$$

Known Information: Given matrix

Solution Approach: Figure E.2.11 includes the calculation procedure in MATLAB's command window. It can be observed from scale of $[A]^{-1}$ in Figure E.2.11 that manually producing the inverse of even this 3×3 matrix will be quite involved.

```
>> syms a11 a12 a13 a21 a22 a23 a31 a32 a33 real;
>> A = [a11, a12, a13
a21, a22, a23
a31, a32, a33];
>> Inverse_A = inv(A)

Inverse_A =

[  (a22*a33 -a23*a32)/(a11*a22*a33 -a11*a23*a32 -a12*a21*a33 +
a12*a23*a31 + a13*a21*a32 - a13*a22*a31), -(a12*a33 -
a13*a32)/(a11*a22*a33- a11*a23*a32 - a12*a21*a33 + a12*a23*a31 +
a13*a21*a32 - a13*a22*a31),  (a12*a23 - a13*a22)/(a11*a22*a33 -
a11*a23*a32 - a12*a21*a33 + a12*a23*a31 + a13*a21*a32 -
a13*a22*a31)]
[ -(a21*a33 - a23*a31)/(a11*a22*a33 - a11*a23*a32 - a12*a21*a33 +
a12*a23*a31 + a13*a21*a32 - a13*a22*a31),  (a11*a33 -
a13*a31)/(a11*a22*a33 - a11*a23*a32 - a12*a21*a33 + a12*a23*a31 +
a13*a21*a32 - a13*a22*a31), -(a11*a23 - a13*a21)/(a11*a22*a33 -
a11*a23*a32 - a12*a21*a33 + a12*a23*a31 + a13*a21*a32 -
a13*a22*a31)]
[  (a21*a32 - a22*a31)/(a11*a22*a33 -a11*a23*a32 - a12*a21*a33 +
a12*a23*a31 + a13*a21*a32 - a13*a22*a31), -(a11*a32 -
a12*a31)/(a11*a22*a33 - a11*a23*a32 - a12*a21*a33 + a12*a23*a31 +
a13*a21*a32 - a13*a22*a31),  (a11*a22 -a12*a21)/(a11*a22*a33 -
a11*a23*a32 - a12*a21*a33 + a12*a23*a31 + a13*a21*a32 -
a13*a22*a31)]

>>
```

FIGURE E.2.11 Example 2.8 solution calculation procedure in MATLAB.

Example 2.9

Problem Statement: Calculate $[A]^{-1}$ and the unknown variables in Equation (2.22).

Known Information: Equation (2.22)

Solution Approach: Figure E.2.12 includes the calculation procedure in MATLAB's command window.

```
>> b = [1, -2, 1, 3]'

b =

     1
    -2
     1
     3

>> A = [1, -2, 1, -3
2, 1, 2, -2
-1, 2, -4, 1
3, 0, 0, -3]

A =

     1    -2     1    -3
     2     1     2    -2
    -1     2    -4     1
     3     0     0    -3

>> inv(A)

ans =

   -0.5000   -0.3750   -0.3125    0.6458
         0    0.5000    0.2500   -0.2500
         0    0.2500   -0.1250   -0.2083
   -0.5000   -0.3750   -0.3125    0.3125

>> x = A\b

x =

    1.8750
   -1.5000
   -1.2500
    0.8750

>>
```

FIGURE E.2.12 Example 2.9 solution calculation procedure in MATLAB.

2.5 INTERMEDIATE AND TOTAL SPATIAL MOTION

Matrix (2.30) is a general spatial angular displacement matrix. In this matrix, the rotation angle is represented by variable δ, and the rotation axis is represented by vector $\mathbf{u}$ (Figure 2.4).[13]

$$\left[R_{\delta,\mathbf{u}}\right]=\begin{bmatrix} u_x^2 \mathrm{v}(\delta)+\cos(\delta) & u_x u_y \mathrm{v}(\delta)-u_z\sin(\delta) & u_x u_z \mathrm{v}(\delta)+u_y\sin(\delta) \\ u_x u_y \mathrm{v}(\delta)+u_z\sin(\delta) & u_y^2 \mathrm{v}(\delta)+\cos(\delta) & u_y u_z \mathrm{v}(\delta)-u_x\sin(\delta) \\ u_x u_z \mathrm{v}(\delta)-u_y\sin(\delta) & u_y u_z \mathrm{v}(\delta)+u_x\sin(\delta) & u_z^2 \mathrm{v}(\delta)+\cos(\delta) \end{bmatrix} \tag{2.30}$$

In Matrix (2.30), $\mathrm{v}(\delta)=1-\cos(\delta)$.

Figure 2.5 illustrates two interconnected members in three-dimensional space. In this system of interconnected members, point $\mathbf{p}_1$ about the axis $\mathbf{u}_0$ by a rotation angle θ and point $\mathbf{q}_1$ rotates about the axis $\mathbf{u}_1$ by angle β.[14] Because both members are interconnected, point $\mathbf{q}_1$ and axis $\mathbf{u}_1$ also rotate about the axis $\mathbf{u}_0$. Point $\mathbf{p}_0$ like axis $\mathbf{u}_0$ are fixed or *grounded* in space. An equation for the rotation of $\mathbf{p}_1$ about $\mathbf{u}_0$-or the *total displacement* of $\mathbf{p}_1$ (which will be labeled $\mathbf{p}$) can be expressed as

$$\mathbf{p}=\left[R_{\theta,\mathbf{u}_0}\right]\left(\mathbf{p}_1-\mathbf{p}_0\right)+\mathbf{p}_0 \tag{2.31}$$

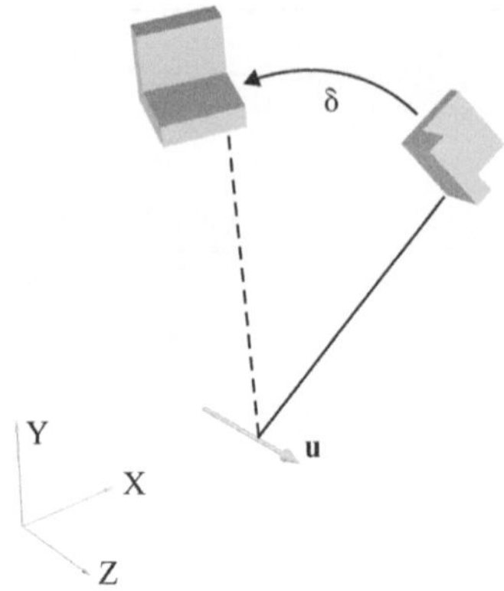

FIGURE 2.4 Spatial rotation of an arbitrary body.

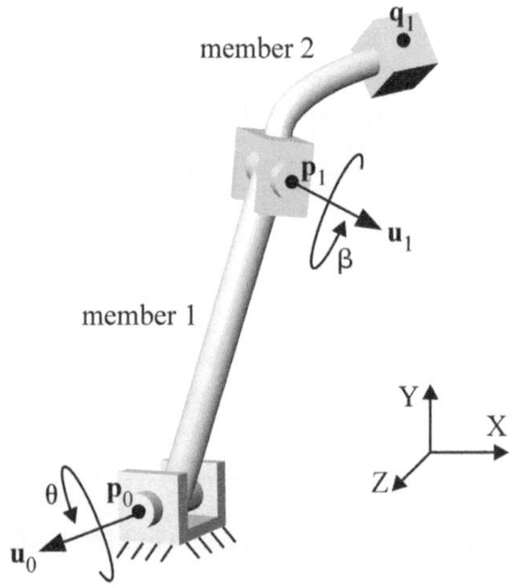

FIGURE 2.5 Spatial motion of a two-body system.

and an equation for the rotation of $\mathbf{u}_1$ about $\mathbf{u}_0$-or the total displacement of $\mathbf{u}_1$ (which will be labeled $\mathbf{u}$) can be expressed as

$$\mathbf{u} = \left[R_{\theta,\mathbf{u}_0}\right]\mathbf{u}_1 \tag{2.32}$$

An equation for the *intermediate displacement* of $\mathbf{q}_1$ about $\mathbf{u}_0$ (which will be labeled $\mathbf{q}'$) can be expressed as

$$\mathbf{q}' = \left[R_{\theta,\mathbf{u}_0}\right](\mathbf{q}_1 - \mathbf{p}_0) + \mathbf{p}_0 \tag{2.33}$$

Equation (2.33) is included in Equation (2.34) to calculate the total displacement of $\mathbf{q}_1$ about $\mathbf{u}_0$ (which will be labeled $\mathbf{q}$).

$$\mathbf{q} = \left[R_{\beta,\mathbf{u}}\right](\mathbf{q}' - \mathbf{p}) + \mathbf{p} \tag{2.34}$$

Matrix (2.35) is a general spatial angular velocity matrix. In this matrix, the angular velocity is represented by variable $\dot{\delta}$, and the rotation axis is represented by vector $\mathbf{u}$.

$$\left[V_{\dot{\delta},\mathbf{u}}\right]=\begin{bmatrix} 0 & -u_z\dot{\delta} & u_y\dot{\delta} \\ u_z\dot{\delta} & 0 & -u_x\dot{\delta} \\ -u_y\dot{\delta} & u_x\dot{\delta} & 0 \end{bmatrix} \tag{2.35}$$

The velocity of $\mathbf{p}_1$ in reference to $\mathbf{u}_0$-or the *total velocity* of $\mathbf{p}_1$ (which will be labeled $\dot{\mathbf{p}}$) can be expressed as

$$\dot{\mathbf{p}}=\left[V_{\dot{\theta},\mathbf{u}_0}\right](\mathbf{p}_1-\mathbf{p}_0) \tag{2.36}$$

An equation for the *intermediate velocity* of $\mathbf{q}_1$ in reference to $\mathbf{u}_0$ (which will be labeled $\dot{\mathbf{q}}'$) can be expressed as

$$\dot{\mathbf{q}}'=\left[V_{\dot{\theta},\mathbf{u}_0}\right](\mathbf{q}-\mathbf{p}_0) \tag{2.37},$$

where variable $\mathbf{q}$ is included from Equation (2.34). Equation (2.37) is included in Equation (2.38) to calculate the total velocity of $\mathbf{q}_1$ in reference to $\mathbf{u}_0$ (which will be labeled $\dot{\mathbf{q}}$)

$$\dot{\mathbf{q}}=\left[V_{\dot{\beta},\mathbf{u}}\right](\mathbf{q}-\mathbf{p})+\dot{\mathbf{q}}' \tag{2.38},$$

where variable $\mathbf{p}$ is included from Equation (2.31).

Matrix (2.39) is a general spatial angular acceleration matrix. In this matrix, the angular velocity and acceleration are represented by variables $\dot{\delta}$ and $\ddot{\delta}$, respectively, and the rotation axis is represented by vector $\mathbf{u}$ from Equation (2.32). In Matrix (2.39), it is assumed that $\dot{\mathbf{u}}=0$ since $\mathbf{u}_1$ typically has a fixed orientation with respect to its associated links [4].[15]

$$\left[A_{\ddot{\delta},\dot{\delta},\mathbf{u}}\right]=\begin{bmatrix} \left(u_x^2-1\right)\dot{\delta}^2 & u_xu_y\dot{\delta}^2-\dot{u}_z\dot{\delta}-u_z\ddot{\delta} & u_xu_z\dot{\delta}^2+\dot{u}_y\dot{\delta}+u_y\ddot{\delta} \\ u_xu_y\dot{\delta}^2+\dot{u}_z\dot{\delta}+u_z\ddot{\delta} & \left(u_y^2-1\right)\dot{\delta}^2 & u_yu_z\dot{\delta}^2-\dot{u}_x\dot{\delta}-u_x\ddot{\delta} \\ u_xu_z\dot{\delta}^2-\dot{u}_y\dot{\delta}-u_y\ddot{\delta} & u_yu_z\dot{\delta}^2-\dot{u}_x\dot{\delta}+u_x\ddot{\delta} & \left(u_y^2-1\right)\dot{\delta}^2 \end{bmatrix} \tag{2.39}$$

The acceleration of $\mathbf{p}_1$ in reference to $\mathbf{u}_0$ or the *total acceleration* of $\mathbf{p}_1$ (which will be labeled $\ddot{\mathbf{p}}$) can be expressed as

$$\ddot{\mathbf{p}}=\left[A_{\ddot{\theta},\dot{\theta},\mathbf{u}_0}\right](\mathbf{p}-\mathbf{p}_0) \tag{2.40}$$

where variable $\mathbf{p}$ is included from Equation (2.31). An equation for the *intermediate acceleration* of $\mathbf{q}_1$ in reference to $\mathbf{u}_0$ (which will be labeled $\ddot{\mathbf{q}}'$) can be expressed as

$$\ddot{\mathbf{q}}'=\left[A_{\ddot{\beta},\dot{\beta},\mathbf{u}_0}\right](\mathbf{q}-\mathbf{p}_0) \tag{2.41}$$

where variable $\mathbf{q}$ is included from Equation (2.34). Equation (2.41) is included in Equation (2.42) to calculate the total acceleration of $\mathbf{q}_1$ in reference to $\mathbf{u}_0$ (which will be labeled $\ddot{\mathbf{q}}$). In Equation (2.42), variables $\mathbf{p}$ and $\mathbf{q}$ are included from Equations (2.31) and (2.34), respectively.

$$\ddot{\mathbf{q}}=\ddot{\mathbf{q}}'+\left[A_{\ddot{\beta},\dot{\beta},\mathbf{u}}\right](\mathbf{q}-\mathbf{p})+2\left[V_{\dot{\theta},\mathbf{u}_0}\right]\left\{\left[V_{\dot{\beta},\mathbf{u}}\right](\mathbf{q}-\mathbf{p})\right\} \tag{2.42}$$

The intermediate and total spatial displacement velocity and acceleration equations appear in the spatial mechanism equation systems in Chapter 8 [3].

Example 2.10

Problem Statement: For the two-body system in Figure 2.5, calculate the displaced values of points $\mathbf{p}_1$, $\mathbf{q}_1$, and axis $\mathbf{u}_1$. The dimensions and rotation angles in this system are as follows:

$\mathbf{p}_0 = (0,0,0)$, $\mathbf{p}_1 = (0,1,0)$, $\mathbf{u}_0 = (0,0,1)$, $\mathbf{u}_1 = (0.7071,0,0.7071)$, $\mathbf{q}_1 = (0.25,1.3536,-0.25)$, $\theta = 30°$, and $\beta = -15°$.

Known Information: Given dimensions, rotation angles, Matrix (2.30), and Equations (2.31) through (2.34)

Solution Approach: Figure E.2.13 includes the calculation procedure in MATLAB's command window.

```
>> p0 = [0, 0, 0]';
>> p1 = [0, 1, 0]';
>> q1 = [0.25, 1.3536, -0.25]';
>> u0 = [0, 0, 1];
>> u0x = u0(1); u0y = u0(2); u0z = u0(3);
>> u1 = [0.7071, 0, 0.7071]';
>> theta = 30*pi/180;
>> beta = -15*pi/180;
>> C = cos(theta);
>> S = sin(theta);
>> V = 1 - C;
>> R_theta_u0 = [...
V*u0x^2 + C, V*u0x*u0y - S*u0z, V*u0x*u0z + S*u0y
V*u0x*u0y + S*u0z, V*u0y^2 + C, V*u0y*u0z - S*u0x
V*u0x*u0z - S*u0y, V*u0y*u0z + S*u0x, V*u0z^2 + C];
>> u = R_theta_u0*u1

u =

    0.6124
    0.3535
    0.7071

>> p = R_theta_u0*(p1 - p0) + p0

p =

   -0.5000
    0.8660
         0

>> q_prime = R_theta_u0*(q1 - p0) + p0;
>> C = cos(beta);
>> S = sin(beta);
>> V = 1 - C;
>> ux = u(1); uy = u(2); uz = u(3);
>> R_beta_u = [...
V*ux^2 + C, V*ux*uy - S*uz, V*ux*uz + S*uy
V*ux*uy + S*uz, V*uy^2 + C, V*uy*uz - S*ux
V*ux*uz -S*uy, V*uy*uz + S*ux, V*uz^2 + C];
>> q = R_beta_u*(q_prime - p) + p

q =

   -0.3599
    1.2357
   -0.3062

>>
```

FIGURE E.2.13 Example 2.10 solution calculation procedure in MATLAB.

2.6 GENERAL TRANSFORMATION MATRIX

Matrices (2.43) through (2.45) are rotation matrices about the x, y, and z-axes of a global coordinate frame, respectively. Given the rotation angle value, the product of any of these matrices and coordinates of a point (in a 3×1 column matrix) are the coordinates of the rotated point.

$$\left[R_{\delta_x}\right] = \begin{bmatrix} 1 & 0 & 0 \\ 0 & \cos\delta_x & -\sin\delta_x \\ 0 & \sin\delta_x & \cos\delta_x \end{bmatrix} \tag{2.43}$$

$$\left[R_{\delta_y}\right] = \begin{bmatrix} \cos\delta_y & 0 & \sin\delta_y \\ 0 & 1 & 0 \\ -\sin\delta_y & 0 & \cos\delta_y \end{bmatrix} \tag{2.44}$$

$$\left[R_{\delta_z}\right] = \begin{bmatrix} \cos\delta_z & -\sin\delta_z & 0 \\ \sin\delta_z & \cos\delta_z & 0 \\ 0 & 0 & 1 \end{bmatrix} \tag{2.45}$$

The product of the three matrices can be expressed as Matrix (2.46)—which can accommodate simultaneous rotations about the *x*, *y*, and *z*-axes (by rotation angles δ_x, δ_y, and δ_z, respectively).

$$\begin{aligned} [R] &= \left[R_{\delta_z}\right]\left[R_{\delta_y}\right]\left[R_{\delta_x}\right] \\ &= \begin{bmatrix} R_{11} & R_{12} & R_{13} \\ R_{21} & R_{22} & R_{23} \\ R_{31} & R_{32} & R_{33} \end{bmatrix} \\ &= \begin{bmatrix} \cos\delta_y\cos\delta_z & (\sin\delta_x\sin\delta_y\cos\delta_z - \cos\delta_x\sin\delta_z) & (\cos\delta_x\sin\delta_y\cos\delta_z + \sin\delta_x\sin\delta_z) \\ \cos\delta_y\sin\delta_z & (\sin\delta_x\sin\delta_y\sin\delta_z + \cos\delta_x\cos\delta_z) & (\cos\delta_x\sin\delta_y\sin\delta_z - \sin\delta_x\cos\delta_z) \\ -\sin\delta_y & \sin\delta_x\cos\delta_y & \cos\delta_x\cos\delta_y \end{bmatrix} \end{aligned} \tag{2.46}$$

Including the elements of Matrix (2.46) into a 4×4 matrix that also considers translations along the *x*, *y*, and *z*-axes (in the fourth matrix column) produces

$${}^{i}_{j}[T] = \begin{bmatrix} R_{11} & R_{12} & R_{13} & \Delta_x \\ R_{21} & R_{22} & R_{23} & \Delta_y \\ R_{31} & R_{32} & R_{33} & \Delta_z \\ 0 & 0 & 0 & 1 \end{bmatrix} \tag{2.47}$$

Matrix (2.47) is a *general transformation matrix* for calculating point coordinates given in one coordinate frame (which we will call frame *j*) in reference to another coordinate frame (which we will call frame *i*) or

$${}^{i}\{\mathbf{p}\} = {}^{i}_{j}[T]\,{}^{j}\{\mathbf{p}\} \tag{2.48}$$

In Equation (2.48), the coordinates in frame *j* or ${}^{j}\{\mathbf{p}\}$ are ${}^{j}\{\mathbf{p}\} = \begin{bmatrix} p_x & p_y & p_z & 1 \end{bmatrix}^T$. Matrix (2.47) is general because it can accommodate all six possible degrees of freedom (*x*-*y*-*z* rotations and *x*-*y*-*z* translations). The rotation angles δ_x, δ_y, and δ_z and translation values Δ_x, Δ_y, and Δ_z in Matrix (2.47) are the angular and linear displacement values required to align frame *i* to frame *j*.

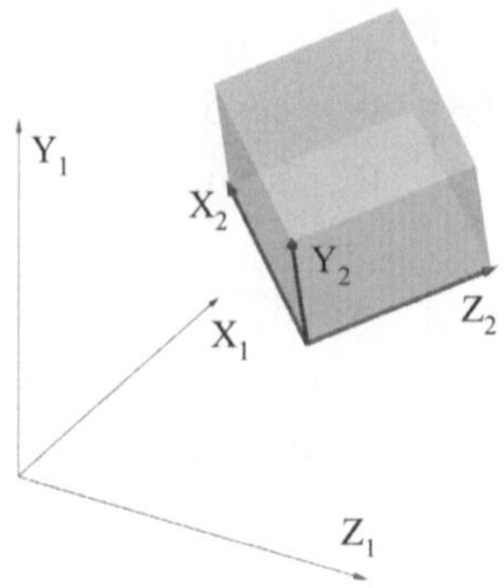

FIGURE 2.6 Coordinate frames i and j (where $i = 1$ and $j = 2$).

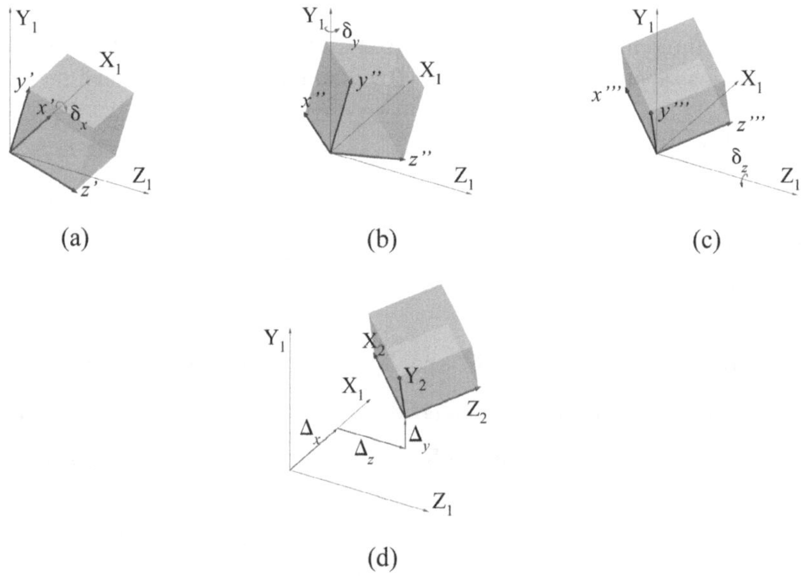

FIGURE 2.7 (a–c) Coordinate frame rotations about and (d) translations along X_1, Y_1, and Z_1-axes.

As an example, Figure 2.6 illustrates a coordinate frame i (where $i = 1$) and arbitrary coordinate frame j (where $j = 2$). Using rotation angles $\delta_x = 20°$, $\delta_y = 40°$, and $\delta_z = 60°$ (Figure 2.7a through c, respectively) and translation values $\Delta_x = 2, \Delta_y = \Delta_z = 1$ (Figure 2.7d), frame 1 is aligned with frame 2. As a result, when these rotation and translation values are used in Equation (2.48), any point coordinates given in reference to frame 2 will be calculated in reference to frame 1. Transformation matrices are used in Chapter 10 for the kinematic analysis of robotic manipulators [5].

Example 2.11

Problem Statement: The coordinates of point $\mathbf{p}_1$ in reference frame 2 of a robotic system are ${}^{2}\{\mathbf{p}_1\} = [2,5,-1,1]^T$ (see Figure E.2.14). Calculate the coordinates of this point in reference frame 1 of the system $\left({}^{1}\{\mathbf{p}_1\}\right)$. The location of the origin of frame 2 with respect to frame 1 is $\Delta = (5,10,-2)$, and the orientation angles of frame 2 with respect to frame 1 are $\delta_x = 0°$, $\delta_y = 15°$, and $\delta_z = 30°$.

Known Information: Given frame rotation and displacement values and Equation (2.48)

Solution Approach: Figure E.2.15 includes the calculation procedure in MATLAB's command window.

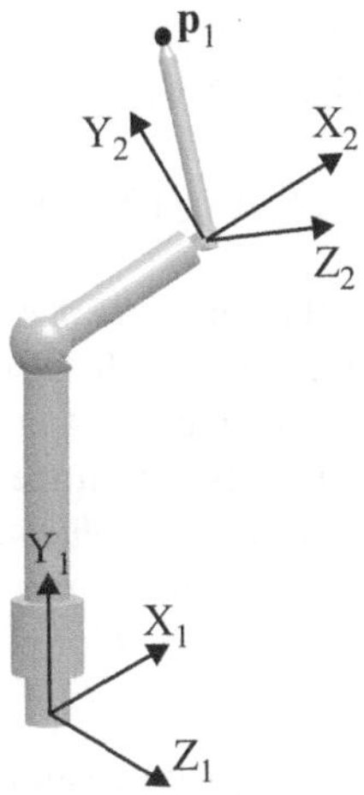

FIGURE E.2.14 Robotic system with reference frames.

```
>> p2 = [2, 5, -1, 1]';
>> Dx = 5;
>> Dy = 10;
>> Dz = -2;
>> delta_x = 0*pi/180;
>> delta_y = 15*pi/180;
>> delta_z = 30*pi/180;
>> Sx = sin(delta_x); Cx = cos(delta_x);
>> Sy = sin(delta_y); Cy = cos(delta_y);
>> Sz = sin(delta_z); Cz = cos(delta_z);
>> T12 = [...
Cy*Cz, Sx*Sy*Cz - Cx*Sz, Cx*Sy*Cz + Sx*Sz, Dx
Cy*Sz, Sx*Sy*Sz + Cx*Cz, Cx*Sy*Sz - Sx*Cz, Dy
-Sy, Sx*Cy, Cx*Cy, Dz
0, 0, 0, 1];
>> p1 = T12*p2

p1 =

    3.9489
   15.1666
   -3.4836
    1.0000

>>
```

FIGURE E.2.15 Example 2.11 solution calculation procedure in MATLAB.

2.7 NONLINEAR EQUATION SYSTEM OPTIMIZATION

2.7.1 Unconstrained Nonlinear Equation System

In comparison to linear simultaneous equations, in *nonlinear simultaneous equations*, at least one unknown variable is nonlinear. Examples of nonlinear variables include x^n (where $n \neq 1$), $\cos(x)$, and $\log(x)$. MATLAB includes *Trust-Region*, *Trust-Region-Dogleg,* and *Levenberg–Marquardt* algorithms for nonlinear simultaneous equations system optimization [6–8]. Linear simultaneous equations are used in Chapters 4, 5, and 6 for planar kinematic analysis, planar motion generation, and planar function generation, respectively. Nonlinear simultaneous equations are used in Chapter 4 and Chapters 6–10 for planar kinematic analysis, planar and spatial dimensional synthesis, and robotic system inverse kinematics.

By moving all equation terms to one side, each equation in the nonlinear simultaneous equation set takes the form $f_i(\mathbf{X}) = 0$, where $\mathbf{X}$ includes all unknown variables (therefore, $\mathbf{X} = (x_1, x_2, x_3, \ldots, x_n)$).

A single function can then be formed, which is the squared sum of each equation in the set. This function, which we will refer to as $f(\mathbf{X})$, is called an *objective function* and takes the form

$$f(\mathbf{X}) = [f_1(\mathbf{X})]^2 + [f_2(\mathbf{X})]^2 + \cdots + [f_n(\mathbf{X})]^2$$

The unknown variable values that solve the simultaneous equation set will also produce an objective function value of 0. In nonlinear optimization (where solutions are approximated), the goal is to calculate the variable values that produce an objective function local minimum.

A problem where an objective function is minimized without any additional conditions is called an *unconstrained nonlinear optimization problem.* MATLAB includes *Trust-Region* and *Quasi-Newton* algorithms for unconstrained nonlinear optimization [7,9,10]. Unconstrained nonlinear optimization models are used in Chapters 6–9 for planar and spatial dimensional synthesis.

2.7.2 Nonlinear Equation System with Nonlinear Equality and Inequality Constraints and Linear Inequality Constraints

Another nonlinear optimization problem considered in this book is one where the objective function is minimized and linear inequality constraints are satisfied. The additional linear inequality constraints limit the minimum and maximum possible values for unknown variables. This type of problem is called a *nonlinear optimization problem with linear inequality constraints.* Unconstrained nonlinear optimization models with linear inequality constraints are used in Chapters 6–9 for planar and spatial dimensional synthesis with order defect elimination.

If additional nonlinear equality constraints and nonlinear inequality constraints are included to further limit the possible values for the unknown variables, this type of problem is called a *nonlinear optimization problem with nonlinear equality and inequality constraints and linear inequality constraints.* MATLAB includes *Interior-Point, Sequential Quadratic Programming* (SQP), and *Active-Set* algorithms for constrained nonlinear optimization of both types [11–14]. Unconstrained nonlinear optimization models with nonlinear equality constraints, nonlinear inequality constraints, and linear inequality constraints are used in Chapters 6–9 for planar and spatial dimensional synthesis with order defect elimination and branch defect elimination.

2.8 GRADIENT VECTOR AND HESSIAN MATRIX

One vector and matrix are commonly employed in nonlinear optimization algorithms due to their effectiveness in ultimately determining the equation system variables for local minima. These matrices are the *gradient vector* and *Hessian matrix.*

For a nonlinear differentiable multivariable function $f(\mathbf{X})$, where $\mathbf{X} = (x_1, x_2, x_3, \ldots, x_n)$, its *gradient* ∇f gives the direction and rate of the fastest increase at a given point. The gradient plays a fundamental role in nonlinear optimization because it is utilized in *gradient descent*—an iterative method for finding the local minima of a differentiable multivariable function. The gradient is a vector of the first partial derivatives of a function and is expressed in Equation (2.49).

$$\nabla f = \begin{pmatrix} \dfrac{\partial f}{\partial x_1} \\ \dfrac{\partial f}{\partial x_2} \\ \vdots \\ \dfrac{\partial f}{\partial x_n} \end{pmatrix} \tag{2.49}$$

While the gradient is a vector of the first partial derivatives of a function, its *Hessian* $\nabla^2 f$ or $\mathbf{H}(f)$ is a matrix of its second partial derivatives and is expressed in Equation (2.50). It describes the curvature of a differentiable multivariable function. The Hessian matrix is useful in nonlinear optimization because it can help determine the saddle points and local extremum of a function within a specific interval.

$$\nabla^2 f = \mathbf{H}(f) = \begin{pmatrix} \frac{\partial^2 f}{\partial x_1^2} & \frac{\partial^2 f}{\partial x_1 \partial x_2} & \cdots & \frac{\partial^2 f}{\partial x_1 \partial x_n} \\ \frac{\partial^2 f}{\partial x_2 \partial x_1} & \frac{\partial^2 f}{\partial x_2^2} & & \frac{\partial^2 f}{\partial x_2 \partial x_n} \\ \vdots & & \ddots & \vdots \\ \frac{\partial^2 f}{\partial x_n \partial x_1} & \frac{\partial^2 f}{\partial x_n \partial x_2} & \cdots & \frac{\partial^2 f}{\partial x_n^2} \end{pmatrix} \tag{2.50}$$

Example 2.12

Problem Statement: Formulate the gradient vector and the Hessian matrix for the function $f(x,y,z) = x^3 + 10xy^2 - \sqrt[3]{z}$.

Known Information: Given function and Equations (2.49) and (2.50)

Solution Approach: Figure E.2.16 includes the calculation procedure in MATLAB's command window.

```
>> syms x y z real
>> f = x^3 + 10*x*y^2 -z^(1/3);
>> Grad_f = gradient(f, [x y z])

Grad_f =

3*x^2 + 10*y^2
        20*x*y
-1/(3*z^(2/3))

>> Hess_f = hessian(f, [x y z])

Hess_f =

[ 6*x, 20*y,              0]
[20*y, 20*x,              0]
[   0,    0, 2/(9*z^(5/3))]

>>
```

FIGURE E.2.16 Example 2.12 solution calculation procedure in MATLAB.

2.9 SUMMARY

Vectors (quantities having both *magnitude* and *direction*) are commonly used in the formulation of equation systems for mechanisms. One way to represent a two-dimensional vector is with a complex number. A complex number is comprised of both a real component and an *imaginary component*. Equation systems for mechanisms can be formulated by producing vector loops for the mechanisms and taking the sum of the individual vector terms in the loops. The first and second derivatives of vector loop equations are taken to calculate mechanism velocities and accelerations.

In addition to vectors, points can also be used to formulate mechanism displacement, velocity, and acceleration equations. In fact, a vector can be produced from the coordinates of two points in 2D or 3D space. Point-based vectors appear in spatial mechanism equation systems such as intermediate and total displacement, velocity, and acceleration equations.

A linear equation is an equation that includes linear or first-order variables. A system of linear equations (or a linear system) is a collection of linear equations including the same variables. If a common solution is sought among a system of linear equations, it is called a set of simultaneous equations.

A set of linear simultaneous equations having *n* equations and *n* unknown variables can be expressed in matrix form. A matrix is a rectangular array of numbers, symbols, or expressions arranged in rows and columns. This form can be particularly convenient when calculating the unknown variables in a simultaneous equation set. Cramer's Rule is a popular matrix-based formula for the solution of linear simultaneous equations having *n* equations and *n* unknown variables.

The transformation matrix, commonly used in the analysis of robotic systems, is used to calculate point coordinates given in one reference frame (frame *j*) in terms of another reference frame (frame *i*). The general spatial transformation matrix can consider up to all six possible degrees of freedom (*x*, *y*, and *z* rotations and translations).

In comparison to linear simultaneous equations which include first-order (or linear) unknown variables, in nonlinear simultaneous equations, at least one unknown variable is nonlinear. MATLAB includes Trust-Region, Trust-Region-Dogleg, and Levenberg–Marquardt algorithms for nonlinear simultaneous equation system optimization.

A problem where an objective function is minimized without any additional conditions is called an unconstrained nonlinear optimization problem. MATLAB includes Trust-Region and Quasi-Newton algorithms for unconstrained nonlinear optimization. Other nonlinear optimization problems considered in this textbook include objective functions with linear inequality constraints, nonlinear inequality constraints, and nonlinear equality constraints. MATLAB includes Interior-Point, Sequential Quadratic Programming (SQP), and Active-Set algorithms for these constrained nonlinear optimization problem types.

Vectors (quantities having both *magnitude* and *direction*) are commonly used in the formulation of equation systems for mechanisms. One way to represent a two-dimensional vector is with a *complex number.* A complex number is comprised of both a *real component* and an *imaginary component.* Equation systems for mechanisms can be formulated by producing vector loops for the mechanisms and taking the sum of the individual vector terms in the loops. The first and second derivatives of vector loop equations are taken to calculate mechanism velocities and accelerations.

In addition to vectors, *points* can also be used to formulate mechanism displacement, velocity, and acceleration equations. In fact, a vector can be produced from the coordinates of two points in 2D or 3D space. Point-based vectors appear in such spatial mechanism equation systems as *intermediate* and *total* displacement, velocity, and acceleration equations.

A *linear equation* is an equation that includes linear or *first-order* variables. A system of linear equations (or a *linear system*) is a collection of linear equations including the same variables. If a common solution is sought among a system of linear equations, it is called a set of *simultaneous equations.*

A set of linear simultaneous equations having *n* equations and *n* unknown variables can be expressed in matrix form. A *matrix* is a rectangular array of numbers, symbols, or expressions arranged in rows and columns. This form can be particularly convenient when calculating the unknown variables in a simultaneous equation set. *Cramer's Rule* is a popular matrix-based formula for the solution of linear simultaneous equations having *n* equations and *n* unknown variables.

The *transformation matrix*, commonly used in the analysis of robotic systems, is used to calculate point coordinates given in one reference frame (frame *j*) in terms of another reference frame (frame *i*). The *general spatial transformation matrix* can consider up to all six possible degrees of freedom (*x*, *y*, and *z* rotations and translations).

Five types of equations are used throughout this book for planar kinematics, planar and spatial dimensional mechanism synthesis, and robotic system inverse kinematics. Two of the five equation types are *linear* and *nonlinear* simultaneous equations. The third type is the *unconstrained nonlinear equation.* The fourth type is the *constrained nonlinear equation system with linear inequality constraints.* The final type is the *constrained nonlinear equation system with linear inequality constraints, nonlinear inequality constraints, and nonlinear equality constraints.*

PROBLEMS

1. Formulate an equation for the vector loop illustrated in Figure P.2.1. Consider that vector $\mathbf{V}_j$ always lies along the real axis.

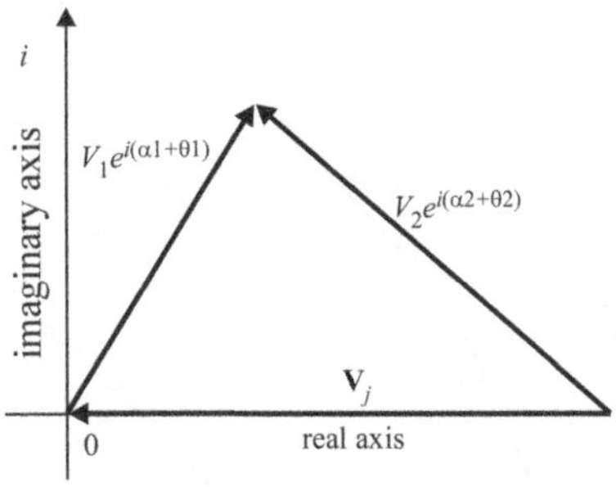

FIGURE P.2.1 Vector loop (three vectors where $\mathbf{V}_j$ changes length) in 2D complex space.

2. Formulate an equation for the vector loop illustrated in Figure P.2.2. Consider that vector $\mathbf{V}_j$ always lies along the real axis and vector $\mathbf{V}_3$ is always perpendicular to the real axis.

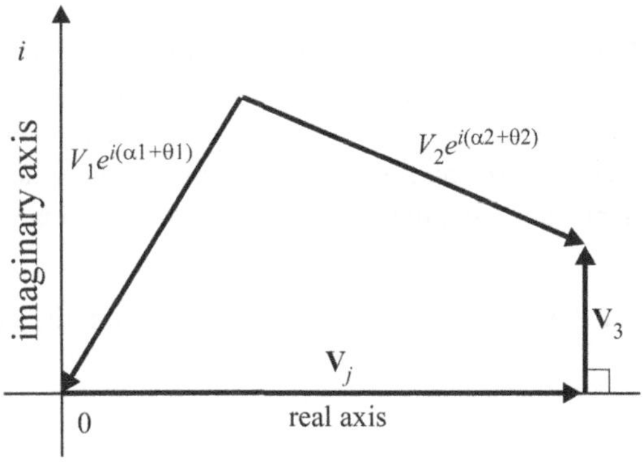

FIGURE P.2.2 Vector loop (four vectors where $\mathbf{V}j$ changes length) in 2D complex space.

3. Calculate the first derivative of the vector loop equation solution from Problem 2. Consider only angles α_1, α_2, and vector $\mathbf{V}_j$ from Problem 2 to be time-dependent.
4. Calculate the second derivative of the vector loop equation solution from Problem 2. Consider only angles α_1, α_2, and vector $\mathbf{V}_j$ from Problem 2 to be time-dependent.
5. Formulate an equation for the vector loop illustrated in Figure P.2.3.

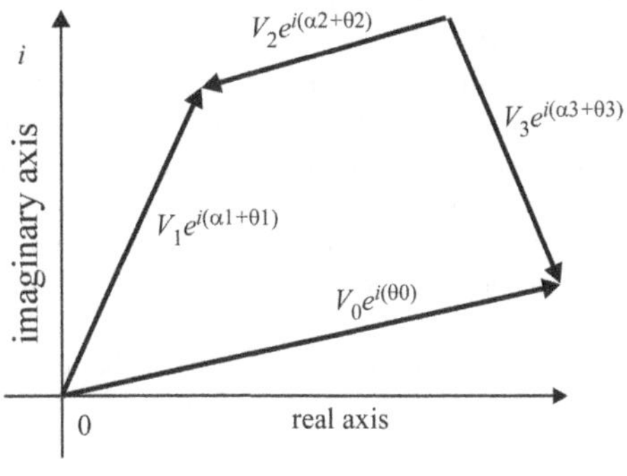

FIGURE P.2.3 Vector loop (four vectors) in 2D complex space.

6. Calculate the first derivative of the vector loop equation solution from Problem 5. Consider only angles α_1, α_2, and α_3 from Problem 5 to be time-dependent.
7. Calculate the second derivative of the vector loop equation solution from Problem 5. Consider only angles α_1, α_2, and α_3 from Problem 5 to be time-dependent.
8. Formulate an equation for the vector loop illustrated in Figure P.2.4.

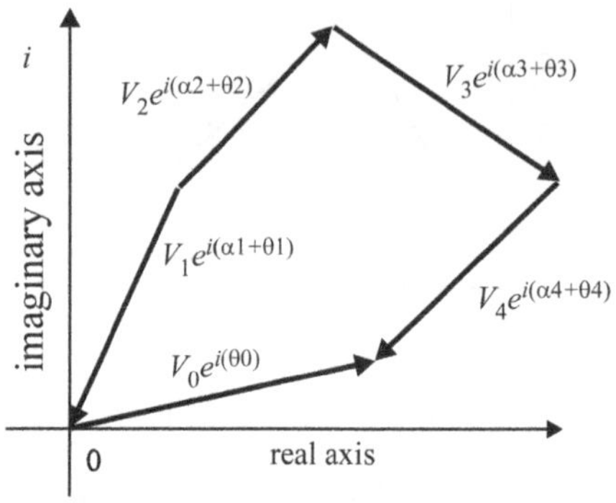

FIGURE P.2.4 Vector loop (five vectors) in 2D complex space.

9. Calculate the first derivative of the vector loop equation solution from Problem 8. Consider only angles α_1, α_2, α_3, and α_4 from Problem 8 to be time-dependent.
10. Calculate the second derivative of the vector loop equation solution from Problem 8. Consider only angles α_1, α_2, α_3, and α_4 from Problem 8 to be time-dependent.
11. Formulate the gradient vector and the Hessian matrix for the function $f(x,y,z) = (3x+y)(y-4z)(2.5x+1.75z)$.
12. Formulate the gradient vector and the Hessian matrix for the function $f(x_1,x_2,x_3) = \frac{\pi x_2 \sin(x_1)}{2} - \frac{1}{x_3}$.

NOTES

1 Therefore, $V_{1x} = V_1 \cos\theta_1$ and $V_{1y} = V_1 \sin\theta_1$.
2 Therefore, $i^2 = -1$, $i^3 = -i$, and $i^4 = 1$ (which are also 90° counter-clockwise rotations each in the 2D complex space).
3 The vector polar forms are the result of *Euler's formula*. In this formula, $e^{i\theta} = \cos\theta + i\sin\theta$.
4 It can be more convenient to use the rectangular form of complex numbers for vector addition.
5 It can be more convenient to use the polar exponential form of complex numbers for vector multiplication (or differentiation).
6 If $\mathbf{V}_1$ represented a link having a length that changes over time, its derivative would also include the vector length derivative term $d\mathbf{V}_1/dt$ or $\dot{\mathbf{V}}_1$.
7 The length of a vector $\mathbf{V}$ is often called the *norm* of $\mathbf{V}$ and is denoted by $\|\mathbf{V}\|$.
8 A first-order variable is one that is of degree 1. Considering a first-order variable, we will arbitrarily define as x, examples of variables that are not first order include x^n (where $n \neq 1$), $\cos(x)$, and $\log(x)$.
9 The transpose of the first matrix is taken to make the number of columns in the first matrix equal to the number of rows in the second matrix—a requirement for the multiplication of any two matrices.
10 Equation (2.26) is the result of the product of a row in matrix $[A]$ and the corresponding column in matrix $[B]$.
11 The main diagonal runs from the top-left matrix corner to the bottom-right matrix corner.
12 *Linear algebra* is the branch of mathematics concerning vector spaces and linear mappings between such spaces.

13 By specifying a rotation axis of $\mathbf{u} = (0,0,1)$, Matrix (2.30) is restricted to rotations in two-dimensional space (as well as Matrix (2.35) and Matrix (2.39)). In this condition, the z-components of the point coordinates could all be specified to 0.
14 Because Matrix (2.30) is a 3×3 matrix, points $\mathbf{p}_0$, $\mathbf{p}_1$, $\mathbf{q}_1$, and axes $\mathbf{u}_0$ and $\mathbf{u}_1$ are 3×1 *column matrices* (where vector rows 1, 2, and 3 include the x, y, and z-components respectively).
15 In the RRSS and 4R spherical mechanisms in Chapter 8, the condition $\dot{\mathbf{u}} = 0$ holds true.

REFERENCES

1. Norton, R. L. 2008. *Design of Machinery*, 4th Edition. pages 186–199. New York: McGraw-Hill.
2. Norton, R. L. 2008. *Design of Machinery*, 4th Edition. pages 309–318. New York: McGraw-Hill.
3. Russell, K., Shen, Q. and Sodhi, R. 2013. *Mechanism Design: Visual and Programmable Approaches.* chapter 7. Boca Raton: CRC Press.
4. Suh, C. H. and Radcliffe, C. W. 1978. *Kinematics and Mechanisms Design.* page 85. New York: John Wiley & Sons.
5. Wilson, C. E. and Peter Sadler, J. 2003. *Kinematics and Dynamics of Machinery*, 3rd Edition, chapter 12. Saddle River: Prentice Hall.
6. "Equation Solving Algorithms," Available at: https://www.mathworks.com. (Accessed on: August 15, 2024).
7. Nocedal, J. and Wright, S. J. 2006. *Numerical Optimization*, 2nd Edition, chapter 4. New York: Springer.
8. Nocedal, J. and Wright, S. J. 2006. *Numerical Optimization*, 2nd Edition, chapter 10. New York: Springer.
9. "Unconstrained Nonlinear Optimization Algorithms," Available at: https://www.mathworks.com. (Accessed on: August 15, 2024).
10. Nocedal, J. and Wright, S. J. 2006. *Numerical Optimization*, 2nd Edition, chapter 6. New York: Springer.
11. "Choosing the Algorithm," Available at: https://www.mathworks.com. (Accessed on: August 15, 2024).
12. Nocedal, J. and Wright, S. J. 2006. *Numerical Optimization*, 2nd Edition, chapter 19. New York: Springer.
13. Nocedal, J. and Wright, S. J. 2006. *Numerical Optimization*, 2nd Edition, chapter 18. New York: Springer.
14. Nocedal, J. and Wright, S. J. 2006. *Numerical Optimization*, 2nd Edition, chapter 16. New York: Springer.

ADDITIONAL READING

Craig, J. J. 2005. *Introduction to Robotics: Mechanics and Control*, 3rd Edition, chapter 2. Upper Saddle River: Pearson Prentice Hall.

Suh, C. H. and Radcliffe, C. W. 1978. *Kinematics and Mechanisms Design.* pages 67–73. New York: John Wiley & Sons.

3 Fundamental Concepts in Kinematics

CONCEPT OVERVIEW

In this chapter, the reader will gain a central understanding regarding the following:

1. Kinematic and design distinctions among select planar and spatial mechanisms
2. Mechanism components and mechanism construction
3. Mechanism *mobility*, *Gruebler's equation*, and *number synthesis*
4. *Grashof Criteria* and the *circuit defect*
5. The *transmission angle* and its relationship with follower link forces
6. *Mechanism inversion*
7. The *passive degree of freedom* and paradoxes to Gruebler's equation

3.1 TYPES OF PLANAR AND SPATIAL MECHANISMS

3.1.1 Planar Four-Bar Mechanism

Figure 3.1a illustrates a *planar four-bar mechanism.* The four interconnected links in this mechanism are the *crank* (the driving link*), coupler, follower,* and *ground.*[1] An input rotation supplied to the crank link compels the motion of the coupler and follower links. Both the crank and follower

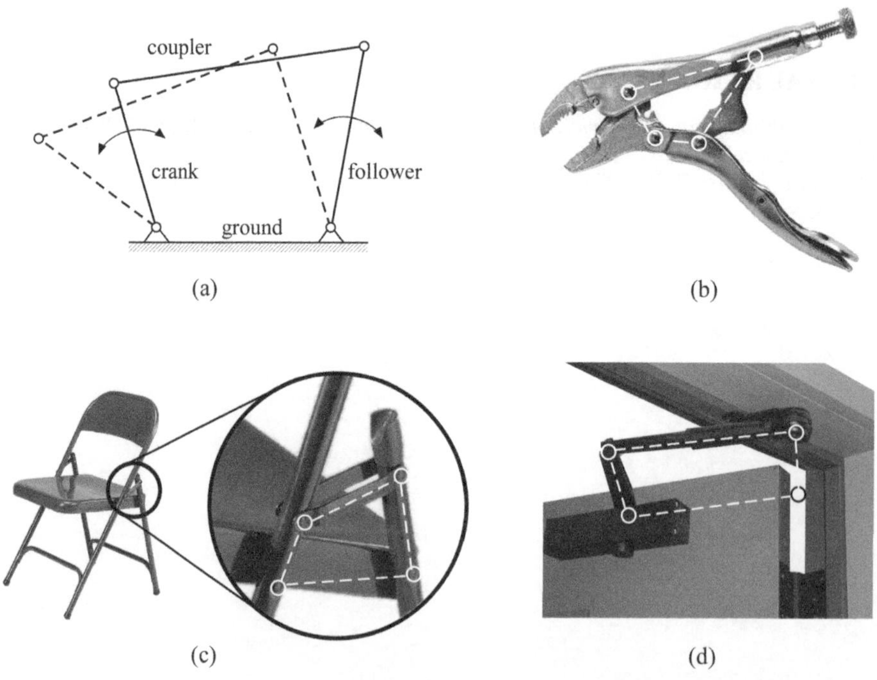

FIGURE 3.1 (a) Planar four-bar mechanism as (b) lock pliers, (c) folding chair, and (d) doorway linkages.

 DOI: 10.1201/9781003591375-3

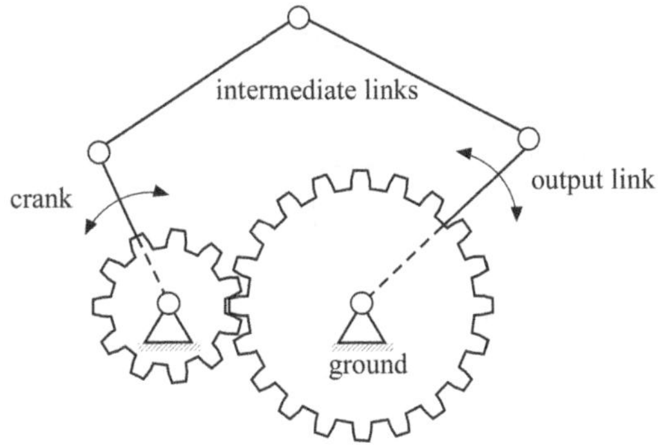

FIGURE 3.2 Geared five-bar mechanism.

links are connected to the ground link (or are *grounded*) and undergo pure rotation (as indicated by the arrows).[2] The planar four-bar mechanism is one of the most widely utilized kinematic chains in everyday devices (Figure 3.1b–d) including locking pliers, folding chairs, and doorways.

3.1.2 Geared Five-Bar Mechanism and Stephenson III Mechanism

Figure 3.2 illustrates a *geared five-bar mechanism.* This mechanism is comprised of five interconnected links where the crank and output link are generally coupled to each other through a gear pair or gear train.[3,4] Because the crank and output links are indirectly interconnected, the input motion supplied to the crank compels the motion of the output and intermediate links. Both the crank and output links of the geared five-bar mechanism undergo pure rotation, whereas the two intermediate links undergo complex motion. This mechanism has been utilized in a range of specialty applications, including vegetable seedling handling, ornithopter design, and forelimb prosthetic design for quadrupeds [1–3].

Though not as commonly utilized in everyday devices as the planar four-bar mechanism, one advantage the planar five-bar mechanism has over the four-bar mechanism is that it can trace paths of higher orders than the latter. This means that the intermediate links of the geared five-bar mechanism have the capacity to trace paths of more complex curvature than the paths traced by the coupler link of the planar four-bar linkage. Using Equation (3.1) (published by Wunderlich in 1963) to calculate the maximum order (*m*) of a link point curve from a planar mechanism that forms a loop of *n* links connected by revolute joints, it can be determined that while a planar four-bar mechanism has a maximum curve order of 6, the planar five-bar mechanism has a maximum order of 10 (or 10.392 to be more exact) [4].

$$m = 2 \cdot 3^{\left(n/2 - 1\right)} \tag{3.1}$$

Attaching a grounded link pair or a *dyad* (the link pair **c**–**d**–**e** in Figure 3.3) to the coupler of the planar four-bar mechanism produces a *planar multi-loop six-bar mechanism.* This particular type of multi-loop planar six-bar mechanism is called a *Stephenson-type III mechanism* [5]. This mechanism includes three links that undergo pure rotation (as indicated by the arrows in Figure 3.3) and the remaining two moving links undergo complex motion. Like the geared five-bar mechanism, the Stephenson III mechanism has been utilized in specialty applications including prosthetic knee design and physical rehabilitation system design [6,7].

The Stephenson III mechanism is one of a class of Stephenson-type mechanisms that includes two others: *Stephenson I* and *Stephenson II* (Figure 3.4). The input motion supplied to the crank (for example, any one of the grounded links in Figure 3.4) compels the motion of the remaining links.

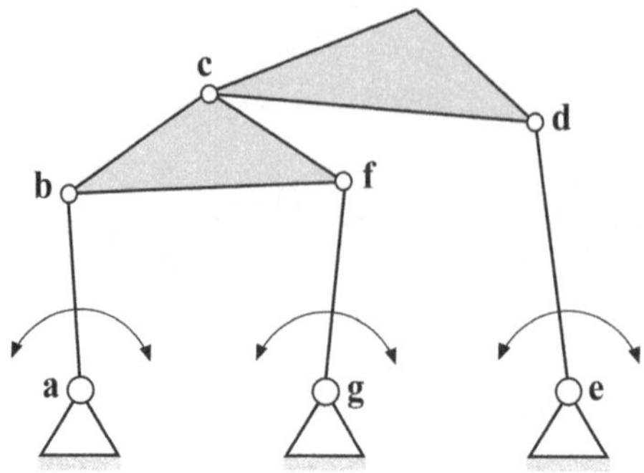

FIGURE 3.3 Planar multi-loop six-bar mechanism (Stephenson III six-bar mechanism).

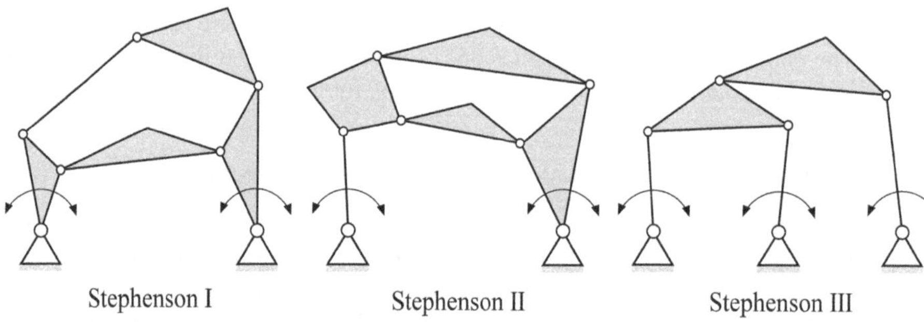

FIGURE 3.4 Stephenson mechanism types.

Stephenson mechanisms include two to three intermediate links that undergo complex motion. This capability enables these mechanisms to deliver both dual and simultaneous *motion* and *path generation* capabilities.[5]

3.1.3 Spatial Four-Bar Mechanisms

Planar mechanisms are restricted to motion in two-dimensional or planar space. *Spatial* mechanisms can exhibit three-dimensional or spatial motion. Spatial mechanism motion is predominantly determined by mechanism joint mobility and joint spatial orientation.

Because spatial mechanisms have the capacity to exhibit spatial motion, they offer a greater variety of possible motions and are structurally more general than planar mechanisms. But because the equations for spatial mechanism analysis and synthesis are often much larger in scale and greater in complexity than those for planar mechanisms, their applications in practice are often limited. In practice, it is not uncommon to find complicated planar mechanism solutions when in fact a simpler spatial mechanism solution is possible. It is, therefore, an ongoing task to devise simple methods of calculation, to produce design aids with diagrams, and to set design standards for spatial mechanisms [8].

Like planar mechanisms, there is also a wide variety of spatial mechanisms. This book considers three types of four-bar spatial mechanisms: the *Revolute–Revolute–Spherical–Spherical* or *RRSS*, the *Revolute–Revolute–Revolute–Revolute Spherical* or *4R Spherical*, and the *Revolute–Spherical–Spherical–Revolute* or *RSSR* mechanisms (Figure 3.5a–c, respectively) [9,10]. The RRSS, 4R Spherical, and RSSR are among the more basic four-bar spatial mechanisms in terms of the types of joints used and the required linkage assembly conditions for motion.

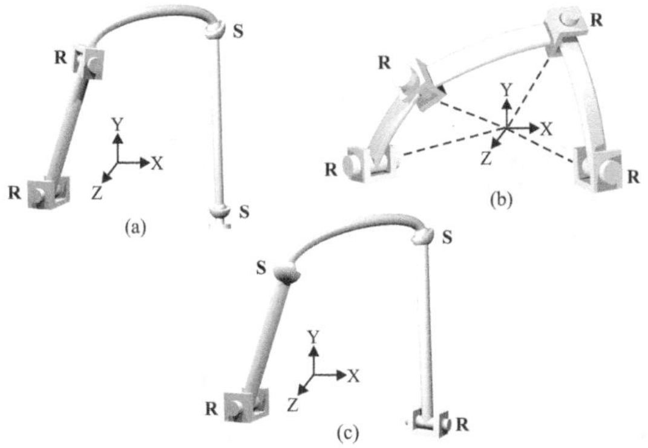

FIGURE 3.5 Spatial (a), RRSS (b), 4R Spherical, and (c) RSSR mechanisms.

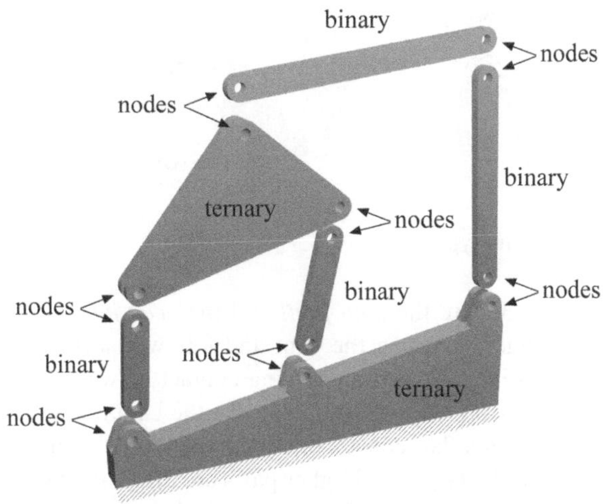

FIGURE 3.6 Link types in the Stephenson III mechanism.

3.2 LINKS, JOINTS, AND MECHANISM MOBILITY

As explained in Chapter 1, a mechanism is an assembly of links and joints. Joints are attached to links at their *nodes*. A link that has two nodes is called a *binary link,* and a *ternary link* has three nodes. The Stephenson III mechanism in Figure 3.6 includes one ternary link as ground and one movable intermediate ternary link. The remaining four links in this mechanism are binary.

Adjacent mechanism links are interconnected at their nodes by *joints*. Joint types differ by their mobility and the type of motion they allow. The joint types used throughout this book are among those illustrated in Figure 3.7. For planar mechanisms, the revolute joint is more commonly used than any other joint. Although the revolute joint is by far the dominant joint type used in practice as well as in the mechanisms in this book, Figure 3.7 includes other common joint types. The *revolute joint* (**R**) has one rotational degrees of freedom (DOF); the *prismatic joint* (**P**) has one translational DOF; the *cylindrical joint* (**C**) has two DOFs—one rotational and one translational; and the *spherical joint* (**S**) has three rotational DOFs. The term *lower pair* describes joints like the R, P, C, and S joints where *surface* contact occurs (e.g., a ball surrounded by a socket for the S joint or a pin surrounded by a hole for the R joint). Lower pairs are also called *full joints* [11].

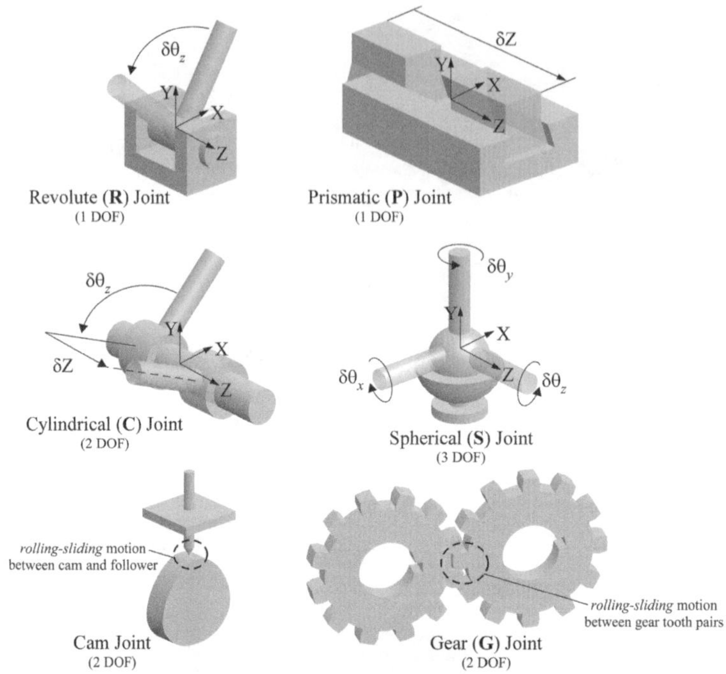

FIGURE 3.7 Six mechanism joint types.

Also illustrated in Figure 3.7 are the *cam joint* and the *gear joint* (**G**). Because the cam joint includes *rolling–sliding* contact (between the cam and follower surfaces), it can have up to two DOF.[6] Because the gear joint includes rolling–sliding contact between meshing gear teeth, it also has two DOF. The term *higher pair* describes joints like the cam and G joints where *line* contact occurs (e.g., a convex surface on a flat, convex or concave surface for the cam joint, or two curved convex surfaces in contact for the **G** joint).[7] Higher pairs are also called *half joints* [7].

Gruebler's equation (Equations 3.2 and 3.3) is used to determine the mobility (the DOF) of a mechanism. Equation (3.2) calculates the mobility of a *planar mechanism*. Since any individual planar mechanism link can have no more than three DOF, the maximum mobility of a planar mechanism with L links is $3L$. Because the ground link is fully constrained, its mobility is subtracted from the maximum mechanism mobility, resulting in $3(L-1)$. Each 1 DOF joint type removes 2 DOF, resulting in $-2J_1$, and each 2 DOF joint type removes 1 DOF, resulting in $-J_2$ (where J_1 and J_2 are the total number 1 and 2 DOF joints, respectively).

Equation (3.3) calculates the mobility of a *spatial* mechanism. Since any individual spatial mechanism link can have no more than six DOF, the maximum mobility of a spatial mechanism with L links is $6L$. Because the ground link is fully constrained, its mobility is subtracted from the maximum mobility, resulting in $6(L-1)$. Each 1 DOF joint type removes 5 DOF, resulting in $-5J_1$. Likewise, each 2 and 3 DOF joint types remove 4 and 3 DOF, respectively. In this textbook, $J_4 = J_5 = 0$ since 4 and 5 DOF joints are not utilized.

$$\mathrm{DOF_{PLANAR}} = 3(L-1) - 2J_1 - J_2 \tag{3.2}$$

$$\mathrm{DOF_{SPATIAL}} = 6(L-1) - 5J_1 - 4J_2 - 3J_3 - 2J_4 - J_5 \tag{3.3}$$

As explained in Section 1.2, the fundamental objective in mechanical system design is to produce specific controlled output motions for supplied input motions. Output motion control is maximized when the mechanical system has a single degree of freedom. Knowing the mobility of a mechanism enables the designer to determine if additional constraints are needed to reach the desired mobility (and if so, how many additional constraints are needed).

Example 3.1

Problem Statement: Determine the mobility of the planar and spatial mechanisms illustrated in Figure E.3.1.

Known Information: Figure E.3.1, Equations (3.2) and (3.3)

Solution Approach: The planar mechanism illustrated in Figure E.3.1 includes eight links, eight revolute joints, and two prismatic joints (therefore, $L = 8$, $J_1 = 10$, and $J_2 = 0$).[8] The spatial linkage illustrated in Figure E.3.1 includes six links, three revolute joints, three spherical joints, and one cylindrical joint (therefore, $L = 6$, $J_1 = 3$, $J_2 = 1$, $J_3 = 3$, and $J_4 = J_5 = 0$). Figure E.3.2 includes the calculation procedure in the MATLAB command window using Gruebler's planar and spatial mechanism equations (3.2 and 3.3).

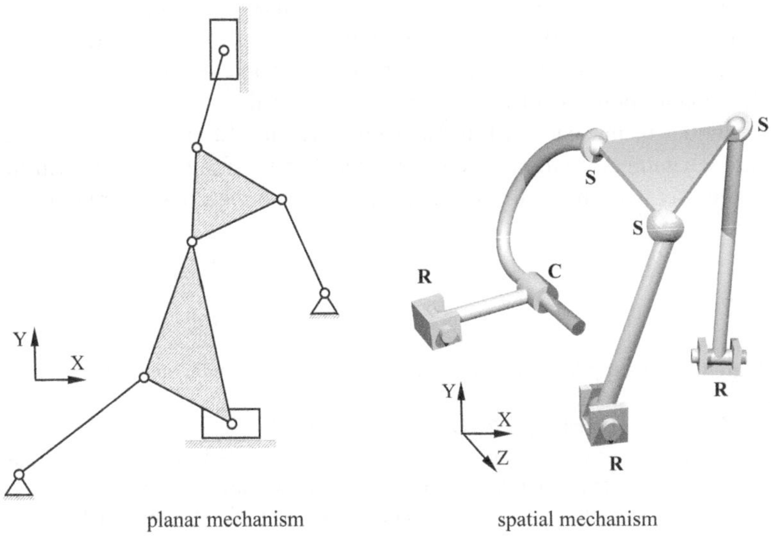

FIGURE E.3.1 (left) Planar and (right) spatial mechanisms.

```
>> L = 8;
>> J1 = 10;
>> J2 = 0;
>> DOF_Planar = 3*(L -1) -2*J1 -J2

DOF_Planar =

     1

>> L = 6;
>> J1 = 3;
>> J2 = 1;
>> J3 = 3;
>> J4 = 0;
>> J5 = 0;
>> DOF_Spatial = 6*(L - 1) - 5*J1 - 4*J2 - 3*J3 - 2*J4 - J5

DOF_Spatial =

     2

>>
```

FIGURE E.3.2 Example 3.1 solution calculation procedure in MATLAB.

3.3 NUMBER SYNTHESIS

Given the number of mechanism links and the number and order of mechanism joints, mechanism mobility is calculated from Gruebler's equation. An inverse application of Gruebler's equation is also possible. For example, Gruebler's equation is useful in determining mechanism link and joint combinations (which result in alternate mechanism solutions) for a given mobility [12]. *Number synthesis* involves the determination of alternate mechanism solutions for a given mobility. Expressing Gruebler's equations for planar and spatial mechanisms as functions of links and joints or as

$$f(L, J_1, J_2) = \text{DOF}_{\text{PLANAR}} \tag{3.4}$$

$$f(L, J_1, J_2, J_3, J_4, J_5) = \text{DOF}_{\text{SPATIAL}} \tag{3.5}$$

shows that for a given mobility, an indefinite number of link and joint combinations-alternate mechanisms-exist. Number synthesis offers not only a means to assist in the creative design of mechanisms, but Equations (3.4) and (3.5) could also be implemented systematically [13,14]. By progressively increasing or decreasing the link and joint variables in Gruebler's equation for a given mobility, tables of concept mechanism solutions are produced.

While the number of mechanism links and joints required to achieve a specified mobility is determined through number synthesis, the specific mechanism design is not determined. With number synthesis, the user determines the specific mechanism design that incorporates the calculated links and joints.

Example 3.2

Problem Statement: Compile a table of single-DOF planar mechanisms having two, three, and four links.

Known Information: Equation (3.2), $\text{DOF}_{\text{PLANAR}} = 1$, $L = 2,3$, and 4

Solution Approach: For each value of variable L, variable J_1 is incrementally increased, and for each value of L and J_1, the corresponding value for the remaining unknown J_2 in Equation (3.2) is calculated. Using this systematic procedure, the resulting mechanism solutions in Table E.3.1 are calculated.

Figure E.3.3 illustrates example mechanism configurations corresponding to mechanism solutions 1, 2, 5, 9, and 10 in Table E.3.1. Each of the mechanism solutions in this table does not necessarily represent a single mechanism configuration. For example, for mechanism solution 9 in Table E.3.1, two mechanism configurations are provided in Figure E.3.3.

TABLE E.3.1
Two-, Three-, and Four-Link Single-DOF Planar Mechanisms

Mechanism Solution	L	J_1	J_1
1	2	0	2
2	2	1	0
3	3	0	5
4	3	1	3
5	3	2	1
6	4	0	8
7	4	1	6
8	4	2	4
9	4	3	2
10	4	4	0

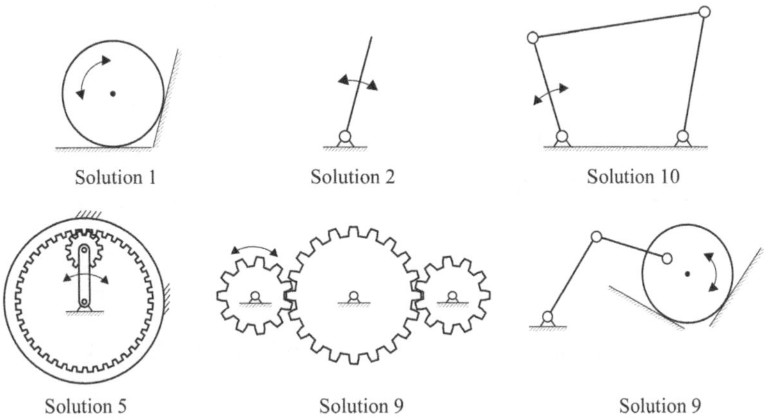

FIGURE E.3.3 Example two-, three-, and four-link single-DOF planar mechanism configurations.

3.4 GRASHOF'S CRITERIA AND TRANSMISSION ANGLE

Grashof's criteria are used to determine the link rotation behavior or link rotatability of four-bar mechanisms. These criteria are based on the lengths of the crank, coupler, follower, and ground links. Table 3.1 includes all of the Grashof and non-Grashof mechanism classifications. For the link length relationships in this table, the sum $S + L$ is the sum of the shortest and longest mechanism link lengths, and the sum $P + Q$ is the sum of the remaining two link lengths.

Figure 3.8 illustrates the link rotations for the *crank-rocker*, *double-crank* (also called *drag-link*), *double-rocker*, and *triple-rocker* mechanisms. As illustrated, only the crank link can undergo a complete rotation in a crank-rocker mechanism, while both the crank and follower links can undergo complete rotations in a double-crank mechanism. In the double-rocker mechanism, the shortest link—the coupler link is the driving link and undergoes a complete rotation, while no link rotates completely in the triple-rocker mechanism.[9] During the motion of a *change-point* mechanism (not illustrated), specifically twice per revolution, all links become simultaneously aligned. During this (theoretical) state, the output behavior of the change point mechanism is unpredictable [15].

Unlike the crank-rocker and drag-link mechanisms, the driving link in the double-rocker mechanism is not grounded. This characteristic can make the double-rocker mechanism less practical for design applications. When considering the double-rocker mechanism, the designer must overcome the technical challenges of affixing a drive system to a movable joint (and operating such a drive system).[10]

TABLE 3.1
Grashof and Non-Grashof Mechanisms

Grashof Type	Link Length Relationship	Shortest Link
Crank-rocker	$S + L < P + Q$	Crank
Double-crank (drag-link)	$S + L < P + Q$	Ground
Double-rocker	$S + L < P + Q$	Coupler
Change-point	$S + L = P + Q$	Any
Non-Grashof Type	**Link Length Relationship**	
Triple-rocker	$S + L > P + Q$	Any

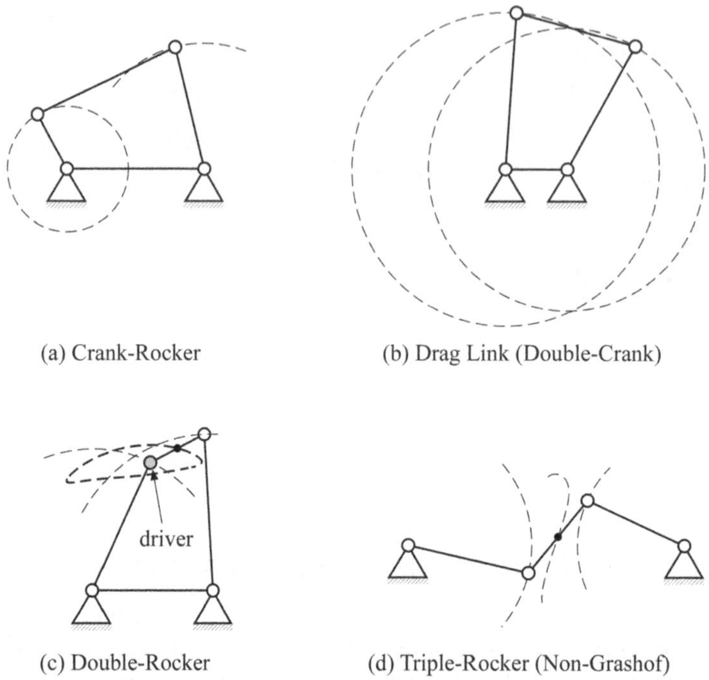

FIGURE 3.8 Link rotations of Grashof mechanisms: (a) crank-rocker, (b) drag-link (double-crank), (c) double-rocker, and (d) non-Grashof triple-rocker

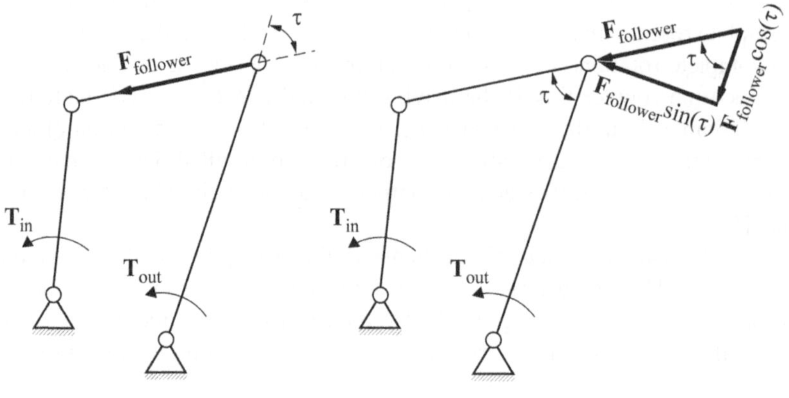

FIGURE 3.9 Planar four-bar mechanism with transmission angle and crank and follower loads.

When force and torque transmission between the crank and follower links are of concern in four-bar mechanism design, knowing the *transmission angle* is critical. The transmission angle is the angle between the coupler and follower links. Figure 3.9a illustrates a planar four-bar mechanism and the transmission angle τ. When an input torque $\mathbf{T}_{\text{in}}$ is applied to the crank link, this link transmits force to the coupler, which subsequently transmits the force to the follower $\mathbf{F}_{\text{follower}}$. One component of this follower force ($\mathbf{F}_{\text{follower}} \sin\tau$ in Figure 3.9b) is normal to the follower link and results in the output torque $\mathbf{T}_{\text{out}}$. The other force component ($\mathbf{F}_{\text{follower}} \cos\tau$) is a columnar load acting along the length of the follower. Both follower force components are functions of the transmission angle. The columnar component of the follower force should be minimized—particularly in applications where the forces and torques transmitted are substantial.[11] A transmission angle of 90° is optimum because the resulting follower force has only a normal component ($\mathbf{F}_{follower}\cos\tau = 0$ when $\tau = 90°$). A transmission angle range of 90° ± 50° is generally preferred by designers [16]. It can be determined from Figure 3.9b that as the transmission angle decreases, the columnar load component increases and the normal load component decreases.

Example 3.3

Problem Statement: The link lengths for three planar four-bar mechanisms are given in Figure E.3.4 (all identical to the link dimensions given in mechanism #1), and the driving links are labeled with rotation arrows. Determine the Grashof type for each mechanism.

Known Information: Figure E.3.4 and Table 3.1

Solution Approach: Because the shortest link in mechanism #1 is the ground link and the Grashof condition $S + L < P + Q$ is true, this mechanism is a *Grashof double-crank*. Because the shortest link in mechanism #2 is the crank link and the Grashof condition $S + L < P + Q$ is true, this mechanism is a *Grashof crank-rocker*. Because the shortest link in mechanism #3 is the coupler link and the Grashof condition $S + L < P + Q$ is true, this mechanism is a *Grashof double-rocker*.

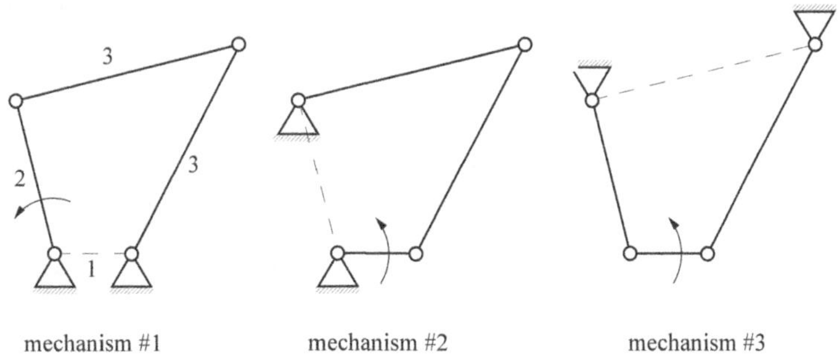

FIGURE E.3.4 Planar four-bar mechanism configurations (with dimensionless link lengths given for mech. #1).

3.5 CIRCUIT DEFECT

While a Grashof mechanism is capable of producing full crank rotation, a non-Grashof mechanism cannot. For the latter mechanism type, the crank can only rotate to the mechanism's *binding position*. When the binding position is reached, the mechanism "locks-up" or is physically precluded from further movement (further movement in the given crank direction). Crank rotation beyond the binding position is only possible through mechanism disassembly.

Figure 3.10 illustrates a non-Grashof triple-rocker and one of its binding positions. The binding position is reached when link $\mathbf{a}_0 - \mathbf{a}_1$ rotates by an angular displacement β. Because the mechanism locks-up at β, further mechanism motion in the counterclockwise direction is only possible when the mechanism is disassembled and reassembled beyond the infeasible region shown in Figure 3.10.[12]

If a mechanism is incapable of achieving a desired crank rotation (as subsequently a desired mechanism position) due to a lock-up condition, it has what is called a *circuit defect*.[13] Circuit defects are fatal to linkage operation (due to mechanism disassembly being required for full crank rotation).

3.6 MECHANISM INVERSION

The motion of a mechanism varies based on which joints are grounded. As illustrated in Figure E.3.4 (see Example 3.3), the planar four-bar mechanisms given are all identical in terms of link length (and even identical in terms of the initial link orientation) but vary only in terms of the

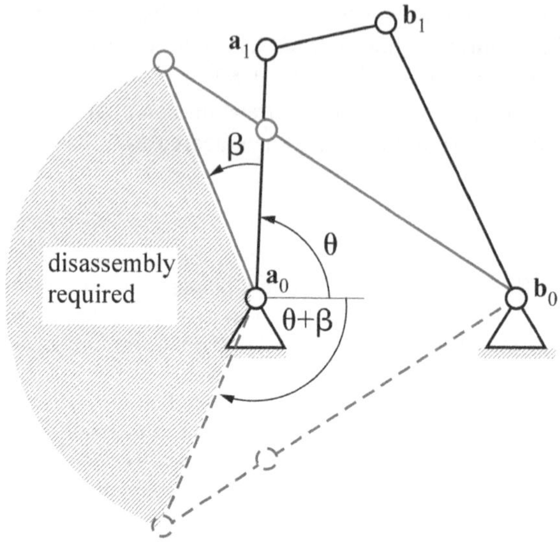

FIGURE 3.10 Planar four-bar mechanism (a non-Grashof triple-rocker) with circuit defect.

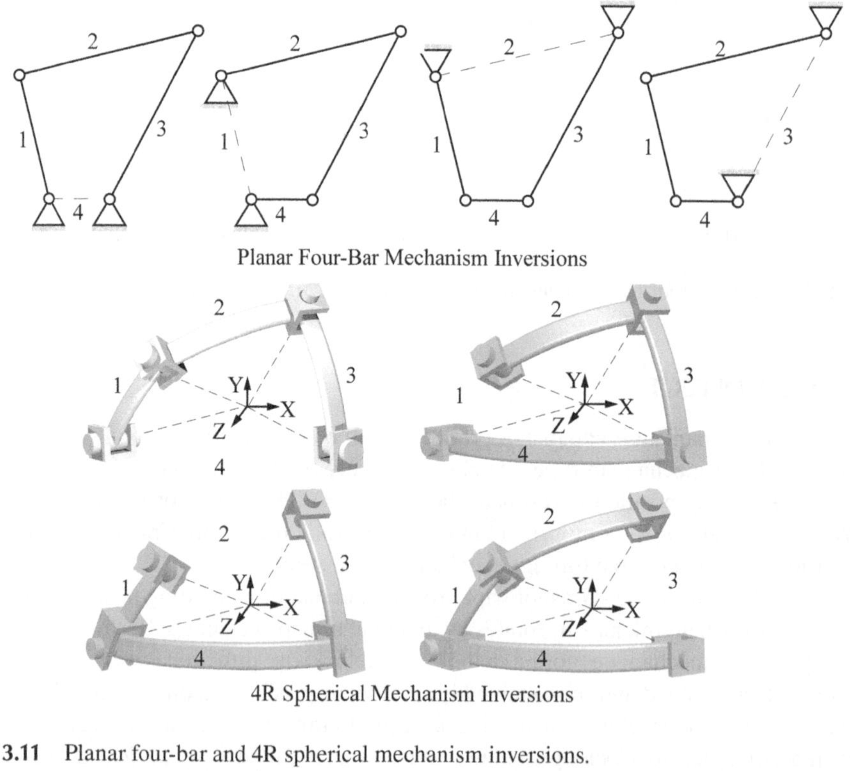

FIGURE 3.11 Planar four-bar and 4R spherical mechanism inversions.

joints that are grounded. The mechanisms in Figure E.3.4 are all *inversions* of each other because they are otherwise identical mechanisms that have different grounded joints.

Figure 3.11 illustrates all the inversions of the planar four-bar and 4R spherical mechanisms. Because the motion of a mechanism varies based on which joints are grounded, producing mechanism inversions is a systematic method to produce mechanisms that have unique motions but include the same general linkage hardware.

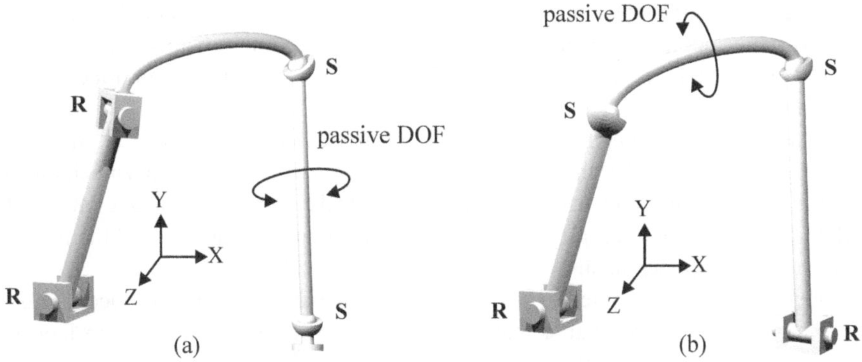

FIGURE 3.12 Passive DOF in spatial (a) RRSS and (b) RSSR mechanisms.

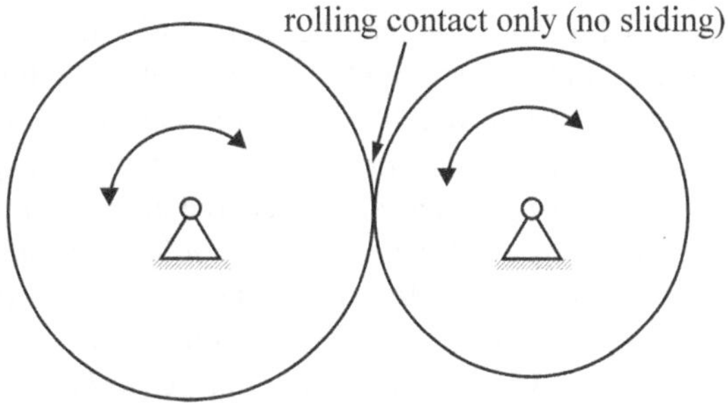

FIGURE 3.13 Rolling cylinder pair (with no sliding contact).

3.7 PASSIVE DEGREE OF FREEDOM AND PARADOXES

There are cases where the mobility values calculated from Equations (3.2) and (3.3) are misleading. For example, although Equation (3.3) produces a mobility value of 2 for the spatial RRSS and RSSR mechanisms, the RRSS mechanism is capable of controlled coupler motion, and the RSSR is capable of controlled follower motion. The extra degree of freedom for these mechanisms is the free rotation of the S-S links about their length axes (Figure 3.12) and is called a *passive degree of freedom*. Because this degree of freedom is highly localized (limited to a single link) for these mechanisms, they have no effect on the kinematics of their links of interest (the coupler link for the RRSS and the follower link for the RSSR).

There are also cases where the mobility values calculated from Equations (3.2) and (3.3) are incorrect. For example, Equation (3.3) produces a mobility of –2 for the 4R spherical mechanism (which has a true mobility of 1), and Equation (3.2) produces a mobility of 0 for the pair of rolling cylinders in Figure 3.13 (which has a true mobility of 1). Mechanisms having true mobility values that violate Gruebler's equation are called *paradoxes* or *maverick mechanisms* [17,18]. Because Gruebler's equation cannot guarantee true mobility results for all mechanisms, it may be necessary for the user to validate the mobility of a mechanism (through kinematic analyses, for example).

3.8 SUMMARY

The *planar four-bar mechanism* is one of the most widely utilized kinematic chains in everyday devices. The four interconnected links in this mechanism are the *crank*, *coupler*, *follower*, and *ground* links. The crank and follower links in the planar four-bar mechanism undergo pure rotation, while the coupler link undergoes *complex motion*. In the *geared five-bar mechanism*, the crank link and output link are coupled by a gear pair or gear train. Though not as commonly utilized in everyday devices as planar four-bar mechanisms, one advantage the planar five-bar mechanism has over the four-bar mechanism is that its intermediate links can trace paths of higher orders than the latter.

Attaching a dyad to the coupler of the planar four-bar mechanism produces a type of *multi-loop planar six-bar mechanism* called the *Stephenson III mechanism*. Although this mechanism is not as commonly utilized as planar four-bar mechanisms in everyday devices, it includes at least two links that simultaneously undergo complex motion, giving it greater capacities in kinematic synthesis than the planar four-bar mechanism.

Planar mechanisms are restricted to motion in two-dimensional or planar space. *Spatial* mechanisms can exhibit three-dimensional or spatial motion. Because spatial mechanisms have the capacity to exhibit spatial motion, they offer a greater variety of possible motions and are structurally more general than planar mechanisms. This book considers three types of four-bar spatial mechanisms: the *Revolute–Revolute–Spherical–Spherical* or *RRSS*, the *Revolute–Revolute–Revolute–Revolute Spherical* or *4R Spherical,* and the *Revolute–Spherical–Spherical–Revolute* or *RSSR* mechanisms. The RRSS, 4R Spherical, and RSSR are among the more basic four-bar spatial mechanisms in terms of the types of joints used and the required linkage assembly conditions for motion.

Mechanism links are interconnected by *joints*. Both the planar four-bar, geared five-bar, and Stephenson III mechanisms include *revolute joints*, a joint having a single rotational degree of freedom. The *prismatic joint* has a single translational degree of freedom and is used where single degree of freedom sliding contact is needed. The *cylindrical joint* and *spherical joint* are other joint types utilized in spatial mechanisms. The cylindrical joint has both a rotational and a translation degree of freedom, while the spherical joint has three rotational DOF. Joints having up to two DOF include the *cam* joint and *gear* joint, which both include *rolling–sliding* contact.

Gruebler's equation is used to calculate mechanism *mobility* or DOF.

Determining the mobility of a mechanism enables the designer to determine if additional constraints are needed to reach the desired mobility, and if so, how many additional constraints are needed. In *number synthesis*, Gruebler's equations are implemented systematically to determine alternate mechanism solutions for a given mobility. *Grashof's Criteria* are used to determine the link rotation behavior of four-bar mechanisms. Determining link rotation behavior is important in design, particularly when coupling a drive system to the crank link.

While a Grashof mechanism can produce full crank rotation, a *non-Grashof* mechanism cannot. For the latter mechanism type, the crank can only rotate to the mechanism's *binding position*. Crank rotation beyond the binding position is only possible through mechanism disassembly. This defect associated with non-Grashof mechanisms is called a *circuit defect*. Circuit defects are considered fatal to linkage operation (due to mechanism disassembly being required for full crank rotation).

Knowing the *transmission angle* (the angle between the coupler and follower links) behavior is important when force and torque transmission between the crank and follower links are of concern in four-bar mechanism design. As the transmission angle decreases, the magnitude of the columnar load on the follower link increases and the magnitude of the load normal to the follower decreases (and vice-versa). A transmission angle range of $90° \pm 50°$ is generally preferred by designers.

The motion of a mechanism can vary based on which joints are grounded. Mechanisms of identical type and link dimensions, but different grounded links, are *inversions* of each other. Because the motion of a mechanism can vary based on which joints are grounded, producing mechanism inversions is a systematic method to produce mechanisms that have unique motions but include the same general linkage hardware.

There are cases where the mobility values calculated using Gruebler's equation are misleading. The follower link of the spatial RRSS mechanism and the coupler link of the spatial RSSR mechanism have a *passive degree of freedom*: the free rotation of these links about their length axes. For example, although Gruebler's equation produces a mobility value of 2 for both mechanisms, the RRSS mechanism is capable of controlled coupler motion, and the RSSR is capable of controlled follower motion.

There are also cases where the mobility values calculated using Gruebler's equation are incorrect. For example, Gruebler's equation produces a mobility of –2 for the 4R Spherical mechanism (which has a true mobility of 1). Mechanisms having true mobility values that violate Gruebler's equation are called *paradoxes* or *maverick mechanisms.* Because Gruebler's equation cannot guarantee true mobility results for all mechanisms, it may be necessary for the user to validate the mobility of a mechanism.

PROBLEMS

1. Planar four-bar linkages have many everyday applications (some are illustrated Figure 3.1). Identify and describe four additional everyday applications for the planar four-bar linkage.
    a. Why is it important to know if a mechanism has a single degree of freedom?
    b. Why is a crank-rocker mechanism more practical than a double-rocker mechanism? Should the transmission angle for the planar four-bar linkage be close to 0°? Explain?
2. For the two linkages illustrated in Figure P.3.1, which (if any) of the links can undergo a complete rotation relative to the other links? How do you know?

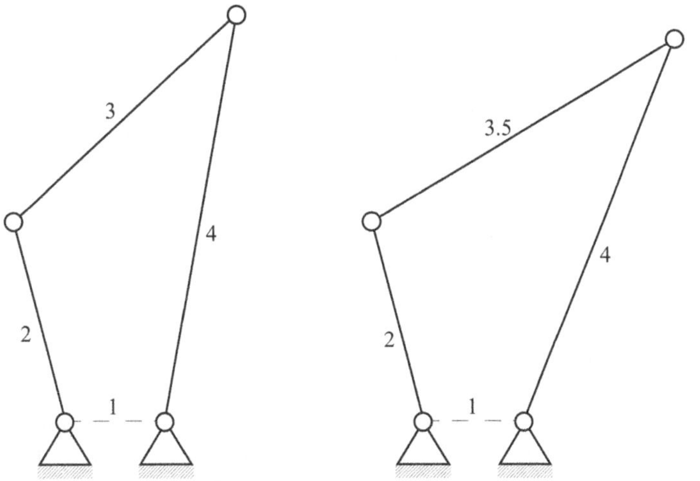

FIGURE P.3.1 Planar four-bar linkages with dimensionless link lengths.

3. Repeat Problem 3, where one linkage has link lengths of 2.35 cm (ground), 3.25 cm (coupler), 2.25 cm, and 2.15 cm and another linkage has link lengths of 2.75 cm (ground), 3.25 cm (coupler), 1.15 cm, and 1.75 cm.

4. Determine the number of links and the mobility of each of the three planar mechanisms in Figure P.3.2.

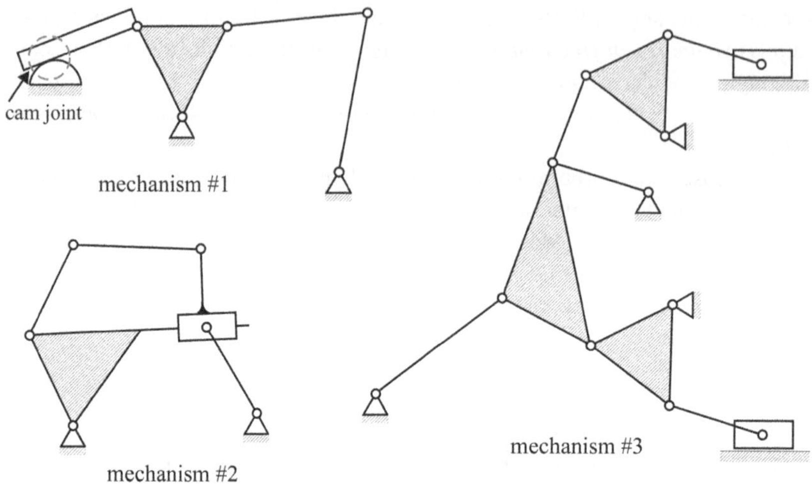

FIGURE P.3.2 Planar mechanisms.

5. For the planar four-bar linkage illustrated in Figure P.3.3, $L_2/L_1 = 1.5$ and $L_3/L_1 = 1.2$. Find the range of L_0/L_1 required for a *drag-link mechanism*.

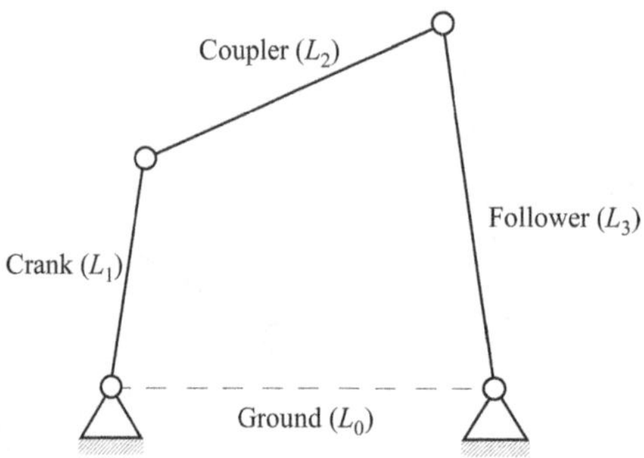

FIGURE P.3.3 Planar four-bar linkage.

6. Repeat Problem 6 for a *crank-rocker mechanism,* where $L_2/L_1 = 2.75$ and $L_3/L_1 = 2.95$.
7. Compile a table of 1 DOF spatial mechanisms having two, three, and four links (let $J_4 = J_5 = 0$ in Equation (3.3)). Illustrate some of these mechanism solutions.
8. Compile a table of 1 DOF spatial mechanisms having five links (let $J_4 = J_5 = 0$ in Equation (3.3)). Illustrate some of these mechanism solutions.

9. Figure P.3.4 illustrates nine spatial mechanisms that include revolute (R), prismatic (P), cylindrical (C), and spherical (S) joints. Calculate the mobility of these mechanisms.

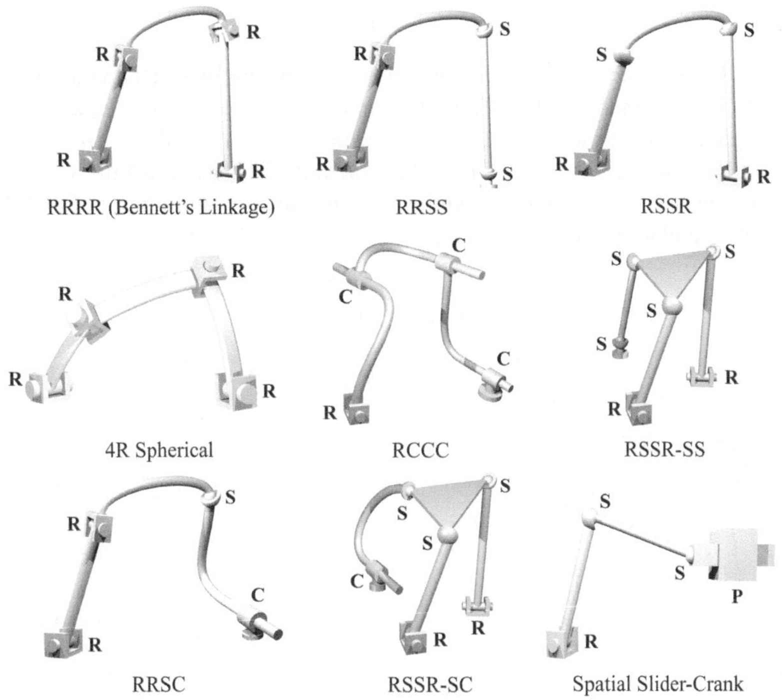

FIGURE P.3.4 Spatial mechanisms comprising R, P, C, and S joints.

10. Figure P.3.5 illustrates five spatial robots that include revolute (R), prismatic (P), and cylindrical (C) joints. Calculate the mobility of these robots.

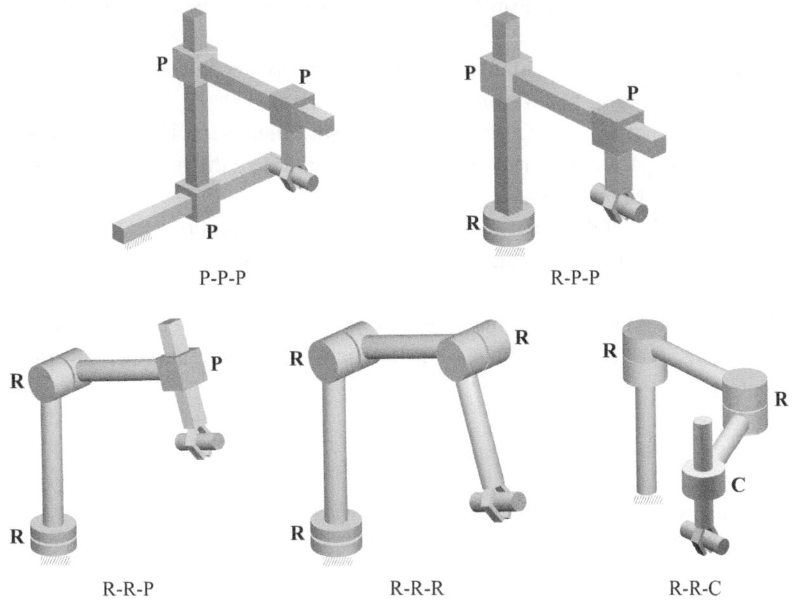

FIGURE P.3.5 Spatial robots comprising R, P, and C joints.

11. Euler's buckling load (F) for a columnar member with pinned ends is

$$F = \frac{\pi^2 EI}{L^2}$$

where E, I, and L are the modulus of elasticity, moment of inertia, and length, respectively, of the columnar member. Formulate an equation from Euler's equation and Figure P.3.6 to calculate the transmission angle corresponding to the follower link buckling load. Calculate the transmission angle for a follower length of 12 in (0.3048 m), a ¼ in (0.635 cm) square follower cross-section, a modulus of elasticity (for aluminum) of 10,000,000 psi (68.05 GPa), and a follower load ($\mathbf{F}_{\text{follower}}$) of 250 lbf (1112.05 N).

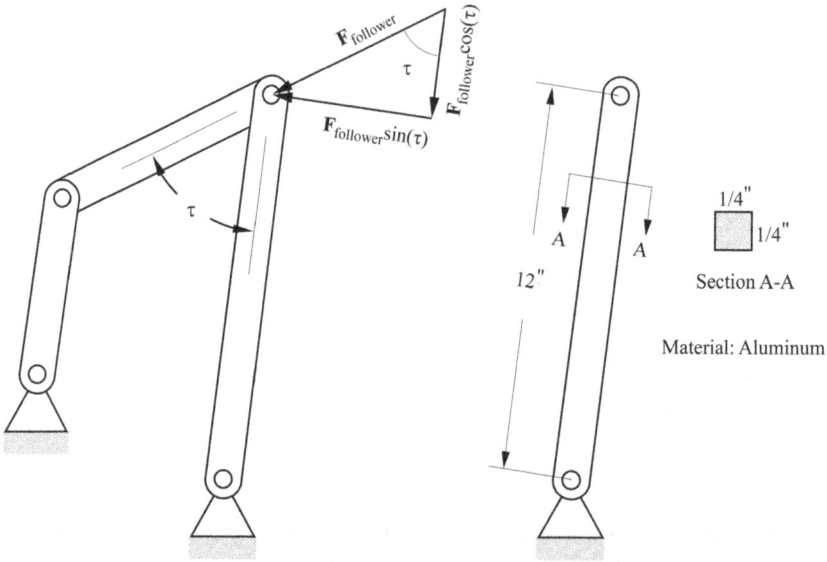

FIGURE P.3.6 Planar four-bar mechanism with the transmission angle and crank and follower loads.

12. *Maverick mechanisms* are mechanisms that defy Gruebler's equation (Gruebler's equation will produce misleading results for maverick mechanisms). *Passive DOFs* are localized DOFs that have no effect on the overall mechanism kinematics. Determine which mechanisms in Figure P.3.4 are maverick mechanisms, and locate the passive DOFs from among the mechanisms in Figure P.3.4.
13. Compile a table of 2 DOF planar mechanisms having four and five links (considering cam or gear joints as J_2). Illustrate one four-bar mechanism solution and one five-bar mechanism solution.
14. Compile a table of 2 DOF planar mechanisms having six and seven links (considering cam or gear joints as J_2). Illustrate one six-bar mechanism solution and one seven-bar mechanism solution.

NOTES

1 Because mechanisms are comprised of links, they are also called *linkages*.
2 The coupler undergoes *complex motion*—a combination of simultaneous rotation and translation (see Section 1.4).
3 Coupling the motion of the crank and output links reduces the resulting mechanism to a single degree of freedom.

4 In addition to using gears for coupling the crank and output links, other options include using *pulley–belt* systems, *chain–sprocket* systems, or *drive motors*.
5 Stephenson mechanism *Motion* generation are introduced in Section 6.8.
6 If no sliding occurs, this joint has one degree of freedom.
7 The term higher pair also describes joints where *point* contact occurs (e.g., two spheres in contact or a sphere on a flat surface).
8 In this planar mechanism, each revolute joint joins only two adjacent links. If more than two adjacent links $\left(L_{adj}\right)$are present, the corresponding number of revolute joints $\left(J_1\right)$ is $J_1 = L_{adj} - 1$.
9 Figure 3.8c illustrates the path achieved by the shortest link in the double-rocker when a driver is applied to the labeled joint (of the shortest link).
10 Affixing a drive system (such as a motor or manual crank) to a grounded link joint (and operating such a system) can be more easily accomplished than to a non-grounded joint.
11 Unlike follower *normal* loads (which links having revolute joints are designed to accommodate), follower *columnar* loads ($\mathbf{F}_{follower} \cos \tau$ in Figure 3.9b) can result in follower buckling or excessive bearing loads in the follower revolute joints.
12 For a given non-Grashof mechanism, the lock-up position can vary based on the designated crank link (for example, link $\mathbf{b}_0 - \mathbf{b}_1$ could be a crank link) as well as the crank rotation direction (clockwise rotation instead of counterclockwise rotation).
13 A mechanism link loop (like loops in electrical and fluid flow systems) is also called a *circuit*.

REFERENCES

1. Chuan Hua, Y., Xian Fa, F., Xin, J., Zhong Jun, L. and Jun, L. 2014. Motion Analysis for Vegetable Potted Seedling Pick-Up Mechanism with Double Crank Geared Linkages. *Applied Mechanics and Materials*. 529: 218–223.
2. Sandhya R., Kadam, M., Balamurugan, G. and Rangavittal, H. K. 2014. Synthesis and Analysis of Geared Five Bar Mechanism for Ornithopter Applications. *2nd International and 17th National Conference on Machines and Mechanisms*. Kanpur, India. December 16–19.
3. Manoharan, G., Udupib, P. R. and Kotian, K. Design of Forelimb Prosthetic for Quadrupeds. 2021. *International Conference on Recent Advances in Modeling and Simulation Techniques in Engineering and Science*. Jaipur, India. December 8–10.
4. Wunderlich, W. 1963. Höhere Koppelkurven. *Österreichisches Ingenieur Archiv*. 17: 162–165.
5. Sandor, G. N. and Erdman, A. G. 1984. *Advanced Mechanism Design: Analysis and Synthesis*, Volume 2. pages 10–11. Englewood Cliffs: Prentice-Hall.
6. Jin, D., Zhang, R., Dimo, H., Wang, R. and Zhang, J. 2003. Kinematic and Dynamic Performance of Prosthetic Knee Joint Using Six-bar Mechanism. *Journal of Rehabilitation Research and Development*. 40: 39–48.
7. Bissell, D. M. 2014. WRIST (Wrist Resonator for Independent Stroke Training). M.S. Thesis. Mechanical and Aerospace Engineering, University of California, Irvine.
8. Capellen, W. M. 1966. Kinematics-A Survey in Retrospect and Prospect. *Mechanism and Machine Theory*. 1: 211–228.
9. Russell, K., Shen, Q. and Sodhi, R. 2013. *Mechanism Design: Visual and Programmable Approaches*. Chapter 7. Boca Raton: CRC Press.
10. Suh, C. H. and Radcliffe, C. W. 1978. *Kinematics and Mechanisms Design*. chapter 4. New York: John Wiley & Sons.
11. Norton, R. L. 2008. *Design of Machinery*, 4th Edition. pages 32–36. New York: McGraw-Hill.
12. Sandor, G. N. and Erdman, A. G. 1984. *Advanced Mechanism Design: Analysis and Synthesis*, Volume 2. page 64. Englewood Cliffs: Prentice-Hall.
13. Sandor, G. N. and Erdman, A. G. 1984. *Advanced Mechanism Design: Analysis and Synthesis*, Volume 2. pages 64–75. Englewood Cliffs: Prentice-Hall.
14. Norton, R. L. 2008. *Design of Machinery*, 4th Edition. pages 42–46. New York: McGraw-Hill.
15. Norton, R. L. 2008. *Design of Machinery*, 4th Edition. page 58. New York: McGraw-Hill.
16. Wilson, C. E. and Peter Sadler, J. 2003. *Kinematics and Dynamics of Machinery*, 3rd Edition. page 33. Saddle River: Prentice Hall.
17. Norton, R. L. 2008. *Design of Machinery*, 4th Edition. pages 46–47. New York: McGraw-Hill.
18. Suh, C. H. and Radcliffe, C. W. 1978. *Kinematics and Mechanisms Design*. pages 109–110. New York: John Wiley & Sons.

4 Kinematic Analysis of Planar Mechanisms

CONCEPT OVERVIEW

In this chapter, the reader will gain a central understanding regarding the following:

1. Link velocity and acceleration components in planar space
2. Vector loop-based displacement, velocity and acceleration equation formulation, and solution calculation
3. Kinematics of mechanism link locations of interest
4. Configurations of closed-loop mechanisms

4.1 INTRODUCTION

In a kinematic analysis, positions, displacements, velocities, and accelerations of mechanism links are determined either *qualitatively* or *quantitatively*. In a quantitative kinematic analysis, equations that fully describe the motion of the mechanism links are used. Qualitative methods include constructing and measuring mechanism schematics and polygons to determine the positions, velocities, and accelerations of mechanism links. As intended by the authors, the kinematic analysis methods presented in this book are all quantitative. Kinematic equations for the planar four-bar, geared five-bar, and Stephenson III mechanisms are formulated in this chapter. Displacement equations are formulated by taking the sum of the closed-vector loop(s) in each mechanism (as introduced in Section 2.2.2). Taking the first and second derivatives of the vector loop displacement equations produces mechanism link velocity and acceleration variables, respectively, and ultimately produces mechanism velocity and acceleration equations, respectively.

As noted in Section 3.1.4, the Stephenson III mechanism (Figure 3.6) is the planar multi-loop mechanisms of choice for analysis in this book. This mechanism has two links that exhibit complex motion. As illustrated in Figure 3.6, the Stephenson III mechanism comprises a planar four-bar mechanism and an additional dyad attached to the coupler link.

Because the Stephenson III mechanism includes the planar four-bar mechanism, the displacement, velocity, and acceleration equations for these planar multi-loop mechanisms will include some of the variables and output from the planar four-bar mechanism equations. Displacement equations for the additional dyads in the Stephenson III mechanism are produced by formulating vector loop equations that include this dyad. Taking the first and second derivatives of the resulting vector loop displacement equations produces velocity and acceleration equations, respectively.

4.2 LINK VELOCITY AND ACCELERATION COMPONENTS IN PLANAR SPACE

Figure 4.1a illustrates an arbitrary grounded rotating link with rotation angle β. Figure 4.1b and c includes the velocity and acceleration components at point $\mathbf{p}_1$ of this link, respectively. Given a link angular velocity (which is labeled $\dot{\beta}$), the velocity vector (which is labeled $\mathbf{V}_{\mathbf{p}_1}$) is produced. This velocity is tangential to the length $\mathbf{p}_0 - \mathbf{p}_1$ and acts in the direction of $\dot{\beta}$.

Given a link angular velocity and angular acceleration (which are labeled $\dot{\beta}$ and $\ddot{\beta}$, respectively), the tangential and normal acceleration vectors (which are labeled $\mathbf{A}_{\mathbf{p}_{1t}}$ and $\mathbf{A}_{\mathbf{p}_{1n}}$, respectively) are produced. The acceleration component $\mathbf{A}_{\mathbf{p}_{1t}}$ is tangential to the length $\mathbf{p}_0 - \mathbf{p}_1$ and acts in the

 DOI: 10.1201/9781003591375-4

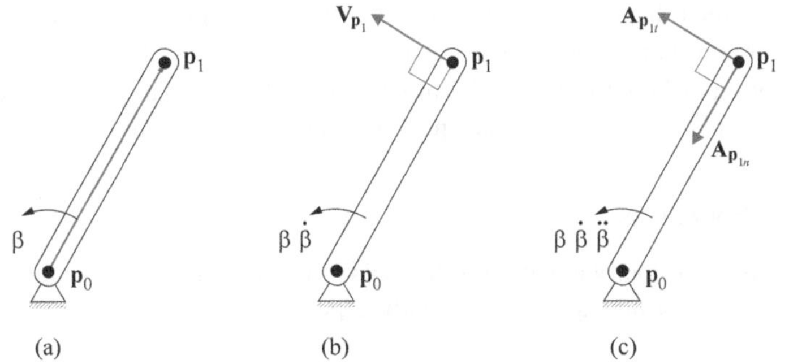

FIGURE 4.1 (a) Rotating link and its (b) velocity and (c) acceleration components.

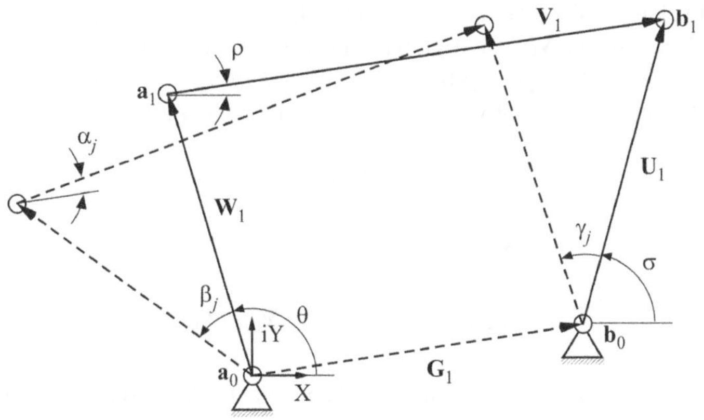

FIGURE 4.2 Planar four-bar mechanism displacement variables.

direction of $\ddot{\beta}$. The acceleration component $\mathbf{A}_{\mathbf{p}_{1n}}$ is along the length $\mathbf{p}_0 - \mathbf{p}_1$ and acts in the direction toward the center of rotation (toward $\mathbf{p}_0$). The total acceleration at $\mathbf{p}_1$ is the sum of acceleration components $\mathbf{A}_{\mathbf{p}_{1t}}$ and $\mathbf{A}_{\mathbf{p}_{1n}}$.

4.3 FOUR-BAR MECHANISM ANALYSIS

4.3.1 Displacement Equations

The planar four-bar mechanism consists of four links interconnected by revolute joints. As calculated from Gruebler's Equation for planar mechanisms (with $L = 4$ and $J_1 = 4$), the planar four-bar mechanism has a single DOF. A single displacement equation is derived for this mechanism by taking the sum of the displaced mechanism vector loop in Figure 4.2. The equation produced from a clockwise sum of the displaced vector loop is

$$W_1 e^{i(\theta+\beta_j)} + V_1 e^{i(\rho+\alpha_j)} - U_1 e^{i(\sigma+\gamma_j)} - \mathbf{G}_1 = 0 \tag{4.1}$$

After expanding Equation (4.1) and grouping its real and imaginary terms as separate equations, the resulting planar four-bar mechanism displacement equations become

$$\begin{aligned} f_1(\alpha_j, \gamma_j) &= W_1 \cos(\theta+\beta_j) + V_1 \cos(\rho+\alpha_j) - U_1 \cos(\sigma+\gamma_j) - G_{1x} = 0 \\ f_2(\alpha_j, \gamma_j) &= W_1 \sin(\theta+\beta_j) + V_1 \sin(\rho+\alpha_j) - U_1 \sin(\sigma+\gamma_j) - G_{1y} = 0 \end{aligned} \tag{4.2}$$

Except for the coupler displacement angle α_j and follower displacement angle γ_j, all other variables in the planar four-bar displacement equations are user-prescribed. Because $f_1(\alpha_j,\gamma_j)$ and $f_2(\alpha_j,\gamma_j)$ in Equation (4.2) both include the unknown displacement angles α_j and γ_j, these equations form a set of nonlinear simultaneous equations from which the angles are calculated.

4.3.2 Velocity Equations

A single planar four-bar velocity equation is derived by differentiating the planar four-bar displacement equation. Differentiating Equation (4.1) with respect to time produces

$$i\dot{\beta}_j W_1 e^{i(\theta+\beta_j)} + i\dot{\alpha}_j V_1 e^{i(\rho+\alpha_j)} - i\dot{\gamma}_j U_1 e^{i(\sigma+\gamma_j)} = 0 \tag{4.3}$$[1]

Equation (4.4) includes the individual velocity variables from Equation (4.3). The velocity $\mathbf{V}_{\mathbf{a}_1}$ is the *global* tangential velocity of the crank link vector—the tangential velocity of $\mathbf{a}_1$ with respect to $\mathbf{a}_0$ (see Figure 4.3). The velocity $\mathbf{V}_{\mathbf{b}_1}$ is the global tangential velocity of the follower link vector—the tangential velocity of $\mathbf{b}_1$ with respect to $\mathbf{b}_0$. Also, the velocity $\mathbf{V}_{\mathbf{b}_1\text{-}\mathbf{a}_1}$ is the *relative* tangential velocity of the coupler link vector—the relative tangential velocity of $\mathbf{b}_1$ with respect to $\mathbf{a}_1$.

$$\begin{aligned}
\mathbf{V}_{\mathbf{a}_1} &= i\dot{\beta}_j W_1 e^{i(\theta+\beta_j)} \\
\mathbf{V}_{\mathbf{b}_1} &= i\dot{\gamma}_j U_1 e^{i(\sigma+\gamma_j)} \\
\mathbf{V}_{\mathbf{b}_1\text{-}\mathbf{a}_1} &= i\dot{\alpha}_j V_1 e^{i(\rho+\alpha_j)} = \mathbf{V}_{\mathbf{b}_1} - \mathbf{V}_{\mathbf{a}_1}
\end{aligned} \tag{4.4}$$

After moving the term $i\dot{\beta}_j W_1 e^{i(\theta+\beta_j)}$ to the right-hand side in Equation (4.3), expanding this equation, and grouping its real and imaginary terms as separate equations, the resulting planar four-bar mechanism velocity equation in the matrix form $[A]\mathbf{x} = \mathbf{b}$ is

$$\begin{bmatrix} -V_1\sin(\rho+\alpha_j) & U_1\sin(\sigma+\gamma_j) \\ V_1\cos(\rho+\alpha_j) & -U_1\cos(\sigma+\gamma_j) \end{bmatrix} \begin{Bmatrix} \dot{\alpha}_j \\ \dot{\gamma}_j \end{Bmatrix} = \dot{\beta}_j \begin{bmatrix} W_1\sin(\theta+\beta_j) \\ -W_1\cos(\theta+\beta_j) \end{bmatrix} \tag{4.5}$$

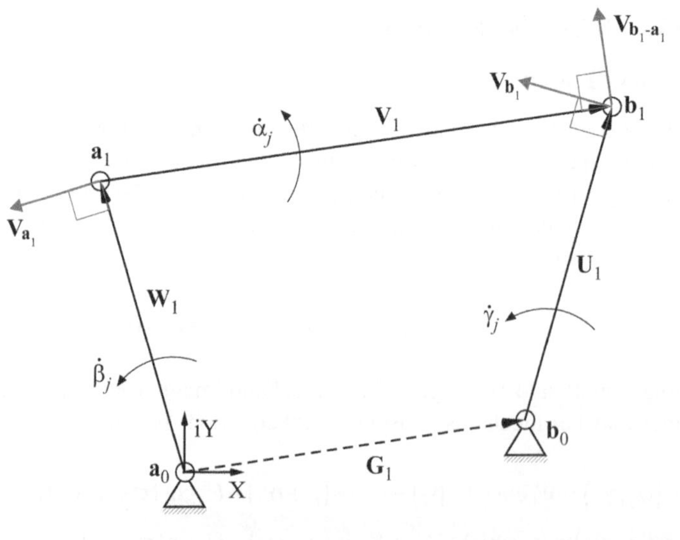

FIGURE 4.3 Planar four-bar mechanism velocity variables.

After including the prescribed mechanism variables and the unknown variables calculated from the planar four-bar displacement equations, Equation (4.5) can be solved (using Cramer's rule, for example) to calculate the coupler angular velocity $\dot{\alpha}_j$ and the follower angular velocity $\dot{\gamma}_j$.

4.3.3 Acceleration Equations

A single planar 4-bar acceleration equation is derived by differentiating the planar four-bar velocity equation. Time differentiation of Equation (4.3) produces

$$-\dot{\beta}_j^2 W_1 e^{i(\theta+\beta_j)} + i\ddot{\beta}_j W_1 e^{i(\theta+\beta_j)} - \dot{\alpha}_j^2 V_1 e^{i(\rho+\alpha_j)} + i\ddot{\alpha}_j V_1 e^{i(\rho+\alpha_j)} + \dot{\gamma}_j^2 U_1 e^{i(\sigma+\gamma_j)} - i\ddot{\gamma}_j U_1 e^{i(\sigma+\gamma_j)} = 0 \quad (4.6)$$

Equation (4.7) includes the individual acceleration variables from Equation (4.6). The accelerations $\mathbf{A}_{\mathbf{a}_{1n}}$ and $\mathbf{A}_{\mathbf{a}_{1t}}$ are the global *normal* and *tangential* accelerations, respectively, of the crank link vector—the normal and tangential accelerations of $\mathbf{a}_1$ with respect to $\mathbf{a}_0$ (see Figure 4.4).[2] The *total* acceleration $\mathbf{A}_{\mathbf{a}_1}$ is the sum of these normal and tangential accelerations. The accelerations $\mathbf{A}_{\mathbf{b}_{1n}}$ and $\mathbf{A}_{\mathbf{b}_{1t}}$ are the global normal and tangential accelerations, respectively, of the follower link vector—the normal and tangential accelerations of $\mathbf{b}_1$ with respect to $\mathbf{b}_0$. The total acceleration $\mathbf{A}_{\mathbf{b}_1}$ is the sum of these normal and tangential accelerations. Also, the accelerations $\mathbf{A}_{(\mathbf{b}_1\text{-}\mathbf{a}_1)_n}$ and $\mathbf{A}_{(\mathbf{b}_1\text{-}\mathbf{a}_1)_t}$ are the relative normal and tangential accelerations, respectively, of $\mathbf{b}_1$ with respect to $\mathbf{a}_1$—the relative normal and tangential accelerations of the coupler link vector. The total relative acceleration $\mathbf{A}_{\mathbf{b}_1\text{-}\mathbf{a}_1}$ is the sum of these normal and tangential accelerations.

$$\begin{aligned}
\mathbf{A}_{\mathbf{a}_1} &= \mathbf{A}_{\mathbf{a}_{1t}} + \mathbf{A}_{\mathbf{a}_{1n}} = i\ddot{\beta}_j W_1 e^{i(\theta+\beta_j)} - \dot{\beta}_j^2 W_1 e^{i(\theta+\beta_j)} \\
\mathbf{A}_{\mathbf{b}_1} &= \mathbf{A}_{\mathbf{b}_{1t}} + \mathbf{A}_{\mathbf{b}_{1n}} = i\ddot{\gamma}_j U_1 e^{i(\sigma+\gamma_j)} - \dot{\gamma}_j^2 U_1 e^{i(\sigma+\gamma_j)} \\
\mathbf{A}_{\mathbf{b}_1-\mathbf{a}_1} &= \mathbf{A}_{(\mathbf{b}_1-\mathbf{a}_1)_t} + \mathbf{A}_{(\mathbf{b}_1-\mathbf{a}_1)_n} = i\ddot{\alpha}_j V_1 e^{i(\rho+\alpha_j)} - \dot{\alpha}_j^2 V_1 e^{i(\rho+\alpha_j)} = \mathbf{A}_{\mathbf{b}_1} - \mathbf{A}_{\mathbf{a}_1}
\end{aligned} \quad (4.7)$$

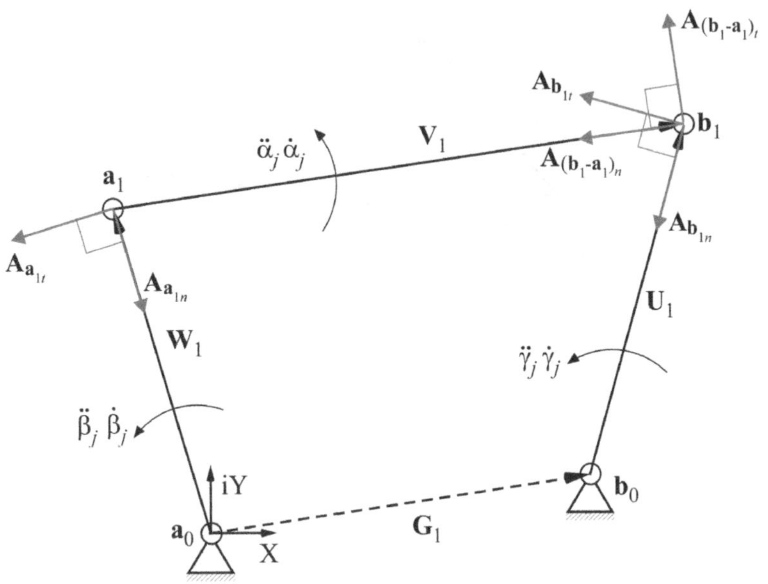

FIGURE 4.4 Planar four-bar mechanism acceleration variables.

After moving the terms $\dot{\beta}_j^2 W_1 e^{i(\theta+\beta_j)}$, $i\ddot{\beta}_j W_1 e^{i(\theta+\beta_j)}$, $\dot{\gamma}_j^2 U_1 e^{i(\sigma+\gamma_j)}$, and $\dot{\alpha}_j^2 V_1 e^{i(\rho+\alpha_j)}$ to the right-hand side in Equation (4.6), expanding this equation, and grouping its real and imaginary terms as separate equations, the resulting planar four-bar mechanism acceleration equation in the matrix form $[A]\mathbf{x} = \mathbf{b}$ is

$$\begin{bmatrix} -V_1 \sin(\rho+\alpha_j) & U_1 \sin(\sigma+\gamma_j) \\ V_1 \cos(\rho+\alpha_j) & -U_1 \cos(\sigma+\gamma_j) \end{bmatrix} \begin{Bmatrix} \ddot{\alpha}_j \\ \ddot{\gamma}_j \end{Bmatrix} = \dot{\beta}_j^2 \begin{bmatrix} W_1 \cos(\theta+\beta_j) \\ W_1 \sin(\theta+\beta_j) \end{bmatrix} - \ddot{\beta}_j \begin{bmatrix} -W_1 \sin(\theta+\beta_j) \\ W_1 \cos(\theta+\beta_j) \end{bmatrix} - \dot{\gamma}_j^2 \begin{bmatrix} U_1 \cos(\sigma+\gamma_j) \\ U_1 \sin(\sigma+\gamma_j) \end{bmatrix} + \dot{\alpha}_j^2 \begin{bmatrix} V_1 \cos(\rho+\alpha_j) \\ V_1 \sin(\rho+\alpha_j) \end{bmatrix} \quad (4.8)$$

After including the prescribed mechanism variables and the unknowns calculated from the planar four-bar displacement and velocity equations, Equation (4.8) can be solved (using Cramer's rule, for example) to calculate the coupler angular acceleration $\ddot{\alpha}_j$ and the follower angular acceleration $\ddot{\gamma}_j$.

Appendix B.1 includes the MATLAB file user instructions for planar four-bar displacement, velocity, and acceleration analysis. Using this MATLAB file (which is available for download at https://www.routledge.com/9781032956800), solutions for Equations (4.2), (4.5), and (4.8) are calculated.

4.3.4 Kinematics of Coupler Locations of Interest

The solutions calculated from the planar four-bar displacement, velocity, and acceleration equations can be used in additional equations to calculate the displacement, velocity, and acceleration of any mechanism link location of interest. Like the planar four-bar kinematic equations, the additional equations are the result of formulating vector loop equations and their derivatives.

For example, Figure 4.5 illustrates a planar four-bar mechanism with a coupler vector $\mathbf{L}_1$ that points to an arbitrary coupler location of interest $\mathbf{p}_1$. Equations (4.9) and (4.10) are produced from the loop $\mathbf{a}_0 - \mathbf{a}_1 - \mathbf{p}_1$ in Figure 4.5 (where Equation (4.10) is the result of expanding and separating Equation (4.9)).

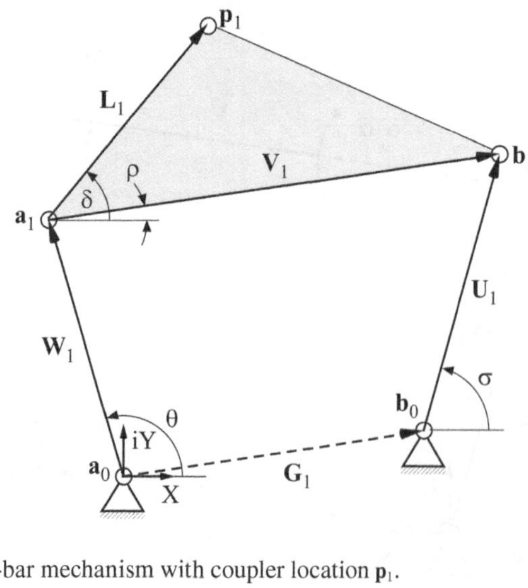

FIGURE 4.5 Planar four-bar mechanism with coupler location $\mathbf{p}_1$.

$$\mathbf{p}_{1j} = W_1 e^{i(\theta+\beta_j)} + L_1 e^{i(\delta+\alpha_j)} \tag{4.9}$$

$$\begin{aligned} p_{1xj} &= W_1 \cos(\theta+\beta_j) + L_1 \cos(\delta+\alpha_j) \\ p_{1yj} &= W_1 \sin(\theta+\beta_j) + L_1 \sin(\delta+\alpha_j) \end{aligned} \tag{4.10}$$

The velocities of the coupler location of interest can be calculated from Equation (4.12)—the expanded form of Equation (4.11) (which is the first derivative of Equation (4.9)).

$$\mathbf{V}_{\mathbf{p}_{1j}} = i\dot{\beta}_j W_1 e^{i(\theta+\beta_j)} + i\dot{\alpha}_j L_1 e^{i(\delta+\alpha_j)} \tag{4.11}$$

$$\begin{aligned} V_{p_{1xj}} &= -\dot{\beta}_j W_1 \sin(\theta+\beta_j) - \dot{\alpha}_j L_1 \sin(\delta+\alpha_j) \\ V_{p_{1yj}} &= \dot{\beta}_j W_1 \cos(\theta+\beta_j) + \dot{\alpha}_j L_1 \cos(\delta+\alpha_j) \end{aligned} \tag{4.12}$$

The accelerations of the coupler location of interest can be calculated from Equation (4.14)—the expanded form of Equation (4.13) (which is the second derivative of Equation (4.9)).

$$\mathbf{A}_{\mathbf{p}_{1j}} = -\dot{\beta}_j^2 W_1 e^{i(\theta+\beta_j)} + i\ddot{\beta}_j W_1 e^{i(\theta+\beta_j)} - \dot{\alpha}_j^2 L_1 e^{i(\delta+\alpha_j)} + i\ddot{\alpha}_j L_1 e^{i(\delta+\alpha_j)} \tag{4.13}$$

$$\begin{aligned} A_{p_{1xj}} &= -\dot{\beta}_j^2 W_1 \cos(\theta+\beta_j) - \ddot{\beta}_j W_1 \sin(\theta+\beta_j) - \dot{\alpha}_j^2 L_1 \cos(\delta+\alpha_j) - \ddot{\alpha}_j L_1 \sin(\delta+\alpha_j) \\ A_{p_{1yj}} &= -\dot{\beta}_j^2 W_1 \sin(\theta+\beta_j) + \ddot{\beta}_j W_1 \cos(\theta+\beta_j) - \dot{\alpha}_j^2 L_1 \sin(\delta+\alpha_j) + \ddot{\alpha}_j L_1 \cos(\delta+\alpha_j) \end{aligned} \tag{4.14}$$

The Appendix B.1 MATLAB file also includes Equations (4.10), (4.12), and (4.14) from which displacement, velocity, and acceleration values are calculated for a coupler link location of interest, respectively.

Example 4.1

Problem Statement: Using the Appendix B.1 MATLAB file, calculate the displaced value of coupler point $\mathbf{p}_1$ for the planar four-bar mechanism configuration in Table E.4.1 for a 25° crank displacement angle.

Known Information: Table E.4.1 and Appendix B.1 MATLAB file

Solution Approach: Figure E.4.1 includes the input specified (in bold text) in the Appendix B.1 MATLAB file. From the calculated output, it can be determined that the displaced coupler point $\mathbf{p}_1$ is $\mathbf{p}_j = 0.2127 + i1.8912$ at a 25° crank displacement angle.

TABLE E.4.1
Planar Four-Bar Mechanism Configuration

W_1, θ	V_1, ρ	U_1, σ	G_{1x}, G_{1y}	L_1, δ
1.75, 90°	1.75, −22.4860°	1.75, 64.5895°	0.8660, −0.5	1, 6.4690°

```
W1 = 1.75*exp(i*90*pi/180);
V1 = 1.75*exp(-i*22.4860*pi/180);
G1 = 0.866 - i*0.5;
U1 = 1.75*exp(i*64.5895*pi/180);
L1 = exp(i*6.469*pi/180);

start_ang   = 0;
step_ang    = 1;
stop_ang    = 25;

angular_vel  = 0 * ones(N+1,1);
angular_acc  = 0 * ones(N+1,1);
```

FIGURE E.4.1 Specified input (in bold text) in the Appendix B.1 MATLAB file for Example 4.1.

Example 4.2

Problem Statement: Using the Appendix B.1 MATLAB file, plot the path traced by point $\mathbf{p}_1$ on the level-luffing crane (Figure E.4.2).[3] The mechanism assembly configuration is given in Table E.4.2. The crank displacement angle range is $\beta_j = 10°, 9°, \cdots - 32°$.

Known Information: Table E.4.2 and Appendix B.1 MATLAB file

Solution Approach: Figure E.4.3 includes the input specified (in bold text) in the Appendix B.1 MATLAB file. In this particular mechanism configuration, the angles for $\mathbf{V}_1$ and $\mathbf{L}_1$ are identical because both vectors are parallel.[4] Figure E.4.4 illustrates the path traced by the coupler point $\mathbf{p}_1$ using the given crank displacement angle range (with a crank rotation increment of −1°). Figure E.4.5 illustrates the level-luffing crane configuration with the calculated path. The lightly shaded section of curve in this figure is useful for leveling applications because this section maintains a near-constant level.

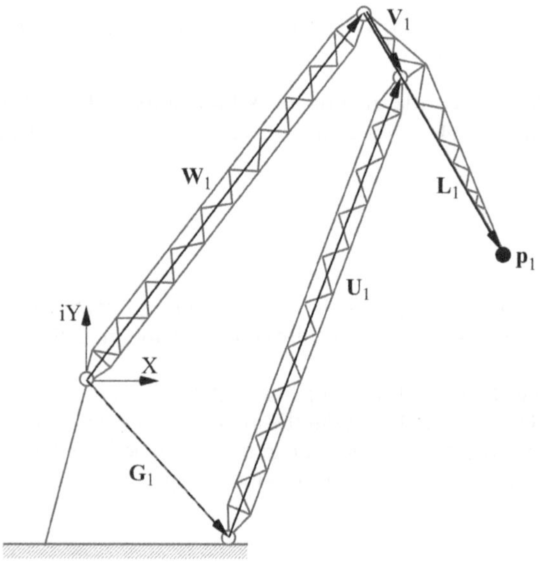

FIGURE E.4.2 Planar four-bar mechanism as a level-luffing crane.

TABLE E.4.2
Level-Luffing Crane Assembly Configuration

W_1, θ	V_1, ρ	U_1, σ	G_{1x}, G_{1y}	L_1, δ
1.96, 47.4041°	0.33, −51.6935°	2.09, 63.7721°	0.6075, −0.6909	1.19, −51.6935°

```
W1 = 1.96*exp(i*47.4041*pi/180);
V1 = 0.33*exp(-i*51.6935*pi/180);
G1 = 0.6075 - i*0.6909;
U1 = 2.09*exp(i*63.7721*pi/180);
L1 = 1.19*exp(-i*51.6935*pi/180);

start_ang    = 10;
step_ang     = -1;
stop_ang     = -32;

angular_vel  = 0 * ones(N+1,1);
angular_acc  = 0 * ones(N+1,1);
```

FIGURE E.4.3 Specified input (in bold text) in the Appendix B.1 MATLAB file for Example 4.2.

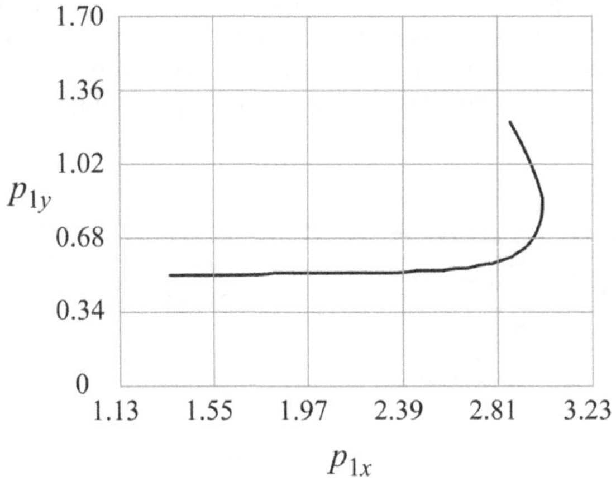

FIGURE E.4.4 Path traced by point $\mathbf{p}_1$ on the level-luffing crane mechanism.

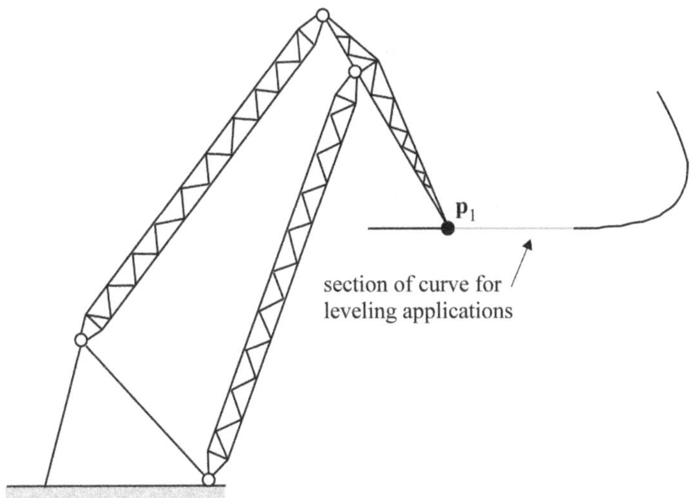

FIGURE E.4.5 Level-luffing crane mechanism with calculated coupler curve.

Example 4.3

Problem Statement: Using the Appendix B.1 MATLAB file, determine if the two door linkage configurations in Table E.4.3 (Figure E.4.6a and b) will operate properly as the door closes.

Known Information: Appendix B.1 MATLAB file, Table E.4.3, and displacement angle range for crank link

Solution Approach: One way to determine the operability of the door linkage configurations is to calculate their coupler link (or follower link) displacement angles throughout the entire crank link rotation range and check for circuit defects.[5] The data calculated from the Appendix B.1 MATLAB file include the coupler and follower link displacement angles α_j and γ_j, respectively. The crank rotation range corresponding to Figure E.4.6b is 0° (door fully open) to 90° (door fully closed).

Figure E.4.7 includes the input specified (in bold text) in the Appendix B.1 MATLAB file. Figure E.4.8 illustrates the calculated coupler link displacement angles with respect to the door (or crank) rotation angle (β) as the door closes. With configuration 1, a discontinuity appears after the door (the crank link of the linkage) exceeds an 84° displacement angle. This discontinuity corresponds to the mechanism exceeding its limiting position and subsequently experiencing a circuit defect (see Section 3.5). Using configuration 1, the door cannot close completely. In contrast, configuration 2 produces a coupler displacement angle curve that is continuous throughout the entire 90° door rotation range. Using configuration 2, the door closes completely.

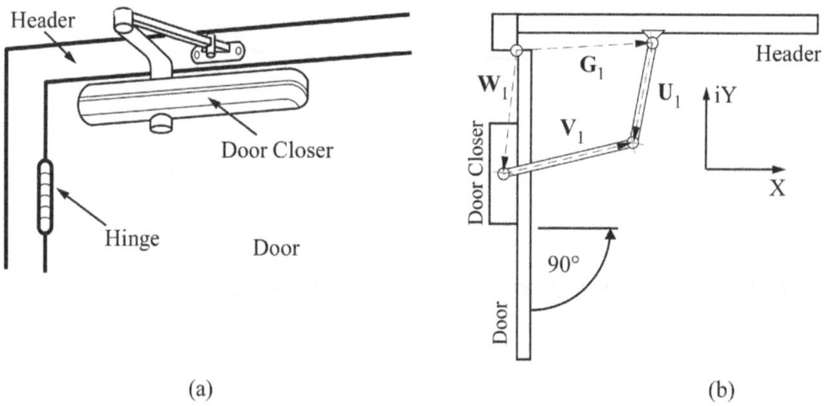

FIGURE E.4.6 (a) Door linkage and (b) kinematic model of four-bar door linkage.

TABLE E.4.3
Door Linkage Assembly Configurations

	W_1, θ	V_1, ρ	U_1, σ	G_{1x}, G_{1y}
Config. 1	10, −90°	8, 8.9743°	9, −103.4799°	10, 0
Config. 2	8, −90°	8, 1.4850°	9, −120.0198°	12.5, 0

```
W1 = 10*exp(-i*90*pi/180);
V1 = 8*exp(i*8.9743*pi/180);
G1 = 10;
U1 = 9*exp(-i*103.4799*pi/180);
L1 = 0;

start_ang   = 0;
step_ang    = 1;
stop_ang    = 90;

angular_vel  = 0 * ones(N+1,1);
angular_acc  = 0 * ones(N+1,1);
```

(MATLAB File Input for Linkage Configuration 1)

```
W1 = 8*exp(-i*90*pi/180);
V1 = 8*exp(i*1.485*pi/180);
G1 = 12.5;
U1 = 9*exp(-i*120.0198*pi/180);
L1 = 0;

start_ang   = 0;
step_ang    = 1;
stop_ang    = 90;

angular_vel  = 0 * ones(N+1,1);
angular_acc  = 0 * ones(N+1,1);
```

(MATLAB File Input for Linkage Configuration 2)

FIGURE E.4.7 Specified input (in bold text) in the Appendix B.1 MATLAB file for Example 4.3.

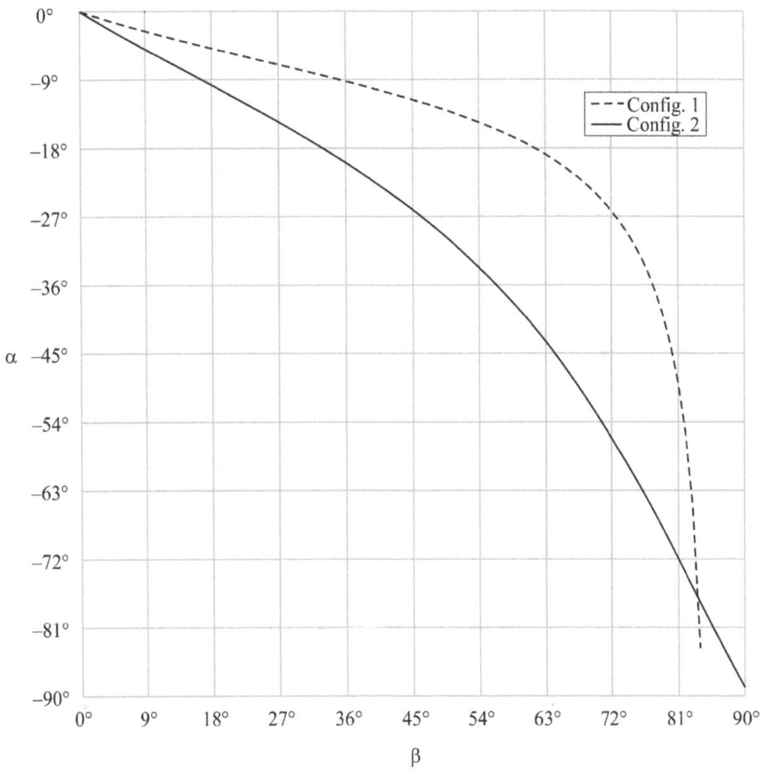

FIGURE E.4.8 Coupler displacement angles for configurations 1 and 2.

4.4 GEARED FIVE-BAR MECHANISM ANALYSIS

4.4.1 Displacement Equations

The geared five-bar mechanism consists of five links interconnected by revolute joints along with a gear pair or gear train that couples both grounded links (see Figure 4.6). As calculated from Gruebler's Equation for planar mechanisms (with $L=6$, $J_1=6$ and $J_2=2$), the geared five-bar mechanism in this figure has a single DOF. A single displacement equation is derived for this mechanism by taking the sum of the displaced mechanism vector loop in Figure 4.6. The equation produced from a clockwise sum of the displaced vector loop is

$$W_1e^{i(\theta+\beta_j)}+V_1e^{i(\rho+\alpha_j)}-S_1e^{i(\psi+\nu_j)}-U_1e^{i(\sigma+\gamma_j)}-\mathbf{G}_1=0 \tag{4.15}$$

The relationship between the displacement angles of the crank $\mathbf{W}_1$ and output link $\mathbf{U}_1$ (angular displacements β_j and γ_j, respectively, in Equation (4.15)) can be expressed as

$$\gamma_j=\pm\beta_j/r \tag{4.16}$$

where the variable r represents the *gear ratio* of the gear pair or train. The gear ratio can be defined as the ratio of the driven gear radius to the driving gear radius ($r=r_{\text{driven}}/r_{\text{driving}}$). If $\mathbf{W}_1$ and $\mathbf{U}_1$ rotate in the same direction (which occurs when an odd number of gears is used), the gear ratio is positive, and it is negative if they rotate in opposite directions (which occurs when an even number of gears is used).

After expanding Equation (4.15) and grouping its real and imaginary terms as separate equations, the resulting geared five-bar mechanism displacement equations become

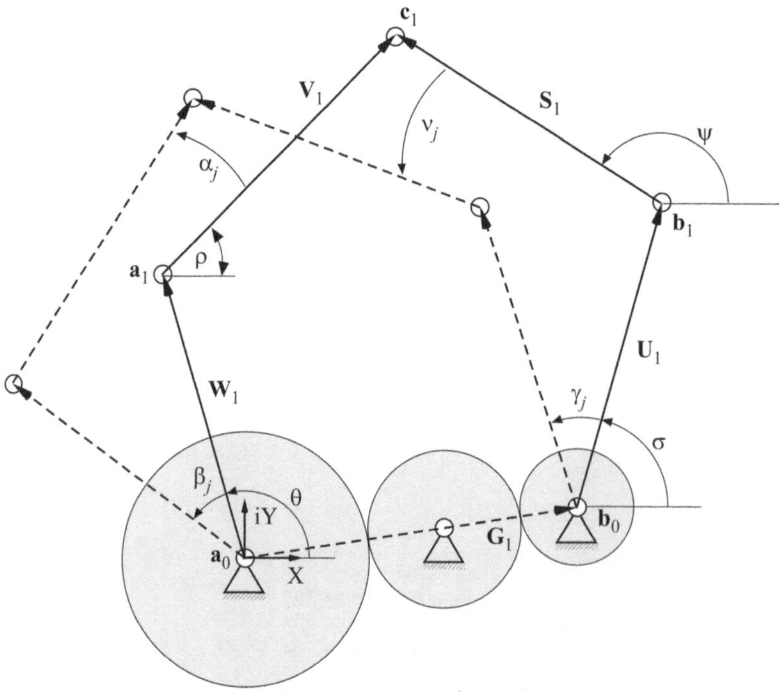

FIGURE 4.6 Geared five-bar mechanism displacement equation variables.

$$f_1(\alpha_j,\nu_j) = W_1\cos(\theta+\beta_j)+V_1\cos(\rho+\alpha_j)-S_1\cos(\psi+\nu_j)-U_1\cos(\sigma+\gamma_j)-G_{1x}=0$$
$$f_2(\alpha_j,\nu_j) = W_1\sin(\theta+\beta_j)+V_1\sin(\rho+\alpha_j)-S_1\sin(\psi+\nu_j)-U_1\sin(\sigma+\gamma_j)-G_{1y}=0 \quad (4.17)$$

Except for the intermediate link displacement angles α_j and ν_j, all other variables in the geared five-bar displacement equations are user-prescribed. Because $f_1(\alpha_j,\nu_j)$ and $f_2(\alpha_j,\nu_j)$ in Equation (4.17) both include the unknown displacement angles α_j and ν_j, these equations form a set of non-linear simultaneous equations from which the angles are calculated.

4.4.2 Velocity Equations

A single geared five-bar velocity equation is derived by differentiating the geared five-bar displacement equation. Differentiating Equation (4.15) with respect to time produces

$$i\dot{\beta}_j W_1 e^{i(\theta+\beta_j)} + i\dot{\alpha}_j V_1 e^{i(\rho+\alpha_j)} - i\dot{\nu}_j S_1 e^{i(\psi+\nu_j)} - i\dot{\gamma}_j U_1 e^{i(\sigma+\gamma_j)} = 0 \quad (4.18)$$

The relationship between the angular velocities of the crank $\mathbf{W}_1$ and output link $\mathbf{U}_1$ (angular velocities $\dot{\beta}_j$ and $\dot{\gamma}_j$, respectively, in Equation (4.18)) can be expressed as

$$\dot{\gamma}_j = \pm\dot{\beta}_j/r \quad (4.19)$$

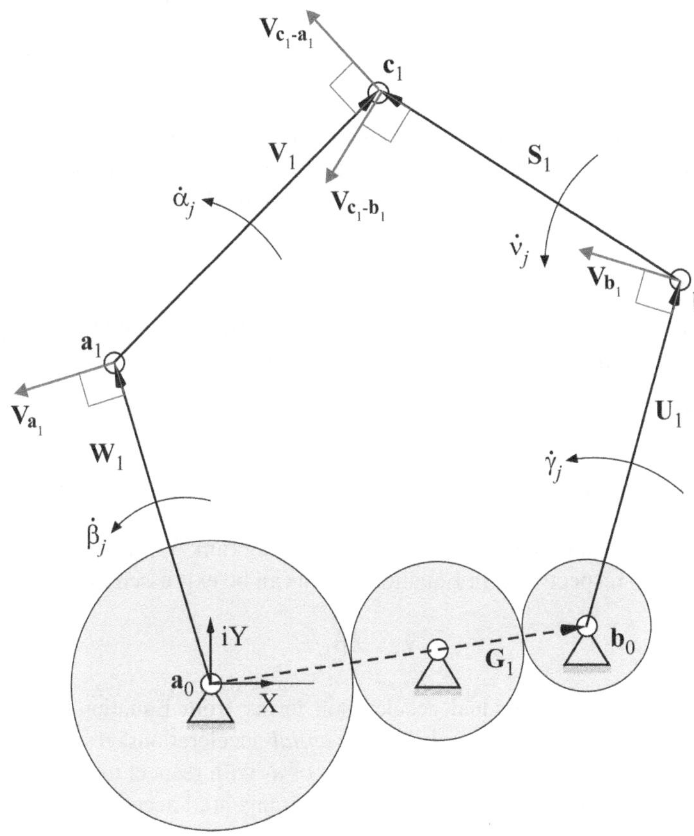

FIGURE 4.7 Geared five-bar mechanism velocity variables.

Equation (4.20) includes the individual velocity terms from Equation (4.18). The velocity $\mathbf{V}_{\mathbf{a}_1}$ is the *global* tangential velocity of the crank link vector—the tangential velocity of $\mathbf{a}_1$ with respect to $\mathbf{a}_0$ (see Figure 4.7). The velocity $\mathbf{V}_{\mathbf{b}_1}$ is the global tangential velocity of the output link vector—the tangential velocity of $\mathbf{b}_1$ with respect to $\mathbf{b}_0$. The velocity $\mathbf{V}_{\mathbf{c}_1\text{-}\mathbf{a}_1}$ is the *relative* tangential velocity of $\mathbf{c}_1$ with respect to $\mathbf{a}_1$—the relative tangential velocity of this intermediate link. Also, the velocity $\mathbf{V}_{\mathbf{c}_1\text{-}\mathbf{b}_1}$ is the *relative* tangential velocity of $\mathbf{c}_1$ with respect to $\mathbf{b}_1$—the relative tangential velocity of this intermediate link. The global velocity of $\mathbf{c}_1$ or $\mathbf{V}_{\mathbf{c}_1}$ can be calculated using $\mathbf{V}_{\mathbf{c}_1} = \mathbf{V}_{\mathbf{a}_1} + \mathbf{V}_{\mathbf{c}_1\text{-}\mathbf{a}_1} = \mathbf{V}_{\mathbf{b}_1} + \mathbf{V}_{\mathbf{c}_1\text{-}\mathbf{b}_1}$.

$$
\begin{aligned}
\mathbf{V}_{\mathbf{a}_1} &= i\dot{\beta}_j W_1 e^{i(\theta+\beta_j)} \\
\mathbf{V}_{\mathbf{b}_1} &= i\dot{\gamma}_j U_1 e^{i(\sigma+\gamma_j)} \\
\mathbf{V}_{\mathbf{c}_1-\mathbf{a}_1} &= i\dot{\alpha}_j V_1 e^{i(\rho+\alpha_j)} = \mathbf{V}_{\mathbf{c}_1} - \mathbf{V}_{\mathbf{a}_1} \\
\mathbf{V}_{\mathbf{c}_1-\mathbf{b}_1} &= i\dot{\nu}_j S_1 e^{i(\psi+\nu_j)} = \mathbf{V}_{\mathbf{c}_1} - \mathbf{V}_{\mathbf{b}_1}
\end{aligned}
\tag{4.20}
$$

After moving the terms $i\dot{\beta}_j W_1 e^{i(\theta+\beta_j)}$ and $i\dot{\gamma}_j U_1 e^{i(\sigma+\gamma_j)}$ to the right-hand side in Equation (4.18), expanding this equation and grouping its real and imaginary terms as separate equations, the resulting geared five-bar mechanism velocity equation in the matrix form $[A]\mathbf{x} = \mathbf{b}$ is

$$
\begin{bmatrix} -V_1 \sin(\rho+\alpha_j) & S_1 \sin(\psi+\nu_j) \\ V_1 \cos(\rho+\alpha_j) & -S_1 \cos(\psi+\nu_j) \end{bmatrix} \begin{Bmatrix} \dot{\alpha}_j \\ \dot{\nu}_j \end{Bmatrix} = \dot{\beta}_j \begin{bmatrix} W_1 \sin(\theta+\beta_j) \\ -W_1 \cos(\theta+\beta_j) \end{bmatrix} + \dot{\gamma}_j \begin{bmatrix} -U_1 \sin(\sigma+\gamma_j) \\ U_1 \cos(\sigma+\gamma_j) \end{bmatrix}
\tag{4.21}
$$

After including the prescribed mechanism variables and the unknowns calculated from the geared five-bar displacement equations, Equation (4.21) can be solved (using Cramer's rule, for example) and the intermediate link angular velocities $\dot{\alpha}_j$ and $\dot{\nu}_j$ are calculated

4.4.3 Acceleration Equations

A single geared five-bar acceleration equation is derived by differentiating the geared five-bar velocity equation. Time differentiation of Equation (4.18) produces

$$
\begin{aligned}
&-\dot{\beta}_j^2 W_1 e^{i(\theta+\beta_j)} + i\ddot{\beta}_j W_1 e^{i(\theta+\beta_j)} - \dot{\alpha}_j^2 V_1 e^{i(\rho+\alpha_j)} + i\ddot{\alpha}_j V_1 e^{i(\rho+\alpha_j)} + \dot{\nu}_j^2 S_1 e^{i(\psi+\nu_j)} - i\ddot{\nu}_j S_1 e^{i(\psi+\nu_j)} \\
&+\dot{\gamma}_j^2 U_1 e^{i(\sigma+\gamma_j)} - i\ddot{\gamma}_j U_1 e^{i(\sigma+\gamma_j)} = 0
\end{aligned}
\tag{4.22}
$$

The relationship between the angular accelerations of the crank $\mathbf{W}_1$ and output link $\mathbf{U}_1$ (angular accelerations $\ddot{\beta}_j$ and $\ddot{\gamma}_j$, respectively, in Equation (4.18)) can be expressed as

$$
\ddot{\gamma}_j = \pm \ddot{\beta}_j / r \tag{4.23}
$$

Equation (4.24) includes the individual acceleration terms from Equation (4.22). The accelerations $\mathbf{A}_{\mathbf{a}_{1n}}$ and $\mathbf{A}_{\mathbf{a}_{1t}}$ are the global *normal* and *tangential* accelerations, respectively, of the crank link vector—the normal and tangential accelerations of $\mathbf{a}_1$ with respect to $\mathbf{a}_0$ (see Figure 4.8). The *total* acceleration of $\mathbf{A}_{\mathbf{a}_1}$ is the sum of these normal and tangential accelerations. The accelerations

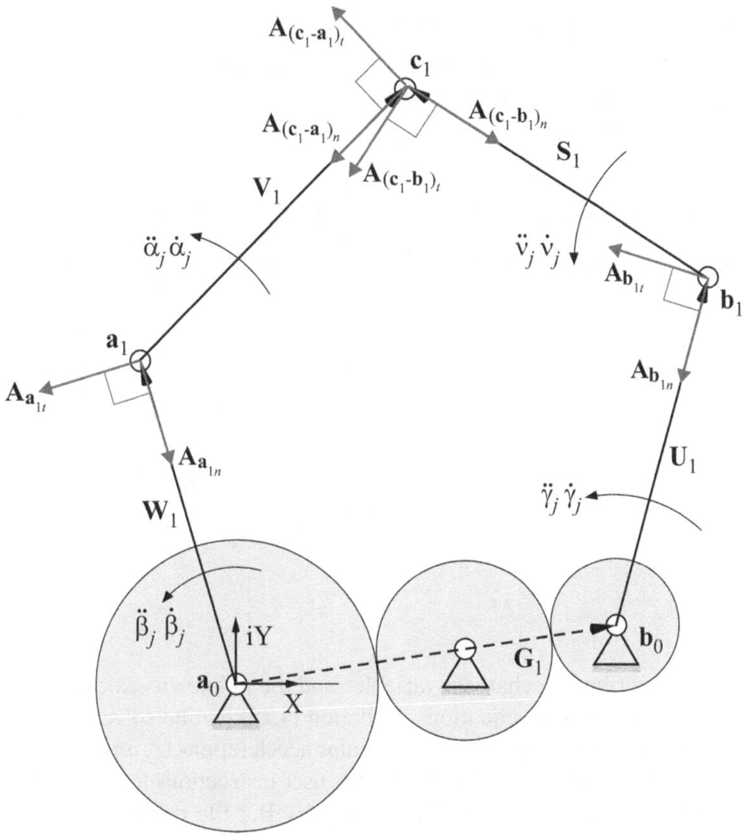

FIGURE 4.8 Geared five-bar mechanism acceleration variables.

$\mathbf{A}_{\mathbf{b}_{1n}}$ and $\mathbf{A}_{\mathbf{b}_{1t}}$ are the global normal and tangential accelerations, respectively, of the output link vector—the normal and tangential accelerations of $\mathbf{b}_1$ with respect to $\mathbf{b}_0$. The total acceleration $\mathbf{A}_{\mathbf{b}_1}$ is the sum of these normal and tangential accelerations. The accelerations $\mathbf{A}_{(\mathbf{c}_1\text{-}\mathbf{a}_1)_n}$ and $\mathbf{A}_{(\mathbf{c}_1\text{-}\mathbf{a}_1)_t}$ are the relative normal and tangential accelerations, respectively, of $\mathbf{c}_1$ with respect to $\mathbf{a}_1$—the relative normal and tangential accelerations of this intermediate link. The total relative acceleration $\mathbf{A}_{\mathbf{c}_1\text{-}\mathbf{a}_1}$ is the sum of these normal and tangential accelerations. Also, the accelerations $\mathbf{A}_{(\mathbf{c}_1\text{-}\mathbf{b}_1)_n}$ and $\mathbf{A}_{(\mathbf{c}_1\text{-}\mathbf{b}_1)_t}$ are the relative normal and tangential accelerations, respectively, of $\mathbf{c}_1$ with respect to $\mathbf{b}_1$—the relative normal and tangential accelerations of this intermediate link. The total relative acceleration $\mathbf{A}_{\mathbf{c}_1\text{-}\mathbf{b}_1}$ is the sum of these normal and tangential accelerations. The global acceleration of $\mathbf{c}_1$ or $\mathbf{A}_{\mathbf{c}_1}$ can be calculated using $\mathbf{A}_{\mathbf{c}_1} = \mathbf{A}_{\mathbf{a}_1} + \mathbf{A}_{\mathbf{c}_1\text{-}\mathbf{a}_1} = \mathbf{A}_{\mathbf{b}_1} + \mathbf{A}_{\mathbf{c}_1\text{-}\mathbf{b}_1}$.

$$\begin{aligned}
\mathbf{A}_{\mathbf{a}_1} &= \mathbf{A}_{\mathbf{a}_{1t}} + \mathbf{A}_{\mathbf{a}_{1n}} = i\ddot{\beta}_j W_1 e^{i(\theta+\beta_j)} - \dot{\beta}_j^2 W_1 e^{i(\theta+\beta_j)} \\
\mathbf{A}_{\mathbf{b}_1} &= \mathbf{A}_{\mathbf{b}_{1t}} + \mathbf{A}_{\mathbf{b}_{1n}} = i\ddot{\gamma}_j U_1 e^{i(\sigma+\gamma_j)} - \dot{\gamma}_j^2 U_1 e^{i(\sigma+\gamma_j)} \\
\mathbf{A}_{\mathbf{c}_1-\mathbf{a}_1} &= \mathbf{A}_{(\mathbf{c}_1-\mathbf{a}_1)_t} + \mathbf{A}_{(\mathbf{c}_1-\mathbf{a}_1)_n} = i\ddot{\alpha}_j V_1 e^{i(\rho+\alpha_j)} - \dot{\alpha}_j^2 V_1 e^{i(\rho+\alpha_j)} = \mathbf{A}_{\mathbf{c}_1} - \mathbf{A}_{\mathbf{a}_1} \\
\mathbf{A}_{\mathbf{c}_1-\mathbf{b}_1} &= \mathbf{A}_{(\mathbf{c}_1-\mathbf{b}_1)_t} + \mathbf{A}_{(\mathbf{c}_1-\mathbf{b}_1)_n} = i\ddot{\nu}_j S_1 e^{i(\psi+\nu_j)} - \dot{\nu}_j^2 S_1 e^{i(\psi+\nu_j)} = \mathbf{A}_{\mathbf{c}_1} - \mathbf{A}_{\mathbf{b}_1}
\end{aligned} \tag{4.24}$$

After moving the terms $\dot{\beta}_j^2 W_1 e^{i(\theta+\beta_j)}$, $i\ddot{\beta}_j W_1 e^{i(\theta+\beta_j)}$, $\dot{\gamma}_j^2 U_1 e^{i(\sigma+\gamma_j)}$, $i\ddot{\gamma}_j U_1 e^{i(\sigma+\gamma_j)}$, $\dot{\alpha}_j^2 V_1 e^{i(\rho+\alpha_j)}$, and $\dot{\nu}_j^2 S_1 e^{i(\psi+\nu_j)}$ to the right-hand side in Equation (4.22), expanding this equation and grouping its real and imaginary terms as separate equations, the resulting geared five-bar mechanism acceleration equation in the matrix form $[A]\mathbf{x} = \mathbf{b}$ is

$$\begin{bmatrix} -V_1\sin(\rho+\alpha_j) & S_1\sin(\psi+\nu_j) \\ V_1\cos(\rho+\alpha_j) & -S_1\cos(\psi+\nu_j) \end{bmatrix} \begin{Bmatrix} \ddot{\alpha}_j \\ \ddot{\nu}_j \end{Bmatrix} = \dot{\beta}_j^2 \begin{bmatrix} W_1\cos(\theta+\beta_j) \\ W_1\sin(\theta+\beta_j) \end{bmatrix} - \ddot{\beta}_j \begin{bmatrix} -W_1\sin(\theta+\beta_j) \\ W_1\cos(\theta+\beta_j) \end{bmatrix}$$
$$- \dot{\gamma}_j^2 \begin{bmatrix} U_1\cos(\sigma+\gamma_j) \\ U_1\sin(\sigma+\gamma_j) \end{bmatrix} + \ddot{\gamma}_j \begin{bmatrix} -U_1\sin(\sigma+\gamma_j) \\ U_1\cos(\sigma+\gamma_j) \end{bmatrix}$$
$$+ \dot{\alpha}_j^2 \begin{bmatrix} V_1\cos(\rho+\alpha_j) \\ V_1\sin(\rho+\alpha_j) \end{bmatrix} - \dot{\nu}_j^2 \begin{bmatrix} S_1\cos(\psi+\nu_j) \\ S_1\sin(\psi+\nu_j) \end{bmatrix} \tag{4.25}$$

After including the prescribed mechanism variables and the unknowns calculated from the geared five-bar displacement and velocity equations, Equation (4.25) can be solved (using Cramer's rule, for example) to calculate the intermediate link angular accelerations $\ddot{\alpha}_j$ and $\ddot{\nu}_j$.

Appendices B.2 and B.3 include the MATLAB file user instructions for geared five-bar displacement, velocity, and acceleration analysis. The Appendix B.2 file considers a mechanism having two gears (or negative gear ratios) and the Appendix B.3 file considers a mechanism having three gears (or positive gear ratios). Using these MATLAB files (which are available for download at https://www.routledge.com/9781032956800), solutions for Equations (4.17), (4.21), and (4.25) are calculated.

4.5.4 Kinematics of Intermediate Link Locations of Interest

Equations (4.10), (4.12), and (4.14) can be directly applied to calculate the displacements, velocities, and accelerations of an arbitrary intermediate link location of interest $\mathbf{p}_1$ of a geared five-bar mechanism (see Figure 4.9). The Appendix B.2 and B.3 MATLAB files include Equations (4.10), (4.12), and (4.14) from which displacement, velocity, and acceleration values are calculated for $\mathbf{p}_1$. As illustrated in Figure 4.9, the point of interest is on the intermediate link that includes vector $\mathbf{V}_1$. One approach to assign the point of interest existing to the intermediate link that includes vector $\mathbf{S}_1$ in the geared five-bar mechanism MATLAB files requires that the *values* assigned to vectors $\mathbf{U}_1$ and $\mathbf{W}_1$ be swapped with each other as well as the values assigned to vectors $\mathbf{S}_1$ and $\mathbf{V}_1$. In addition, the value assigned to vector $\mathbf{G}_1$ must be multiplied by –1 (which moves the coordinate system origin to point $\mathbf{b}_0$ in Figure 4.9). If this approach is used in the geared five-bar mechanism MATLAB files, the value assigned to vector $\mathbf{L}_1$ will correspond to a point of interest on the intermediate link that includes vector $\mathbf{S}_1$.

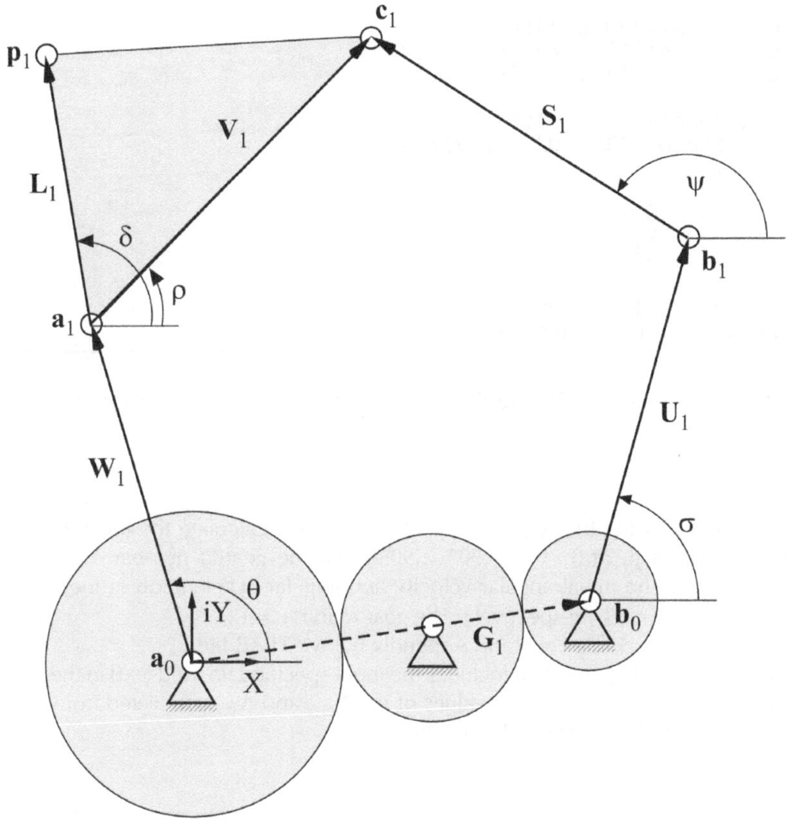

FIGURE 4.9 Geared five-bar mechanism with intermediate link location $\mathbf{p}_1$.

Example 4.4

Problem Statement: Using the Appendix B.3 MATLAB file, calculate the displaced value of link point $\mathbf{p}_1$ and its velocity for the geared five-bar mechanism configuration in Table E.4.4 for a 45° crank displacement angle and a constant rotation speed of 1.5 rad/s. The gear ratio is $r = +2$.

Known Information: Appendix B.3 MATLAB file, Table E.4.4, and β and $\dot{\beta}$

Solution Approach: Figure E.4.9 includes the input specified (in bold text) in the Appendix B.3 MATLAB file. From the calculated output, it can be determined that the displaced coupler point $\mathbf{p}_1$ is $\mathbf{p}_j = (0.0972 + i0.1865)$ m and its velocity is $\mathbf{V}_{\mathbf{p}_j} = (-0.3480 - i0.2398)$ m/s at a 45° crank displacement angle.

TABLE E.4.4
Geared Five-Bar Mechanism Configuration (with Link Lengths in Meters)

W_1, θ	V_1, ρ	U_1, σ	S_1, ψ	G_{1x}, G_{1y}	L_1, δ
0.35, 90°	0.525, 54.7643°	0.35, 60°	0.525, 115.0279°	0.35, 0	0.35, −15.7645°

```
W1 = 0.35*exp(i*90*pi/180);
V1 = 0.525*exp(i*54.7643*pi/180);
L1 = 0.35*exp(-i*15.7645*pi/180);
G1 = 0.35;
U1 = 0.35*exp(i*60*pi/180);
S1 = 0.525*exp(i*115.0279*pi/180);
ratio = 2;

start_ang   = 0;
step_ang    = 1;
stop_ang    = 45;

angular_vel  = 1.5 * ones(N+1,1);
angular_acc  = 0 * ones(N+1,1);
```

FIGURE E.4.9 Specified input (in bold text) in the Appendix B.3 MATLAB file for Example 4.4.

Example 4.5

Problem Statement: Using the Appendix B.3 MATLAB file, calculate the location, velocity, and acceleration values of $\mathbf{p}_1$ at $\beta = 0°, 15°, 30°, \ldots, 90°$ given the geared five-bar mechanism dimensions in Table E.4.5. The initial angular velocity and angular acceleration of the driving link are $\dot{\beta}_0 = 1$ rad/s and $\ddot{\beta} = 0.1$ rad/s^2, respectively. The gear ratio is $r = +2$.

Known Information: Table E.4.5 and Appendix B.3 MATLAB file

Solution Approach: Figure E.4.10 includes the input specified (in bold text) in the Appendix B.3 MATLAB file. Table E.4.6 includes the values of $\mathbf{p}_{1j}$, $\mathbf{V}_{\mathbf{p}_{1j}}$, and $\mathbf{A}_{\mathbf{p}_{1j}}$ calculated from the Appendix B.3 MATLAB file over the given crank rotation range.

TABLE E.4.5
Geared Five-Bar Mechanism Dimensions (with Lengths in Meters)

W_1, θ	V_1, ρ	U_1, σ	S_1, ψ	G_{1x}, G_{1y}	L_1, δ
1, 90°	1.5, 32.7304°	1.5, 45°	1.5, 149.9837°	1.5, 0	1, 74.1400°

```
W1 = 1.0*exp(i*90*pi/180);
V1 = 1.5*exp(i*32.7304*pi/180);
L1 = 1.0*exp(i*74.1400*pi/180);
G1 = 1.5;
U1 = 1.5*exp(i*45*pi/180);
S1 = 1.5*exp(i*149.9847*pi/180);
ratio = 2;

start_ang   = 0;
step_ang    = 1;
stop_ang    = 90;

angular_vel  = 1 * ones(N+1,1);
angular_acc  = 0.1 * ones(N+1,1);
```

FIGURE E.4.10 Specified input (in bold text) in the Appendix B.3 MATLAB file for Example 4.5.

TABLE E.4.6
Calculated Geared Five-Bar Point Positions, Velocities, and Accelerations

β	$\mathbf{p}_{1j}$ (m)	$\mathbf{V}_{p1_j}$ (m/s)	$\mathbf{A}_{p1_j}$ (m/s²)
0°	0.2733, 1.9619	−0.8980, −0.0290	−0.2352, −0.9733
15°	0.0360, 1.9215	−0.9273, −0.2851	0.0102, −0.9959
30°	−0.1949, 1.8184	−0.8956, −0.5302	0.2336, −0.9326
45°	−0.4068, 1.6609	−0.8202, −0.7419	0.3516, −0.7683
60°	−0.5944, 1.4624	−0.7440, −0.8976	0.2326, −0.5178
75°	−0.7657, 1.2386	−0.7199, −0.9973	0.0079, −0.3769
90°	−0.9304, 0.9976	−0.6957, −1.0978	0.2980, −0.5166

4.5 STEPHENSON III MECHANISM ANALYSIS

4.5.1 Displacement Equations

The Stephenson III mechanism consists of six links interconnected by revolute joints. As calculated from Gruebler's Equation for planar mechanisms (with $L = 6$ and $J_1 = 7$), the Stephenson III mechanism has a single DOF.

Because the Stephenson III mechanism loop $\mathbf{W}_1 - \mathbf{V}_1 - \mathbf{U}_1 - \mathbf{G}_1$ is a planar four-bar mechanism, a displacement analysis for this mechanism loop can be conducted using the planar four-bar displacement equations in Section 4.3.1. In this section, the displacement equations for Stephenson III mechanism dyad $\mathbf{U}_1^* - \mathbf{V}_1^*$ (Figure 4.10) are presented.

To analyze the $\mathbf{U}_1^* - \mathbf{V}_1^*$ dyad (Figure 4.10b), a vector loop must be formed that includes this dyad. The equation produced from the clockwise vector loop sum $\mathbf{W}_1 - \mathbf{L}_1 - \mathbf{V}_1^* - \mathbf{U}_1^* - \mathbf{G}_1^* - \mathbf{G}_1$ is

$$W_1e^{i(\theta+\beta_j)} + L_1e^{i(\delta+\alpha_j)} + V_1^*e^{i(\rho^*+\alpha_j^*)} - U_1^*e^{i(\sigma^*+\gamma_j^*)} - \mathbf{G}_1^* - \mathbf{G}_1 = 0 \tag{4.26}$$

After expanding Equation (4.26) and grouping its real and imaginary terms as separate equations, the resulting Stephenson III mechanism displacement equations become

$$\begin{aligned} f_1(\alpha_j^*,\gamma_j^*) &= W_1\cos(\theta+\beta_j) + L_1\cos(\delta+\alpha_j) + V_1^*\cos(\rho^*+\alpha_j^*) - U_1^*\cos(\sigma^*+\gamma_j^*) - G_{1x}^* - G_{1x} = 0 \\ f_2(\alpha_j^*,\gamma_j^*) &= W_1\sin(\theta+\beta_j) + L_{11}\sin(\delta+\alpha_j) + V_1^*\sin(\rho^*+\alpha_j^*) - U_1^*\sin(\sigma^*+\gamma_j^*) - G_{1y}^* - G_{1y} = 0 \end{aligned} \tag{4.27}$$

Except for the $\mathbf{U}_1^* - \mathbf{V}_1^*$ dyad displacement angles α_j^* and γ_j^*, all other variables in the Stephenson III mechanism displacement equations are either user-prescribed or calculated from Equation (4.2). Because $f_1(\alpha_j^*,\gamma_j^*)$ and $f_2(\alpha_j^*,\gamma_j^*)$ in Equation (4.27) both include the unknown $\mathbf{U}_1^* - \mathbf{V}_1^*$ dyad displacement angles α_j^* and γ_j^*, these equations form a set of nonlinear simultaneous equations from which the angles are calculated.

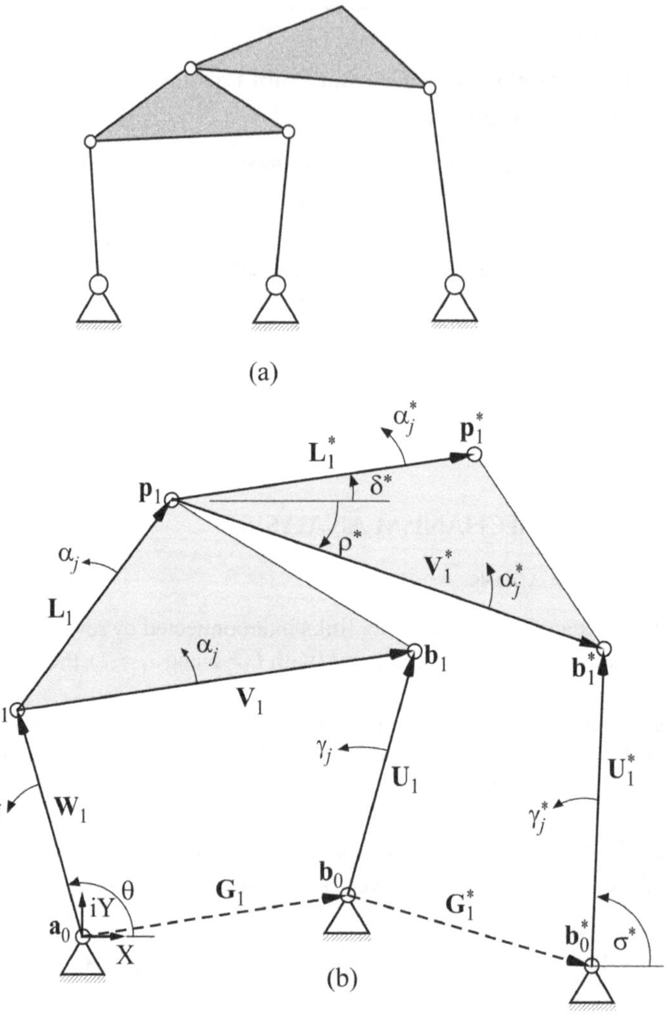

FIGURE 4.10 (a) Stephenson III mechanism and (b) mechanism displacement variables

Example 4.6

Problem Statement: Using the Appendix B.4 MATLAB file, calculate the displacement angle of vector $\mathbf{V}_1^*$ for the Stephenson III mechanism configuration in Tables E.4.1 (see Example 4.1) and E.4.7 for a 50° crank displacement angle.

Known Information: Tables E.4.1, E.4.7, and Appendix B.4 MATLAB file

Solution Approach: Figure E.4.11 includes the input specified (in bold text) in the Appendix B.4 MATLAB file. From the calculated output, it can be determined that the displacement angle of vector $\mathbf{V}_1^*$ is $\alpha_j^* = 7.1215°$ at a 50° crank displacement angle.

TABLE E.4.7
Stephenson III Mechanism Configuration

V_1^*, ρ^*	U_1^*, σ^*	G_{1x}^*, G_{1y}^*
1.75, −8.8397°	1.75, 65.6031°	1.1340, 0.5

```
W1 = 1.75*exp(i*90*pi/180);
V1 = 1.75*exp(-i*22.4860*pi/180);
G1 = 0.866 - i*0.5;
U1 = 1.75*exp(i*64.5895*pi/180);
L1 = exp(i*6.469*pi/180);

V1s = 1.75*exp(-i*8.8397*pi/180);
G1s = 1.1340 + i*0.5;
U1s = 1.75*exp(i*65.6031*pi/180);
L1s = 0;

start_ang   = 0;
step_ang    = 1;
stop_ang    = 50;

angular_vel  = 0 * ones(N+1,1);
angular_acc  = 0 * ones(N+1,1);
```

FIGURE E.4.11 Specified input (in bold text) in the Appendix B.4 MATLAB file for Example 4.6.

4.5.2 Velocity Equations

Because the Stephenson III mechanism loop $\mathbf{W}_1 - \mathbf{V}_1 - \mathbf{U}_1 - \mathbf{G}_1$ is a planar four-bar mechanism, a velocity analysis for this mechanism loop can be conducted using the planar four-bar velocity equations in Section 4.3.2. In this section, the velocity equations for Stephenson III mechanism dyad $\mathbf{U}_1^* - \mathbf{V}_1^*$ (Figure 4.11) are presented.

A single Stephenson III velocity equation is derived by differentiating the Stephenson III displacement equation. Differentiating Equation (4.26) with respect to time produces

$$i\dot{\beta}_j W_1 e^{i(\theta+\beta_j)} + i\dot{\alpha}_j L_1 e^{i(\delta+\alpha_j)} + i\dot{\alpha}_j^* V_1^* e^{i(\rho^*+\alpha_j^*)} - i\dot{\gamma}_j^* U_1^* e^{i(\sigma^*+\gamma_j^*)} = 0 \tag{4.28}$$

Equation (4.29) includes the individual velocity terms from Equation (4.28). The velocity $\mathbf{V}_{\mathbf{b}_1^*}$ is the *global* tangential velocity of vector $\mathbf{U}_1^*$—the tangential velocity of $\mathbf{b}_1^*$ with respect to $\mathbf{b}_0^*$ (see Figure 4.11). The velocity $\mathbf{V}_{\mathbf{p}_1\text{-}\mathbf{b}_1^*}$ is the *relative* tangential velocity of $\mathbf{p}_1$ with respect to $\mathbf{b}_1^*$—the

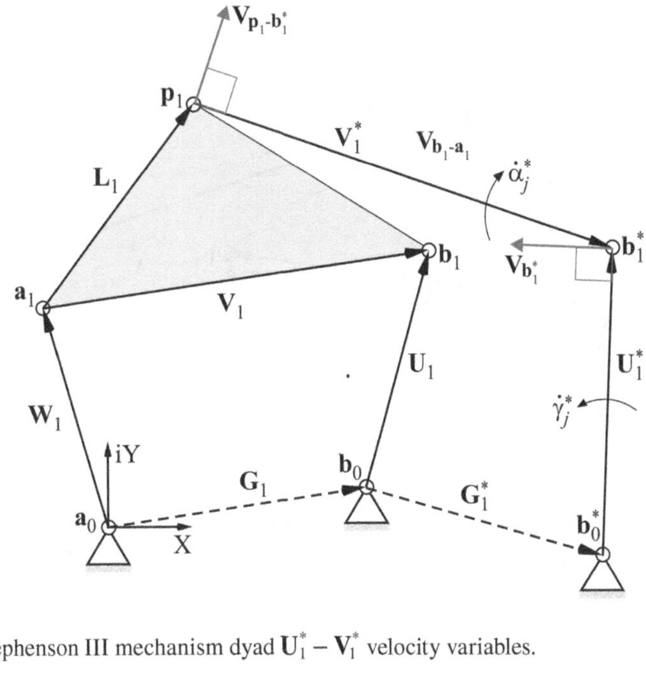

FIGURE 4.11 Stephenson III mechanism dyad $\mathbf{U}_1^* - \mathbf{V}_1^*$ velocity variables.

relative tangential velocity of vector $\mathbf{V}_1^*$. The global velocity of $\mathbf{p}_1$ or $\mathbf{V}_{\mathbf{p}_1}$ can be calculated using $\mathbf{V}_{\mathbf{p}_1} = \mathbf{V}_{\mathbf{b}_1^*} + \mathbf{V}_{\mathbf{p}_1\text{-}\mathbf{b}_1^*}$ and is identical to the Equation (4.11) value for the planar four-bar mechanism loop in the Stephenson III mechanism.

$$\begin{aligned} \mathbf{V}_{\mathbf{b}_1^*} &= i\dot{\gamma}_j^* U_1^* e^{i(\sigma^* + \gamma_j^*)} \\ \mathbf{V}_{\mathbf{p}_1\text{-}\mathbf{b}_1^*} &= i\dot{\alpha}_j^* V_1^* e^{i(\rho^* + \alpha_j^*)} = \mathbf{V}_{\mathbf{p}_1} - \mathbf{V}_{\mathbf{b}_1^*} \end{aligned} \tag{4.29}$$

After moving the terms $i\dot{\beta}_j W_1 e^{i(\theta+\beta_j)}$ and $i\dot{\alpha}_j L_1 e^{i(\delta+\alpha_j)}$ to the right-hand side in Equation (4.28), expanding this equation and grouping its real and imaginary terms as separate equations, the resulting Stephenson III mechanism velocity equation in the matrix form $[A]\mathbf{x} = \mathbf{b}$ is

$$\begin{bmatrix} -V_1^* \sin(\rho^* + \alpha_j^*) & U_1^* \sin(\sigma^* + \gamma_j^*) \\ V_1^* \cos(\rho^* + \alpha_j^*) & -U_1^* \cos(\sigma^* + \gamma_j^*) \end{bmatrix} \begin{Bmatrix} \dot{\alpha}_j^* \\ \dot{\gamma}_j^* \end{Bmatrix} = \dot{\beta}_j \begin{bmatrix} W_1 \sin(\theta + \beta_j) \\ -W_1 \cos(\theta + \beta_j) \end{bmatrix} + \dot{\alpha}_j \begin{bmatrix} L_1 \sin(\delta + \alpha_j) \\ -L_1 \cos(\delta + \alpha_j) \end{bmatrix} \tag{4.30}$$

After including the prescribed mechanism variables and the unknowns calculated from the Stephenson III displacement equations, Equation (4.30) can be solved (using Cramer's rule, for example) to calculate the $\mathbf{U}_1^*$ - $\mathbf{V}_1^*$ dyad angular velocities $\dot{\alpha}_j^*$ and $\dot{\gamma}_j^*$.

4.5.3 Acceleration Equations

Because the Stephenson III mechanism loop $\mathbf{W}_1 - \mathbf{V}_1 - \mathbf{U}_1 - \mathbf{G}_1$ is a planar four-bar mechanism, an acceleration analysis for this mechanism loop can be conducted using the planar four-bar acceleration equations in Section 4.3.3. In this section, the acceleration equations for Stephenson III mechanism dyad $\mathbf{U}_1^* - \mathbf{V}_1^*$ (Figure 4.12) are presented.

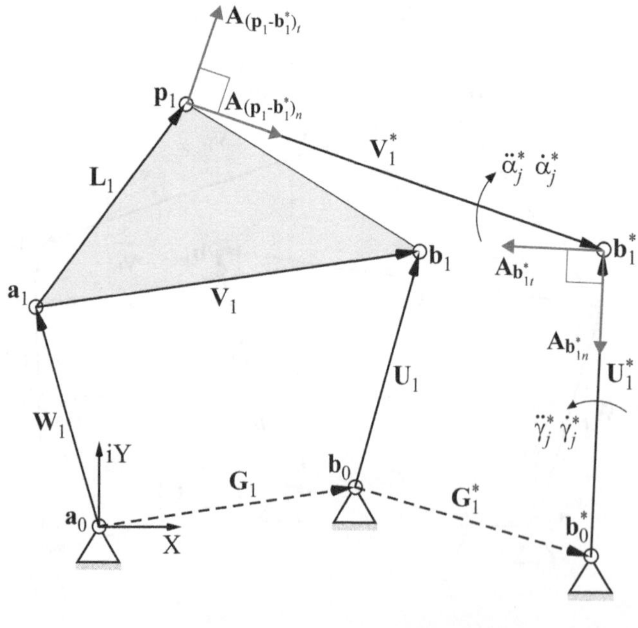

FIGURE 4.12 Stephenson III mechanism dyad $\mathbf{U}_1^* - \mathbf{V}_1^*$ acceleration variables.

A single planar four-bar acceleration equation is derived by differentiating the planar four-bar velocity equation. Time differentiation of the Equation (4.28) produces

$$
\begin{aligned}
&-\dot{\beta}_j^2 W_1 e^{i(\theta+\beta_j)} + i\ddot{\beta}_j W_1 e^{i(\theta+\beta_j)} - \dot{\alpha}_j^2 L_1 e^{i(\delta+\alpha_j)} + i\ddot{\alpha}_j L_1 e^{i(\delta+\alpha_j)} - \left(\dot{\alpha}_j^*\right)^2 V_1^* e^{i\left(\rho^*+\alpha_j^*\right)} + i\ddot{\alpha}_j^* V_1^* e^{i\left(\rho^*+\alpha_j^*\right)} \\
&+\left(\dot{\gamma}_j^*\right)^2 U_1^* e^{i\left(\sigma^*+\gamma_j^*\right)} - i\ddot{\gamma}_j^* U_1^* e^{i\left(\sigma^*+\gamma_j^*\right)} = 0
\end{aligned} \tag{4.31}
$$

Equation (4.32) includes the individual acceleration terms from Equation (4.31). The accelerations $\mathbf{A}_{\mathbf{b}_{1n}^*}$ and $\mathbf{A}_{\mathbf{b}_{1t}^*}$ are the global *normal* and *tangential* accelerations, respectively, of vector $\mathbf{U}_1^*$—the normal and tangential accelerations $\mathbf{b}_1^*$ with respect to $\mathbf{b}_0^*$ (see Figure 4.12). The *total* acceleration of $\mathbf{A}_{\mathbf{b}_1^*}$ is the sum of these normal and tangential accelerations. The accelerations $\mathbf{A}_{\left(\mathbf{p}_1\text{-}\mathbf{b}_1^*\right)_n}$ and $\mathbf{A}_{\left(\mathbf{p}_1\text{-}\mathbf{b}_1^*\right)_t}$ are the relative normal and tangential accelerations, respectively, of $\mathbf{p}_1$ with respect to $\mathbf{b}_1^*$—the relative normal and tangential accelerations of vector $\mathbf{V}_1^*$. The total relative acceleration $\mathbf{A}_{\mathbf{p}_1\text{-}\mathbf{b}_1^*}$ is the sum of these normal and tangential accelerations. The global acceleration of $\mathbf{p}_1$ or $\mathbf{A}_{\mathbf{p}_1}$ can be calculated using $\mathbf{A}_{\mathbf{p}_1} = \mathbf{A}_{\mathbf{b}_1^*} + \mathbf{A}_{\mathbf{p}_1\text{-}\mathbf{b}_1^*}$ and is identical to the Equation (4.13) value for the planar four-bar mechanism loop in the Stephenson III mechanism.

$$
\begin{aligned}
&\mathbf{A}_{\mathbf{b}_1^*} = \mathbf{A}_{\mathbf{b}_{1t}^*} + \mathbf{A}_{\mathbf{b}_{1n}^*} = i\ddot{\gamma}_j^* U_1^* e^{i\left(\sigma^*+\gamma_j^*\right)} - \left(\dot{\gamma}_j^*\right)^2 U_1^* e^{i\left(\sigma^*+\gamma_j^*\right)} \\
&\mathbf{A}_{\mathbf{p}_1\text{-}\mathbf{b}_1^*} = \mathbf{A}_{\left(\mathbf{p}_1\text{-}\mathbf{b}_1^*\right)_t} + \mathbf{A}_{\left(\mathbf{p}_1\text{-}\mathbf{b}_1^*\right)_n} = i\ddot{\alpha}_j^* V_1^* e^{i\left(\rho^*+\alpha_j^*\right)} - \left(\dot{\alpha}_j^*\right)^2 V_1^* e^{i\left(\rho^*+\alpha_j^*\right)} = \mathbf{A}_{\mathbf{p}_1} - \mathbf{A}_{\mathbf{b}_1^*}
\end{aligned} \tag{4.32}
$$

After moving all of the terms, except $i\ddot{\alpha}_j^* V_1^* e^{i\left(\rho^*+\alpha_j^*\right)}$ and $i\ddot{\gamma}_j^* U_1^* e^{i\left(\sigma^*+\gamma_j^*\right)}$, to the right-hand side in Equation (4.31), expanding this equation and grouping its real and imaginary terms as separate equations, the resulting Stephenson III mechanism acceleration equation in the matrix form $[A]\mathbf{x} = \mathbf{b}$ is

$$
\begin{aligned}
&\begin{bmatrix} -V_1^* \sin\left(\rho^*+\alpha_j^*\right) & U_1^* \sin\left(\sigma^*+\gamma_j^*\right) \\ V_1^* \cos\left(\rho^*+\alpha_j^*\right) & -U_1^* \cos\left(\sigma^*+\gamma_j^*\right) \end{bmatrix} \begin{Bmatrix} \ddot{\alpha}_j^* \\ \ddot{\gamma}_j^* \end{Bmatrix} \\
&= \dot{\beta}_j^2 \begin{bmatrix} W_1 \cos\left(\theta+\beta_j\right) \\ W_1 \sin\left(\theta+\beta_j\right) \end{bmatrix} - \ddot{\beta}_j \begin{bmatrix} -W_1 \sin\left(\theta+\beta_j\right) \\ W_1 \cos\left(\theta+\beta_j\right) \end{bmatrix} \\
&+ \dot{\alpha}_j^2 \begin{bmatrix} L_1 \cos\left(\delta+\alpha_j\right) \\ L_1 \sin\left(\delta+\alpha_j\right) \end{bmatrix} - \ddot{\alpha}_j \begin{bmatrix} -L_1 \sin\left(\delta+\alpha_j\right) \\ L_1 \cos\left(\delta+\alpha_j\right) \end{bmatrix} \\
&+ \left(\dot{\alpha}_j^*\right)^2 \begin{bmatrix} V_1^* \cos\left(\rho^*+\alpha_j^*\right) \\ V_1^* \sin\left(\rho^*+\alpha_j^*\right) \end{bmatrix} - \left(\dot{\gamma}_j^*\right)^2 \begin{bmatrix} U_1^* \cos\left(\sigma^*+\gamma_j^*\right) \\ U_1^* \sin\left(\sigma^*+\gamma_j^*\right) \end{bmatrix}
\end{aligned} \tag{4.33}
$$

After including the prescribed mechanism variables and the unknowns calculated from the Stephenson III displacement and velocity equations, Equation (4.33) can be solved (using Cramer's rule, for example) to calculate the $\mathbf{U}_1^* - \mathbf{V}_1^*$ dyad angular accelerations $\ddot{\alpha}_j^*$ and $\ddot{\gamma}_j^*$, respectively.

Appendix B.4 includes the MATLAB file user instructions for Stephenson III displacement, velocity, and acceleration analysis. Using this MATLAB files (which are available for download at https://www.routledge.com/9781032956800), solutions for Equations (4.27), (4.30), and (4.33) are calculated.

4.5.4 Kinematics of Intermediate Link Locations of Interest

The results calculated from the Stephenson III displacement, velocity, and acceleration equations can be used in additional equations to calculate the displacement, velocity, and acceleration of intermediate link locations of interest. These equations are the result of formulating vector loop equation and their derivatives.

For example, Figure 4.10b illustrates a Stephenson III mechanism with an intermediate link vector $\mathbf{L}_1^*$ that points to an arbitrary intermediate link location of interest $\mathbf{p}_1^*$. Equations (4.34) and (4.35) are produced from the loop $\mathbf{a}_0 - \mathbf{a}_1 - \mathbf{p}_1 - \mathbf{p}_1^*$ in Figure 4.10b (where Equation (4.35) is the result of expanding and separating Equation (4.34)).

$$\mathbf{p}_{1j}^* = W_1 e^{i(\theta+\beta_j)} + L_1 e^{i(\delta+\alpha_j)} + L_1^* e^{i(\delta^*+\alpha_j^*)} \tag{4.34}$$

$$\begin{aligned} p_{1xj}^* &= W_1 \cos(\theta+\beta_j) + L_1 \cos(\delta+\alpha_j) + L_1^* \cos(\delta^*+\alpha_j^*) \\ p_{1yj}^* &= W_1 \sin(\theta+\beta_j) + L_1 \sin(\delta+\alpha_j) + L_1^* \sin(\delta^*+\alpha_j^*) \end{aligned} \tag{4.35}$$

The velocities of the intermediate link location of interest can be calculated from Equation (4.37)—the expanded form of Equation (4.36) (which is the first derivative of Equation (4.34)).

$$\mathbf{V}_{\mathbf{p}_{1j}^*} = i\dot{\beta}_j W_1 e^{i(\theta+\beta_j)} + i\dot{\alpha}_j L_1 e^{i(\delta+\alpha_j)} + i\dot{\alpha}_j^* L_1^* e^{i(\delta^*+\alpha_j^*)} \tag{4.36}$$

$$\begin{aligned} V_{p_{1xj}^*} &= -\dot{\beta}_j W_1 \sin(\theta+\beta_j) - \dot{\alpha}_j L_1 \sin(\delta+\alpha_j) - \dot{\alpha}_j^* L_1^* \sin(\delta^*+\alpha_j^*) \\ V_{p_{1yj}^*} &= \dot{\beta}_j W_1 \cos(\theta+\beta_j) + \dot{\alpha}_j L_1 \cos(\delta+\alpha_j) + \dot{\alpha}_j^* L_1^* \cos(\delta^*+\alpha_j^*) \end{aligned} \tag{4.37}$$

The accelerations of the intermediate link location of interest can be calculated from Equation (4.39)—the expanded form of Equation (4.38) (which is the second derivative of Equation (4.34)).

$$\mathbf{A}_{\mathbf{p}_{1j}^*} = -\dot{\beta}_j^2 W_1 e^{i(\theta+\beta_j)} + i\ddot{\beta}_j W_1 e^{i(\theta+\beta_j)} - \dot{\alpha}_j^2 L_1 e^{i(\delta+\alpha_j)} + i\ddot{\alpha}_j L_1 e^{i(\delta+\alpha_j)} - \left(\dot{\alpha}_j^*\right)^2 L_1^* e^{i(\delta^*+\alpha_j^*)} + i\ddot{\alpha}_j^* L_1^* e^{i(\delta^*+\alpha_j^*)} \tag{4.38}$$

$$\begin{aligned} A_{p_{1xj}^*} &= -\dot{\beta}_j^2 W_1 \cos(\theta+\beta_j) - \ddot{\beta}_j W_1 \sin(\theta+\beta_j) - \dot{\alpha}_j^2 L_1 \cos(\delta+\alpha_j) - \ddot{\alpha}_j L_1 \sin(\delta+\alpha_j) \\ &\quad -\left(\dot{\alpha}_j^*\right)^2 L_1^* \cos(\delta^*+\alpha_j^*) - \ddot{\alpha}_j^* L_1^* \sin(\delta^*+\alpha_j^*) \\ A_{p_{1yj}^*} &= -\dot{\beta}_j^2 W_1 \sin(\theta+\beta_j) + \ddot{\beta}_j W_1 \cos(\theta+\beta_j) - \dot{\alpha}_j^2 L_1 \sin(\delta+\alpha_j) + \ddot{\alpha}_j L_1 \cos(\delta+\alpha_j) \\ &\quad -\left(\dot{\alpha}_j^*\right)^2 L_1^* \sin(\delta^*+\alpha_j^*) + \ddot{\alpha}_j^* L_1^* \cos(\delta^*+\alpha_j^*) \end{aligned} \tag{4.39}$$

The Appendix B.4 MATLAB file also includes Equations (4.35), (4.37), and (4.39) from which displacement, velocity, and acceleration values are calculated for the intermediate link locations of interest.

Example 4.7

Problem Statement: Using the Appendix B.4 MATLAB file, calculate the path traced by point $\mathbf{p}_1^*$ of the Stephenson III mechanism given in Table E.4.8 for a complete crank rotation.

Known Information: Table E.4.8 and Appendix B.4 MATLAB file

Solution Approach: Figure E.4.12 includes the input specified (in bold text) in the Appendix B.4 MATLAB file. Figure E.4.13 includes the Stephenson III mechanism and the $\mathbf{p}_1^*$ path calculated from the Appendix B.4 MATLAB file.

TABLE E.4.8
Stephenson III Mechanism Dimensions

W_1, θ	V_1, ρ	U_1, σ	G_{1x}, G_{1y}	L_1, δ
1, 90°	1.5, 19.3737°	1.5, 93.2461°	1.5, 0	1, 60.7834°
L_1^*, δ^*	V_1^*, ρ^*	U_1^*, σ^*	G_{1x}^*, G_{1y}^*	
1, 63.7091°	2, 17.1417°	2, 76.4844°	0.4318, 0.5176	

```
W1 = exp(i*90*pi/180);
V1 = 1.5*exp(i*19.3737*pi/180);
G1 = 1.5;
U1 = 1.5*exp(i*93.2461*pi/180);
L1 = exp(i*60.7834*pi/180);

V1s = 2*exp(i*17.1417*pi/180);
G1s = 0.4318 + i*0.5176;
U1s = 2*exp(i*76.4844*pi/180);
L1s = exp(i*63.7091*pi/180);

start_ang    = 0;
step_ang     = 1;
stop_ang     = 360;

angular_vel  = 0 * ones(N+1,1);
angular_acc  = 0 * ones(N+1,1);
```

FIGURE E.4.12 Specified input (in bold text) in the Appendix B.4 MATLAB file for Example 4.7.

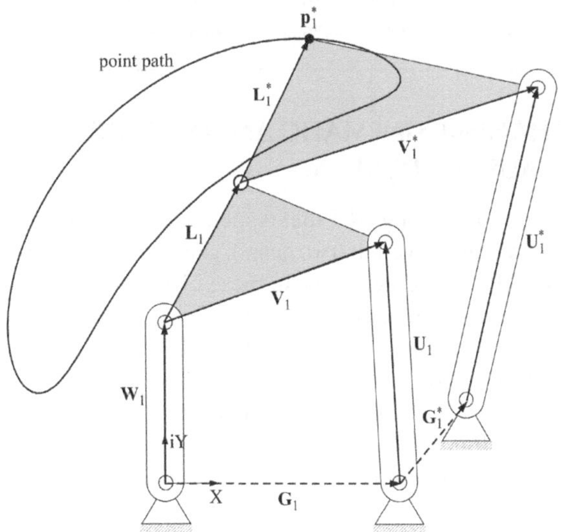

FIGURE E.4.13 Stephenson III mechanism and $\mathbf{p}_1^*$ path.

4.6 TIME AND DRIVER ANGULAR VELOCITY

In the Appendix B (and Appendix I) MATLAB files, the driving link angular velocity $\dot{\beta}$ is calculated according to $\dot{\beta} = \dot{\beta}_0 + \ddot{\beta}t$, where $\dot{\beta}_0$, $\ddot{\beta}$, and t are the initial angular velocity, angular acceleration, and time, respectively.[6] Under a constant-velocity condition (where $\ddot{\beta} = 0$ and subsequently $\dot{\beta} = \dot{\beta}_0$), the time increment becomes $t = \beta/\dot{\beta}$. Under a condition where both an initial angular velocity and angular acceleration are present, the time increment becomes

$$t = \frac{-\dot{\beta}_0 \pm \sqrt{\dot{\beta}_0^2 + 2\ddot{\beta}\beta}}{\ddot{\beta}}$$

where only the smallest positive value of the two solutions is valid.

Example 4.8

Problem Statement: For the problem in Example 4.5, include $\dot{\beta}$ in Table E.4.6.

Known Information: Example 4.5 and Appendix B.3 MATLAB file

Solution Approach: Table E.4.9 includes the $\dot{\beta}$ values calculated using $\dot{\beta}_0$ and $\ddot{\beta}$ from Example 4.5, the t values from the Appendix B.3 MATLAB file, and the angular velocity equation $\dot{\beta} = \dot{\beta}_0 + \ddot{\beta}t$ presented in Section 4.6.

TABLE E.4.9
Geared Five-Bar Driver Velocities and Point Positions, Velocities, and Accelerations

β	$\dot{\beta}$ (rad/s)	p_{1j} (m)	$V_{p_{1j}}$ (m/s)	$A_{p_{1j}}$ (m/s^2)
0°	1	0.2733, 1.9619	−0.8980, −0.0290	−0.2352, −0.9733
15°	1.0258	0.0360, 1.9215	−0.9273, −0.2851	0.0102, −0.9959
30°	1.0511	−0.1949, 1.8184	−0.8956, −0.5302	0.2336, −0.9326
45°	1.0757	−0.4068, 1.6609	−0.8202, −0.7419	0.3516, −0.7683
60°	1.0997	−0.5944, 1.4624	−0.7440, −0.8976	0.2326, −0.5178
75°	1.1233	−0.7657, 1.2386	−0.7199, −0.9973	0.0079, −0.3769
90°	1.1464	−0.9304, 0.9976	−0.6957, −1.0978	0.2980, −0.5166

4.7 PLANAR MECHANISM KINEMATIC ANALYSIS AND MODELING IN SIMSCAPE MULTIBODY™

It has been noted in Sections 4.3 through 4.5 that Appendices B.1 through B.4 include user instructions for the planar four-bar, geared five-bar (two gears), geared five-bar (three gears), and Stephenson III mechanisms in MATLAB files, respectively. In these files, the displacement, velocity, and acceleration equations formulated in Sections 4.3 through 4.5 are solved. These MATLAB files provide a means for the user to efficiently conduct planar four-bar, geared five-bar, and Stephenson III kinematic analyses by solving their displacement, velocity, and acceleration equations.

This book also utilizes Simscape Multibody as an alternate approach for simulation-based kinematic analysis. Simscape Multibody (formerly SimMechanics) is the latest MATLAB toolbox that provides a physical modeling environment for the mechanical modeling and simulation of rigid, multibody systems. The MATLAB and Simscape Multibody files that accompany this book use distinct approaches for kinematic analysis. In the Appendix B MATLAB files, for example, solutions are calculated from the expanded–separated vector loop equations in Sections 4.3 through 4.5. In Simscape Multibody, however, *Newtonian equations* (produced by building mechanism links and joints in a physical modeling environment) are solved using *ordinary differential equation* (ODE) solvers [6].

A library of Simscape Multibody files is available for download at https://www.routledge.com/9781032956800 to conduct displacement, velocity, and acceleration analyses on planar four-bar, geared five-bar, and Stephenson III mechanisms.[7] With these files, the user specifies the mechanism link dimensions and driving link parameters (e.g. crank displacements, velocities, and/or accelerations) and the displacements, velocities, and/or accelerations of the mechanism locations of interest are measured. Additionally, the motion of the mechanism itself is simulated. The Simscape Multibody file user instructions for the planar four-bar, geared five-bar (two gears), geared five-bar (three gears), and Stephenson III mechanisms are given in Appendices C.1 through C.4, respectively.

Example 4.9

Problem Statement: Using the Appendix C.1 Simscape Multibody files, plot the transmission angle (see Section 3.4) versus crank displacement angle plot for the planar four-bar mechanism in Table E.4.10 for a complete crank rotation.

Known Information: Table E.4.10 and Appendix C.1 Simscape Multibody files.

Solution Approach: Figure E.4.14 includes the input specified (in bold text) in the Appendix C.1 Simscape Multibody file. Figure E.4.15 illustrates a plot of the transmission angle data measured in the Appendix C.1 Simscape Multibody files. Figure E.4.16 is an initial position snapshot of the mechanism animation generated from the Appendix C.1 Simscape Multibody files.

TABLE E.4.10
Planar Four-Bar Mechanism Dimensions

W_1, θ	V_1, ρ	U_1, σ	G_{1x}, G_{1y}	L_1, δ
1, 90°	1.5, 4.2451°	1.5, 88.2046°	1.4489, –0.3882	0.75, 45°

```
W1 = 1*exp(i*90*pi/180);
V1 = 1.5*exp(i*4.2451*pi/180);
G1 = 1.4489 - i*0.3882;
U1 = 1.5*exp(i*88.2046*pi/180);
L1 = 0.75*exp(i*45*pi/180);

start_ang   = 0;
step_ang    = 1;
stop_ang    = 360;

angular_vel  = 0;
angular_acc  = 0;
```

FIGURE E.4.14 Specified input (in bold text) in the Appendix C.1 Simscape Multibody file for Example 4.9.

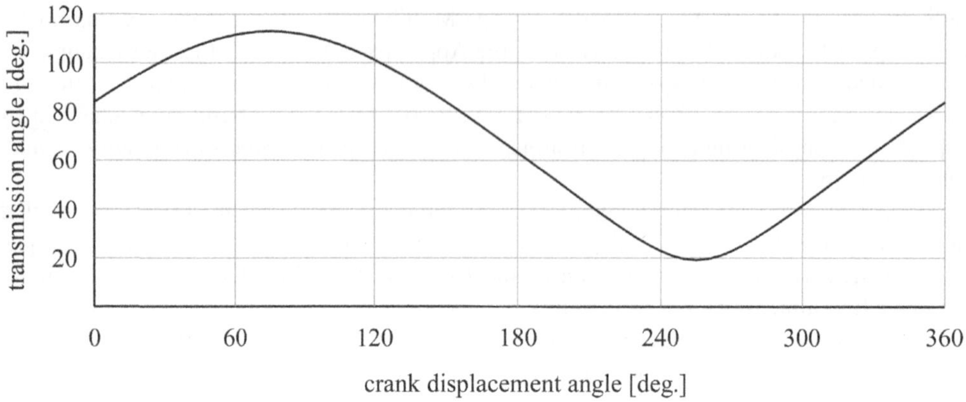

FIGURE E.4.15 Transmission angle vs. crank displacement angle plot.

FIGURE E.4.16 Initial position of the planar four-bar mechanism in the Appendix C.1 Simscape Multibody animation for Example 4.9.

Example 4.10

Problem Statement: Repeat Example 4.5 using the Appendix C.2 Simscape Multibody files. In this example, $\beta = 0°, -15°, -30°, \ldots, -90°$, $\dot{\beta}_0 = -1$ rad/s, and $\ddot{\beta} = -0.1$ rad/s^2.

Known Information: Table E.4.5 (from Example 4.5) and Appendix C.2 Simscape Multibody files,

Solution Approach: Because the Appendix C.2 Simscape Multibody files consider geared five-bar mechanisms, where the crank and output links rotate in opposite directions, the gear ratio for this problem is $r = -2$.[8] Figure E.4.17 includes the input specified (in bold text) in the Appendix C.2 Simscape Multibody file. Table E.4.11 includes the values of $\mathbf{p}_{1j}$, $\mathbf{V}_{p_{1j}}$, and $\mathbf{A}_{p_{1j}}$ measured from Appendix C.2 Simscape Multibody files. Figure E.4.18 is an initial position snapshot of the mechanism animation generated from the Appendix C.2 Simscape Multibody files.

```
W1 = 1.0*exp(i*90*pi/180);
V1 = 1.5*exp(i*32.7304*pi/180);
L1 = 1.0*exp(i*74.1400*pi/180);
G1 = 1.5;
U1 = 1.5*exp(i*45*pi/180);
S1 = 1.5*exp(i*149.9847*pi/180);
ratio = -2;

start_ang   = 0;
step_ang    = -15;
stop_ang    = -90;

angular_vel  = -1;
angular_acc  = -0.1;
```

FIGURE E.4.17 Specified input (in bold text) in the Appendix C.2 Simscape Multibody file for Example 4.10.

TABLE E.4.11
Measured Geared Five-Bar Point Positions, Velocities, and Accelerations

c	$\mathbf{p}_{1j}$ (m)	$\mathbf{V}_{p_{1j}}$ (m/s)	$\mathbf{A}_{p_{1j}}$ (m/s^2)
0°	0.2733, 1.9619	−0.1473, 0.3259	0.4798, −2.5866
−14.984°	0.2391, 1.9658	−0.1530, −0.2874	−0.3755, −2.3556
−29.977°	0.1849, 1.8154	−0.2855, −0.9217	−0.5639, −2.7119
−44.981°	0.1021, 1.5037	−0.3576, −1.6097	0.1516, −2.6833
−59.973°	0.0315, 1.0517	−0.1901, −2.0725	1.0790, −0.8210
−74.968°	0.0135, 0.5647	0.0060, −1.9746	0.3157, 1.4206
−90.001°	0.0119, 0.1535	−0.0653, −1.5665	−0.8276, 1.8546

FIGURE E.4.18 Initial position of the geared five-bar mechanism in the Appendix C.2 Simscape Multibody animation for Example 4.10.

Example 4.11

Problem Statement: Using the Appendix C.4 Simscape Multibody files for the mechanism in Example 4.7, measure the position of vector $\mathbf{V}_1^*$ over a complete crank rotation (at 45° rotation increments).

Known Information: Table E.4.8 (from Example 4.7) and Appendix C.4 Simscape Multibody files

Solution Approach: One way to define the position of vector $\mathbf{V}_1^*$ of the Stephenson III mechanism is by point $\mathbf{p}_1$ and angle α^*. Figure E.4.19 includes the input specified (in bold text) in the Appendix C.4 Simscape Multibody file. Table E.4.12 includes the $\mathbf{V}_1^*$ position data measured from the Appendix C.4 Simscape Multibody files. Figure E.4.20 is an initial position snapshot of the mechanism animation generated from the Appendix C.4 Simscape Multibody files.

```
W1 = exp(i*90*pi/180);
V1 = 1.5*exp(i*19.3737*pi/180);
G1 = 1.5;
U1 = 1.5*exp(i*93.2461*pi/180);
L1 = exp(i*60.7834*pi/180);

V1s = 2*exp(i*17.1417*pi/180);
G1s = 0.4318 + i*0.5176;
U1s = 2*exp(i*76.4844*pi/180);
L1s = exp(i*63.7091*pi/180);

start_ang   = 0;
step_ang    = 45;
stop_ang    = 360;

angular_vel  = 0;
angular_acc  = 0;
```

FIGURE E.4.19 Specified input (in bold text) in the Appendix C.4 Simscape Multibody file for Example 4.11.

TABLE E.4.12
Measured Position Data for Stephenson III Mechanism Dyad Vector $\mathbf{V}_1^*$

β	$\mathbf{p}_1$	α^*
0°	0.4881, 1.8728	–0.0017°
45.004°	–0.2542, 1.5986	8.9811°
89.982°	–0.7405, 0.9661	20.6900°
135.020°	–0.8565, 0.2815	33.3070°
179.990°	–0.6180, –0.2137	47.3220°
225.030°	–0.1788, –0.2435	58.5700°
270.000°	0.4727, 0.8497	38.0690°
315.000°	1.0664, 1.6404	–0.2744°
360.000°	0.4881, 1.8728	–0.0016°

FIGURE E.4.20 Initial position of the Stephenson III mechanism in the Appendix C.4 Simscape Multibody animation for Example 4.11.

4.8 SUMMARY

The method of vector loop closure is employed in this chapter to formulate kinematic equations for the planar four-bar mechanism, geared five-bar mechanism, and Stephenson III mechanism. With this method, a sum of the closed loop of mechanism link vectors is taken, expanded, grouped into real and imaginary components, and expressed as a system of two equations to calculate two unknown mechanism variables. With this approach, equations that fully describe the position, displacement, velocity, and acceleration of each mechanism link are formulated. Taking the first and second derivatives of the mechanism displacement equations produce velocity and acceleration equations, respectively. The unknown variables calculated from these kinematic equations can be used in additional vector loop equations to calculate the positions, displacements, velocities, and accelerations of additional mechanism locations of interest.

The displacement equations presented in this chapter form sets of two *nonlinear* simultaneous equations. Unlike linear simultaneous equations, nonlinear simultaneous equations cannot be solved algebraically. Using a *nonlinear optimization method*, such equation sets can be solved numerically.

The *Newton–Raphson method* is one of the most nonlinear optimization methods. In the Appendix B.1 through B.4 MATLAB files, the displacement, velocity, and acceleration equations for the planar four-bar, geared five-bar, and Stephenson III mechanisms, respectively, are solved numerically.

This book also utilizes Simscape Multibody as an alternate approach for simulation-based kinematic analyses. Using the Appendix C.1 through C.4 Simscape Multibody files, the user can conduct displacement, velocity, and acceleration analyses on the planar four-bar, geared five-bar, and Stephenson III mechanisms, respectively, as well as simulate mechanism motion.

PROBLEMS

1. Figure P.4.1 illustrates a planar four-bar mechanism used to guide a hatch from the closed-hatch position to the opened-hatch position. The dimensions for the illustrated mechanism are included in Table P.4.1. Using a crank rotation increment of –1°, determine the crank rotation range required to reach the opened-hatch position as well as the value of point $\mathbf{p}_1$ at the opened-hatch position (using the Appendix B.1 or C.1 files).

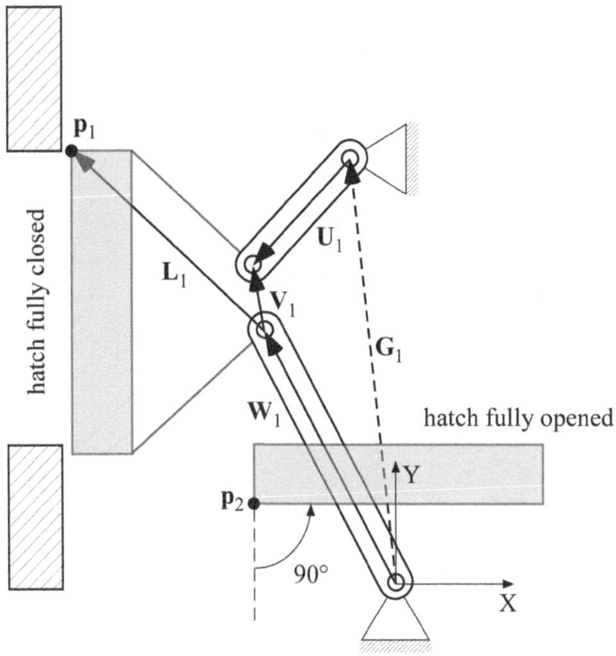

FIGURE P.4.1 Hatch mechanism.

TABLE P.4.1
Hatch Mechanism Dimensions

W_1, θ	V_1, ρ	U_1, σ	G_{1x}, G_{1y}	L_1, δ
4.4127, 118.7982°	1.0214, 101.2268°	2.2807, 225.1319°	–0.7156, 6.4851	4.1345, 139.4559°

2. Figure P.4.2 illustrates a planar four-bar mechanism used to guide a bucket from the loading position to the unloading position. The dimensions for the illustrated mechanism are included in Table P.4.2. Considering a crank rotation increment of –0.1°, determine the crank rotation range required to reach the unloading position as well as the value of point $\mathbf{p}_1$ at the unloading position (using the Appendix B.1 or C.1 files).

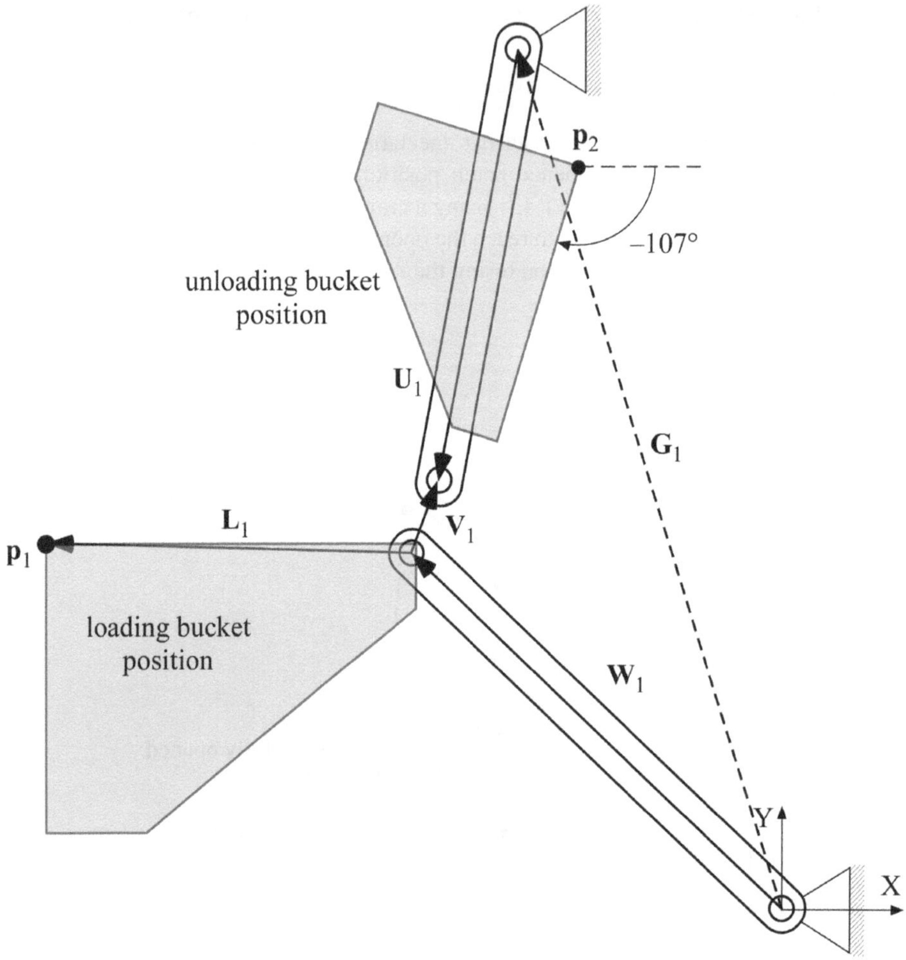

FIGURE P.4.2 Loading–unloading mechanism.

TABLE P.4.2
Loading–Unloading Mechanism Dimensions

W_1, θ	V_1, ρ	U_1, σ	G_{1x}, G_{1y}	L_1, δ
2.9777, 136.588°	0.4032, 70.2408°	2.513, 259.5885°	–1.5726, 4.8975	2.1995, 179.5331°

3. Using the Appendix B.1 or C.1 files, produce a velocity magnitude versus crank angular displacement plot and an acceleration magnitude versus crank angular displacement plot for point $\mathbf{p}_1$ of the leveling crane illustrated in Figure P.4.3 over a –35° crank rotation range. The initial crank angular velocity and angular acceleration are –1 rad/s and –0.1 rad/s^2, respectively.

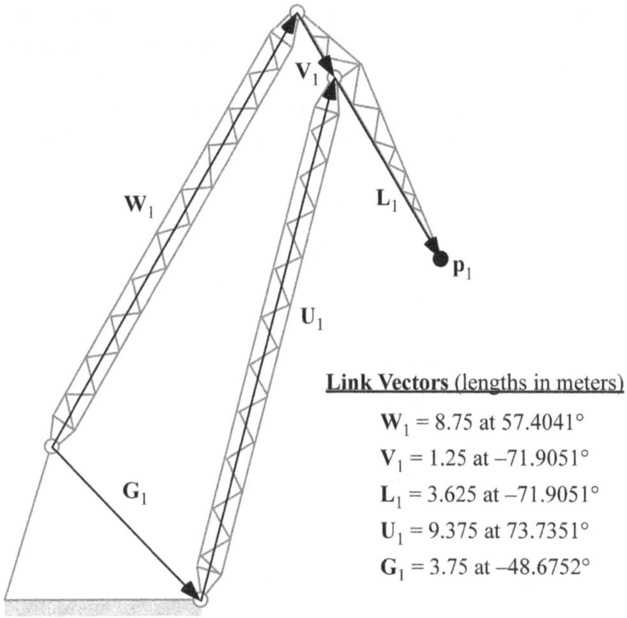

FIGURE P.4.3 Leveling crane.

4. Figure P.4.4 illustrates a planar four-bar mechanism used to guide a component from the initial position to the assembled position. The dimensions for the illustrated mechanism are included in Table P.4.3. Considering a crank rotation increment of +0.025°, determine the crank rotation range required to reach the assembled position as well as the value of point $\mathbf{p}_1$ at the assembled position (using the Appendix B.1 or C.1 files).

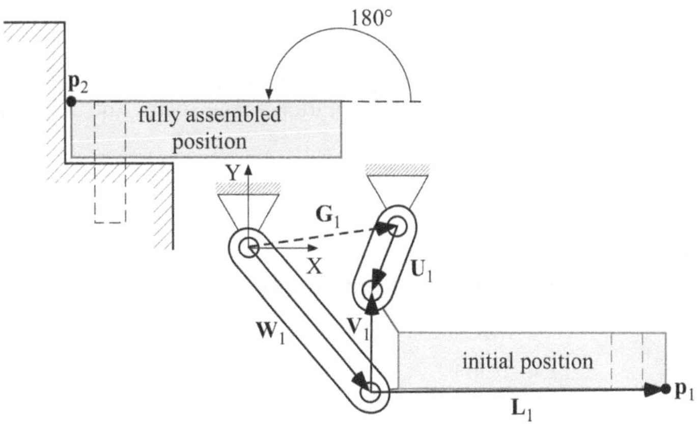

FIGURE P.4.4 Component assembly mechanism.

TABLE P.4.3
Component Assembly Mechanism Dimensions

W_1, θ	V_1, ρ	U_1, σ	G_{1x}, G_{1y}	L_1, δ
3.0645, 310.6493°	1.6039, 89.216°	1.1359, –112.147°	2.4465, 0.3309	4.8525, 0.7042°

5. Figure P.4.5 illustrates a planar four-bar mechanism used to guide a digging bucket from the initial position to the final position. The dimensions for the illustrated mechanism are included in Table P.4.4. Considering a crank rotation increment of –0.2°, determine the crank rotation range required to reach the final digging position as well as the value of point $\mathbf{p}_1$ at the final digging position (using the Appendix B.1 or C.1 files).

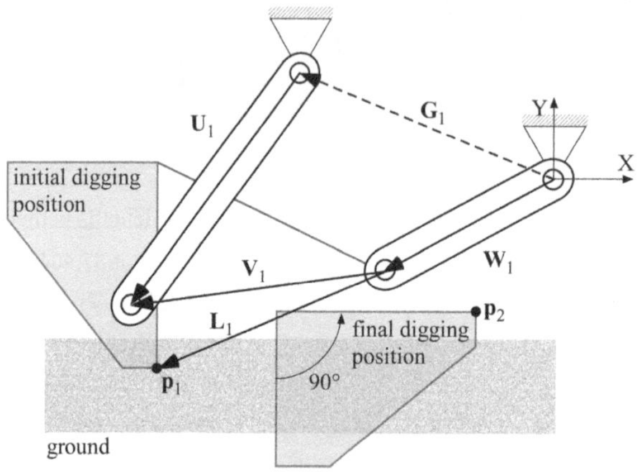

FIGURE P.4.5 Digging mechanism.

TABLE P.4.4
Digging Mechanism Dimensions

W_1, θ	V_1, ρ	U_1, σ	G_{1x}, G_{1y}	L_1, δ
4.1332, 207.829°	5.3857, 187.8282°	6.0475, 233.487°	–5.3924, 2.1975	5.1839, 202.3921°

6. For the two planar four-bar mechanism configurations given in Table P.4.5, plot the paths traced by point **p** over a complete crank rotation range (using the Appendix B.1 or C.1 files).

TABLE P.4.5
Planar Four-Bar Mechanism Configurations

W_1, θ	V_1, ρ	U_1, σ	G_{1x}, G_{1y}	L_1, δ
2.1, 90°	2.4, –16.0138°	2.4, 58.1173°	1.0392, –0.6	2.4, 12.9412°
0.7361, –0.425	1.7, 58.1173°	1.7, –16.0138°	1.4875, 90°	1.7, 12.9412°

7. Using the Appendix B.1 or C.1 files, calculate the rotation range of the designated coupler link in the planar four-bar folding chair linkage from the fully opened position to the fully closed position illustrated in Figure P.4.6.

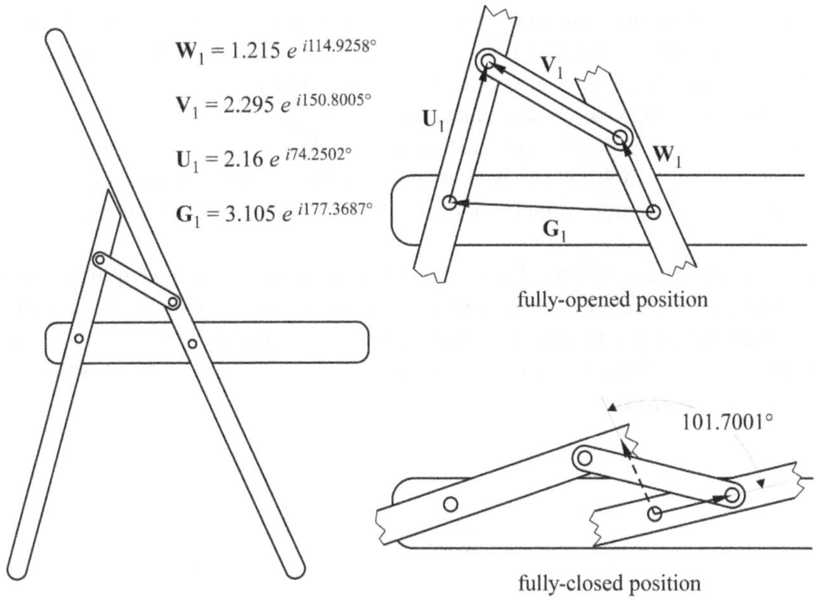

FIGURE P.4.6 Folding chair linkage in (left) fully opened and (right) fully closed position.

8. Using the Appendix B.1 or C.1 files, determine if the two folding chair linkage designs illustrated in Figure P.4.7 will produce properly folded chairs (a chair that permits a crank rotation of –101.7°) and if not, why?

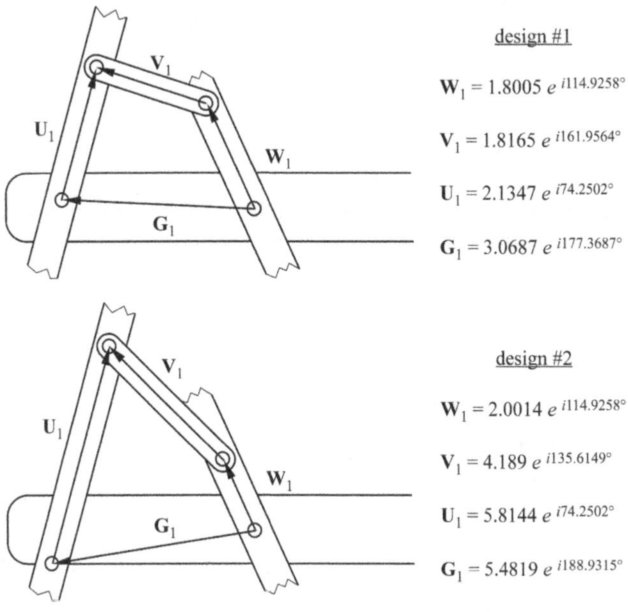

FIGURE P.4.7 Folding chair linkage designs.

9. For the two planar four-bar mechanism configurations given in Problem 8, produce the transmission angle (see Section 3.4) versus crank displacement angle plots over a complete crank rotation range (using the Appendix B.1 or C.1 files).
10. For the two planar four-bar mechanism configurations given in Problem 8 (with link lengths in meters), produce tables of the velocity and acceleration vectors of $\mathbf{p}_1$ over a complete crank rotation range at 30° crank rotation increments (using the Appendix B.1 or C.1 files). The initial crank angular velocity and angular acceleration are 1 rad/s and 0.25 rad/s^2, respectively.
11. Figure P.4.8 illustrates a planar four-bar mechanism used to guide a wiping blade. For this mechanism, produce a table of the angular displacements, velocities, and accelerations of the follower link over a 45° crank rotation range at 5° crank rotation increments (using the Appendix B.1 or C.1 files). The crank angular velocity and acceleration are 1.25 rad/s and 0.15 rad/s^2, respectively.

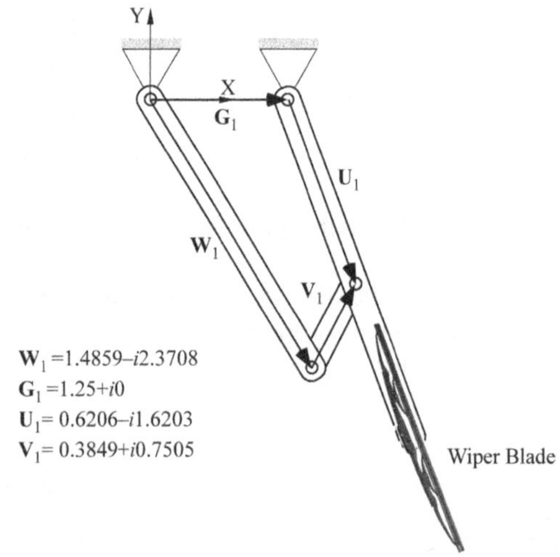

FIGURE P.4.8 Wiping blade mechanism.

12. Using the Appendix B.3 or C.3 files, plot the path traced by point $\mathbf{p}_1$ over a complete crank rotation range for the geared five-bar mechanism configuration given in Table P.4.6. The gear ratio is +2.

TABLE P.4.6
Geared Five-Bar Mechanism Configuration

W_1, θ	V_1, ρ	U_1, σ	S_1, ψ	G_{1x}, G_{1y}	L_1, δ
1.3, 60°	9.1, 173.6421°	5.2, 150°	11.7, 182.2848°	7.8, 0	7.8, 203.6421°

13. For the geared five-bar mechanism configuration given in Table P.4.6, plot the angular displacements of $\mathbf{V}_1$ and $\mathbf{S}_1$ (versus crank displacement) over a complete crank rotation range (using the Appendix B.3 or C.3 files). The gear ratio is +2.

14. For the geared five-bar mechanism configuration given in Table P.4.7, produce tables of the velocity and acceleration of point $\mathbf{p}_1$ over a complete crank rotation range at 30° crank rotation increments (using the Appendix B.2 or C.2 files).

TABLE P.4.7
Geared Five-Bar Mechanism Configuration

W_1, θ	V_1, ρ	U_1, σ	S_1, ψ	G_1, angle
3.302 cm, 90°	13.208 cm, 140.2031°	4.953 cm, 75°	19.812 cm, 154.0128°	6.604 cm, –15°
L_1, δ	$\dot{\beta}$ (rad/s)	$\ddot{\beta}$ (rad/s²)	**Gear Ratio**	
9.906 cm, 180°	1	0.5	–2	

15. For the geared five-bar mechanism configuration given in Table P.4.7, produce tables of the angular velocity and acceleration of $\mathbf{V}_1$ and $\mathbf{S}_1$ over a complete crank rotation range at 30° crank rotation increments (using the Appendix B.2 or C.2 files).
16. Figure P.4.9 illustrates a Stephenson III mechanism used to guide a gripping tool. The dimensions for the illustrated mechanism are included in Table P.4.8. Considering a crank rotation increment of 1°, determine the crank rotation range required to simultaneously achieve a –75° upper jaw rotation and a –15° lower jaw rotation (using the Appendix B.4 or Appendix C.4 files).

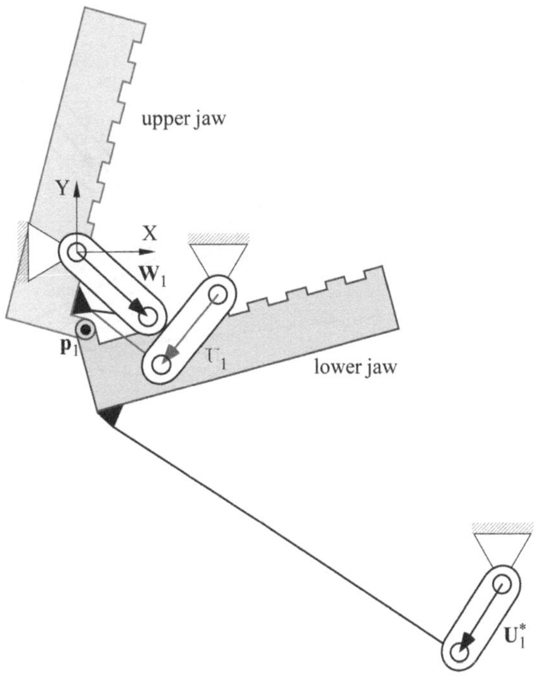

FIGURE P.4.9 Stephenson III gripper mechanism.

TABLE P.4.8
Stephenson III Gripper Mechanism Dimensions

W_1, θ	V_1, ρ	U_1, σ	G_{1x}, G_{1y}	L_1, δ
1.7995, 318.4178°	0.9091, 286.4362°	1.6482, 230.6143°	2.6491, −0.7924	1.2304, 189.1904°
V_1^*, ρ^*	U_1^*, σ^*	L_1^*, δ^*	G_{1x}^*, G_{1y}^*	
9.2542, 319.9647°	1.5166, 236.6216°	5.7298, 11.8113°	5.4021, −5.2847	

17. For the Stephenson III mechanism configuration given in Table P.4.8, produce plots of the angular displacement of the upper jaw $\mathbf{V}_1$ and upper jaw $\mathbf{V}_1^*$ versus the crank rotation range calculated in Problem 16 at 1° crank rotation increments (using the Appendix B.4 or Appendix C.4 files).
18. Figure P.4.10 illustrates a Stephenson III mechanism used to guide a digging tool. The dimensions for the illustrated mechanism are included in Table P.4.9. Considering a crank rotation increment of 1°, determine the crank rotation range required to simultaneously achieve a 65° digging arm rotation and a 75° digging bucket rotation (using the Appendix B.4 or Appendix C.4 files).

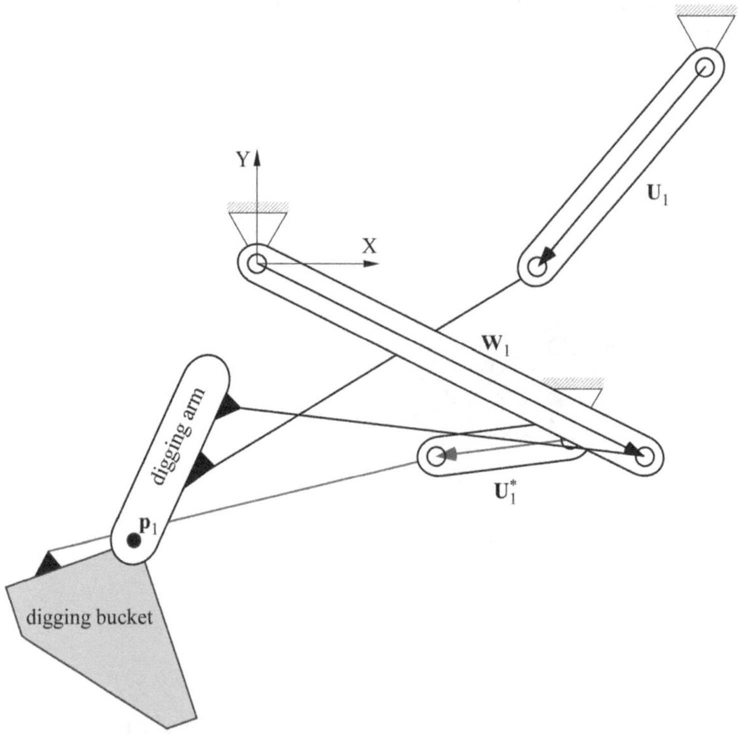

FIGURE P.4.10 Stephenson III digger mechanism.

19. For the Stephenson III mechanism configuration given in Table P.4.10, produce a table of the location, velocity, and acceleration of $\mathbf{p}_1^*$ over a complete crank rotation range at 30° crank rotation increments (using the Appendix B.4 or Appendix C.4 files). The initial crank angular velocity and angular acceleration are 1.35 rad/s and 0.6 rad/s^2, respectively.

TABLE P.4.9
Stephenson III Digging Mechanism Dimensions

W_1, θ	V_1, ρ	U_1, σ	G_{1x}, G_{1y}	L_1, δ
4.2366, –25.6775°	2.0805, 120.3504°	2.5252, 229.6156°	4.4031, 1.8831	5.1069, 189.0097°
V_1^*, ρ^*	U_1^*, σ^*	L_1^*, δ^*	G_{1x}^*, G_{1y}^*	
3.0994, 14.6953°	1.3340, 187.0724°	1.9944, 65°	–1.3069, –3.5681	

20. For the Stephenson III mechanism configuration given in Table P.4.10, produce a table of the angular position, velocity, and acceleration of $\mathbf{V}_1^*$ over a complete crank rotation range at –30° crank rotation increments (using the Appendix B.4 or Appendix C.4 files). The initial crank angular velocity and angular acceleration are –0.35 rad/s and –0.05 rad/s^2, respectively.

TABLE P.4.10
Stephenson III Mechanism Dimensions

W_1, θ	V_1, ρ	U_1, σ	G_{1x}, G_{1y}	L_1, δ
1.45 m, 90°	2.175 m, 19.3737°	2.175 m, 93.2461°	2.175 m, 0 m	4.35 m, –5.1593°
V_1^*, ρ^*	U_1^*, σ^*	G_{1x}^*, G_{1y}^*	L_1^*, δ^*	
2.9 m, –46.9725°	2.9 m, –6.6563°	1.2557 m, –0.725 m	2.175 m, 1.2172°	

NOTES

1 Because the complex coefficient is fully distributed in Equation (4.3) (as well as in the velocity equations for the forthcoming planar mechanisms), it can be canceled from the equation if preferred.
2 *Tangential acceleration* (which points along or opposite the direction of the velocity vector) is the result of a change in the velocity vector magnitude, while *normal acceleration* or *centripetal acceleration* (which points towards the center of path curvature) is the result of a change in the velocity vector direction.
3 A level-luffing crane is a crane designed to trace a horizontal path for leveling applications.
4 The angles for $\mathbf{V}_1$ and $\mathbf{L}_1$ (angles ρ and δ, respectively) are identical in this example problem. Identical values for these angles, however, are not required for four-bar linkages in general.
5 During circuit defects, no coupler and follower displacement angles appear in the Appendix B.1 MATLAB file. As a result, circuit defects will appear as discontinuities in coupler and follower displacement angle plots.
6 In Chapter 8, the crank rotation variables $\dot{\beta}_0$, $\dot{\beta}$, and $\ddot{\beta}$ become $\dot{\theta}_0$, $\dot{\theta}$, and $\ddot{\theta}$, respectively.
7 This library can also be downloaded from the authors' server http://www.softalink.com/kruss/mech_des_vis_prog_app_2nd_ed/MATLAB_SIMMULT_FILES.zip.
8 The Appendix C.3 Simscape Multibody file considers geared five-bar mechanism where the crank and output links rotate in the same direction, while the Appendix C.2 file considers links that rotate in the opposite direction.

5 Analytical Planar Four-Bar, Geared Five-Bar, and Stephenson III Motion Generation

CONCEPT OVERVIEW

In this chapter, the reader will gain a central understanding regarding the following:

1. *Dimensional Synthesis* and its categories
2. Defects in dimensional synthesis and defect elimination methods
3. Analytical planar four-bar, geared five-bar, and Stephenson III motion generation for three, four, and five precision positions

5.1 INTRODUCTION

In *kinematic synthesis,* the objective is to calculate the mechanism dimensions required to achieve or approximate the prescribed mechanism output motion.[1] Calculated mechanism dimensions include link lengths, link positions, and joint coordinates.[2] The prescribed mechanism output motion includes link positions, path points, and displacement angles. Because the calculated output in kinematic synthesis includes mechanism dimensions, kinematic synthesis is also called *dimensional synthesis* [1]. In contrast to *kinematic analysis*-where mechanism dimensions are known and the resulting mechanism output motion is calculated—in kinematic synthesis, mechanism output parameters are known and the corresponding mechanism dimensions are calculated (Figure 5.1) [2].

Kinematic synthesis includes three distinct subcategories: *motion generation*, *path generation,* and *function generation* (Figure 5.2). In motion generation, mechanism dimensions are calculated to achieve or approximate prescribed rigid-body positions, while the prescribed output motion is rigid-body path points in path generation.[3] In function generation, mechanism dimensions are calculated to achieve or approximate prescribed link displacement angles. Although this chapter considers planar motion generation exclusively, Chapters 6 and 7 consider planar motion path and function generation, respectively.

An overview of published research in kinematic synthesis will reveal a substantial assortment of *qualitative* and *quantitative* methods for motion, path, and function generation [3]. Qualitative methods include graphical techniques which can provide a wealth of information with virtually no computational effort [4]. Quantitative methods include mathematical models that can be solved analytically or numerically by way of solution algorithms and optimization methods [5–7]. As intended by the authors, quantitative kinematic synthesis methods are presented throughout this book.

DOI: 10.1201/9781003591375-5

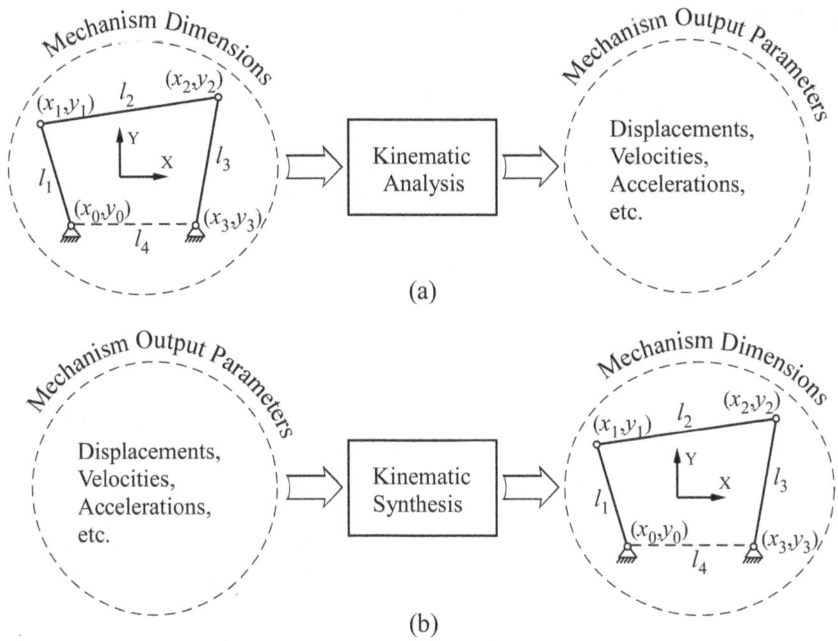

FIGURE 5.1 Kinematic analysis versus kinematic synthesis.

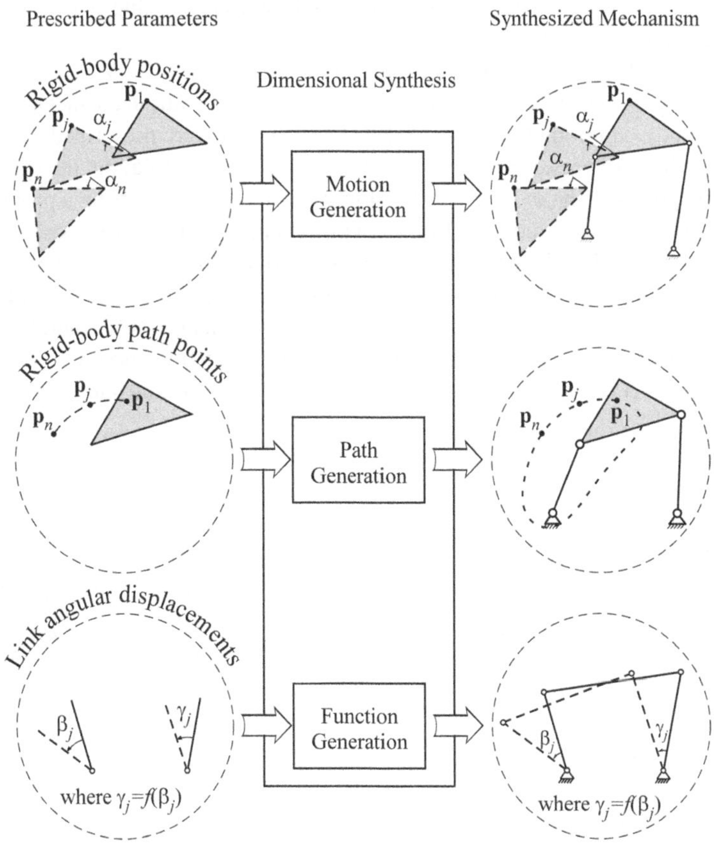

FIGURE 5.2 Subcategories of kinematic synthesis.

5.2 BRANCH AND ORDER DEFECTS

Although motion and path generation ensure that the synthesized mechanisms will achieve or approximate prescribed rigid-body output motion, they do not guarantee the following:

- The synthesized mechanism will achieve or approximate the prescribed rigid-body output motion without a change in its original assembly configuration (thus requiring mechanism disassembly).
- The synthesized mechanism will achieve or approximate the prescribed rigid-body output motion in the intended order.

These two uncertainties are defects inherent in motion and path generation: mechanism synthesis requiring prescribed rigid-body positions and rigid-body path points, respectively.

The first noted defect is commonly called a *branch defect* [8,9]. Figure 5.3a illustrates the two assembly configurations of a planar four-bar mechanism: configurations a-b-c-d and a-b-c*-d.[4] In kinematic synthesis, a *branch* represents the rigid-body positions or path points that are achieved by a single mechanism assembly configuration. When both mechanism assembly configurations are required to achieve all the rigid-body positions or path points, this can introduce a potential design problem because the mechanism would require disassembly and reassembly from one assembly configuration to the other during operation to achieve all of the prescribed rigid-body output motion.[5] Figure 5.3b illustrates a planar four-bar branch defect. In this figure, rigid-body positions 1-2-3 are achieved by the planar four-bar configuration a-b-c-d, while position 1* is achieved by the configuration a-b-c*-d. So, if the design application requires that this mechanism achieves the positions in the order 1*-1-2-3 continuously, this would not be possible with a single mechanism assembly configuration due to the given branch defect.

The second noted defect is commonly called an *order defect* [9,10]. When rigid-body positions or path points are prescribed, it is typically desired that the synthesized mechanism achieve the prescribed rigid-body output motion in the order in which they were prescribed. If, for example, the rigid-body path points are prescribed in the order 1-2-3-4-1 (Figure 5.4a) but the synthesized mechanism achieves the points in the order 1-2-4-3-1 (Figure 5.4b), this can present a problem—especially if the prescribed point order is required to trace a particular coupler curve. As shown in Figure 5.4, the order in which the rigid-body path points are achieved determines the profile of the path achieved.

Branch and order defect elimination by way of constraint equations, graphical methods, or prescribed crank displacement angles are often employed in motion and path generation models to produce motion and path generator solutions that are branch defect-free and order defect-free. Several branch and order defect elimination approaches and constraints are introduced in Chapters 6 and 8 and in Section 5.6.

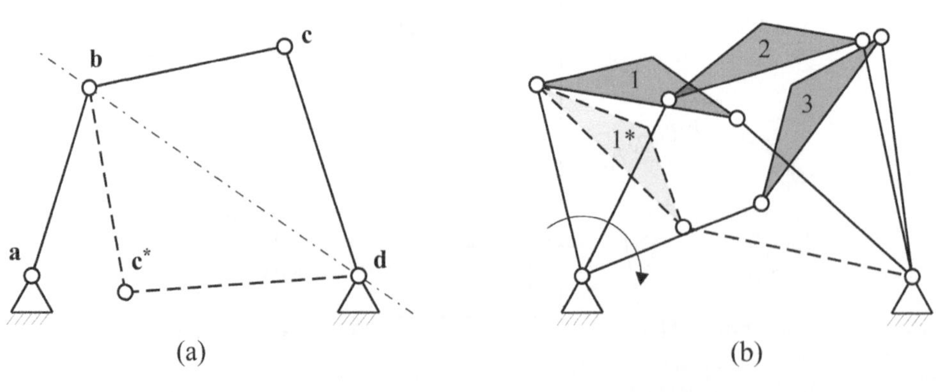

FIGURE 5.3 (a) Planar four-bar assembly configurations and (b) branch defect.

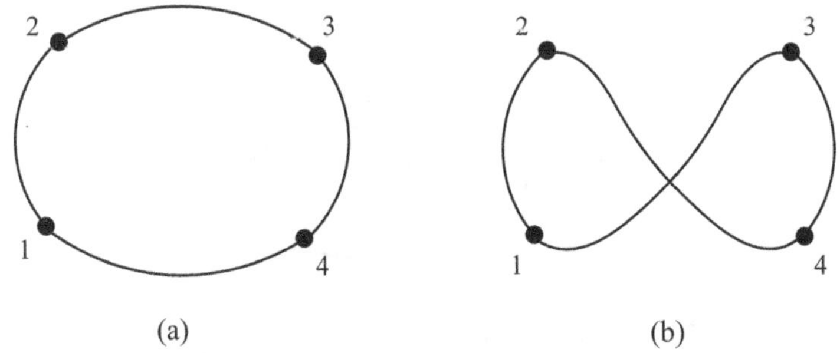

FIGURE 5.4 Paths associated with point orders: (a) 1-2-3-4-1 and (b) 1-2-4-3-1.

5.3 FOUR-BAR MOTION GENERATION: THREE, FOUR, AND FIVE PRECISION POSITIONS

Figure 5.5a illustrates a planar four-bar mechanism and the three coupler positions it achieves. The location and orientation or the *position* of the coupler are defined by the coordinates of coupler point $\mathbf{p}_j$ and coupler displacement angle α_j. In motion generation, the mechanism dimensions required to achieve or approximate *precision positions* are calculated.[6] Figure 5.5b illustrates the operation sequence of an aircraft landing gear, which further conveys motion generation. For effective operation, the aircraft landing gear should be guided from within the airframe to a position where the wheels can properly contact the ground. The planar four-bar motion generation equations presented in this section are useful for calculating the four-bar mechanism dimensions required to achieve a group of three, four, or five precision positions.

Figure 5.6 illustrates both dyads of a planar four-bar mechanism in a starting position (Position 1) and a displaced position (Position *j*). The left- and right-side dyads include the vector chains $\mathbf{W}-\mathbf{Z}$ and $\mathbf{U}-\mathbf{S}$, respectively. By taking the vector sum between the starting and displaced positions for each dyad, vector loop closure equations are derived [11,12]. After taking the counterclockwise vector sum for each dyad (starting with $\mathbf{W}_1$ for the left-side dyad and $\mathbf{U}_1$ for the right-side dyad), the vector loop closure equations for the four-bar mechanism dyads become

$$W_1e^{i\theta} + Z_1e^{i\phi} + P_{j1}e^{i\delta_j} - Z_1e^{i(\phi+\alpha_j)} - W_1e^{i(\theta+\beta_j)} = 0 \tag{5.1}$$

$$U_1e^{i\sigma} + S_1e^{i\psi} + P_{j1}e^{i\delta_j} - S_1e^{i(\psi+\alpha_j)} - U_1e^{i(\sigma+\gamma_j)} = 0 \tag{5.2}$$

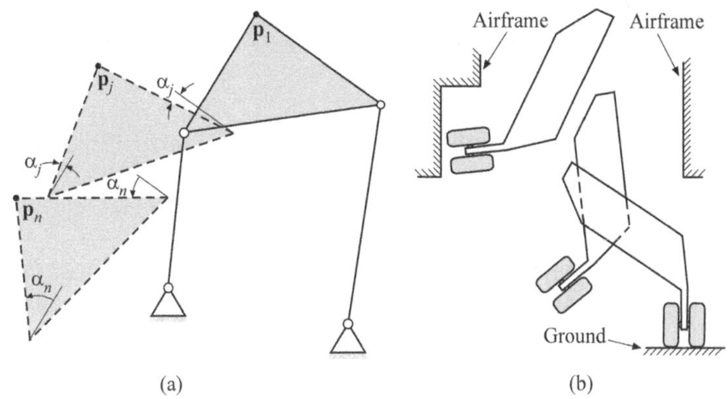

FIGURE 5.5 (a) Four-bar mechanism and coupler positions and (b) aircraft landing gear example.

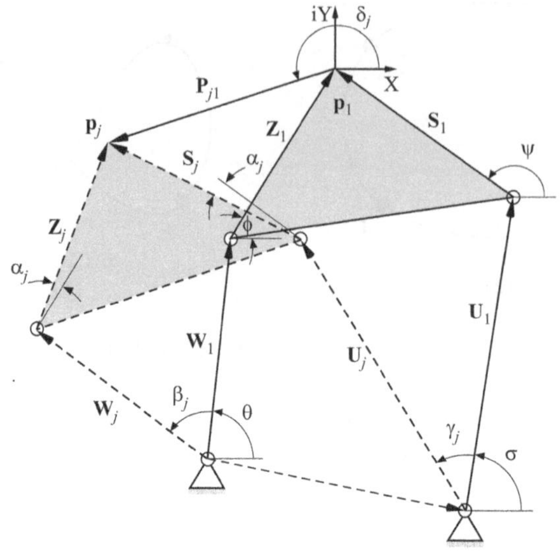

FIGURE 5.6 Four-bar mechanism in starting and (dashed) displaced positions.

where $P_{j1}e^{i\delta_j} = \mathbf{p}_j - \mathbf{p}_1$. Because $\mathbf{p}_1$ is the coordinate frame origin ($\mathbf{p}_1 = (0,0)$), $p_{jx} = P_{j1}\cos\delta_j$, and $p_{jy} = P_{j1}\sin\delta_j$.

After factoring the terms for the starting dyad position, the resulting *standard form* loop closure equations become

$$W_1e^{i\theta}\left(e^{i\beta_j} - 1\right) + Z_1e^{i\phi}\left(e^{i\alpha_j} - 1\right) = P_{j1}e^{i\delta_j} \tag{5.3}$$

$$U_1e^{i\sigma}\left(e^{i\gamma_j} - 1\right) + S_1e^{i\psi}\left(e^{i\alpha_j} - 1\right) = P_{j1}e^{i\delta_j} \tag{5.4}$$

Expanding Equation (5.3) and grouping the real and imaginary terms as separate equations produce the equation set

$$\begin{aligned} W_1\cos\theta\left(\cos\beta_j - 1\right) - W_1\sin\theta\sin\beta_j + Z_1\cos\phi\left(\cos\alpha_j - 1\right) - Z_1\sin\phi\sin\alpha_j &= P_{j1}\cos\delta_j \\ W_1\sin\theta\left(\cos\beta_j - 1\right) + W_1\cos\theta\sin\beta_j + Z_1\sin\phi\left(\cos\alpha_j - 1\right) + Z_1\cos\phi\sin\alpha_j &= P_{j1}\sin\delta_j \end{aligned} \tag{5.5}$$

After specifying $W_1\cos\theta = W_{1x}$, $W_1\sin\theta = W_{1y}$, $Z_1\cos\phi = Z_{1x}$, and $Z_1\sin\phi = Z_{1y}$, Equation (5.5) becomes

$$\begin{aligned} W_{1x}\left(\cos\beta_j - 1\right) - W_{1y}\sin\beta_j + Z_{1x}\left(\cos\alpha_j - 1\right) - Z_{1y}\sin\alpha_j &= P_{j1}\cos\delta_j \\ W_{1y}\left(\cos\beta_j - 1\right) + W_{1x}\sin\beta_j + Z_{1y}\left(\cos\alpha_j - 1\right) + Z_{1x}\sin\alpha_j &= P_{j1}\sin\delta_j \end{aligned} \tag{5.6}$$

Likewise, after separating the real and imaginary terms in the right-side dyad Equation (5.4) and specifying $U_1\cos\sigma = U_{1x}$, $U_1\sin\sigma = U_{1y}$, $S_1\cos\psi = S_{1x}$, and $S_1\sin\psi = S_{1y}$ the equation becomes

$$\begin{aligned} U_{1x}\left(\cos\gamma_j - 1\right) - U_{1y}\sin\gamma_j + S_{1x}\left(\cos\alpha_j - 1\right) - S_{1y}\sin\alpha_j &= P_{j1}\cos\delta_j \\ U_{1y}\left(\cos\gamma_j - 1\right) + U_{1x}\sin\gamma_j + S_{1y}\left(\cos\alpha_j - 1\right) + S_{1x}\sin\alpha_j &= P_{j1}\sin\delta_j \end{aligned} \tag{5.7}$$

When expressed in matrix form for N precision positions (where $j = 2, 3, \ldots, N$), Equations (5.6) and (5.7) become the following equations, respectively:

$$\begin{bmatrix} \cos\beta_j - 1 & -\sin\beta_j & \cos\alpha_j - 1 & -\sin\alpha_j \\ \sin\beta_j & \cos\beta_j - 1 & \sin\alpha_j & \cos\alpha_j - 1 \\ & \vdots & & \\ \cos\beta_N - 1 & -\sin\beta_N & \cos\alpha_N - 1 & -\sin\alpha_N \\ \sin\beta_N & \cos\beta_N - 1 & \sin\alpha_N & \cos\alpha_N - 1 \end{bmatrix} \begin{Bmatrix} W_{1x} \\ W_{1y} \\ Z_{1x} \\ Z_{1y} \end{Bmatrix} = \begin{bmatrix} P_{j1}\cos\delta_j \\ P_{j1}\sin\delta_j \\ \vdots \\ P_{N1}\cos\delta_N \\ P_{N1}\sin\delta_N \end{bmatrix} \tag{5.8}$$

$$\begin{bmatrix} \cos\gamma_j - 1 & -\sin\gamma_j & \cos\alpha_j - 1 & -\sin\alpha_j \\ \sin\gamma_j & \cos\gamma_j - 1 & \sin\alpha_j & \cos\alpha_j - 1 \\ & \vdots & & \\ \cos\gamma_N - 1 & -\sin\gamma_N & \cos\alpha_N - 1 & -\sin\alpha_N \\ \sin\gamma_N & \cos\gamma_N - 1 & \sin\alpha_N & \cos\alpha_N - 1 \end{bmatrix} \begin{Bmatrix} U_{1x} \\ U_{1y} \\ S_{1x} \\ S_{1y} \end{Bmatrix} = \begin{bmatrix} P_{j1}\cos\delta_j \\ P_{j1}\sin\delta_j \\ \vdots \\ P_{N1}\cos\delta_N \\ P_{N1}\sin\delta_N \end{bmatrix} \tag{5.9}$$

Although as written, Table 5.1 includes the design variables for the left-side dyad only; this table also corresponds to the right-side dyad (replacing variables W_{1x}, W_{1y}, Z_{1x}, Z_{1y}, and β_j with U_{1x}, U_{1y}, S_{1x}, S_{1y}, and γ_j, respectively).

With three precision positions ($N = 3$ in Table 5.1), Equation (5.8) forms a set of four linear simultaneous equations with four scalar unknown variables that can be calculated using Cramer's rule. Angles β_2 and β_3 are the two "free choices" that can be prescribed along with the precision positions (for the right-side dyad, the corresponding free choices are angles γ_2 and γ_3). With three precision positions, the resulting mechanism not only achieves the precision positions precisely but does so according to the prescribed dyad displacement angles β and γ. This is known as motion generation with *prescribed timing.* Prescribed timing offers a solution to the order defect problem presented in Section 5.2 because by prescribing the dyad displacement angles along with the precision positions, the intended order in which the synthesized mechanism is to achieve the precision positions is ensured.

Beyond three precision positions, however, the number of simultaneous equations exceeds the number of scalar dyad unknowns in Equations (5.8) and (5.9). Beyond three precision positions, these equations are no longer square and subsequently cannot be solved by Cramer's rule. For motion generation with four precision positions, this book includes an algorithm (described in Appendix D.1) to analytically calculate β_3 and β_4 for a prescribed value of β_2 [13]. Given the four

TABLE 5.1
Maximum Number of Solutions for Equation (5.8) for Three, Four, and Five Precision Positions

Num. of Precision Positions (N)	Num. of Equations	Total Number of Unknowns	Number of Free Choices	Number of Solutions
3	4	(6) W_{1x}, W_{1y}, Z_{1x}, Z_{1y}, β_2, and β_3	2	Infinite
4	6	(7) W_{1x}, W_{1y}, Z_{1x}, Z_{1y}, β_2, β_3, and β_4	1	Infinite
5	8	(8) W_{1x}, W_{1y}, Z_{1x}, Z_{1y}, β_2, β_3, β_4, and β_5	0	Finite

precision positions, β_2 and the algorithm-calculated values for β_3 and β_4, any four of the resulting six simultaneous equations from Equation (5.8) can be used to solve for W_{1x}, W_{1y}, Z_{1x}, and Z_{1y}.

Given a range of β_2 values (where $0 \le \beta_2 \le 2\pi$), a corresponding range of values for $\mathbf{W}_1$ and $\mathbf{Z}_1$ can be calculated using the algorithm described in Section D.1. Knowing that the ungrounded mechanism joints or *moving pivots* of a planar four-bar mechanism correspond to $-\mathbf{Z}_1$ and the grounded mechanism joints or *fixed pivots* correspond to $-(\mathbf{Z}_1 + \mathbf{W}_1)$, a locus of moving and fixed pivots can be produced from a range of $\mathbf{W}_1$ and $\mathbf{Z}_1$ solutions (calculated from a range of β_2 values). The resulting moving and fixed pivot loci are known as *Burmester circle and center points,* respectively, and confirm the existence of an infinite number of planar four-bar mechanism solutions given four precision positions, as shown in Table 5.1 [14]. A planar four-bar mechanism can be constructed from any two pairs of circle points and center points.[7] Section D.2 includes user instructions for the MATLAB file to calculate the *Burmester circle* and *center points*. The MATLAB file itself is available for download at https://www.routledge.com/9781032956800.

With five precision positions, not only is Cramer's rule inapplicable for Equation (5.8), but also a finite number of solutions exist, as shown in Table 5.1. Section E.1 includes an algorithm to analytically calculate $\beta_2 \sim \beta_5$ [15]. As conveyed in Section E.1, for motion generation with five precision positions, four, two, or no real values exist for $\beta_2 \sim \beta_5$. This corresponds to the existence of four, two, or no circle-center point pairs (mechanism dyad solutions) or subsequently six, two, or no motion generator solutions, respectively. Section E.2 includes user instructions for the MATLAB file to calculate circle and center points for five-position synthesis. The MATLAB file itself is available for download at https://www.routledge.com/9781032956800.

Example 5.1

Problem Statement: Synthesize a planar four-bar mechanism to guide the landing gear through the three precision positions in Figure E.5.1.

Known Information: Equation (5.8), (5.9), and Table E.5.1

Solution Approach: Since the first precision point $\mathbf{p}_1$ lies on the coordinate system origin, the x- and y-components of $\mathbf{p}_j$ correspond to $P_{j1}\cos\delta_j$ and $P_{j1}\sin\delta_j$, respectively, in Equations (5.8) and (5.9). Table E.5.1 includes the precision position and dyad displacement data used in Equations (5.8) and (5.9) to calculate the components of the dyad vectors for the planar four-bar mechanism (W_{1x}, W_{1y}, Z_{1x}, Z_{1y}, U_{1x}, U_{1y}, S_{1x}, and S_{1y}).

Figure E.5.2 includes the calculation procedure in the MATLAB command window and

Figure E.5.3 illustrates the resulting planar four-bar mechanism. Being an analytically calculated result, this mechanism achieves the precision positions precisely. A kinematic analysis of this solution is conducted in Example 5.8 in Section 5.6.

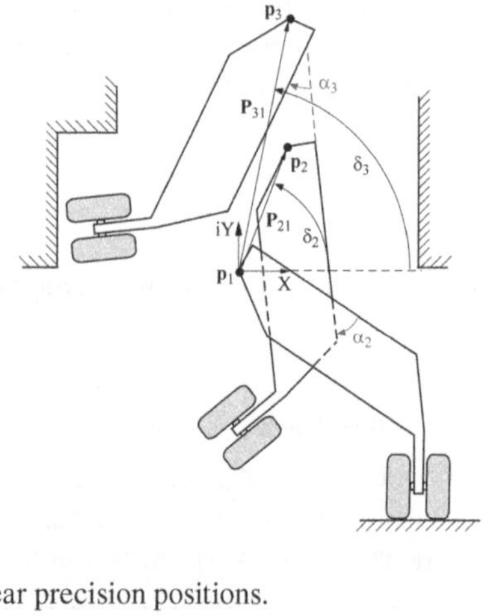

FIGURE E.5.1 Three landing gear precision positions.

TABLE E.5.1
Landing Gear Precision Position Parameters and Dyad Displacement Angles

Precision Position	$\mathbf{p}_j$	α_j (deg)	β_j (deg)	γ_j (deg)
1	0, 0			
2	0.292, 0.734	–51.7124	18	–40
3	0.299, 1.461	–84.9734	38	–87

```
>> p1 = [0, 0]';
>> p2 = [0.292, 0.734]';
>> p3 = [0.299, 1.461]';
>> A2 = -51.7124*pi/180; A3 = -84.9734*pi/180;
>> B2 = 18*pi/180; B3 = 38*pi/180;
>> G2 = -40*pi/180; G3 = -87*pi/180;

>> P21 = p2 - p1;
>> P31 = p3 - p1;
>> CA2 = cos(A2); SA2 = sin(A2); CA3 = cos(A3); SA3 = sin(A3);
>> CB2 = cos(B2); SB2 = sin(B2); CB3 = cos(B3); SB3 = sin(B3);
>> CG2 = cos(G2); SG2 = sin(G2); CG3 = cos(G3); SG3 = sin(G3);

>> WZ = inv([CB2 - 1, -SB2, CA2 -1, -SA2
SB2, CB2 - 1, SA2, CA2 - 1
CB3 - 1, -SB3, CA3 - 1, -SA3
SB3, CB3 - 1, SA3, CA3 - 1])*[P21(1), P21(2), P31(1), P31(2)]'

WZ =

    2.0580
   -0.8054
   -0.1324
    0.1191

>> US = inv([CG2 - 1, -SG2, CA2 - 1, -SA2
SG2, CG2 - 1, SA2, CA2 - 1
CG3 - 1, -SG3, CA3 - 1, -SA3
SG3, CG3 -1, SA3, CA3 -1])*[P21(1), P21(2), P31(1), P31(2)]'

US =

    0.5808
   -1.8615
   -1.5053
    1.3400

>>
```

FIGURE E.5.2 Example 5.1 $\mathbf{W}_1 - \mathbf{Z}_1$ and $\mathbf{U}_1 - \mathbf{S}_1$ calculation procedure in the MATLAB command window.

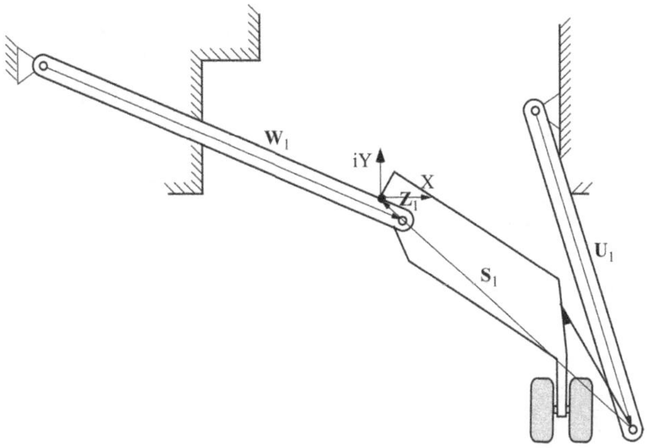

FIGURE E.5.3 Synthesized planar four-bar motion generator.

Example 5.2

Problem Statement: Synthesize a planar four-bar mechanism to guide the gripper jaw through the four precision positions in Figure E.5.4.

Known Information: Appendix D MATLAB file and Table E.5.2

Solution Approach: Table E.5.2 includes the precision position data used as input in the Appendix D MATLAB file. Vectors $\mathbf{P}_{j1}$ are the magnitudes of the coupler points $\mathbf{p}_j$, and angles δ_j are the angles $\mathbf{P}_{j1}$ that make with the *x*-axis. Figure E.5.5 includes the Table E.5.2 input specified in the Appendix D MATLAB file. The circle and center points illustrated in Figure E.5.6 correspond to the β_2-β_3-β_4 solution for the first mechanism branch (as opposed to the alternate β_2-$\tilde{\beta}_3$-$\tilde{\beta}_4$ solution for the second mechanism branch).[8]

Table E.5.3 and Figure E.5.7 include the mechanism dyad components for the circle and center point pairs selected from the calculated circle and center point curves in Figure E.5.6. The $\mathbf{W}_1$ and $\mathbf{Z}_1$ values in Table E.5.3 are for $\beta_2 = -21°$ in the Appendix D MATLAB file output for the first branch.

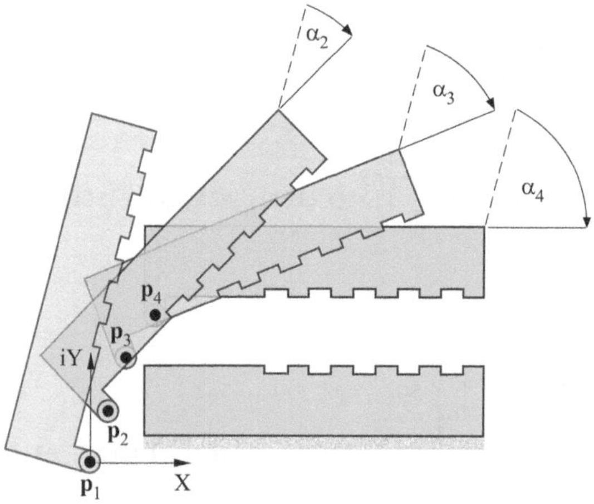

FIGURE E.5.4 Four gripper jaw precision positions.

TABLE E.5.2
Gripper Jaw Precision Position Parameters

Precision Position	$\mathbf{p}_j$	$\mathbf{P}_{j1}$	δ_j (deg)	α_j (deg)
1	0, 0	0		
2	0.1815, 0.4882	0.5208	69.6062	–30
3	0.3700, 0.9775	1.0452	69.2675	–53
4	0.6647, 1.4078	1.5568	64.7254	–75

```
alpha2 = -30*pi/180;
alpha3 = -53*pi/180;
alpha4 = -75*pi/180;

P21 = 0.5208;
P31 = 1.0452;
P41 = 1.5568;

delta2 = 69.6062*pi/180;
delta3 = 69.2675*pi/180;
delta4 = 64.7254*pi/180;
```

FIGURE E.5.5 Specified input (in bold type) in the Appendix D MATLAB file for Example 5.2.

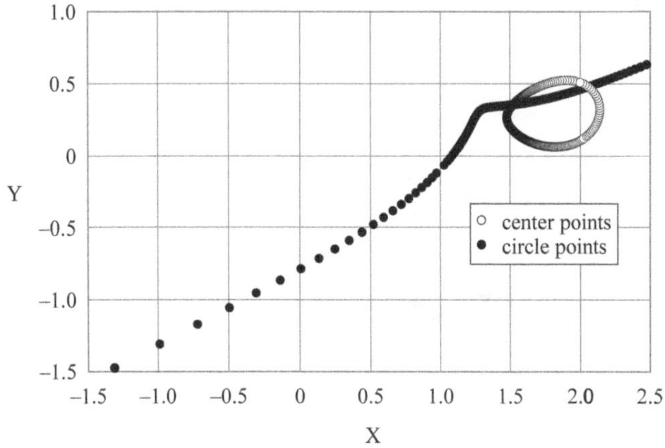

FIGURE E.5.6 Calculated circle and center points.

TABLE E.5.3
Selected Four-Bar Mechanism Dyad Variables

$\mathbf{W}_1$	$\mathbf{Z}_1$	$\mathbf{U}_1$	$\mathbf{S}_1$
−3.4300, −1.6978	1.3122, 1.4758	−1.6042, −0.8795	−0.5211, 0.4801

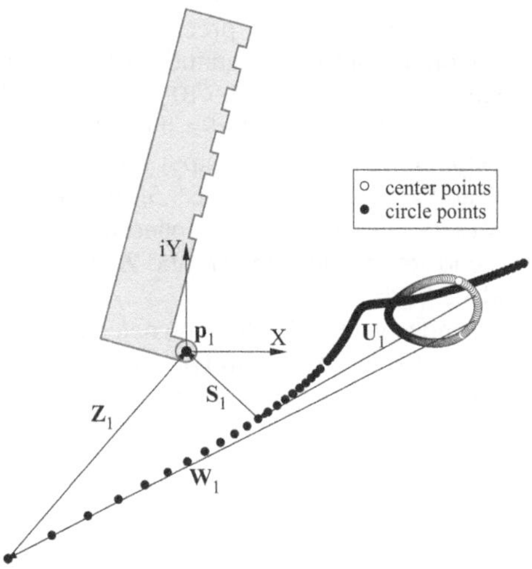

FIGURE E.5.7 Four-bar mechanism dyads selected from circle and center points.

The Appendix D MATLAB file output provides values for $-\mathbf{Z}_1$ and $-\mathbf{Z}_1 - \mathbf{W}_1$. Therefore, to produce the $\mathbf{W}_1$ and $\mathbf{Z}_1$ values in Table E.5.3, the value for $-\mathbf{Z}_1$ is multiplied by −1, and this value is added to $-\mathbf{Z}_1 - \mathbf{W}_1$ and the sum multiplied by −1. The $\mathbf{U}_1$ and $\mathbf{S}_1$ values in Table E.5.3 are for $\beta_2 = -10°$ in the Appendix D MATLAB file output for the first branch. To produce the values for $\mathbf{U}_1$ and $\mathbf{S}_1$ in Table E.5.3, the second set of values for $\mathbf{W}_1$ and $\mathbf{Z}_1$ from the Appendix D MATLAB file output are assigned to $\mathbf{U}_1$ and $\mathbf{S}_1$, respectively.

To achieve positions 2, 3, and 4, vector $\mathbf{W}_1$ (the designated driver link) is rotated by −21°, −40.1692°, and −59.1006° (β_2,β_3, and β_4), respectively. These angles were calculated in the Appendix D MATLAB file for the first branch. Figure E.5.8 illustrates the resulting planar four-bar mechanism. Because the circle and center points are calculated analytically, the resulting mechanism achieves the precision positions precisely.

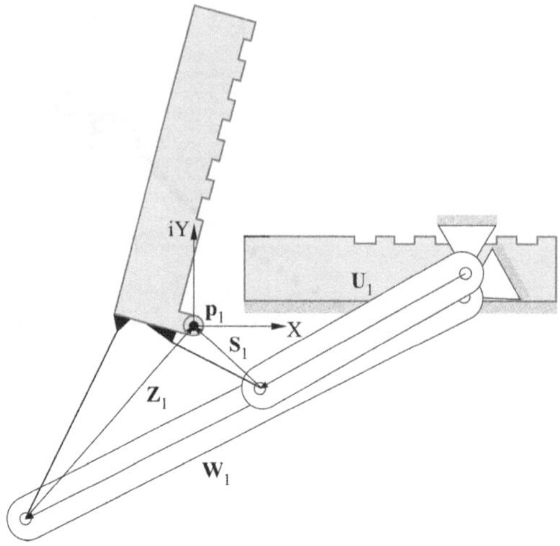

FIGURE E.5.8 Synthesized planar four-bar motion generator.

Example 5.3

Problem Statement: Synthesize a planar four-bar mechanism to guide the guide the gripper jaw through the five precision positions in Figure E.5.9.

Known Information: Appendix E MATLAB file and Table E.5.4

Solution Approach: Table E.5.4 includes the precision position data used as input in the Appendix E MATLAB file. Vectors $\mathbf{P}_{j1}$ are the magnitudes of the coupler points $\mathbf{p}_j$, and angles δ_j are the angles $\mathbf{P}_{j1}$ make with the *x*-axis. Figure E.5.10 includes the Table E.5.4 input specified in the Appendix E MATLAB file. Table E.5.5 includes the selected planar four-bar mechanism dyad components. The Appendix E MATLAB file output provides values for $-\mathbf{Z}_1$ and $-\mathbf{Z}_1 - \mathbf{W}_1$. Therefore, to produce the $\mathbf{W}_1$ and $\mathbf{Z}_1$ values in Table E.5.5, the value for $-\mathbf{Z}_1$ is multiplied by −1, and this value is added to $-\mathbf{Z}_1 - \mathbf{W}_1$ and the sum multiplied by −1. To produce the values for $\mathbf{U}_1$ and $\mathbf{S}_1$ in Table E.5.5, the second set of values for $\mathbf{W}_1$ and $\mathbf{Z}_1$ from the Appendix E MATLAB file output are assigned to $\mathbf{U}_1$ and $\mathbf{S}_1$.

To achieve positions 2, 3, 4, and 5, vector $\mathbf{W}_1$ (the designated driver link) is rotated by −9.2361°, −21.0897°, −40.3030°, and −59.2610° (β_2, β_3, β_4, and β_5), respectively. These angles were calculated in the Appendix E MATLAB file for the first branches. Figure E.5.11 illustrates the synthesized

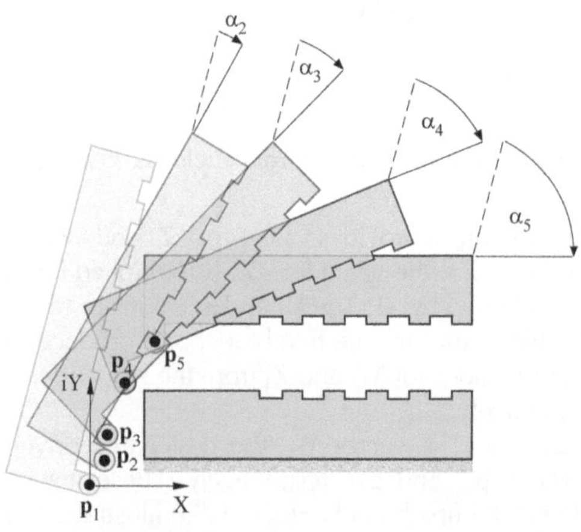

FIGURE E.5.9 Five gripper jaw precision positions.

TABLE E.5.4
Gripper Jaw Precision Position Parameters

Precision Position	$\mathbf{p}_j$	$\mathbf{P}_{j1}$	δ_j (deg)	α_j (deg)
1	0, 0	0		
2	0.1104, 0.1789	0.2102	58.3224	−15
3	0.1815, 0.4882	0.5208	69.6062	−30
4	0.3700, 0.9775	1.0452	69.2675	−53
5	0.6647, 1.4078	1.5568	64.7254	−75

```
alpha2 = -15*pi/180;
alpha3 = -30*pi/180;
alpha4 = -53*pi/180;
alpha5 = -75*pi/180;

P21 = 0.2102;
P31 = 0.5208;
P41 = 1.0452;
P51 = 1.5568;

delta2 = 58.3224*pi/180;
delta3 = 69.6062*pi/180;
delta4 = 69.2675*pi/180;
delta5 = 64.7254*pi/180;
```

FIGURE E.5.10 Specified input (in bold type) in the Appendix E MATLAB file for Example 5.3.

TABLE E.5.5
Selected Four-Bar Mechanism Dyad Variables

$\mathbf{W}_1$	$\mathbf{Z}_1$	$\mathbf{U}_1$	$\mathbf{S}_1$
−3.4627, −1.7135	1.3455, 1.4928	−1.5664, −0.8643	−0.5553, 0.4575

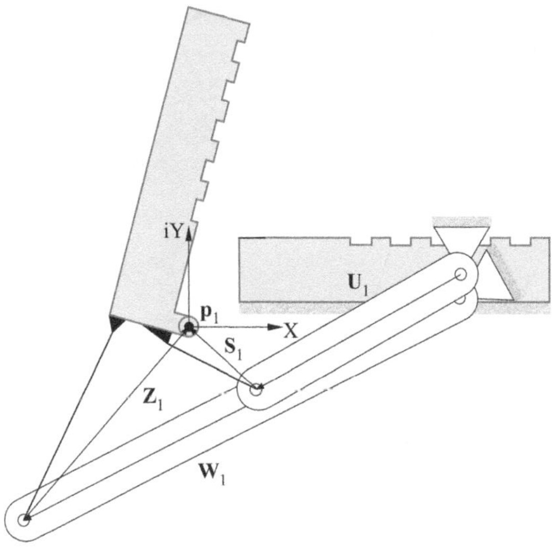

FIGURE E.5.11 Synthesized planar four-bar motion generator.

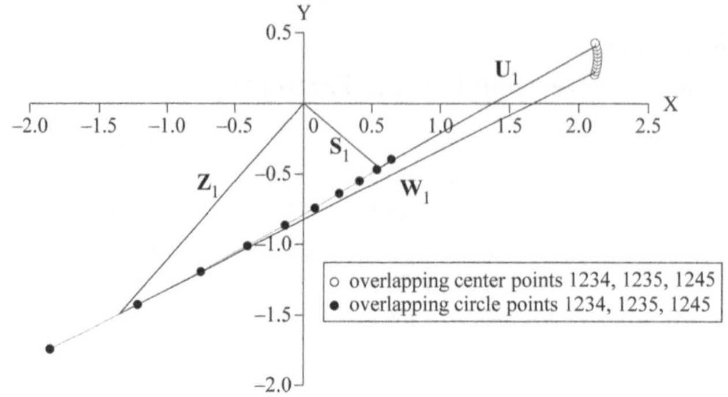

FIGURE E.5.12 Center and circle point solutions for the five-position synthesis problem in Example 5.3.

planar four-bar mechanism. Because the circle and center points are calculated analytically, the resulting mechanism achieves the precision positions precisely.

Another solution option for planar four-bar synthesis with five precision positions is to divide a five-precision position problem into two or more four-precision position problems, calculate the circle and center point loci for these problems and find common intersection points or overlapping regions among the curves. If they exist, a common solution that achieves all five precision positions exists. Dividing the five precision positions in Example 5.3 into three four-position synthesis problems (1-2-3-4, 1-2-3-5, and 1-2-4-5) and plotting the center point loci reveal a common overlapping region which coincides with the moving pivot solutions for Example 5.3 (Figure E.5.12). Plotting the circle point loci reveals a common overlapping region which coincides with the moving pivot solutions for Example 5.3 (Figure E.5.12).

5.4 GEARED FIVE-BAR MOTION GENERATION: THREE PRECISION POSITIONS

While the planar four-bar mechanism comprises two dyads (dyads $\mathbf{W}-\mathbf{Z}$ and $\mathbf{U}-\mathbf{S}$ in Figure 5.6), the geared five-bar mechanism includes a $\mathbf{W}-\mathbf{Z}$ dyad and a $\mathbf{U}$-$\mathbf{S}$-$\mathbf{T}$ triad (Figure 5.7).[9] Equation (5.8) can be applied calculate the $\mathbf{W}-\mathbf{Z}$ dyad components in geared five-bar motion generation for

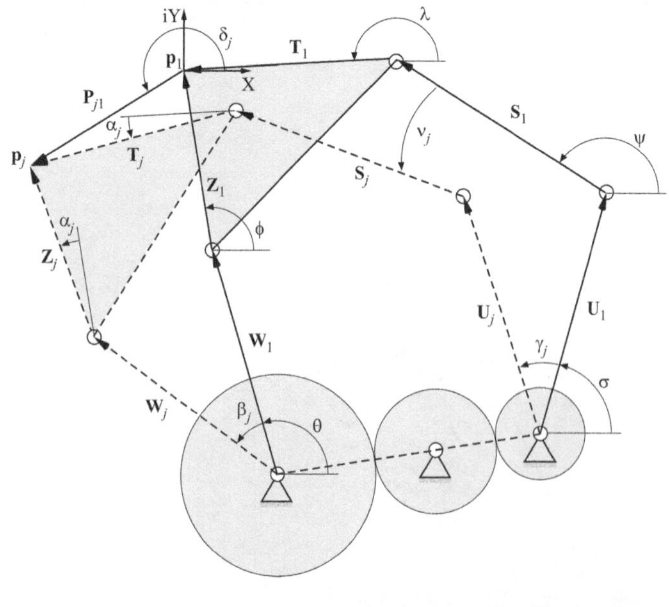

FIGURE 5.7 Geared five-bar mechanism in starting and (dashed) displaced positions.

three precision positions. An equation is formulated here to calculate the **U** - **S** - **T** triad components for three-precision position motion generation.

By taking the vector sum between the starting and displaced positions for the **U** - **S** - **T** triad, a vector loop closure equation is derived. After taking the counterclockwise vector sum for the triad (starting with $\mathbf{U}_1$), the vector loop closure equation becomes

$$U_1 e^{i\sigma} + S_1 e^{i\psi} + T_1 e^{i\lambda} + P_{j1} e^{i\delta_j} - T_1 e^{i(\lambda+\alpha_j)} - S_1 e^{i(\psi+\nu_j)} - U_1 e^{i(\sigma+\gamma_j)} = 0 \tag{5.10}$$

where $P_{j1} e^{i\delta_j} = \mathbf{p}_j - \mathbf{p}_1$. Because $\mathbf{p}_1$ is the coordinate frame origin ($\mathbf{p}_1 = (0,0)$), $p_{jx} = P_{j1}\cos\delta_j$ and $p_{jy} = P_{j1}\sin\delta_j$.

After factoring the terms for the starting dyad position, the resulting standard form loop closure equations become

$$S_1 e^{i\psi}\left(e^{i\nu_j} - 1\right) + T_1 e^{i\lambda}\left(e^{i\alpha_j} - 1\right) = P_{j1} e^{i\delta_j} + U_1 e^{i\sigma} - U_1 e^{i(\sigma+\gamma_j)} \tag{5.11}$$

Expanding Equation (5.11) and grouping the real and imaginary terms as separate equations produce the equation set

$$\begin{aligned}
&S_1\cos\psi\left(\cos\nu_j - 1\right) - S_1\sin\psi\sin\nu_j + T_1\cos\lambda\left(\cos\alpha_j - 1\right) - T_1\sin\lambda\sin\alpha_j \\
&= P_{j1}\cos\delta_j + U_1\left(\cos\sigma - \cos\left(\sigma+\gamma_j\right)\right) \\
&S_1\sin\psi\left(\cos\nu_j - 1\right) + S_1\cos\psi\sin\nu_j + T_1\sin\lambda\left(\cos\alpha_j - 1\right) + T_1\cos\lambda\sin\alpha_j \\
&= P_{j1}\sin\delta_j + U_1\left(\sin\sigma - \sin\left(\sigma+\gamma_j\right)\right)
\end{aligned} \tag{5.12}$$

After specifying $U_1\cos\sigma = U_{1x}$, $U_1\,\mathrm{Sin}\,\sigma = U_{1y}$, $T_1\cos\lambda = T_{1x}$ and $T_1\sin\lambda = T_{1y}$ Equation (5.12) becomes

$$\begin{aligned}
&S_{1x}\left(\cos\nu_j - 1\right) - S_{1y}\sin\nu_j + T_{1x}\left(\cos\alpha_j - 1\right) - T_{1y}\sin\alpha_j \\
&= P_{j1}\cos\delta_j + U_1\left(\cos\sigma - \cos\left(\sigma+\gamma_j\right)\right) \\
&S_{1y}\left(\cos\nu_j - 1\right) + S_{1x}\sin\nu_j + T_{1y}\left(\cos\alpha_j - 1\right) + T_{1x}\sin\alpha_j \\
&= P_{j1}\sin\delta_j + U_1\left(\sin\sigma - \sin\left(\sigma+\gamma_j\right)\right)
\end{aligned} \tag{5.13}$$

When expressed in matrix form for three precision positions, Equations (5.13) becomes

$$\begin{aligned}
&\begin{bmatrix}
\cos\nu_2 - 1 & -\sin\nu_2 & \cos\alpha_2 - 1 & -\sin\alpha_2 \\
\sin\nu_2 & \cos\nu_2 - 1 & \sin\alpha_2 & \cos\alpha_2 - 1 \\
\cos\nu_3 - 1 & -\sin\nu_3 & \cos\alpha_3 - 1 & -\sin\alpha_3 \\
\sin\nu_3 & \cos\nu_3 - 1 & \sin\alpha_3 & \cos\alpha_3 - 1
\end{bmatrix}
\begin{Bmatrix} U_{1x} \\ U_{1y} \\ T_{1x} \\ T_{1y} \end{Bmatrix} \\
&= \begin{bmatrix}
P_{21}\cos\delta_2 + U_1\left(\cos\sigma - \cos\left(\sigma+\gamma_2\right)\right) \\
P_{21}\sin\delta_2 + U_1\left(\sin\sigma - \sin\left(\sigma+\gamma_2\right)\right) \\
P_{31}\cos\delta_3 + U_1\left(\cos\sigma - \cos\left(\sigma+\gamma_3\right)\right) \\
P_{31}\sin\delta_3 + U_1\left(\sin\sigma - \sin\left(\sigma+\gamma_3\right)\right)
\end{bmatrix}
\end{aligned} \tag{5.14}$$

Example 5.4

Problem Statement: Calculate the components of $\mathbf{S}_1$ and $\mathbf{T}_1$ for the $\mathbf{U}_1$ - $\mathbf{S}_1$ - $\mathbf{T}_1$ triad of a geared five-bar mechanism to guide the landing gear through the three precision positions in Example 5.1. Use the $\mathbf{W}_1 - \mathbf{Z}_1$ dyad solution from Example 5.1 along with $\gamma_j = 3\beta_j/4$, $U_1 = 2.5$, $\sigma = 30°$, $\nu_2 = 25°$, and $\nu_3 = 40°$.

Known Information: Equation (5.14) and Table E.5.1

Solution Approach: Since the first precision point $\mathbf{p}_1$ lies on the coordinate system origin, the *x*- and *y*-components of $\mathbf{p}_j$ correspond to $P_{j1}\cos\delta_j$ and $P_{j1}\sin\delta_j$, respectively, in Equation (5.14). Table E.5.1 includes the precision position and dyad displacement data used in Equation (5.14) to calculate the components of the triad vectors for the geared five-bar mechanism (S_{1x}, S_{1y}, T_{1x}, and T_{1y}).

Figure E.5.13 includes the calculation procedure in the MATLAB command window, and Figure E.5.14 illustrates the resulting geared five-bar mechanism. Being an analytically calculated result, this mechanism achieves the precision positions precisely. A kinematic analysis of this solution is shown in Example 5.9 in Section 5.6.

```
>> p1 = [0, 0]';
>> p2 = [0.292, 0.734]';
>> p3 = [0.299, 1.461]';
>> A2 = -51.7124*pi/180; A3 = -84.9734*pi/180;
>> B2 = 18*pi/180; B3 = 38*pi/180;
>> V2 = 25*pi/180; V3 = 40*pi/180;
>> G2 = (3/4)*B2; G3 = (3/4)*B3;
>> U1 = 2.5; SIGMA = 30*pi/180;

>> P21 = p2 - p1;
>> P31 = p3 - p1;
>> CA2 = cos(A2); SA2 = sin(A2); CA3 = cos(A3); SA3 = sin(A3);
>> CV2 = cos(V2); SV2 = sin(V2); CV3 = cos(V3); SV3 = sin(V3);

>> ST = inv([CV2 - 1,-SV2, CA2 - 1, -SA2
>> SV2, CV2 - 1, SA2, CA2 - 1
>> CV3 - 1, -SV3, CA3 - 1,-SA3
>> SV3, CV3 -1, SA3, CA3 -1])*...
>> [P21(1) + U1*(cos(SIGMA) - cos(SIGMA + G2)),...
P21(2) + U1*(sin(SIGMA) - sin(SIGMA + G2)),...
P31(1) + U1*(cos(SIGMA) - cos(SIGMA + G3)),...
P31(2) + U1*(sin(SIGMA) - sin(SIGMA + G3))]'

ST =

   -0.4553
   -1.6385
   -0.2658
   -0.2454
>>
```

FIGURE E.5.13 Example 5.4 $\mathbf{S}_1 - \mathbf{T}_1$ calculation procedure in the MATLAB command window.

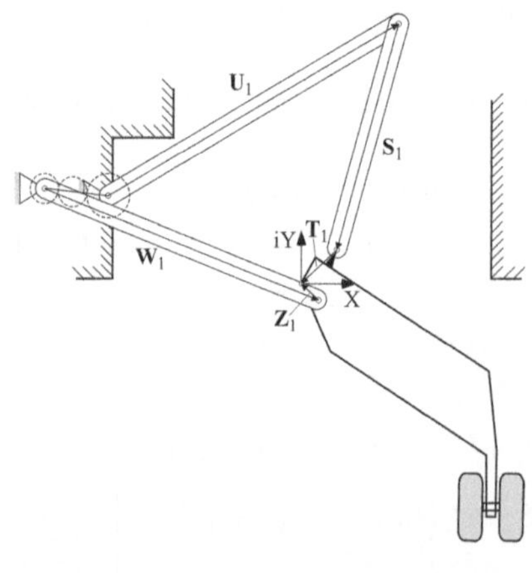

FIGURE E.5.14 Synthesized geared five-bar motion generator.

5.5 STEPHENSON III MOTION GENERATION: THREE, FOUR, AND FIVE PRECISION POSITIONS

As noted in Section 3.1.2, the Stephenson III mechanism is a planar four-bar mechanism with an additional dyad attached to the coupler link. The additional dyad is the $\mathbf{U}^* - \mathbf{S}^*$ dyad in Figure 5.8. Because the Stephenson III mechanism includes the planar four-bar mechanism, Equations (5.8) and (5.9) can be used to calculate the $\mathbf{W} - \mathbf{Z}$ and $\mathbf{U} - \mathbf{S}$ dyad vector components ($W_{1x}, W_{1y}, Z_{1x}, Z_{1y}, U_{1x}, U_{1y}, S_{1x}$, and S_{1y}) for motion generation with three precision positions given dyad displacement angles β_j and γ_j.

When modifying Equation (5.9) to reflect the $\mathbf{U}^* - \mathbf{S}^*$ vector components U^*_{1x}, U^*_{1y}, S^*_{1x}, and S^*_{1y} and displacement angles γ^*_j, it becomes

$$\begin{bmatrix} \cos\gamma^*_j - 1 & -\sin\gamma^*_j & \cos\alpha^*_j - 1 & -\sin\alpha^*_j \\ \sin\gamma^*_j & \cos\gamma^*_j - 1 & \sin\alpha^*_j & \cos\alpha^*_j - 1 \\ & & \vdots & \\ \cos\gamma^*_N - 1 & -\sin\gamma^*_N & \cos\alpha^*_N - 1 & -\sin\alpha^*_N \\ \sin\gamma^*_N & \cos\gamma^*_N - 1 & \sin\alpha^*_N & \cos\alpha^*_N - 1 \end{bmatrix} \begin{Bmatrix} U^*_{1x} \\ U^*_{1y} \\ S^*_{1x} \\ S^*_{1y} \end{Bmatrix} = \begin{bmatrix} P_{j1}\cos\delta_j \\ P_{j1}\sin\delta_j \\ \vdots \\ P_{N1}\cos\delta_N \\ P_{N1}\sin\delta_N \end{bmatrix} \quad (5.15)$$

While the displacement angle for vector $\mathbf{S}^*_1\,(\alpha^*_j)$ is not the same as the displacement angle for vectors $\mathbf{S}_1$ and $\mathbf{Z}_1\,(\alpha_j)$, all three vectors share the same link point $\mathbf{p}$.

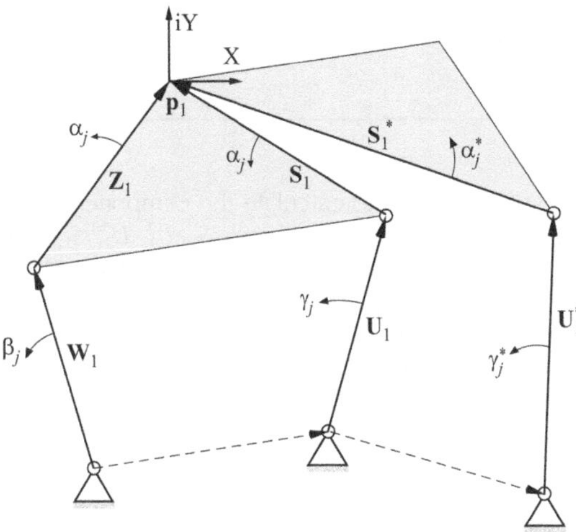

FIGURE 5.8 Stephenson III mechanism dyads.

Example 5.5

Problem Statement: Synthesize a Stephenson III mechanism to guide the gripper through the three precision positions in Figure E.5.15.

Known Information: Equations (5.8), (5.9), and (5.10) and Table E.5.6

Solution Approach: Since the first precision point $\mathbf{p}_1$ lies on the coordinate system origin, the x- and y-components of $\mathbf{p}_j$ correspond to $P_{j1}\cos\delta_j$ and $P_{j1}\sin\delta_j$, respectively, in Equations (5.8), (5.9), and (5.10). Table E.5.6 includes the precision position and dyad displacement data

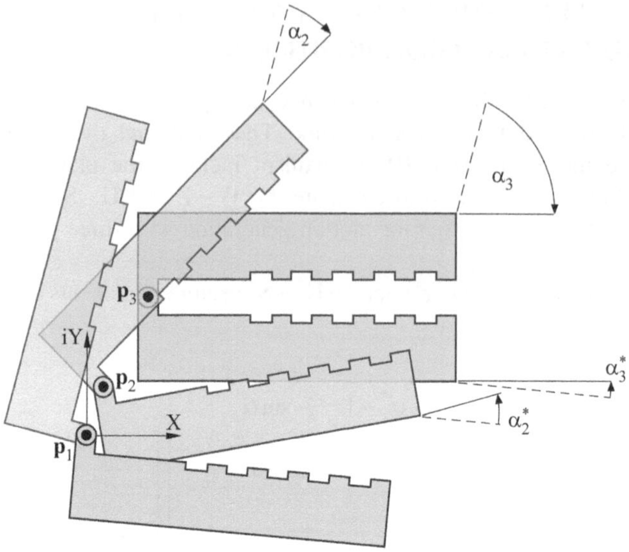

FIGURE E.5.15 Three gripper precision positions.

TABLE E.5.6
Gripper Precision Position Parameters and Dyad Displacement Angles

Precision Position	$\mathbf{p}_j$	α_j (deg)	β_j (deg)	γ_j (deg)	α^*_j (deg)	γ^*_j (deg)
1	0, 0					
2	0.1815, 0.4882	–30	15	–20	15	10
3	0.6647, 1.4078	–75	40	–35	5	30

used in Equations (5.8), (5.9), and (5.10) to calculate the components of the dyad vectors for the Stephenson III mechanism ($W_{1x}, W_{1y}, Z_{1x}, Z_{1y}, U_{1x}, U_{1y}, S_{1x}, S_{1y}, U^*_{1x}, U^*_{1y}, S^*_{1x}$, and S^*_{1y}).

Figures E.5.16 and E.5.17 include the calculation procedure in the MATLAB command window, and Figure E.5.18 illustrates the resulting Stephenson III mechanism. Being an analytically calculated result, this mechanism achieves the precision positions precisely. A kinematic analysis of this solution is conducted in Example 5.10 in Section 5.6.

Example 5.6

Problem Statement: Synthesize a Stephenson III mechanism to guide the gripper through the four precision positions in Figure E.5.19.

Known Information: Appendix D MATLAB file and Table E.5.7

Solution Approach: Table E.5.7 includes the precision position data used as input in the Appendix D MATLAB file. Vectors $\mathbf{P}_{j1}$ are the magnitudes of the coupler points $\mathbf{p}_j$, and angles δ_j are the angles $\mathbf{P}_{j1}$ make with the *x*-axis. Figure E.5.20 includes the Table E.5.7 input specified in the Appendix D MATLAB file. The circle and center points illustrated in Figures E.5.21 and E.5.22 correspond to the β_2-β_3-β_4 solutions for the first mechanism branch (as opposed to the alternate β_2-$\tilde{\beta}_3$-$\tilde{\beta}_4$ solution for the second mechanism branch).[10]

Table E.5.8 includes the mechanism dyad components for the circle and center point pairs selected from the calculated circle and center point curves. The selected dyads are also illustrated in Figures E.5.21 and E.5.22. The $\mathbf{W}_1$ and $\mathbf{Z}_1$ values in Table E.5.8 are for $\beta_2 = -21°$ in the Appendix D MATLAB file output for the first branch. The $\mathbf{U}_1$ and $\mathbf{S}_1$ values in Table E.5.8 are for $\beta_2 = -10°$ in

```
>> p1 = [0, 0]';
>> p2 = [0.1815, 0.4882]';
>> p3 = [0.6647, 1.4078]';
>> A2 = -30*pi/180;A3 = -75*pi/180;
>> B2 = 15*pi/180; B3 = 40*pi/180;
>> G2 = -20*pi/180; G3 = -35*pi/180;

>> P21 = p2 - p1;
>> P31 = p3 - p1;
>> CA2 = cos(A2); SA2 = sin(A2); CA3 = cos(A3); SA3 = sin(A3);
>> CB2 = cos(B2); SB2 = sin(B2); CB3 = cos(B3); SB3 = sin(B3);
>> CG2 = cos(G2); SG2 = sin(G2); CG3 = cos(G3); SG3 = sin(G3);

>> WZ = inv([CB2 -1,-SB2, CA2 - 1, -SA2
SB2, CB2 - 1, SA2, CA2 - 1
CB3 - 1, -SB3, CA3 - 1, -SA3
SB3, CB3 -1, SA3, CA3 - 1])*[P21(1), P21(2), P31(1), P31(2)]'

WZ =

    0.5854
   -1.2248
   -0.4926
   -0.3631

>> US = inv([CG2 -1,-SG2, CA2 - 1, -SA2
SG2, CG2 - 1, SA2, CA2 - 1
CG3 - 1,-SG3, CA3 - 1,-SA3
SG3, CG3 - 1, SA3, CA3 - 1])*[P21(1), P21(2), P31(1), P31(2)]'

US =

    0.1905
    1.8923
   -1.0184
   -1.1813

>>
```

FIGURE E.5.16 Example 5.4 $\mathbf{W}_1 - \mathbf{Z}_1$ and $\mathbf{U}_1 - \mathbf{S}_1$ calculation procedure in the MATLAB command window.

```
>> p1 = [0, 0]';
>> p2 = [0.1815, 0.4882]';
>> p3 = [0.6647, 1.4078]';
>> A2s = 15*pi/180;A3s = 5*pi/180;
>> G2s = 10*pi/180; G3s = 30*pi/180;

>> P21 = p2 - p1;
>> P31 = p3 - p1;
>> CA2s = cos(A2s); SA2s = sin(A2s);
>> CA3s = cos(A3s); SA3s = sin(A3s);
>> CG2s = cos(G2s); SG2s = sin(G2s);
>> CG3s = cos(G3s); SG3s = sin(G3s);

>> UsSs= inv([CG2s - 1, -SG2s, CA2s - 1,-SA2s
SG2s, CG2s -1 , SA2s, CA2s - 1
CG3s - 1,-SG3s, CA3s - 1,-SA3s
SG3s, CG3s - 1, SA3s, CA3s - 1])*[P21(1), P21(2), P31(1), P31(2)]'

UsSs =

    2.2215
   -2.0095
    0.3400
    0.4718

>>
```

FIGURE E.5.17 Example 5.4 $\mathbf{U}_1^* - \mathbf{S}_1^*$ calculation procedure in the MATLAB command window.

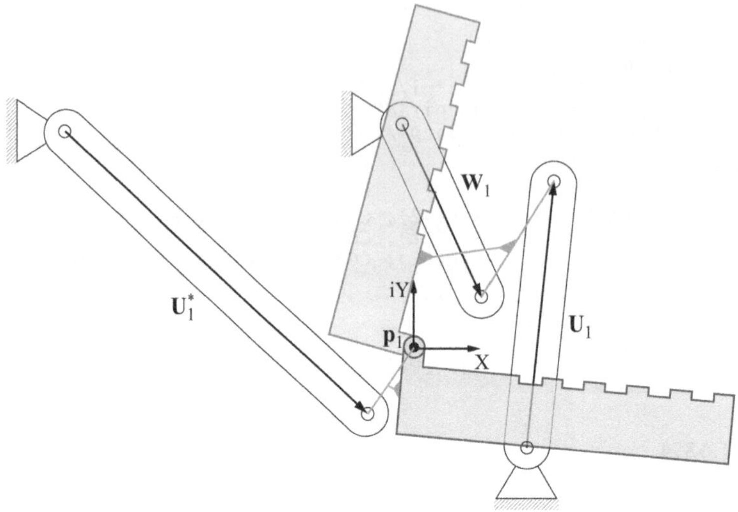

FIGURE E.5.18 Synthesized Stephenson III motion generator.

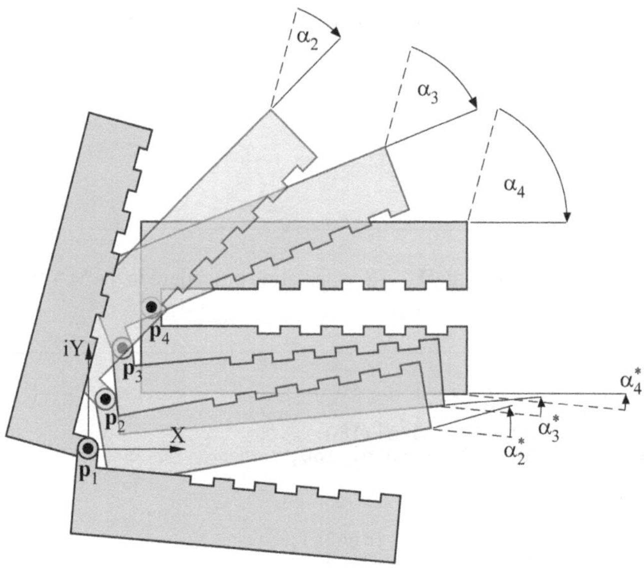

FIGURE E.5.19 Four gripper precision positions.

TABLE E.5.7
Gripper Precision Position Parameters

Precision Position	$\mathbf{p}_j$	$\mathbf{P}_{j1}$	δ_j (deg)	α_j(deg)	α_j^* (deg)
1	0, 0	0			
2	0.1815, 0.4882	0.5208	69.6062	–30	15
3	0.3700, 0.9775	1.0452	69.2675	–53	10
4	0.6647, 1.4078	1.5568	64.7254	–75	5

```
alpha2 = -30*pi/180;
alpha3 = -53*pi/180;
alpha4 = -75*pi/180;

P21 = 0.5208;
P31 = 1.0452;
P41 = 1.5568;

delta2 = 69.6062*pi/180;
delta3 = 69.2675*pi/180;
delta4 = 64.7254*pi/180;
```

```
alpha2 = 15*pi/180;
alpha3 = 10*pi/180;
alpha4 = 5*pi/180;

P21 = 0.5208;
P31 = 1.0452;
P41 = 1.5568;

delta2 = 69.6062*pi/180;
delta3 = 69.2675*pi/180;
delta4 = 64.7254*pi/180;
```

FIGURE E.5.20 Specified input (in bold type) in the Appendix D MATLAB file for Example 5.5 for (top) $\mathbf{W}_1 - \mathbf{Z}_1$ and $\mathbf{U}_1 - \mathbf{S}_1$ dyads and (bottom) $\mathbf{U}_1^* - \mathbf{S}_1^*$ dyad.

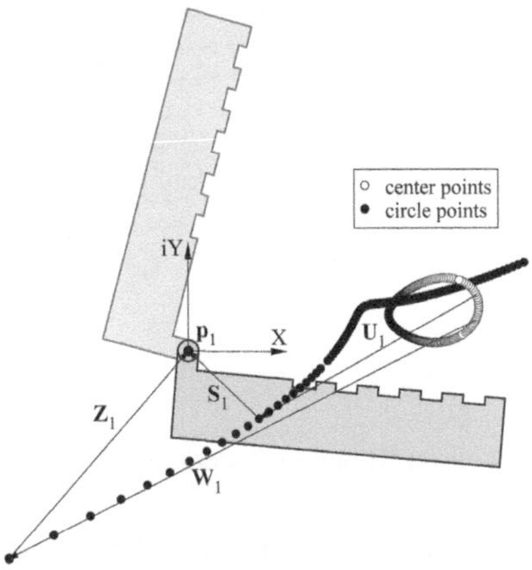

FIGURE E.5.21 Stephenson III mechanism $\mathbf{W}_1 - \mathbf{Z}_1$ and $\mathbf{U}_1 - \mathbf{S}_1$ dyads selected from circle and center points.

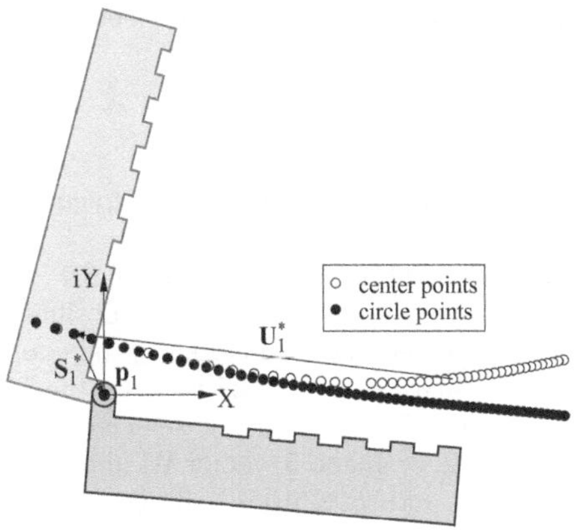

FIGURE E.5.22 Stephenson III mechanism $\mathbf{U}_1^* - \mathbf{S}_1^*$ dyads selected from circle and center points.

TABLE E.5.8
Selected Stephenson III Mechanism Dyad Variables

$\mathbf{W}_1$	$\mathbf{Z}_1$	$\mathbf{U}_1$	$\mathbf{S}_1$
−3.4300, −1.6978	1.3122, 1.4758	−1.6042, −0.8795	−0.5211, 0.4801
$\mathbf{U}_1^*$	$\mathbf{S}_1^*$		
−2.4757, 0.2867	0.1908, −0.3886		

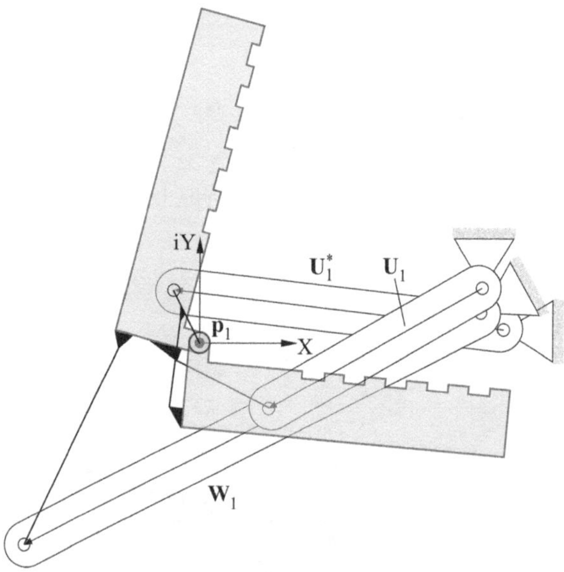

FIGURE E.5.23 Synthesized Stephenson III motion generator.

the Appendix D MATLAB file output for the first branch. The $\mathbf{U}_1^*$ and $\mathbf{S}_1^*$ values in Table E.5.8 are for $\beta_2 = -10°$ in the Appendix D MATLAB file output for the first branch. To achieve positions 2, 3, and 4, vector $\mathbf{W}_1$ (the designated driver link) is rotated by −21°, −40.1692°, and −59.1006° (β_2, β_3, and β_4), respectively. These angles were also calculated in the Appendix D MATLAB file for the first branch. Figure E.5.23 illustrates the resulting Stephenson III mechanism. Because the circle and center points are calculated analytically, the resulting mechanism achieves the precision positions precisely.

Example 5.7

Problem Statement: Synthesize a Stephenson III mechanism to guide the gripper through the five precision positions in Figure E.5.24.

Known Information: Appendix E MATLAB file and Table E.5.9

Solution Approach: Table E.5.9 includes the precision position data used as input in the Appendix E MATLAB file. Vectors $\mathbf{P}_{j1}$ are the magnitudes of the coupler points $\mathbf{p}_j$, and angles δ_j are the angles $\mathbf{P}_{j1}$ make with the x-axis. Figure E.5.25 includes the Table E.5.9 input specified in the Appendix E MATLAB file. Table E.5.10 includes the selected Stephenson III mechanism dyad components. To achieve positions 2, 3, 4, and 5, vector $\mathbf{W}_1$ (the designated driver link) is rotated by −9.2361°, −21.0897°, −40.3030°, and −59.2610° (β_2, β_3, β_4, and β_5), respectively. These angles were calculated in the Appendix E MATLAB file for the first branches. Figure E.5.26 illustrates the synthesized Stephenson III mechanism. Because the circle and center points are calculated analytically, the resulting mechanism achieves the precision positions precisely.

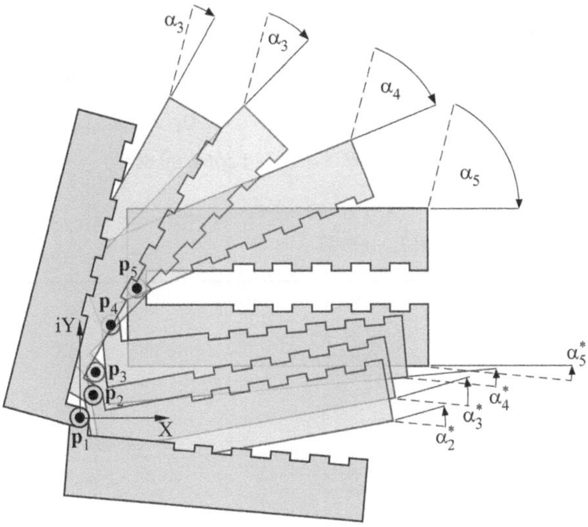

FIGURE E.5.24 Five gripper precision positions.

TABLE E.5.9
Gripper Precision Position Parameters

Precision Position	$\mathbf{p}_j$	$\mathbf{P}_{j1}$	δ_j (deg)	α_j(deg)	α_j^* (deg)
1	0, 0	0			
2	0.1104, 0.1789	0.2102	58.3239	–15	15
3	0.1815, 0.4882	0.5209	69.6073	–30	15
4	0.3700, 0.9776	1.0453	69.2691	–53	10
5	0.6647, 1.4079	1.5569	64.7270	–75	5

```
alpha2 = -15*pi/180;
alpha3 = -30*pi/180;
alpha4 = -53*pi/180;
alpha5 = -75*pi/180;

P21 = 0.2102;
P31 = 0.5208;
P41 = 1.0452;
P51 = 1.5568;

delta2 = 58.3224*pi/180;
delta3 = 69.6062*pi/180;
delta4 = 69.2675*pi/180;
delta5 = 64.7254*pi/180;
```

```
alpha2 = 15*pi/180;
alpha3 = 15*pi/180;
alpha4 = 10*pi/180;
alpha5 = 5*pi/180;

P21 = 0.2102;
P31 = 0.5208;
P41 = 1.0452;
P51 = 1.5568;

delta2 = 58.3224*pi/180;
delta3 = 69.6062*pi/180;
delta4 = 69.2675*pi/180;
delta5 = 64.7254*pi/180;
```

FIGURE E.5.25 Specified input (in bold type) in the Appendix D MATLAB file for Example 5.7 for (top) $\mathbf{W}_1 - \mathbf{Z}_1$ and $\mathbf{U}_1 - \mathbf{S}_1$ dyads and (bottom) $\mathbf{U}_1^* - \mathbf{S}_1^*$ dyad.

TABLE E.5.10
Selected Stephenson III Mechanism Dyad Variables

W_1	Z_1	U_1	S_1
−3.4627, −1.7135	1.3455, 1.4928	−1.5664, −0.8643	−0.5553, 0.4575
U_1^*	S_1^*		
−2.4790, 0.2887	0.1967, −0.3903		

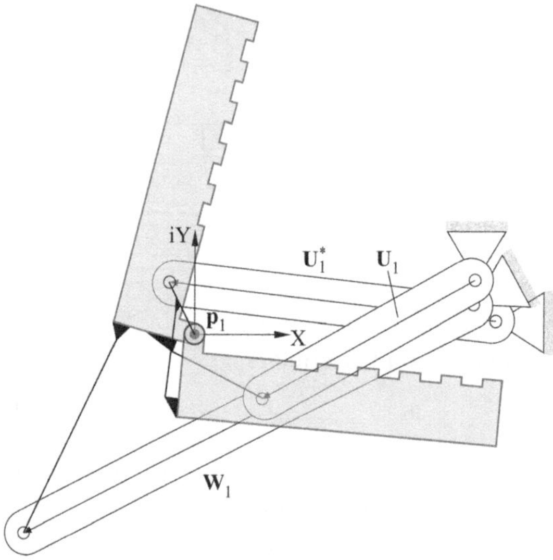

FIGURE E.5.26 Synthesized Stephenson III motion generator.

5.6 FROM MOTION GENERATION TO KINEMATIC DISPLACEMENT ANALYSIS

After synthesizing planar four-bar, geared five-bar, and Stephenson III motion generators, kinematic displacement analyses should be conducted to determine if the mechanisms achieve their precision positions and do so without encountering any circuit, branch, or order defects. In Chapter 4, kinematic equations and corresponding MATLAB and Simscape Multibody files are presented for these mechanisms from which the user can conduct displacement analyses. All mechanism link vectors in kinematic analysis, however, do not explicitly appear in the motion generator solutions. Also, the reference frames for motion generation and kinematic analysis are not in the same location on the mechanism. This section presents approaches to resolve both issues, and as a result, smoothly transition from motion generation to kinematic analysis in MATLAB and Simscape Multibody.

5.6.1 Planar Four-Bar Mechanism

Figure 5.9a includes the dyad vectors calculated in planar four-bar motion generation, and Figure 5.9b includes the mechanism vectors specified in planar four-bar displacement analysis. But as shown in this figure, vectors $\mathbf{V}_1$, $\mathbf{L}_1$, and $\mathbf{G}_1$ do not explicitly appear in a motion generator solution. These vectors, however, can be calculated or directly assigned values using dyad vectors from a motion generator solution.

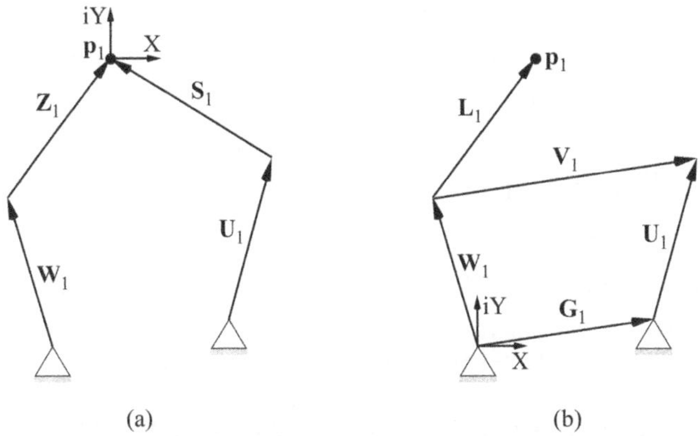

FIGURE 5.9 Vectors for planar four-bar mechanism: (a) motion generation and (b) kinematic displacement analysis.

TABLE 5.2
Vectors for Planar Four-Bar Motion Generation and Kinematic Displacement Analysis

Motion Generation		Kinematic Displacement Analysis	
Vector	**Vector Expression**	**Vector**	**Vector Expression**
$\mathbf{W}_1$	$\mathbf{W}_1 = W_{1x} + iW_{1y}$	$\mathbf{W}_1$	$\mathbf{W}_1 = W_{1x} + iW_{1y}$
$\mathbf{Z}_1$	$\mathbf{Z}_1 = Z_{1x} + iZ_{1y}$	$\mathbf{L}_1$	$\mathbf{L}_1 = Z_{1x} + iZ_{1y}$
$\mathbf{U}_1$	$\mathbf{U}_1 = U_{1x} + iU_{1y}$	$\mathbf{U}_1$	$\mathbf{U}_1 = U_{1x} + iU_{1y}$
$\mathbf{S}_1$	$\mathbf{S}_1 = S_{1x} + iS_{1y}$	$\mathbf{V}_1$	$\mathbf{V}_1 = \mathbf{Z}_1 - \mathbf{S}_1$
		$\mathbf{G}_1$	$\mathbf{G}_1 = \mathbf{W}_1 + \mathbf{Z}_1 - \mathbf{S}_1 - \mathbf{U}_1$

Table 5.2 includes the dyad vectors calculated in planar four-bar motion generation. Because vector components are calculated (see Equations (5.8) and (5.9)), the dyad vectors can be expressed in rectangular form, as shown. Table 5.2 also includes the vectors specified in planar four-bar displacement analysis. In displacement analysis, vectors $\mathbf{W}_1$ and $\mathbf{U}_1$ can be specified as they appear in motion generation. Also, the calculated components of $\mathbf{Z}_1$can be assigned to $\mathbf{L}_1$ since both vectors join $\mathbf{W}_1$ and $\mathbf{p}_1$, as shown in Figure 5.9. From this figure, it can also be determined that $\mathbf{V}_1 = \mathbf{Z}_1 - \mathbf{S}_1$ and $\mathbf{G}_1 = \mathbf{W}_1 + \mathbf{Z}_1 - \mathbf{S}_1 - \mathbf{U}_1$.

As illustrated in Figure 5.9, the reference frames for motion generation and kinematic analysis are not in the same position. Because of this, the $\mathbf{p}_j$ values calculated in kinematic analysis cannot be compared as is to the $\mathbf{p}_j$ values specified in motion generation. To resolve this, $\mathbf{p}_1$ should be subtracted from each $\mathbf{p}_j$ value calculated in kinematic analysis.

Example 5.8

Problem Statement: Using the Appendix B.1 or C.1 files, calculate the precision positions achieved by the motion generator solution in Example 5.1.

Known Information: Example 5.1 solution and Appendix B.1 MATLAB file

Solution Approach: Figure E.5.27 includes the input specified (in bold text) in the Appendix B.1 MATLAB file. Table E.5.11 includes the landing gear positions achieved. While the values of α_j and γ_j can be directly compared to the prescribed values in Example 1 (Table E.5.1), $\mathbf{p}_1$ must be subtracted from each $\mathbf{p}_j$ value to produce results that can also be compared to those in Table E.5.1.

```
W1 = 2.0580 - i*0.8054;
Z1 = -0.1324 + i*0.1191;
U1 = 0.5808 - i*1.8615;
S1 = -1.5053 + i*1.3400;

L1 = Z1;
V1 = Z1 - S1;
G1 = W1 + Z1 - S1 - U1;

start_ang   = 0;
step_ang    = 1;
stop_ang    = 38;

angular_vel  = 0 * ones(N+1,1);
angular_acc  = 0 * ones(N+1,1);
```

FIGURE E.5.27 Specified input (in bold text) in the Appendix B.1 MATLAB file for Example 5.8.

TABLE E.5.11
Landing Gear Positions Achieved

Position	β_j (deg)	$\mathbf{p}_j$	α_j (deg)	γ_j (deg)	$\mathbf{p}_j - \mathbf{p}_1$
1		1.9256, –0.6863			0, 0
2	18	2.2176, 0.0477	–51.7160	–40.0020	0.2920, 0.7340
3	38	2.2246, 0.7747	–84.9800	–87.0040	0.2990, 1.4610

5.6.2 Geared Five-Bar Mechanism

Figure 5.10a includes the dyad and triad vectors calculated in geared five-bar motion generation, and Figure 5.10b includes the mechanism vectors specified in geared five-bar displacement analysis. Like the planar four-bar motion generator, vectors $\mathbf{V}_1$, $\mathbf{L}_1$, and $\mathbf{G}_1$ do not explicitly appear in a geared five-bar motion generator solution. These vectors, however, can be calculated or directly assigned values using dyad vectors from a motion generator solution.

Table 5.3 includes the dyad vectors calculated in geared five-bar motion generation. Because vector components are calculated (see Equations (5.8) and (5.14)), the dyad and triad vectors can be expressed in rectangular form, as shown. Table 5.3 also includes the vectors specified in geared five-bar displacement analysis. In displacement analysis, vectors $\mathbf{W}_1$ and $\mathbf{S}_1$ can be specified as they appear in motion generation. Also, the calculated components of $\mathbf{Z}_1$ can be assigned to $\mathbf{L}_1$ since both vectors join $\mathbf{W}_1$ and $\mathbf{p}_1$, as shown in Figure 5.10. From this figure, it can also be determined that $\mathbf{V}_1 = \mathbf{Z}_1 - \mathbf{T}_1$ and $\mathbf{G}_1 = \mathbf{W}_1 + \mathbf{Z}_1 - \mathbf{T}_1 - \mathbf{S}_1 - \mathbf{U}_1$.

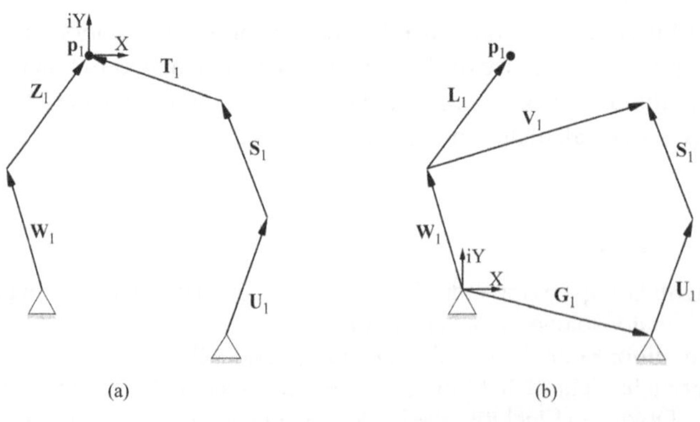

FIGURE 5.10 Vectors for geared five-bar mechanism: (a) motion generation and (b) kinematic displacement analysis.

TABLE 5.3
Vectors for Geared Five-Bar Motion Generation and Kinematic Displacement Analysis

Motion Generation		Kinematic Displacement Analysis	
Vector	Vector Expression	Vector	Vector Expression
$\mathbf{W}_1$	$\mathbf{W}_1 = W_{1x} + iW_{1y}$	$\mathbf{W}_1$	$\mathbf{W}_1 = W_{1x} + iW_{1y}$
$\mathbf{Z}_1$	$\mathbf{Z}_1 = Z_{1x} + iZ_{1y}$	$\mathbf{L}_1$	$\mathbf{L}_1 = Z_{1x} + iZ_{1y}$
$\mathbf{S}_1$	$\mathbf{S}_1 = S_{1x} + iS_{1y}$	$\mathbf{S}_1$	$\mathbf{S}_1 = S_{1x} + iS_{1y}$
$\mathbf{T}_1$	$\mathbf{T}_1 = T_{1x} + iT_{1y}$	$\mathbf{U}_1$	$\mathbf{U}_1 = U_1 e^{i\sigma}$
		$\mathbf{V}_1$	$\mathbf{V}_1 = \mathbf{Z}_1 - \mathbf{T}_1$
		$\mathbf{G}_1$	$\mathbf{G}_1 = \mathbf{W}_1 + \mathbf{Z}_1 - \mathbf{T}_1 - \mathbf{S}_1 - \mathbf{U}_1$

As illustrated in Figure 5.10, the reference frames for motion generation and kinematic analysis are not in the same position. Like with planar four-bar kinematic analysis, $\mathbf{p}_1$ should be subtracted from each $\mathbf{p}_j$ value calculated in geared five-bar kinematic analysis. Doing this will allow the user to make a direct comparison between the points specified in motion generation and those calculated in a kinematic displacement analysis.

Example 5.9

Problem Statement: Using the Appendix B.2 or C.2 files, calculate the precision positions achieved by the motion generator solution in Example 5.4.

Known Information: Example 5.4 solution and Appendix B.2 MATLAB file

Solution Approach: Figure E.5.28 includes the input specified (in bold text) in the Appendix B.2 MATLAB file. The gear ratio specified here is the inverted γ_j/β_j value specified in Example 5.4. Table E.5.12 includes the landing gear positions achieved. While the values of α_j and ν_j can be directly compared to their prescribed values, $\mathbf{p}_1$ must be subtracted from each $\mathbf{p}_j$ value to produce results that can also be compared to those in Table E.5.1.

```
W1 = 2.0580 - i*0.8054;
Z1 = -0.1324 + i*0.1191;
U1 = 2.5*exp(i*30*pi/180);
S1 = -0.4553 - i*1.6385;
T1 = -0.2658 - i*0.2454;

L1 = Z1;
V1 = Z1 - T1;
G1 = W1 + Z1 - T1 - S1 - U1;
ratio = 4/3;

start_ang   = 0;
step_ang    = 1;
stop_ang    = 38;

angular_vel  = 0 * ones(N+1,1);
angular_acc  = 0 * ones(N+1,1);
```

FIGURE E.5.28 Specified input (in bold text) in the Appendix B.2 MATLAB file for Example 5.9.

5.6.3 Stephenson III Mechanism

Figure 5.11a includes the dyad vectors calculated in Stephenson III motion generation, and Figure 5.11b includes the mechanism vectors specified in Stephenson III displacement analysis. But as shown in this figure, vectors $\mathbf{V}_1$, $\mathbf{L}_1$, $\mathbf{V}_1^*$, $\mathbf{G}_1$, and $\mathbf{G}_1^*$ do not explicitly appear in a Stephenson III

TABLE E.5.12
Landing Gear Positions Achieved

Position	β_j (deg)	$\mathbf{p}_j$	α_j (deg)	ν_j (deg)	$\mathbf{p}_j - \mathbf{p}_1$
1		1.9256, –0.6863			0, 0
2	18	2.2176, 0.0477	–51.7170	25.0010	0.2920, 0.7340
3	38	2.2246, 0.7747	–84.9760	40.0020	0.2990, 1.4610

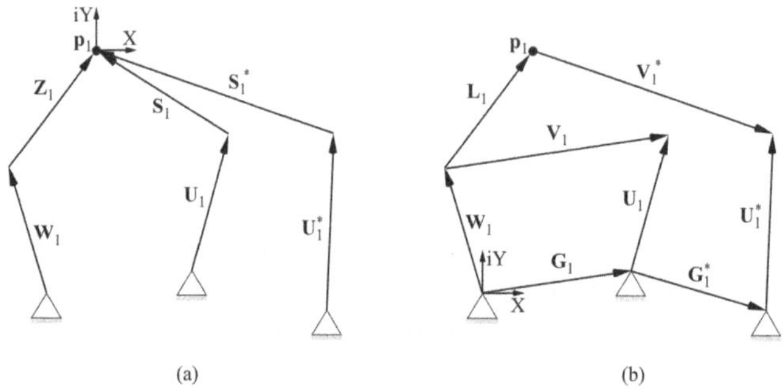

FIGURE 5.11 Vectors for Stephenson III mechanism: (a) motion generation and (b) kinematic displacement analysis.

TABLE 5.4
Vectors for Stephenson III Motion Generation and Kinematic Displacement Analysis

Motion Generation		Kinematic Displacement Analysis	
Vector	Vector Expression	Vector	Vector Expression
$\mathbf{W}_1$	$\mathbf{W}_1 = W_{1x} + iW_{1y}$	$\mathbf{W}_1$	$\mathbf{W}_1 = W_{1x} + iW_{1y}$
$\mathbf{Z}_1$	$\mathbf{Z}_1 = Z_{1x} + iZ_{1y}$	$\mathbf{L}_1$	$\mathbf{L}_1 = Z_{1x} + iZ_{1y}$
$\mathbf{U}_1$	$\mathbf{U}_1 = U_{1x} + iU_{iy}$	$\mathbf{U}_1$	$\mathbf{U}_1 = U_{1x} + iU_{1y}$
$\mathbf{S}_1$	$\mathbf{S}_1 = S_{1x} + iS_{1y}$	$\mathbf{V}_1$	$\mathbf{V}_1 = \mathbf{Z}_1 - \mathbf{S}_1$
$\mathbf{U}_1^*$	$\mathbf{U}_1^* = U_{1x}^* + iU_{1y}^*$	$\mathbf{V}_1^*$	$\mathbf{V}_1^* = -\mathbf{S}_1^*$
$\mathbf{S}_1^*$	$\mathbf{S}_1^* = S_{1x}^* + iS_{1y}^*$	$\mathbf{U}_1^*$	$\mathbf{U}_1^* = U_{1x}^* + iU_{1y}^*$
		$\mathbf{G}_1$	$\mathbf{G}_1 = \mathbf{W}_1 + \mathbf{Z}_1 - \mathbf{S}_1 - \mathbf{U}_1$
		$\mathbf{G}_1^*$	$\mathbf{G}_1^* = \mathbf{W}_1 + \mathbf{Z}_1 - \mathbf{S}_1^* - \mathbf{U}_1^* - \mathbf{G}_1$

motion generator solution. These vectors, however, can be calculated or directly assigned values using dyad vectors from a motion generator solution.

Table 5.4 includes the dyad vectors calculated in Stephenson III motion generation. Because vector components are calculated (see Equations (5.8), (5.9), and (5.15)), the dyad vectors can be expressed in rectangular form as shown. Table 5.4 also includes the vectors specified in Stephenson III displacement analysis. In displacement analysis, vectors $\mathbf{W}_1$, $\mathbf{U}_1$, and $\mathbf{U}_1^*$ can be specified as they appear in motion generation. Also, the calculated components of $\mathbf{Z}_1$ can be assigned to $\mathbf{L}_1$ since both vectors join $\mathbf{W}_1$ and $\mathbf{p}_1$, as shown in Figure 5.11. From this figure, it can also be determined that $\mathbf{V}_1 = \mathbf{Z}_1 - \mathbf{T}_1$, $\mathbf{V}_1^* = -\mathbf{S}_1^*$, $\mathbf{G}_1 = \mathbf{W}_1 + \mathbf{Z}_1 - \mathbf{T}_1 - \mathbf{S}_1 - \mathbf{U}_1$, and $\mathbf{G}_1^* = \mathbf{W}_1 + \mathbf{Z}_1 - \mathbf{S}_1^* - \mathbf{U}_1^* - \mathbf{G}_1$.

As illustrated in Figure 5.11, the reference frames for motion generation and kinematic analysis are not in the same position. Like with planar four-bar and geared five-bar kinematic analysis,

$\mathbf{p}_1$ should be subtracted from each $\mathbf{p}_j$ value calculated in Stephenson III kinematic analysis. Doing this will allow the user to make a direct comparison between the points specified in motion generation and those calculated in a kinematic displacement analysis.

Example 5.10

Problem Statement: Using the Appendix B.3 or C.3 files, calculate the precision positions achieved by the motion generator solution in Example 5.5.

Known Information: Example 5.5 solution and Appendix B.3 MATLAB file

Solution Approach: Figure E.5.29 includes the input specified (in bold text) in the Appendix B.3 MATLAB file. Table E.5.13 includes the gripper positions achieved. While the values of α_j, γ_j, α^*_j, and α^*_j can be directly compared to their prescribed values, $\mathbf{p}_1$ must be subtracted from each $\mathbf{p}_j$ value to produce results that can also be compared to those in Table E.5.6.

```
W1 = 0.5854 - i*1.2248;
Z1 = -0.4926 - i*0.3631;
U1 = 0.1905 + i*1.8923;
S1 = -1.0184 - i*1.1813;
U1s = 2.2215 - i*2.0095;
S1s = 0.3400 + i*0.4718;

L1 = Z1;
V1 = Z1 - S1;
G1 = W1 + Z1 - S1 - U1;
V1s = -S1s;
G1s = W1 + Z1 - S1s - U1s - G1;
L1s = 1;

start_ang   = 0;
step_ang    = 1;
stop_ang    = 38;

angular_vel  = 0 * ones(N+1,1);
angular_acc  = 0 * ones(N+1,1);
```

FIGURE E.5.29 Specified input (in bold text) in the Appendix B.3 MATLAB file for Example 5.10.

TABLE E.5.13
Gripper Positions Achieved

Position	β_j (deg)	$\mathbf{p}_j$	α_j (deg)	γ_j (deg)	α^*_j (deg)	α^*_j (deg)	$\mathbf{p}_j - \mathbf{p}_1$
1	0	0.0928,–1.5879					0, 0
2	15	0.2743,–1.0997	–30.0000	–20.0000	14.9990	9.9997	0.1815, 0.4882
3	40	0.7575,–0.1801	–75.0010	–35.0000	4.9995	29.9990	0.6647, 1.4078

5.7 ORDER AND BRANCH DEFECT ELIMINATION

Although branch and order defects are inherent in motion and path generation, several conventional practices and construction methods are used to produce motion and path generators that are branch defect- and order defect-free. These practices and construction methods include *Prescribed Timing, Filemon's Construction,* and *Waldron's Construction.*

Motion and path generation with prescribed timing prevents order defects because with this practice, each precision position and its corresponding dyad displacement angles are prescribed. For example, in planar four-bar three-position motion generation (see Example 5.1), the user specifies the precision positions along with displacement angles β_j for dyad $\mathbf{W}-\mathbf{Z}$ and γ_j for dyad $\mathbf{U}-\mathbf{S}$. In Example 5.4, the user specifies the precision positions along with displacement angles β_j for dyad

$\mathbf{W}-\mathbf{Z}$ and γ_j and ν_j for triad $\mathbf{U}-\mathbf{S}-\mathbf{T}$. The prescribed timing is also used in Example 5.5. By specifying both the precision positions and the corresponding dyad displacement angles, the order in which the precision positions are achieved is maintained.

In the early 1970s, Filemon introduced a construction method to ensure that planar four-bar motion generator solutions selected from circle and center points are branch defect-free [16,17]. Filemon's work shows that as long as the moving pivot of the mechanism driving link is selected outside the wedge-shaped region produced using her construction method, the resulting planar four-bar motion generator will pass through all of the precision positions without disassembly (it is a non-branching solution).[11]

Filemon's construction assumes that an output link has already been selected. The follower link moving pivot and fixed pivot will be defined as $\mathbf{b}_1$ and $\mathbf{b}_0$, respectively, where $\mathbf{b}_1=-\mathbf{S}_1$ and $\mathbf{b}_0=-(\mathbf{S}_1+\mathbf{U}_1)$. Relative inverse displacements of the follower link are taken to sweep a planar wedge-shaped region (with angles $\angle\mathbf{b}_{0j}\mathbf{b}_1\mathbf{b}_0$).[12] The positions that the follower fixed pivot can take relative to the follower moving pivot are computed as

$$\mathbf{b}_{0j}=\left[M_j\right]\mathbf{b}_0 \tag{5.16}$$

where

$$\left[M_j\right]=\begin{bmatrix}1&0&0\\0&1&0\\0&0&1\end{bmatrix}\begin{bmatrix}\cos\alpha_j&-\sin\alpha_j&P_{j1x}\\\sin\alpha_j&\cos\alpha_j&P_{j1y}\\0&0&1\end{bmatrix}^{-1}\quad j=2,3,\ldots \tag{5.17}$$

and $\mathbf{b}_0=\left(b_{0x},b_{0y},1\right)^T$.

Given the precision position variables α_j and $\mathbf{P}_{j1}$, Equation (5.16) calculates the ground link displacements of the inverted four-bar mechanism. When selecting the input link (having already constructed the wedge-shaped region) the moving pivot for the input link should be chosen from outside the region. Filemon's construction can be applied in motion generation with 3, 4, and 5 precision positions (and beyond).

Example 5.11

Problem Statement: Using Filemon's construction, calculate the wedge region using the precision positions in Example 5.1.

Known Information: Example 5.1 precision positions and Equation (5.16)

Solution Approach: Figure E.5.30 includes the calculation procedure in the MATLAB command window. The calculated follower $(\mathbf{U}_1-\mathbf{S}_1)$ in Example 5.1 is the selected output link in this example. From Table E.5.1 in Example 5.1, the coupler displacement angles α_j and the *x*- and *y*-components of the coupler point vectors $\mathbf{P}_{j1}$ are known. Using the data in Equation (5.16), the two positions of the fixed pivot $\mathbf{b}_0$ ($\mathbf{b}_{02}$ and $\mathbf{b}_{03}$) are calculated (Figures E.5.30 and E.5.31). The line passing through $\mathbf{b}_1$ and $\mathbf{b}_{02}$ and the line passing through $\mathbf{b}_1$ and $\mathbf{b}_{03}$ form the borders for the planar wedge-shaped region (Figure E.5.32).

The moving pivot for the driving link dyad $(\mathbf{W}_1-\mathbf{Z}_1)$ must lie outside the wedge-shaped region, and the resulting four-bar mechanism must be either a Grashof crank-rocker or drag-link. Figure E.5.32 illustrates the $\mathbf{W}_1-\mathbf{Z}_1$ solution from Equation (5.8) with $\beta_2=10°$ and $\beta_3=40°$. In this solution, $\mathbf{W}_1=1.4236+i0.1298$ and $\mathbf{Z}_1=-0.6724+i0.1024$. The resulting planar four-bar motion generator in this figure, a Grashof crank-rocker, is branch defect-free.

In the mid-1970s, Waldron introduced a feasible region construction method for selecting non-branching planar four-bar motion generators for three precision positions [18,19]. Waldron's

```
>> alpha2 = -51.7124*pi/180; alpha3 = -84.9734*pi/180;
>> P21 = 0.7899; P31 = 1.4913;
>> delta2 = 68.3063*pi/180; delta3 = 78.4339*pi/180;
>> US = [0.5808, -1.8615, -1.5053, 1.3400];
>> b0 = [-US(1) - US(3), -US(2) - US(4), 1]'

b0 =

    0.9245
    0.5215
    1.0000

>> A2 = alpha2; P2x = P21*cos(delta2); P2y = P21*sin(delta2);
>> A3 = alpha3; P3x = P31*cos(delta3); P3y = P31*sin(delta3);
>> M2 = [1, 0, 0; 0, 1, 0; 0, 0, 1]*...
inv([cos(A2), -sin(A2), P2x; sin(A2), cos(A2), P2y; 0, 0, 1]);
>> M3 = [1, 0, 0; 0, 1, 0; 0, 0, 1]*...
inv([cos(A3), -sin(A3), P3x; sin(A3), cos(A3), P3y; 0, 0, 1]);
>> b02 = M2*b0

b02 =

    0.5587
    0.3648
    1.0000

>> b03 = M3*b0

b03 =

    0.9907
    0.5408
    1.0000

>>
```

FIGURE E.5.30 Example 5.11 $\mathbf{b}_{02}$ and $\mathbf{b}_{03}$ calculation procedure in the MATLAB command window.

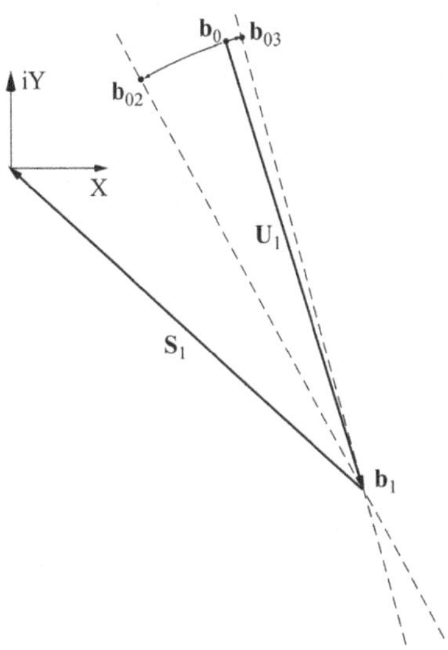

FIGURE E.5.31 Wedge-shaped region borders formed by $\mathbf{b}_{02}$ and $\mathbf{b}_{03}$ (form $\mathbf{U}_1$).

work shows that follower link moving pivots selected outside the three circles produced using his construction method as well as follower link moving pivots selected in regions where any two circles overlap result in non-branching motion generator solutions. Once a suitable follower link dyad is calculated according to Waldron's construction method, a corresponding suitable crank link dyad is calculated according to Filemon's construction.

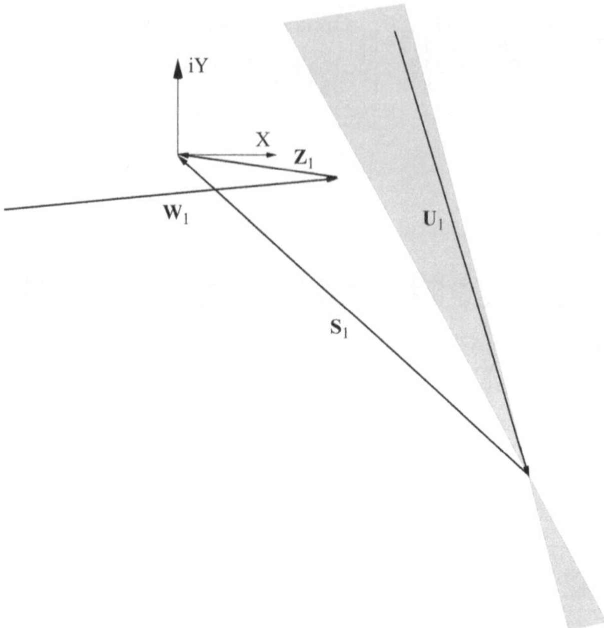

FIGURE E.5.32 Wedge-shaped region with calculated dyads $\mathbf{W}_1 - \mathbf{Z}_1$ and $\mathbf{U}_1 - \mathbf{S}_1$.

Given three precision positions, three rotation centers or *poles* exist to rotate from positions 1 to 2 (pole $\mathbf{p}_{12}$), positions 1 to 3 (pole $\mathbf{p}_{13}$), and positions 2 to 3 (pole $\mathbf{p}_{23}$) [20]. These poles form a triangle (called a *pole triangle)* $\Delta\mathbf{p}_{12}\mathbf{p}_{13}\mathbf{p}_{23}$ [21]. The reflection of pole $\mathbf{p}_{23}$ about the triangle side $\mathbf{p}_{12}\mathbf{p}_{13}$ produces a new pole (called an *image pole*) $\mathbf{p}'_{23}$. In Waldron's construction, each side of the triangle $\Delta\mathbf{p}_{12}\mathbf{p}_{13}\mathbf{p}'_{23}$ is equal to a diameter of each circle.

The rotation pole equation corresponding to the two rigid-body positions in Figure 5.6 is defined as

$$\text{pole } \mathbf{p}_{jk} = P_{j1}e^{i\delta_j} - \left[\frac{P_{k1}e^{i\delta_k} - P_{j1}e^{i\delta_j}}{e^{i(\alpha_k - \alpha_j)} - 1}\right] \tag{5.18}$$

where $P_{11} = \delta_1 = \alpha_1 = 0$ [22]. Equation (5.18) is used to calculate pole $\mathbf{p}_{12}$, pole $\mathbf{p}_{13}$, and pole $\mathbf{p}_{23}$.

Example 5.12

Problem Statement: Calculate a non-branching landing gear mechanism for Example 5.1 using Waldron's construction and Filemon's construction.

Known Information: Example 5.1 precision positions and Equation (5.18)

Solution Approach: Figure E.5.33 includes the calculation procedure in the MATLAB command window, and Figure E.5.34 includes the calculated pole triangle and image pole. Figure E.5.35 includes the three-circle diagram resulting from Waldron's construction along with the $\mathbf{U}_1 - \mathbf{S}_1$ and $\mathbf{W}_1 - \mathbf{Z}_1$ dyads from Example 5.1. Suitable follower link moving pivots lie either outside the three circles or in regions where any two of the three circles overlap.

It can be determined from Figure E.5.35 that the $\mathbf{U}_1 - \mathbf{S}_1$ follower link dyad solution from Example 5.1 is satisfactory since its moving pivot lies outside the three circles. The $\mathbf{W}_1 - \mathbf{Z}_1$ crank link dyad solution from Example 5.1 is also satisfactory since its moving pivot lies outside the wedge-shaped region, as illustrated in Figure E.5.36. The resulting four-bar motion generator is branch defect-free.

```
>> alpha1 = 0; alpha2 = -51.7124*pi/180; alpha3 = -84.9734*pi/180;
>> P11 = 0; P21 = 0.7899; P31 = 1.4913;
>> delta1=0; delta2 = 68.3063*pi/180; delta3 = 78.4339*pi/180;

>> Pole_12 = P11*exp(i*delta1) - (P21*exp(i*delta2) - ...
P11*exp(i*delta1))/(exp(i*(alpha2 - alpha1)) - 1)

Pole_12 =

   0.9032 + 0.0657i

>> Pole_13 = P11*exp(i*delta1) - (P31*exp(i*delta3) -...
P11*exp(i*delta1))/(exp(i*(alpha3 - alpha1)) - 1)

Pole_13 =

   0.9471 + 0.5673i

>> Pole_23 = P21*exp(i*delta2) - (P31*exp(i*delta3) -...
P21*exp(i*delta2))/(exp(i*(alpha3 - alpha2)) - 1)

Pole_23 =

   1.5126 + 1.0857i

>> V1 = [real(Pole_23 - Pole_13), imag(Pole_23 - Pole_13)];
>> V2 = [real(Pole_12 - Pole_13), imag(Pole_12 - Pole_13)];

>> Angle = 2*(pi - acos(dot(V1,V2)/(norm(V1)*norm(V2))));
>> Rotation = [cos(Angle), -sin(Angle), real(Pole_13)*(1 -
cos(Angle)) + ...
imag(Pole_13)*sin(Angle); sin(Angle), cos(Angle), imag(Pole_13)*...
(1 - cos(Angle)) - real(Pole_13)*sin(Angle); 0, 0, 1]*...
[real(Pole_23), imag(Pole_23), 1]';

>> Pole_23p = Rotation(1) + i*Rotation(2)

Pole_23p =

   0.4802 + 1.1760i

>>
```

FIGURE E.5.33 Example 5.12 pole $\mathbf{p}_{12}$, pole $\mathbf{p}_{13}$, pole $\mathbf{p}_{23}$, and pole $\mathbf{p}'_{23}$ calculation procedure in the MATLAB command window.

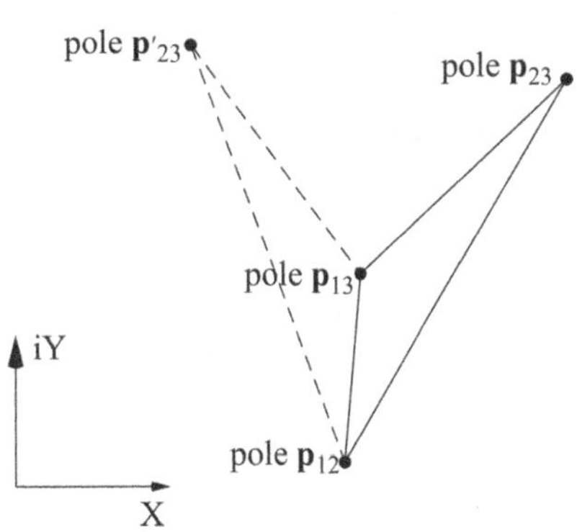

FIGURE E.5.34 Calculated pole triangle with an image pole.

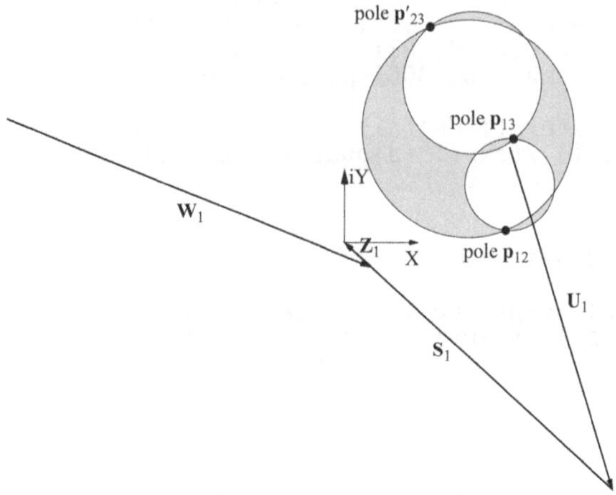

FIGURE E.5.35 Three-circle diagram with $\mathbf{U}_1 - \mathbf{S}_1$ and $\mathbf{W}_1 - \mathbf{Z}_1$ dyads from Example 5.1.

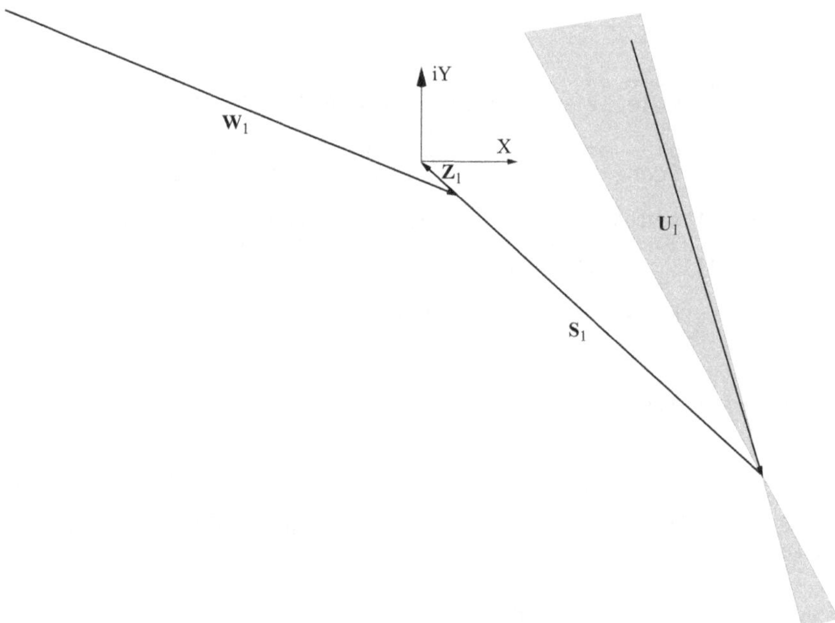

FIGURE E.5.36 Wedge-shaped region with calculated dyads with $\mathbf{U}_1 - \mathbf{S}_1$ and $\mathbf{W}_1 - \mathbf{Z}_1$ dyads from Example 5.1.

5.8 SUMMARY

While in kinematic analysis, mechanism link dimensions are known and the mechanism motion is calculated, in *kinematic synthesis*, mechanism motion is known and the corresponding link dimensions are calculated. Kinematic synthesis (also called dimensional synthesis) includes three subcategories: *motion generation, path generation,* and *function generation*. In motion generation, mechanism link positions are prescribed (specifically coupler link positions for the planar four-bar mechanism). In path generation, link path points are prescribed, while crank and follower displacement angles are prescribed in planar four-bar function generation. Prescribed coupler positions are commonly called *precision positions*.

Branch defects and *order defects* are inherent in planar four-bar motion generation. A mechanism having a branch defect cannot achieve its precision positions in a single-assembly configuration (called a *branch*). A mechanism having an order defect cannot achieve its precision positions in the intended order. *Prescribed timing*, *Filemon's construction*, and *Waldron's construction* are commonly employed methods to eliminate order and branch defects.

Linear simultaneous equation systems for planar four-bar, geared five-bar, and Stephenson III motion generation are formulated using vector loop closure for the starting and displaced positions of a mechanism dyad (and triad for the geared five-bar mechanism). These equations are applicable for three, four, and five precision positions. Cramer's rule is applicable in three-position synthesis, while solution algorithms for four- and five-position synthesis have been codified in the Appendix D and E MATLAB files, respectively.

A locus of fixed and moving pivots (called *Burmester center and circle points*) can be produced given four precision positions. These loci confirm the existence of an infinite number of planar dyad solutions for four-position synthesis. Center and circle point loci can also be used to produce four-bar mechanism solutions for five-position synthesis by dividing it into three four-position problems and locating common intersection points or overlapping regions of their Burmester center and circle points.

PROBLEMS

1. Synthesize a branch defect-free planar four-bar mechanism for the three hatch positions in Figure P.5.1. Branch defect elimination can be verified through *Filemon's Construction*, *Waldron's Construction*, or a displacement analysis (using the Appendix B.1 or C.1 files).

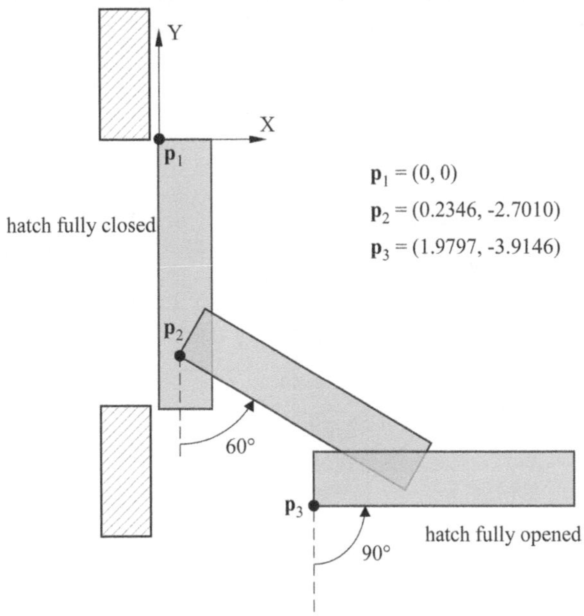

FIGURE P.5.1 Three hatch positions.

2. Synthesize a branch defect-free planar four-bar mechanism for the three load–unload bucket positions in Figure P.5.2. Branch defect elimination can be verified through *Filemon's Construction*, *Waldron's Construction*, or a displacement analysis (using the Appendix B.1 or C.1 files).

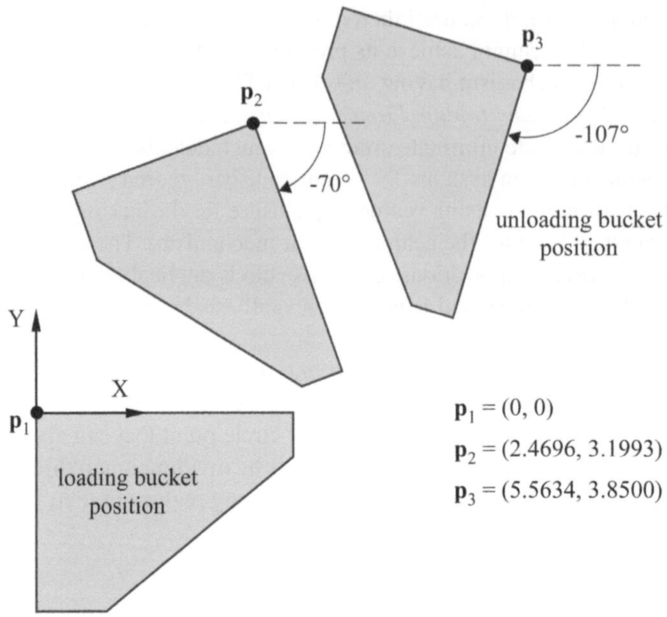

FIGURE P.5.2 Three load–unload bucket positions.

3. Synthesize a branch defect-free planar four-bar mechanism for the three stamping tool positions in Figure P.5.3. Branch defect elimination can be verified through *Filemon's Construction*, *Waldron's Construction,* or a displacement analysis (using the Appendix B.1 or C.1 files).

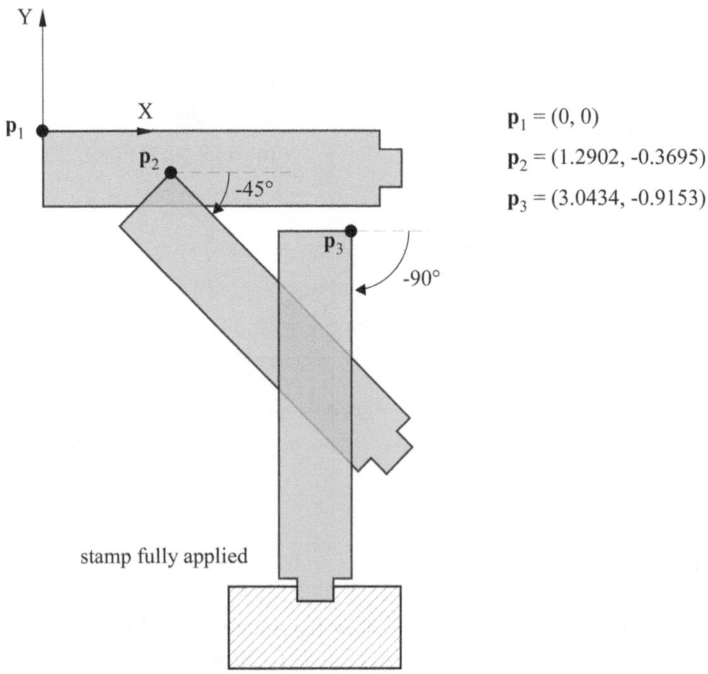

FIGURE P.5.3 Three stamping tool positions.

4. Synthesize a branch defect-free planar four-bar mechanism for the three folding wing positions in Figure P.5.4. Branch defect elimination can be verified through *Filemon's Construction*, *Waldron's Construction,* or a displacement analysis (using the Appendix B.1 or C.1 files).

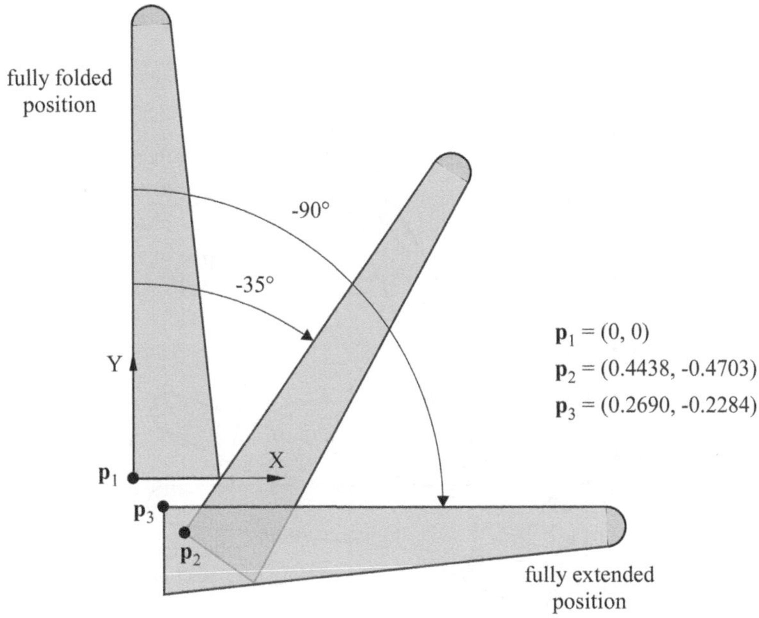

FIGURE P.5.4 Three folding wing positions.

5. Synthesize a branch defect-free planar four-bar mechanism for the three lower blade positions in Figure P.5.5. Branch defect elimination can be verified through *Filemon's Construction*, *Waldron's Construction,* or a displacement analysis (using the Appendix B.1 or C.1 files).

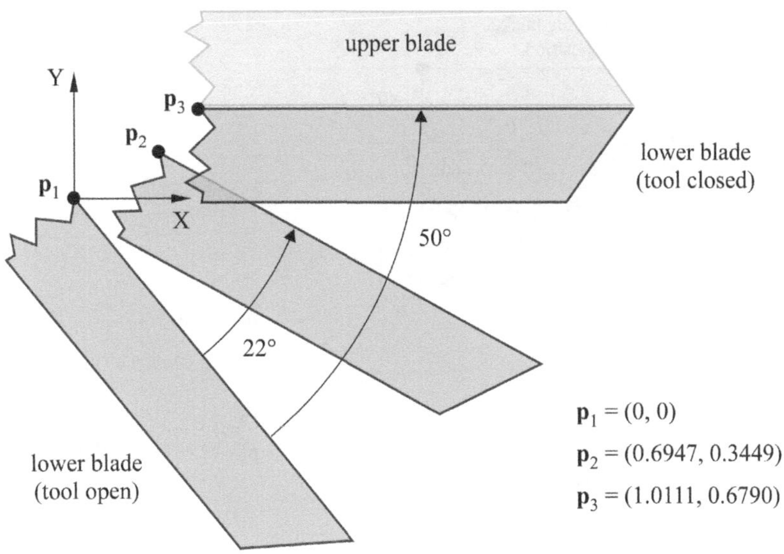

FIGURE P.5.5 Three lower cutting blade positions.

6. Synthesize an order- and branch defect-free planar four-bar mechanism (for the first mechanism branch) for the four folding wing positions in Figure P.5.6. Branch defect elimination can be verified through *Filemon's Construction* or a displacement analysis (using the Appendix B.1 or C.1 files).

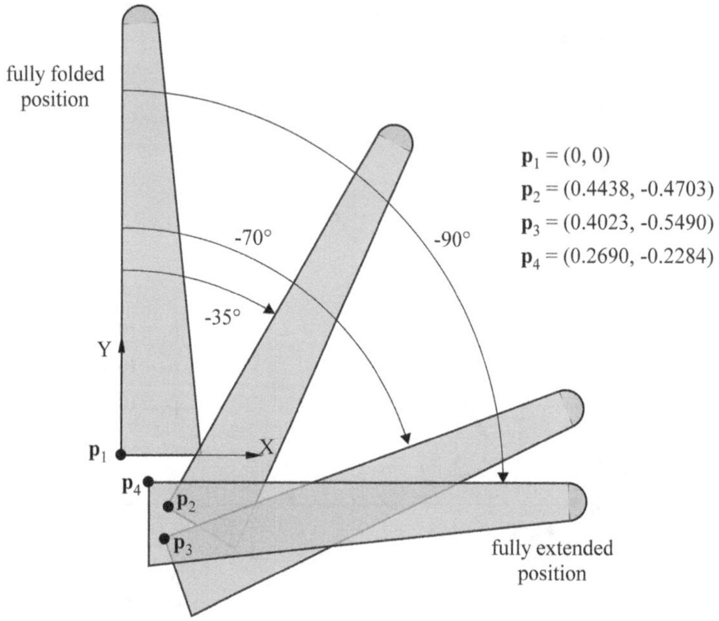

FIGURE P.5.6 Four folding wing positions.

7. Synthesize an order- and branch defect-free planar four-bar mechanism (for the first mechanism branch) for the four loading–unloading bucket positions in Figure P.5.7. Branch defect elimination can be verified through *Filemon's Construction* or a displacement analysis (using the Appendix B.1 or C.1 files).

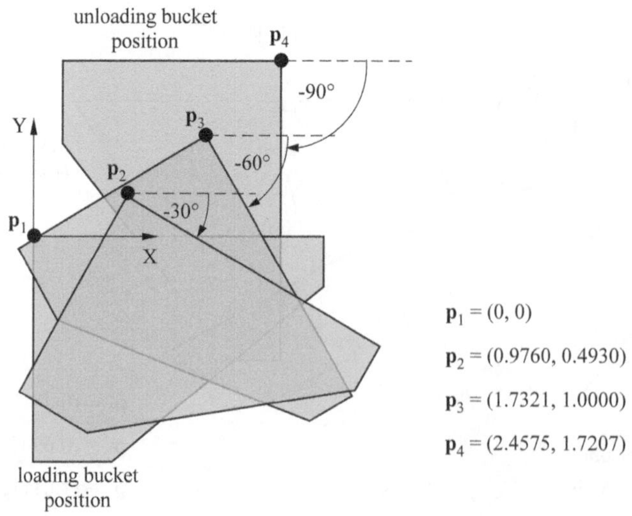

FIGURE P.5.7 Four loading–unloading bucket positions.

8. Synthesize an order- and branch defect-free planar four-bar mechanism (for the first mechanism branch) for the four hatch positions in Figure P.5.8. Branch defect elimination can be verified through *Filemon's Construction* or a displacement analysis (using the Appendix B.1 or C.1 files).

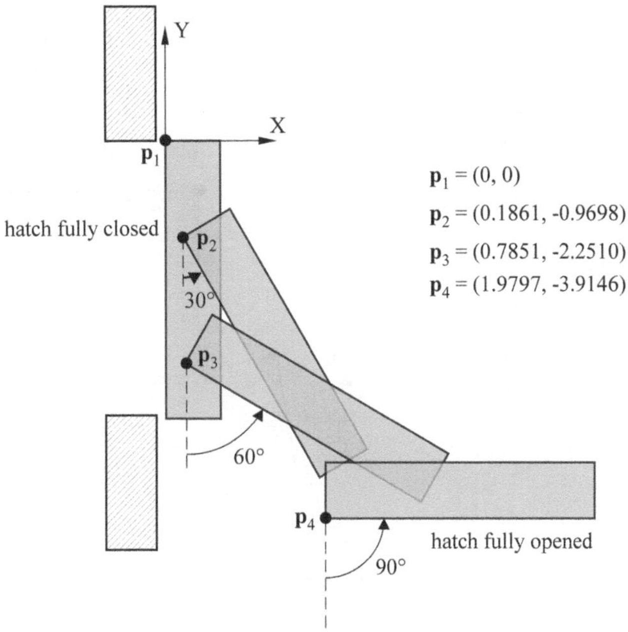

FIGURE P.5.8 Four hatch positions.

9. Synthesize an order- and branch defect-free planar four-bar mechanism (for the first mechanism branch) for the four stamping tool positions in Figure P.5.9. Branch defect elimination can be verified through *Filemon's Construction* or a displacement analysis (using the Appendix B.1 or C.1 files).

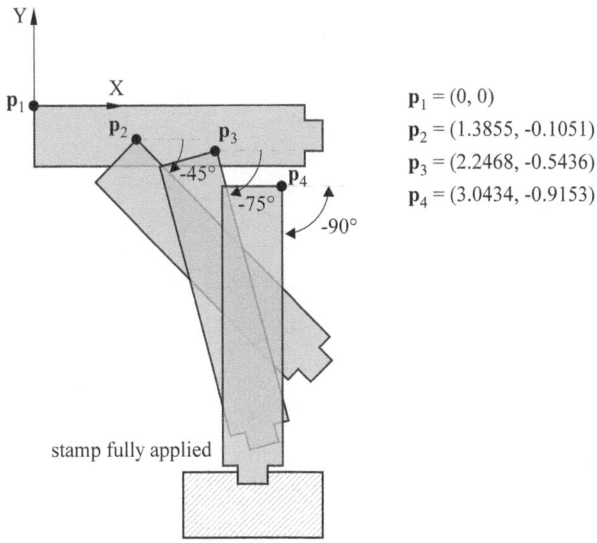

FIGURE P.5.9 Four stamping tool positions.

10. As noted in Appendix E, either zero, two, or four motion generator solutions are calculated in analytical planar four-bar motion generation for five precision positions. Determine the number of motion generator solutions that exist (for the first mechanism branch) for the five digger bucket positions in Figure P.5.10.

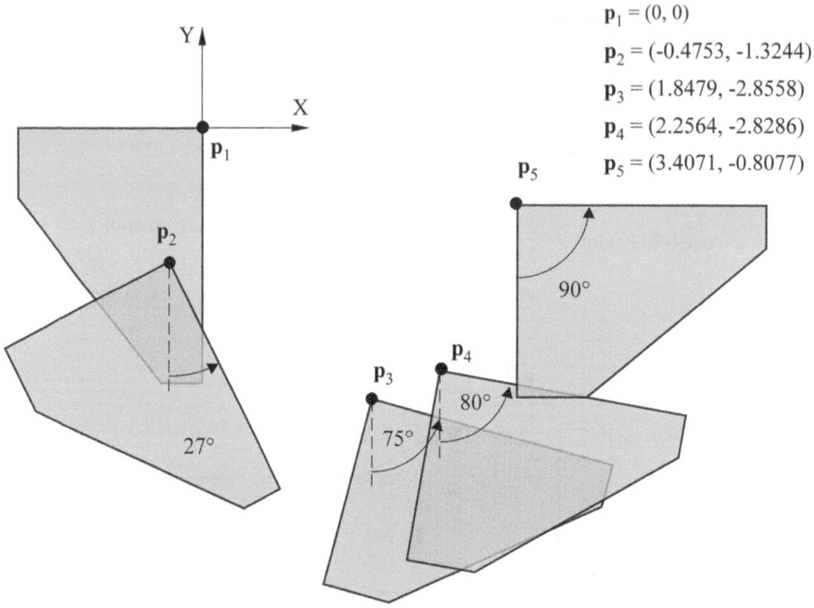

FIGURE P.5.10 Five digger bucket positions.

11. Synthesize an order- and branch defect-free planar four-bar mechanism (for the first mechanism branch) for the five gripper jaw positions in Figure P.5.11. Branch defect elimination can be verified through *Filemon's Construction* or a displacement analysis (using the Appendix B.1 or C.1 files).

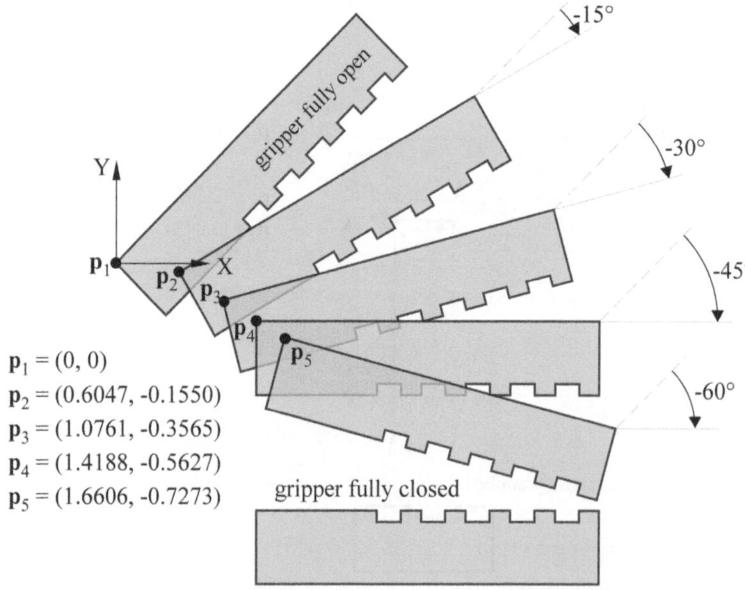

FIGURE P.5.11 Five gripper jaw positions.

12. Synthesize an order- and branch defect-free planar four-bar mechanism (for the first mechanism branch) for the five loading–unloading bucket positions in Figure P.5.12. Branch defect elimination can be verified through *Filemon's Construction* or a displacement analysis (using the Appendix B.1 or C.1 files).

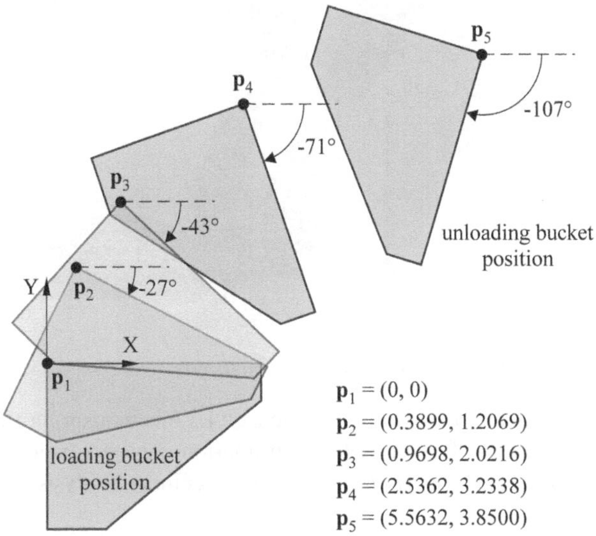

FIGURE P.5.12 Five loading–unloading bucket positions.

13. Calculate the components of $\mathbf{S}_1$ and $\mathbf{T}_1$ for the $\mathbf{U}_1 - \mathbf{S}_1 - \mathbf{T}_1$ triad of a geared five-bar mechanism to guide the landing gear through the three hatch positions in Problem 1. Use the $\mathbf{W}_1 - \mathbf{Z}_1$ and β_j solutions from Problem 1 along with $\gamma_j = 17\beta_j/20$, $U_1 = 3$, $\sigma = 35°$, $\nu_2 = 25°$, and $\nu_3 = 40°$.
14. Calculate the components of $\mathbf{S}_1$ and $\mathbf{T}_1$ for the $\mathbf{U}_1 - \mathbf{S}_1 - \mathbf{T}_1$ triad of a geared five-bar mechanism to guide the landing gear through the three load–unload bucket positions in Problem 2. Use the $\mathbf{W}_1 - \mathbf{Z}_1$ and β_j solutions from Problem 2 along with $\gamma_j = -16\beta_j/20$, $U_1 = 1$, $\sigma = 0°$, $\nu_2 = 10°$, and $\nu_3 = 25°$.
15. Calculate the components of $\mathbf{S}_1$ and $\mathbf{T}_1$ for the $\mathbf{U}_1 - \mathbf{S}_1 - \mathbf{T}_1$ triad of a geared five-bar mechanism to guide the landing gear through the three load–unload bucket positions in Problem 3. Use the $\mathbf{W}_1 - \mathbf{Z}_1$ and β_j solutions from Problem 3 along with $\gamma_j = 13\beta_j/20$, $U_1 = 2.5$, $\sigma = 45°$, $\nu_2 = 13°$, and $\nu_3 = 26°$.
16. Calculate the components of $\mathbf{S}_1$ and $\mathbf{T}_1$ for the $\mathbf{U}_1 - \mathbf{S}_1 - \mathbf{T}_1$ triad of a geared five-bar mechanism to guide the landing gear through the three folding wing positions in Problem 4. Use the $\mathbf{W}_1 - \mathbf{Z}_1$ and β_j solutions from Problem 4 along with $\gamma_j = -11\beta_j/20$, $U_1 = 1.25$, $\sigma = 15°$, $\nu_2 = 10°$, and $\nu_3 = 20°$.
17. Calculate the components of $\mathbf{S}_1$ and $\mathbf{T}_1$ for the $\mathbf{U}_1 - \mathbf{S}_1 - \mathbf{T}_1$ triad of a geared five-bar mechanism to guide the landing gear through the three folding wing positions in Problem 5. Use the $\mathbf{W}_1 - \mathbf{Z}_1$ and β_j solutions from Problem 5 along with $\gamma_j = \beta_j/2$, $U_1 = 1.75$, $\sigma = 90°$, $\nu_2 = 7°$, and $\nu_3 = 20°$.
18. Synthesize a branch defect-free planar Stephenson III mechanism for the three adjustable seat positions in Figure P.5.13. Branch defect elimination can be verified through *Filemon's Construction*, *Waldron's Construction,* or a displacement analysis (using the Appendix B.4 or C.4 files).

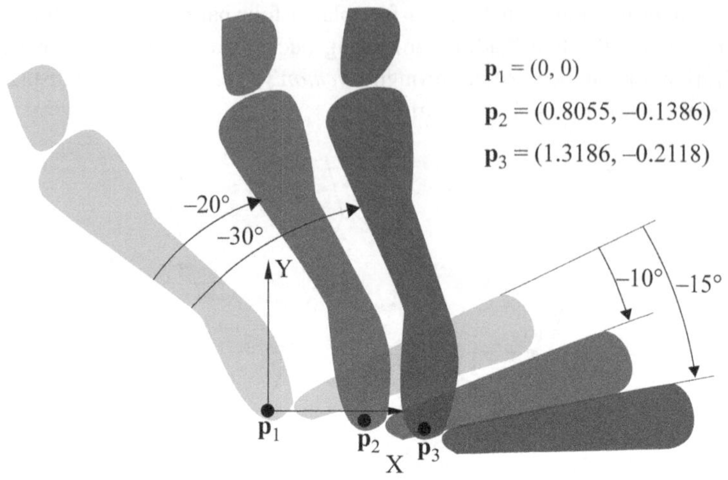

FIGURE P.5.13 Three adjustable seat positions.

19. Synthesize a branch defect-free planar Stephenson III mechanism for the three digger arm positions in Figure P.5.14. Branch defect elimination can be verified through *Filemon's Construction*, *Waldron's Construction,* or a displacement analysis (using the Appendix B.4 or C.4 files).

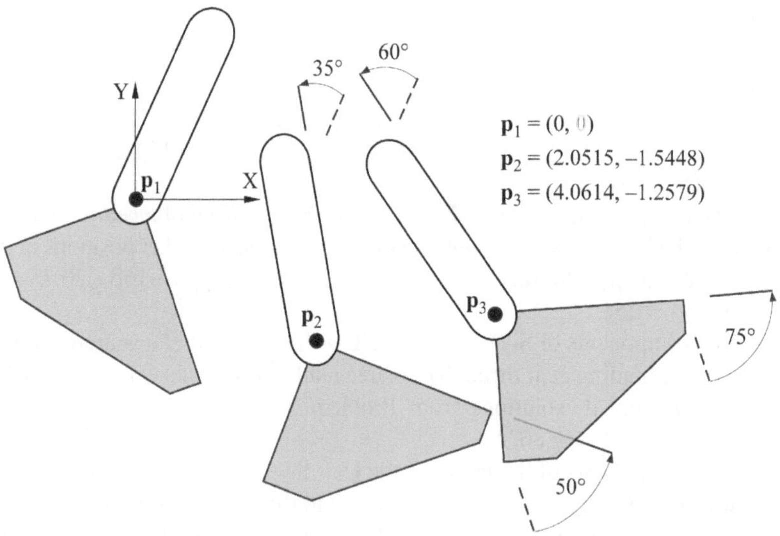

FIGURE P.5.14 Three digger arm positions.

20. Synthesize a branch defect-free Stephenson III mechanism for the three cutting tool positions in Figure P.5.15. Branch defect elimination can be verified through *Filemon's Construction*, *Waldron's Construction,* or a displacement analysis (using the Appendix B.4 or C.4 files).

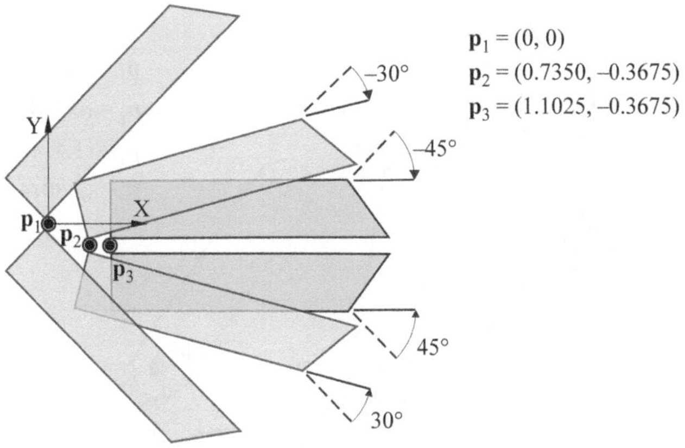

FIGURE P.5.15 Three cutting tool positions.

21. Repeat Problem 5.18 with $\mathbf{p}_2 = (2.9, -1.45)$ and $\mathbf{p}_3 = (4.35, -0.3625)$. Branch defect elimination can be verified through *Filemon's Construction*, *Waldron's Construction,* or a displacement analysis (using the Appendix B.4 or C.4 files).
22. Synthesize an order- and branch defect-free Stephenson III mechanism (for the first mechanism branch) for the four adjustable seat positions in Figure P.5.16. Branch defect elimination can be verified through *Filemon's Construction* or a displacement analysis (using the Appendix B.4 or C.4 files).

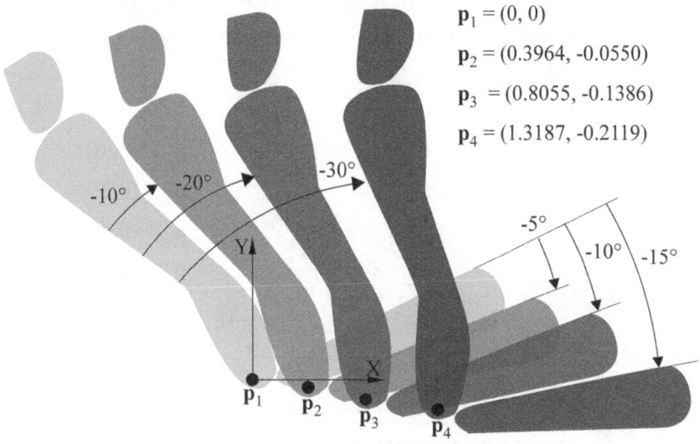

FIGURE P.5.16 Four adjustable seat positions.

23. Synthesize an order- and branch defect-free Stephenson III mechanism (for the first mechanism branch) for the four digger arm positions in Figure P.5.17. Branch defect elimination can be verified through *Filemon's Construction* or a displacement analysis (using the Appendix B.4 or C.4 files).

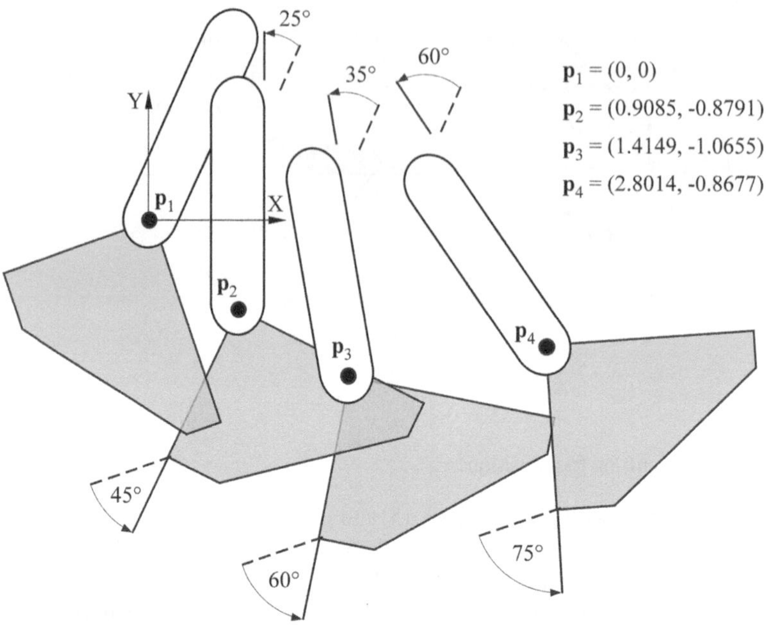

FIGURE P.5.17 Four digger arm positions.

24. Synthesize an order- and branch defect-free Stephenson III mechanism (for the first mechanism branch) for the four cutting tool positions in Figure P.5.18. Branch defect elimination can be verified through *Filemon's Construction* or a displacement analysis (using the Appendix B.4 or C.4 files).

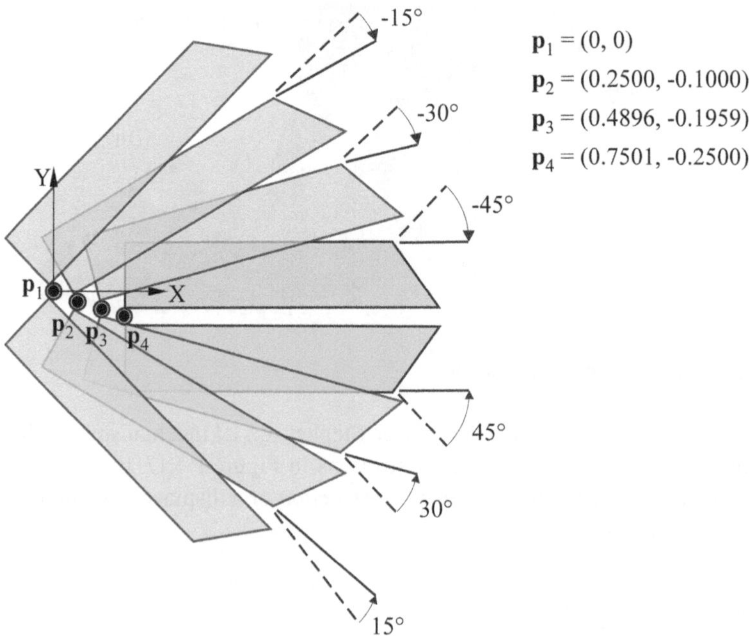

FIGURE P.5.18 Four cutting tool positions.

25. Synthesize an order- and branch defect-free Stephenson III mechanism (for the first mechanism branch) for the five adjustable seat precision positions in Figure P.5.1. Branch defect elimination can be verified through *Filemon's Construction* or a displacement analysis (using the Appendix B.4 or C.4 files) (Table P.5.1).

TABLE P.5.1
Adjustable Seat Precision Position Parameters

Precision Position	$\mathbf{p}_j$	α_j (deg)	α_j^* (deg)
1	0.1161, 2.2930		
2	0.5125, 2.2380	−10	−5
3	0.7956, 2.1814	−17	−8.5
4	1.0534, 2.1267	−23	−11.5
5	1.4349, 2.0808	−30	−15

NOTES

1 Another basic description of kinematic synthesis is *the design of a mechanism to produce a desired output motion for a given input motion.*
2 Mechanism links are also called "rigid-bodies" because mechanism links are generally assumed rigid (non-deforming) in kinematic synthesis.
3 In four-bar motion and path generation, the coupler link is the rigid-body for which positions and path points are prescribed.
4 The assembly configurations a-b-c-d and a-b-c*-d of the planar four-bar mechanism are called the *open* and *crossed* configurations, respectively.
5 The *branch defect* is distinct from the *circuit defect* (see Section 3.5 and Example 4.3), although the branch and circuit defects both require mechanism disassembly. With the circuit defect, the mechanism is reassembled in another position of the same assembly configuration (and not from one assembly configuration to another as with the branch defect).
6 Prescribed rigid-body positions are also called *precision positions.*
7 One circle point–center point pair corresponds to the $\mathbf{W}_1 - \mathbf{Z}_1$ dyad, and the $\mathbf{U}_1 - \mathbf{S}_1$ dyad corresponds to the other pair.
8 The Appendix D MATLAB file calculates circle and center point solutions for both mechanism branches.
9 While a dyad is a grounded two-link pair, a grounded three-link chain is called a *triad.*
10 The Appendix D MATLAB file calculates circle and center point solutions for both mechanism branches.
11 Filemon's construction ensures that there will be no change of branch if the motion generator solution is a Grashof crank-rocker or drag-link mechanism.
12 If the angle of the wedge-shaped region is 180° or greater, the region will fill the entire 2D space (and no solution will be available under to Filemon's construction).

REFERENCES

1. Sandor, G. N. and Erdman, A. G. 1984. *Advanced Mechanism Design: Analysis and Synthesis*, Volume 2. page 51. Englewood Cliffs: Prentice-Hall.
2 Martin, G. H. 1969. *Kinematics and Dynamics of Machines.* Page 319. New York: McGraw-Hill.
3 Erdman, A. G. 1993. *Modern Kinematics-Developments in the Last Forty Years.* chapters 4–5. New York: John Wiley & Sons.
4 Kimbrell, J. T. 1991. *Kinematics Analysis and Synthesis.* chapter 7. New York: McGraw-Hill.
5 Kimbrell, J. T. 1991. *Kinematics Analysis and Synthesis.* chapter 8. New York: McGraw-Hill.
6 Lee, W. and Russell, K. 2018. Developments in Quantitative Dimensional Synthesis (1970–Present): Four-Bar Motion Generation. *Inverse Problems in Science and Engineering.* 26: 133–148.

7 Lee, W. and Russell, K. 2018. Developments in Quantitative Dimensional Synthesis (1970–Present): Four-Bar Path and Function Generation. *Inverse Problems in Science and Engineering*. 26: 1280–1304.
8. Mallik, A. K., Amitabha, G. and Dittrich, G. 1994. *Kinematic Analysis and Synthesis of Mechanisms*. pages 306–308. Boca Raton: CRC Press.
9. Balli, S. S. and Chand, S. 2002. Defects in Link Mechanisms and Solution Rectification. *Mechanism and Machine Theory*. 37: 851–876.
10. Mallik, A. K., Amitabha, G. and Dittrich, G. 1994. *Kinematic Analysis and Synthesis of Mechanisms*. pages 307–308. Boca Raton: CRC Press.
11. Norton, R. L. 2008. *Design of Machinery*, 4th Edition. chapter 5. New York: McGraw-Hill.
12. Sandor, G. N. and Erdman, A. G. 1984. *Advanced Mechanism Design: Analysis and Synthesis*, Volume 2. pages 179–180. Englewood Cliffs: Prentice-Hall.
13. Sandor, G. N. and Erdman, A. G. 1984. *Advanced Mechanism Design: Analysis and Synthesis*, Volume 2. pages 180–183. Englewood Cliffs: Prentice-Hall.
14. Kimbrell, J. T. 1991. *Kinematics Analysis and Synthesis*. pages 189–192. New York: McGraw-Hill.
15. Sandor, G. N. and Erdman, A. G. 1984. *Advanced Mechanism Design: Analysis and Synthesis*, Volume 2. pages 199–204, 261–263. Englewood Cliffs: Prentice-Hall.
16. McCarthy, M. J. 2000. *Geometric Design of Linkages*. pages 110–111. New York: Springer.
17. Filemon, E. 1972. Useful Ranges of Centerpoint Curves for Design of Crank-and-Rocker Linkages. *Mechanism and Machine Theory*. 7: 47–53.
18. McCarthy, M. J. 2000. *Geometric Design of Linkages*. pages 111–112. New York: Springer.
19. Waldron, K. L. 1976. Elimination of the Branch Problem in Graphical Burmester Mechanism Synthesis for Four Finitely Separated Positions. *Transactions of the ASME, Journal of Engineering for Industry*. 98: 176–182.
20. Sandor, G. N. and Erdman, A. G. 1984. *Advanced Mechanism Design: Analysis and Synthesis*, Volume 2. page 76. Englewood Cliffs: Prentice-Hall.
21. McCarthy, M. J. 2000. *Geometric Design of Linkages*. pages 51–52. New York: Springer.
22. Sandor, G. N. and Erdman, A. G. 1984. *Advanced Mechanism Design: Analysis and Synthesis*, Volume 2. pages 198–199. Englewood Cliffs: Prentice-Hall.

6 Planar Four-Bar, Geared Five-Bar, and Stephenson III Path and Motion Generation

CONCEPT OVERVIEW

In this chapter, the reader will gain a central understanding regarding the following:

1. The distinctions between motion and path generation
2. Dwell motion in coupler curves
3. Constrained and unconstrained nonlinear optimization for planar four-bar, geared five-bar, and Stephenson III mechanism dyad dimensional synthesis for motion and path generation
4. Constrained nonlinear optimization for planar four-bar mechanism loop dimensional synthesis for motion and path generation

6.1 INTRODUCTION

Figure 6.1a illustrates a planar four-bar mechanism and the curve traced by coupler point $\mathbf{p}_1$. In quantitative path generation, the mechanism dimensions required to achieve or approximate prescribed coupler path points are calculated.[1] To further convey the concept of path generation, Figure 6.1b illustrates a *level-luffing crane*, a mechanism including a hook that remains at a constant level during motion. For effective operation, the extremity of the level-luffing crane should follow a horizontal path. The planar four-bar path generation equations presented in this chapter are useful for calculating the planar four-bar mechanism dimensions required to approximate precision points and, subsequently, the paths that include them.

There is a distinct difference between path generation and motion generation. In motion generation, *precision positions* are achieved or approximated, while *precision points* are achieved or approximated in path generation. Because coupler path points rather than coupler positions are

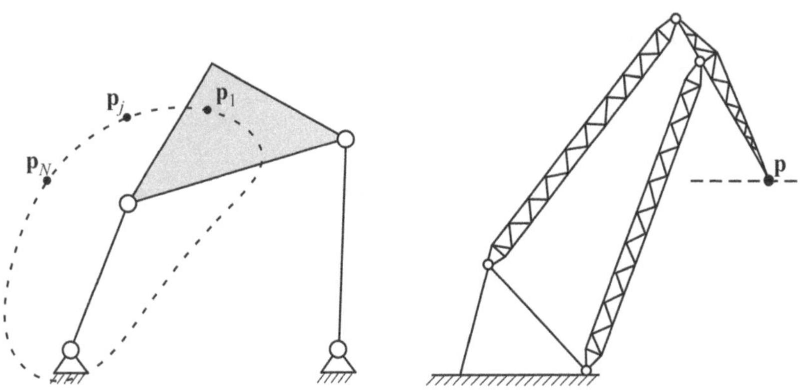

(a) Four-bar mechanism & coupler path points (b) Level-luffing crane mechanism

FIGURE 6.1 Path generation concepts.

DOI: 10.1201/9781003591375-6

prescribed in path generation, coupler displacement angles are not of particular concern. The planar four-bar motion generation equations presented in Chapter 5 can be directly applied for path generation for three, four, and five precision points since the mechanisms calculated from these equations achieve prescribed coupler points in addition to prescribed coupler displacement angles. Because coupler displacement angles are required in the equations mentioned in Chapter 5, the user is free to specify the coupler displacement angles arbitrarily. Path generation beyond five precision points, however, requires the use of an algorithm for nonlinear optimization problems for solution calculation. One reason for this is because no analytical solution method is available for Equations (5.8) and (5.9) beyond five precision positions. Another reason is because the inclusion of the coupler displacement angles among the unknown mechanism variables to be calculated (rather than being specified variables in motion generation) contributes to the non-linearity of Equations (5.8) and (5.9). In this chapter, path generation equations where coupler displacement angles are included among the unknowns to be calculated are presented.

6.2 COUPLER CURVES AND DWELL MOTION

Unlike the crank and follower links of the planar four-bar mechanism which undergo pure rotation, the coupler link of this mechanism undergoes complex motion: a combination of simultaneous rotation and translation. The complex motion of the coupler link enables it to trace curves of higher orders than the crank and follower links. Planar four-bar mechanisms can produce coupler curves of up to order six [1].[2]

Although the specific features of the coupler curve desired depend on the particular mechanical design application, there are certain coupler curve features that are more common over a range of design applications. *Straight lines* and *circular arcs* are among the most basic but widely used curve features in design applications. Figure 6.2 includes the *Roberts Straight-Line* linkage as well as a linkage that traces a circular arc [2].[3]

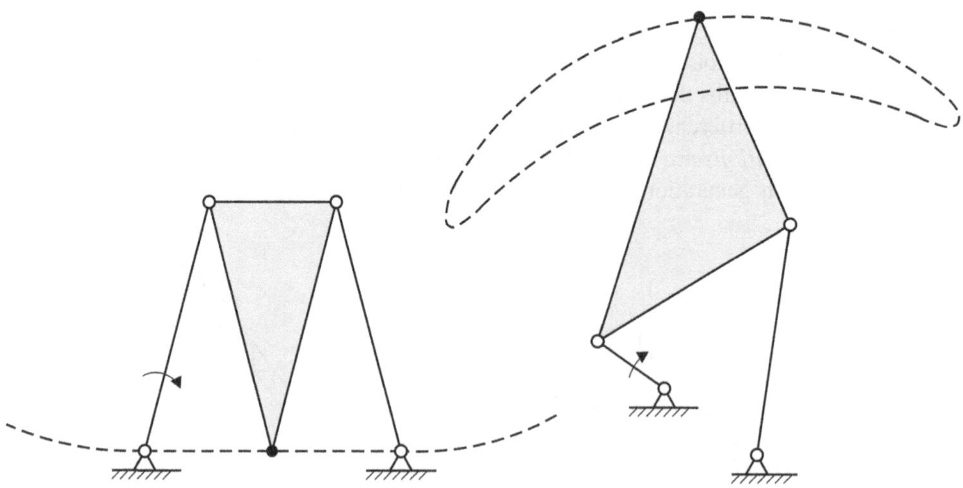

(a) Robert s approximate straight-line mechanism (b) Circular arc mechanism

FIGURE 6.2 Straight line and circular arc tracing four-bar mechanisms.

The shape of a coupler curve is determined by the mechanism link length proportions and the specific coupler point location from which the curve is traced. As a result, a planar four-bar linkage can produce an infinite variety of distinct coupler curves. A *Coupler Curve Atlas* includes a variety of coupler curves and their corresponding four-bar mechanism link proportions [3]. Coupler curve atlases can provide designers with judicious mechanism dimension starting points for design applications. Figure 6.3 illustrates several planar four-bar coupler curve variations with coupler point locations (Figure 6.3a) and link length proportions (Figure 6.3b).

One coupler curve feature of interest is the *dwell region* [4]. This region is distinct because it has *constant curvature*—a curve of constant radius. A dwell region could either be a circular arc or a straight-line segment on a coupler curve.[4] The four-bar mechanism in Figure 6.4 traces a curve having both dwell region types. Dwell regions are often of interest to designers because as they have a constant curvature, they produce no radial displacement. This zero radial output motion for

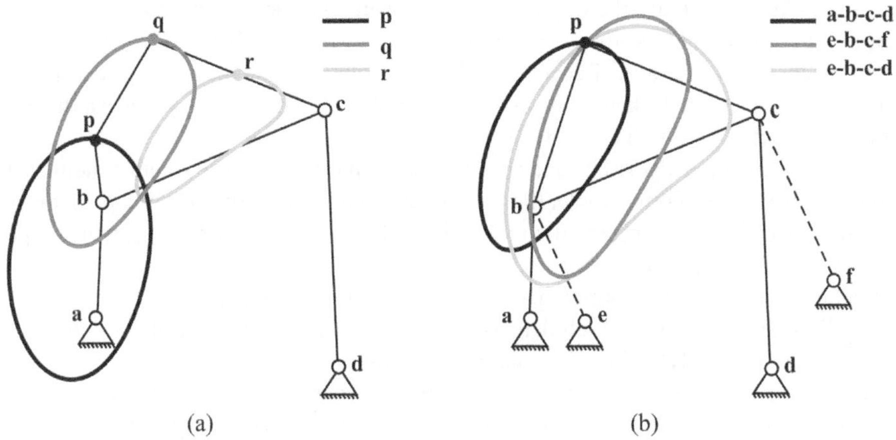

FIGURE 6.3 Coupler curve variation with (a) coupler point location and (b) link length ratio.

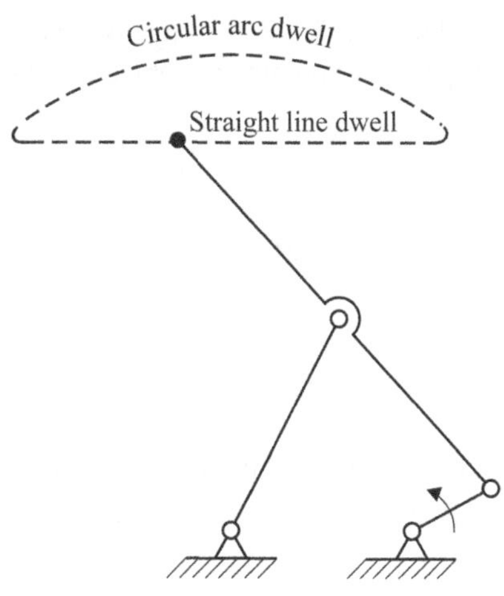

FIGURE 6.4 Planar four-bar coupler curve dell regions.

a nonzero rotational input motion (supplied by the crank link) is called *dwell motion* and is particularly useful in many design applications. A common example of dwell motion is observed in cam-follower systems [5–7]. Although dwell motion is commonly produced through cam-follower systems, linkage-based mechanism systems offer a design alternative for achieving dwell motion.

6.3 PLANAR FOUR-BAR PATH AND MOTION GENERATION

As shown in Table 5.1, the user can specify a maximum of five precision positions using Equations (5.8) and (5.9). There will be occasions however when five precision positions are not enough to adequately describe the desired coupler position sequence. As noted in Section 6.1, beyond five precision positions, Equations (5.8) and (5.9) require the use of an algorithm for nonlinear optimization problems to calculate the unknown mechanism dyad variables. Using analytical dimensional synthesis methods, mechanism solutions are calculated algebraically, and prescribed coupler motion data are achieved *precisely* by the calculated mechanism solutions. Using nonlinear optimization algorithms for dimensional synthesis, prescribed coupler motion data are *approximated* by the calculated mechanism solutions.[5] Because mechanism solutions calculated via the latter approach approximate the prescribed link motion, *structural error*—finite differences between the prescribed and achieved link motion—will exist [8].

Nonlinear optimization algorithms may be more favorable as analytical solution methods become more complex or limited. For example, the analytical solution algorithm for five-position synthesis in Appendix E.1 includes a substantial computation schedule, and beyond five precision positions, no analytical solution algorithms exist.

Nonlinear optimization algorithms may also be more favorable because they can include constraints to eliminate order and branch defects. As explained in Section 5.2, these defects are commonly encountered in analytical motion and path generation.

The dimension synthesis models presented from this chapter through the rest of the book are either *unconstrained* or *constrained nonlinear optimization models*. The *Quasi-Newton* algorithm included in MATLAB was implemented in this work for unconstrained nonlinear optimization [9,10].[6] The *Interior-Point* algorithm included in MATLAB was implemented in this work for constrained nonlinear optimization [11].

An objective function is derived for the $\mathbf{W}-\mathbf{Z}$ dyad by taking the sum of the squares of Equation (5.6). For path generation with N precision points, this objective function is expressed as

$$f(\mathbf{X})=\sum_{j=2}^{N}\left\{\begin{array}{l}\left(W_{1x}\left(\cos\beta_j-1\right)-W_{1y}\sin\beta_j+Z_{1x}\left(\cos\alpha_j-1\right)-Z_{1y}\sin\alpha_j-P_{j1}\cos\delta_j\right)^2\\+\left(W_{1y}\left(\cos\beta_j-1\right)+W_{1x}\sin\beta_j+Z_{1y}\left(\cos\alpha_j-1\right)+Z_{1x}\sin\alpha_j-P_{j1}\sin\delta_j\right)^2\end{array}\right\} \tag{6.1}$$

where $\mathbf{X}=(W_{1x},W_{1y},Z_{1x},Z_{1y},\alpha_2\ldots\alpha_4,\beta_2\ldots\beta_4)$ with four precision points and $\mathbf{X}=(W_{1x},W_{1y},Z_{1x},Z_{1y},\alpha_2\ldots\alpha_5,\beta_2\ldots\beta_5)$ with five precision points. With N precision points, $\mathbf{X}=(W_{1x},W_{1y},Z_{1x},Z_{1y},\alpha_2\ldots\alpha_N,\beta_2\ldots\beta_N)$.[7]

Likewise, an objective function is derived for the $\mathbf{U}-\mathbf{S}$ dyad by taking the sum of the squares of Equation (5.7). With N precision points, this objective function is expressed as

$$f(\mathbf{X})=\sum_{j=2}^{N}\left\{\begin{array}{l}\left(U_{1x}\left(\cos\gamma_j-1\right)-U_{1y}\sin\gamma_j+S_{1x}\left(\cos\alpha_j-1\right)-S_{1y}\sin\alpha_j-P_{j1}\cos\delta_j\right)^2\\+\left(U_{1y}\left(\cos\gamma_j-1\right)+U_{1x}\sin\gamma_j+S_{1y}\left(\cos\alpha_j-1\right)+S_{1x}\sin\alpha_j-P_{j1}\sin\delta_j\right)^2\end{array}\right\} \tag{6.2}$$

where $\mathbf{X} = (U_{1x}, U_{1y}, S_{1x}, S_{1y}, \gamma_2 \ldots \gamma_4)$ with four precision points and $\mathbf{X} = (U_{1x}, U_{1y}, S_{1x}, S_{1y}, \gamma_2 \ldots \gamma_5)$ with five precision points. With N precision points, $\mathbf{X} = (U_{1x}, U_{1y}, S_{1x}, S_{1y}, \gamma_2 \ldots \gamma_N)$. In path generation, the α_j values calculated from Equation (6.1)—the $\mathbf{W} - \mathbf{Z}$ dyad are used as prescribed values in Equation (6.2)—form the $\mathbf{U} - \mathbf{S}$ dyad.

As presented, Equations (6.1) and (6.2) are for path generation. By prescribing both the coupler points and coupler displacement angles, Equations (6.1) and (6.2) can be used for motion generation. Using Equation (6.1) for motion generation, where the dyad displacement angles β_j are calculated, $\mathbf{X} = (W_{1x}, W_{1y}, Z_{1x}, Z_{1y}, \beta_2 \ldots \beta_4)$ with four precision positions and $\mathbf{X} = (W_{1x}, W_{1y}, Z_{1x}, Z_{1y}, \beta_2 \ldots \beta_5)$ with five precision positions. With N precision positions, $\mathbf{X} = (W_{1x}, W_{1y}, Z_{1x}, Z_{1y}, \beta_2 \ldots \beta_N)$.

To eliminate order defects, Inequality (6.3) should accompany Equation (6.1) as an inequality constraint for the driver dyad.

$$\begin{cases} \beta_j > \beta_{j-1} \\ \beta_N < 2\pi \end{cases} \quad j = 2,3,\ldots \tag{6.3}$$

As expressed, Inequality (6.3) constrains the crank-to-counter-clockwise rotation. To constrain the crank-to-clockwise rotation, the inequality $\beta_j < \beta_{j-1}$ should be used instead of $\beta_j > \beta_{j-1}$.

Equation (6.1) used in combination with Inequality (6.3) produces a *constrained nonlinear optimization model with inequality constraints.* Equation (6.2) produces an *unconstrained nonlinear optimization model.* Appendices F.1 and F.2 include MATLAB file user instructions for these optimization models for planar four-bar motion generation and path generation, respectively.

Example 6.1

Problem Statement: Using the Appendix F.1 MATLAB files, synthesize a planar four-bar mechanism to approximate the landing gear precision positions in Table E.6.1.

Known Information: Table E.6.1 and Appendix F.1 MATLAB files

Solution Approach: Figure E.6.1 includes the input in bold type in the Appendix F.1 MATLAB file *W_Z_interior_point.m*. Figure E.6.2 includes the output from the same file. Figure E.6.3 includes the input in bold type in the Appendix F.1 MATLAB file *U_S_quasi_newton.m*. Figure E.6.4 includes the output from the same file. Figure E.6.5 illustrates the resulting planar four-bar mechanism. A kinematic analysis of this solution is shown in Example 6.9.

TABLE E.6.1
Landing Gear Precision Positions

Precision Position	$\mathbf{p}_j$	α_j(deg)
1	0, 0	
2	0.203, 0.415	–25
3	0.397, 0.848	–40
4	0.385, 1.182	–70
5	0.182, 1.460	–90

```
N = 5;

alpha2 = -25 *pi/180;
alpha3 = -40 *pi/180;
alpha4 = -70 *pi/180;
alpha5 = -90 *pi/180;

P1 = [0, 0];
P2 = [0.203, 0.415];
P3 = [0.397, 0.848];
P4 = [0.385, 1.182];
P5 = [0.182, 1.460];

W1x = -1;
W1y = -1;
Z1x = -1;
Z1y = -1;

beta2 = 15 *pi/180;
beta3 = 30 *pi/180;
beta4 = 45 *pi/180;
beta5 = 60 *pi/180;
```

FIGURE E.6.1 Input in bold type in the Appendix F.1 MATLAB file *W_Z_interior_point.m* for Example 6.1.

```
Iterations: 42
Func Evals: 69
Algorithm: interior-point
exit flag = 2
error = ( 2.290903e-03 )
W1x = ( 0.648748 )
W1y = ( -0.561522 )
Z1x = ( -0.283731 )
Z1y = ( 0.136138 )
beta2 = ( 22.590697 )
beta3 = ( 49.544840 )
beta4 = ( 72.536784 )
beta5 = ( 102.190603 )
Elapsed time is 0.701511 seconds.
Solution calculation is complete.
>>
```

FIGURE E.6.2 Output from the Appendix F.1 MATLAB file *W_Z_interior_point.m* for Example 6.1.

```
N = 5;

alpha2 = -25 *pi/180;
alpha3 = -40 *pi/180;
alpha4 = -70 *pi/180;
alpha5 = -90 *pi/180;

P1 = [0, 0];
P2 = [0.203, 0.415];
P3 = [0.397, 0.848];
P4 = [0.385, 1.182];
P5 = [0.182, 1.460];

U1x = 1;
U1y = -1;
S1x = -1;
S1y = -1;

gamma2 = -10 *pi/180;
gamma3 = -40 *pi/180;
gamma4 = -60 *pi/180;
gamma5 = -70 *pi/180;
```

FIGURE E.6.3 Input in bold type in the Appendix F.1 MATLAB file *U_S_quasi_newton.m* for Example 6.1.

```
Iterations: 100
Func Evals: 114
Algorithm: quasi-newton
exit flag = 5
error = ( 7.259571e-33 )
U1x = ( -0.178559 )
U1y = ( -0.450057 )
S1x = ( -0.839784 )
S1y = ( -0.582417 )
gamma2 = ( 44.989920 )
gamma3 = ( 76.582544 )
gamma4 = ( 46.190653 )
gamma5 = ( -9.990035 )
Elapsed time is 0.187881 seconds.
Solution calculation is complete.
>>
```

FIGURE E.6.4 Output from the Appendix F.1 MATLAB file *U_S_quasi_newton.m* for Example 6.1.

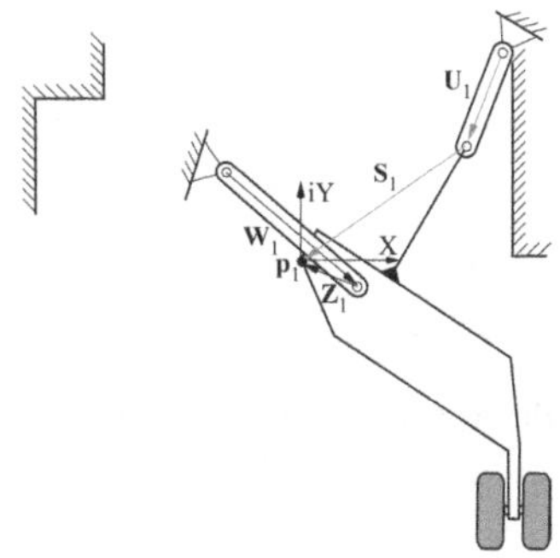

FIGURE E.6.5 Synthesized planar four-bar motion generator.

Example 6.2

Problem Statement: Using the Appendix F.2 MATLAB files, synthesize a planar four-bar mechanism to approximate the dwell path precision points in Table E.6.2.

Known Information: Table E.6.2 and Appendix F.2 MATLAB files

Solution Approach: Figure E.6.6 includes the input in bold type in the Appendix F.2 MATLAB file *W_Z_interior_point.m*. Figure E.6.7 includes the output from the same file.

Figure E.6.8 includes the input in bold type in the Appendix F.2 MATLAB file *U_S_quasi_newton.m*. Figure E.6.9 includes the output from the same file. Figure E.6.10 illustrates the resulting planar four-bar mechanism. A kinematic analysis of this solution is shown in Example 6.10.

TABLE E.6.2
Dwell Path Precision Points

Precision Point	$\mathbf{p}_j$
1	0, 0
2	–1, 0
3	–2, 0
4	–3, 0
5	–4, 0
6	–5, 0
7	–6, 0
8	–7, 0

```
N = 8;

P1 = [0, 0];
P2 = [-1, 0];
P3 = [-2, 0];
P4 = [-3, 0];
P5 = [-4, 0];
P6 = [-5, 0];
P7 = [-6, 0];
P8 = [-7, 0];

W1x = 1;
W1y = 1;
Z1x = -10;
Z1y = -10;

alpha2 = 10 *pi/180;
alpha3 = 10 *pi/180;
alpha4 = 10 *pi/180;
alpha5 = 10 *pi/180;
alpha6 = 10 *pi/180;
alpha7 = 10 *pi/180;
alpha8 = 10 *pi/180;

beta2 = 10 *pi/180;
beta3 = 20 *pi/180;
beta4 = 30 *pi/180;
beta5 = 40 *pi/180;
beta6 = 50 *pi/180;
beta7 = 60 *pi/180;
beta8 = 70 *pi/180;
```

FIGURE E.6.6 Input in bold type in the Appendix F.2 MATLAB file *W_Z_interior_point.m* for Example 6.2.

```
Iterations: 351
Func Evals: 1686
Algorithm: interior-point
exit flag = 2
error = ( 5.026590e-20 )
W1x = ( 1.579630 )
W1y = ( 1.581584 )
Z1x = ( -12.386988 )
Z1y = ( -19.454246 )
alpha2 = ( 2.945129 )
alpha3 = ( 1.848532 )
alpha4 = ( 0.282210 )
alpha5 = ( -1.629721 )
alpha6 = ( -3.902549 )
alpha7 = ( -6.699715 )
alpha8 = ( -13.716011 )
beta2 = ( 56.219638 )
beta3 = ( 73.108150 )
beta4 = ( 87.679965 )
beta5 = ( 101.842678 )
beta6 = ( 116.892642 )
beta7 = ( 134.872387 )
beta8 = ( 193.678608 )
Elapsed time is 1.220771 seconds.
Solution calculation is complete.
>>
```

FIGURE E.6.7 Output from the Appendix F.2 MATLAB file *W_Z_interior_point.m* for Example 6.2.

```
N = 8;

alpha2 = 2.945129 *pi/180;
alpha3 = 1.848532 *pi/180;
alpha4 = 0.282210 *pi/180;
alpha5 = -1.629721 *pi/180;
alpha6 = -3.902549 *pi/180;
alpha7 = -6.699715 *pi/180;
alpha8 = -13.716011 *pi/180;

P1 = [0, 0];
P2 = [-1, 0];
P3 = [-2, 0];
P4 = [-3, 0];
P5 = [-4, 0];
P6 = [-5, 0];
P7 = [-6, 0];
P8 = [-7, 0];

U1x = 1;
U1y = 1;
S1x = -10;
S1y = -10;

gamma2 = -10 *pi/180;
gamma3 = -20 *pi/180;
gamma4 = -30 *pi/180;
gamma5 = -40 *pi/180;
gamma6 = -50 *pi/180;
gamma7 = -60 *pi/180;
gamma8 = -70 *pi/180;
```

FIGURE E.6.8 Input in bold type in the Appendix F.2 MATLAB file *U_S_quasi_newton.m* for Example 6.2.

```
Iterations: 145
Func Evals: 179
Algorithm: quasi-newton
exit flag = 5
error = ( 4.649542e-06 )
U1x = ( 1.582902 )
U1y = ( -9.010446 )
S1x = ( 1.943278 )
S1y = ( -20.536207 )
gamma2 = ( -12.906191 )
gamma3 = ( -16.734804 )
gamma4 = ( -19.517774 )
gamma5 = ( -21.516600 )
gamma6 = ( -22.684408 )
gamma7 = ( -22.644479 )
gamma8 = ( -13.048697 )
Elapsed time is 0.241147 seconds.
Solution calculation is complete.
>>
```

FIGURE E.6.9 Output from the Appendix F.2 MATLAB file *U_S_quasi_newton.m* for Example 6.2.

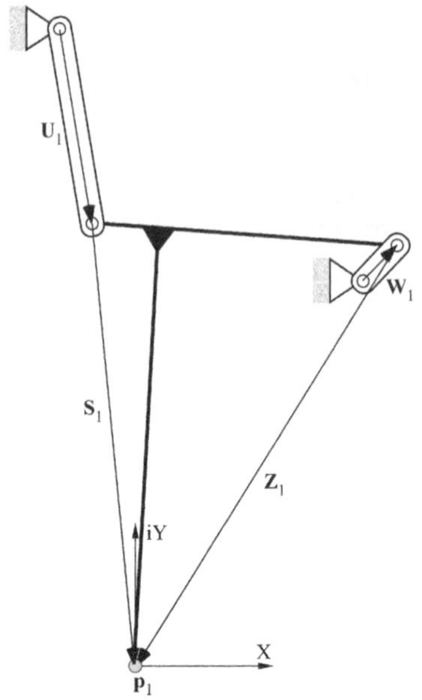

FIGURE E.6.10 Synthesized planar four-bar path generator.

6.4 GEARED FIVE-BAR MOTION AND PATH GENERATION

Equation (6.1) and Inequality (6.3) can be directly applied for the $\mathbf{W}-\mathbf{Z}$ dyad or the geared five-bar mechanism for motion and path generation, respectively. Like the four-bar mechanism, $\mathbf{X}=(W_{1x},W_{1y},Z_{1x},Z_{1y},\alpha_2\ldots\alpha_N,\beta_2\ldots\beta_N)$ for geared five-bar path generation with N precision points and $\mathbf{X}=(W_{1x},W_{1y},Z_{1x},Z_{1y},\beta_2\ldots\beta_N)$ for geared five-bar motion generation with N precision positions.

An objective function is derived for the $\mathbf{U}-\mathbf{S}-\mathbf{T}$ triad (see Figure 5.7) by taking the sum of the squares of Equation (5.13). With N precision points or precision positions, this objective function is expressed as

$$f(\mathbf{X})=\sum_{j=2}^{N}\left\{\begin{bmatrix}\begin{pmatrix}S_{1x}(\cos\nu_j-1)-S_{1y}\sin\nu_j+T_{1x}(\cos\alpha_j-1)-T_{1y}\sin\alpha_j+\\ U_{1x}(\cos\gamma_j-1)-U_{1y}\sin\gamma_j+P_{j1}\cos\delta_j\end{pmatrix}^2\\ +\begin{pmatrix}S_{1y}(\cos\nu_j-1)+S_{1x}\sin\nu_j+T_{1y}(\cos\alpha_j-1)+T_{1x}\sin\alpha_j+\\ U_{1y}(\cos\gamma_j-1)+U_{1x}\sin\gamma_j-P_{j1}\sin\delta_j\end{pmatrix}^2\end{bmatrix}\right\} \tag{6.4}$$

With N precision points or precision positions, $\mathbf{X}=(U_{1x},U_{1y},S_{1x},S_{1y},T_{1x},T_{1y},\nu_2\ldots\nu_N)$. In path generation, the α_j values calculated from Equation (6.1)—the $\mathbf{W}-\mathbf{Z}$ dyad are used as prescribed values in Equation (6.4)—form the $\mathbf{U}-\mathbf{S}-\mathbf{T}$ triad. The displacement angles $\gamma_2\ldots\gamma_N$ are specified according to the gear ratio used.

Like Equation (6.2), Equation (6.4) produces an *unconstrained nonlinear optimization model.* Appendices F.3 and F.4 include MATLAB file user instructions for these optimization models for geared five-bar motion generation and path generation, respectively.

Example 6.3

Problem Statement: Using the Appendix F.3 MATLAB files, synthesize a geared five-bar mechanism to approximate the landing gear precision positions given in Example 6.1. Use the $\boldsymbol{W}_1 - \boldsymbol{Z}_1$ dyad solution from Example 6.1 along with $\gamma_j = 2\beta_j/3$.

Known Information: Example 6.1 $\boldsymbol{W}_1 - \boldsymbol{Z}_1$ dyad solution and Appendix F.3 MATLAB files

Solution Approach: Figure E.6.11 includes the input in bold type in the Appendix F.3 MATLAB file *U_S_T_quasi_newton.m*. Figure E.6.12 includes the output from the same file. Figure E.6.13 illustrates the resulting geared five-bar mechanism. A kinematic analysis of this solution is presented in Example 6.11.

```
N = 5;

alpha2 = -25 *pi/180;
alpha3 = -40 *pi/180;
alpha4 = -70 *pi/180;
alpha5 = -90 *pi/180;

gamma2 = (2/3)*22.590697 *pi/180;
gamma3 = (2/3)*49.544840 *pi/180;
gamma4 = (2/3)*72.536784 *pi/180;
gamma5 = (2/3)*102.190603 *pi/180;

P1 = [0, 0];
P2 = [0.203, 0.415];
P3 = [0.397, 0.848];
P4 = [0.385, 1.182];
P5 = [0.182, 1.460];

U1x = -1;
U1y = -1;
S1x = -1;
S1y = -1;
T1x = -1;
T1y = -1;

nu2 = 10 *pi/180;
nu3 = 20 *pi/180;
nu4 = 30 *pi/180;
nu5 = 30 *pi/180;
```

FIGURE E.6.11 Input in bold type in the Appendix F.3 MATLAB file *U_S_T_quasi_newton.m* for Example 6.3.

```
Iterations: 67
Func Evals: 69
Algorithm: quasi-newton
exit flag = 5
error = ( 8.117893e-33 )
U1x = ( 0.657244 )
U1y = ( -0.546383 )
S1x = ( 0.200861 )
S1y = ( -0.613504 )
T1x = ( -0.383144 )
T1y = ( -0.153482 )
nu2 = ( 10.918639 )
nu3 = ( 21.909577 )
nu4 = ( 5.849295 )
nu5 = ( -12.925750 )
Elapsed time is 0.223029 seconds.
Solution calculation is complete.
>>
```

FIGURE E.6.12 Output from the Appendix F.3 MATLAB file *U_S_T_quasi_newton.m* for Example 6.3.

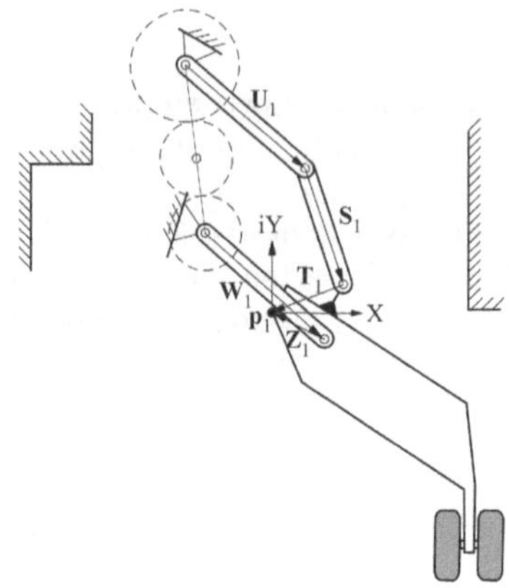

FIGURE E.6.13 Synthesized geared five-bar motion generator.

Example 6.4

Problem Statement: Using the Appendix F.4 MATLAB files, synthesize a geared five-bar mechanism to approximate the dwell path precision positions in Example 6.2. Use the $\mathbf{W}_1 - \mathbf{Z}_1$ dyad solution from Example 6.2 along with $\gamma_j = -4\beta_j/5$.

Known Information: Example 6.2 $\mathbf{W}_1 - \mathbf{Z}_1$ dyad solution and Appendix F.4 MATLAB files

Solution Approach: Figure E.6.14 includes the input in bold type in the Appendix F.4 MATLAB file *U_S_T_quasi_newton.m*. Figure E.6.15 includes the output from the same file. Figure E.6.16 illustrates the resulting geared five-bar mechanism. A kinematic analysis of this solution is presented in Example 6.12.

```
N = 8;

alpha2 = 2.945129 *pi/180;
alpha3 = 1.848532 *pi/180;
alpha4 = 0.282210 *pi/180;
alpha5 = -1.629721 *pi/180;
alpha6 = -3.902549 *pi/180;
alpha7 = -6.699715 *pi/180;
alpha8 = -13.716011 *pi/180;

gamma2 = -(4/5)*56.219638  *pi/180;
gamma3 = -(4/5)*73.108150 *pi/180;
gamma4 = -(4/5)*87.679965 *pi/180;
gamma5 = -(4/5)*101.842678 *pi/180;
gamma6 = -(4/5)*116.892642 *pi/180;
gamma7 = -(4/5)*134.872387 *pi/180;
gamma8 = -(4/5)*193.678608 *pi/180;

P1 = [0, 0];
P2 = [-1, 0];
P3 = [-2, 0];
P4 = [-3, 0];
P5 = [-4, 0];
P6 = [-5, 0];
P7 = [-6, 0];
P8 = [-7, 0];

U1x = -1;
U1y = 1;
S1x = 1;
S1y = 1;
T1x = 1;
T1y = 1;

nu2 = 10 *pi/180;
nu3 = 10 *pi/180;
nu4 = 10 *pi/180;
nu5 = 10 *pi/180;
nu6 = 10 *pi/180;
nu7 = 10 *pi/180;
nu8 = 10 *pi/180;
```

FIGURE E.6.14 Input in bold type in the Appendix F.4 MATLAB file *U_S_T_quasi_newton.m* for Example 6.4.

```
Iterations: 2935
Func Evals: 3700
Algorithm: quasi-newton
exit flag = 5
error = ( 1.748046e-08 )
U1x = ( -2.794764 )
U1y = ( -2.116332 )
S1x = ( -21.287198 )
S1y = ( 19.108909 )
T1x = ( -9.360478 )
T1y = ( -27.544693 )
nu2 = ( 5.554088 )
nu3 = ( 7.895161 )
nu4 = ( 9.984747 )
nu5 = ( 12.000753 )
nu6 = ( 14.043266 )
nu7 = ( 16.219317 )
nu8 = ( 19.543789 )
Elapsed time is 0.996637 seconds.
Solution calculation is complete.
>>
```

FIGURE E.6.15 Output from the Appendix F.4 MATLAB file *U_S_T_quasi_newton.m* for Example 6.4.

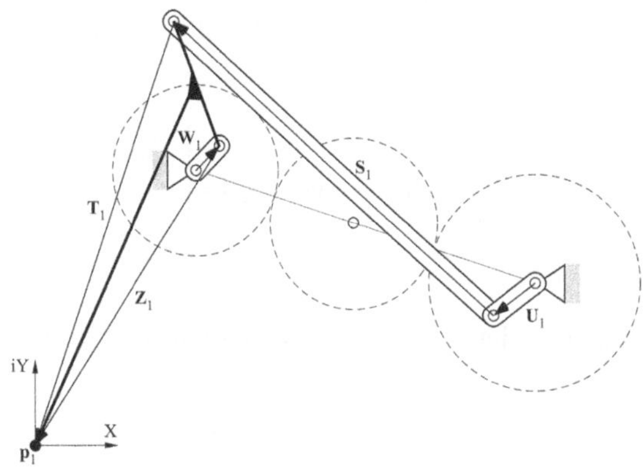

FIGURE E.6.16 Synthesized geared five-bar path generator.

6.5 ALGEBRAIC PLANAR FOUR-BAR DISPLACEMENT MODEL

The planar four-bar displacement model in Section 4.3.1 is a nonlinear simultaneous equation system and as a result is numerically solved in MATLAB using the algorithms given in Section 2.7.1. Suh and Radcliffe, however, introduced a displacement model for the planar four-bar mechanism that is solved algebraically [12].[8] This alternate kinematic model is based on Equation (6.5)—the constant-length equation for the follower link.

$$(\mathbf{b}-\mathbf{b}_0)^T(\mathbf{b}-\mathbf{b}_0)-(\mathbf{b}_1-\mathbf{b}_0)^T(\mathbf{b}_1-\mathbf{b}_0)=0 \tag{6.5}$$

The rotation of the crank link (link $\mathbf{a}_0$ - $\mathbf{a}_1$ Figure 6.5) is given by

$$\mathbf{a}=[R_\theta](\mathbf{a}_1-\mathbf{a}_0)+\mathbf{a}_0 \tag{6.6}$$

The global displacement of the coupler link (and follower link) pivot $\mathbf{b}_1$ is expressed as

$$\mathbf{b} = [R_\alpha](\mathbf{b}_1' - \mathbf{a}) + \mathbf{a} \quad (6.7)$$

where

$$\mathbf{b}_1' = [R_\theta](\mathbf{b}_1 - \mathbf{a}_0) + \mathbf{a}_0 \quad (6.8)$$

Matrices $[R_\theta]$ and $[R_\alpha]$ are planar rotation matrices with crank and coupler link displacement angles θ and α, respectively. Substituting Equation (6.7) into Equation (6.5) produces

$$E\cos(\alpha) + F\sin(\alpha) + G = 0 \quad (6.9)$$

where

$$E = (\mathbf{a} - \mathbf{b}_0)^T \{[I - Q](\mathbf{b}_1' - \mathbf{a})\}^9 \quad (6.10)$$

$$F = (\mathbf{a} - \mathbf{b}_0)^T \{[P](\mathbf{b}_1' - \mathbf{a})\} \quad (6.11)$$

$$G = (\mathbf{a} - \mathbf{b}_0)^T \{[Q](\mathbf{b}_1' - \mathbf{a})\} + \frac{1}{2}\begin{Bmatrix} (\mathbf{b}_1' - \mathbf{a})^T (\mathbf{b}_1' - \mathbf{a}) + (\mathbf{a} - \mathbf{b}_0)^T (\mathbf{a} - \mathbf{b}_0) \\ -(\mathbf{b}_1 - \mathbf{b}_0)^T (\mathbf{b}_1 - \mathbf{b}_0) \end{Bmatrix} \quad (6.12)$$

$$[P] = \begin{bmatrix} 0 & -1 & 0 \\ 1 & 0 & 0 \\ 0 & 0 & 0 \end{bmatrix} \quad \text{and} \quad [Q] = \begin{bmatrix} 0 & 0 & 0 \\ 0 & 0 & 0 \\ 0 & 0 & 1 \end{bmatrix}$$

Two coupler link rotation angle solutions exist for Equation (6.9). These two coupler rotation angle solutions (α) are

$$\alpha_{1,2} = 2\tan^{-1} \frac{-F \pm \sqrt{E^2 + F^2 - G^2}}{G - E}{}^{10} \quad (6.13)$$

The coupler displacement angle α calculated from Equations (6.13) is the difference between the angles $\angle \mathbf{b} - \mathbf{a} - \mathbf{a}_0$ and $\angle \mathbf{b}_1 - \mathbf{a}_1 - \mathbf{a}_0$ (angle α_0 in Figure 6.5).

The global displacement of an arbitrary coupler point $\mathbf{p}_1$ can be expressed as

$$\mathbf{p} = [R_\alpha](\mathbf{p}' - \mathbf{a}) + \mathbf{a} \quad (6.14)$$

where

$$\mathbf{p}' = [R_\theta](\mathbf{p}_1 - \mathbf{a}_0) + \mathbf{a}_0 \quad (6.15)$$

Mechanism variables $\mathbf{a}_0$, $\mathbf{a}_1$, $\mathbf{b}_0$, $\mathbf{b}_1$, and $\mathbf{p}_1$ are 3×1 vectors containing x- and y-components (with the z-components all being zero).

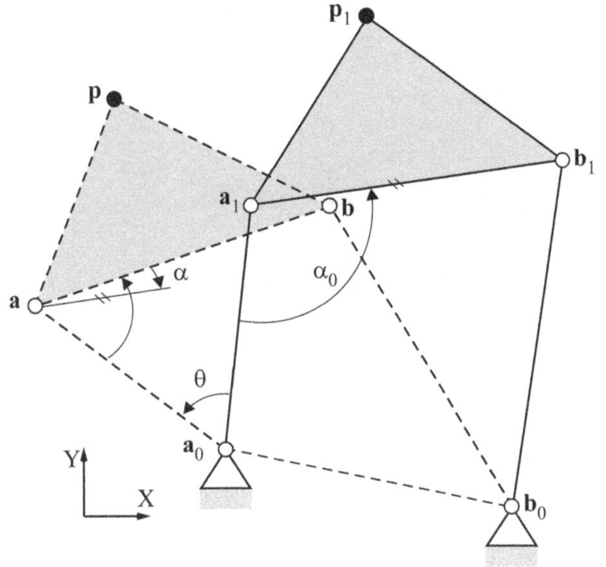

FIGURE 6.5 Four-bar mechanism in starting and displaced positions (algebraic model).

6.6 FOUR-BAR MOTION GENERATION: UNCONSTRAINED OPTIMIZATION USING ALTERNATE DYAD MODEL

Until now, the motion generation equations presented in this book were derived using the vector loop closure method for mechanism dyads. The unknown variables to be solved using these models are the dyad vector components and displacement angles.

Suh and Radcliffe presented a set of nonlinear equations for the synthesis of planar four-bar motion generators [13]. The unknown variables to be solved in these equations are the mechanism dyad fixed and moving pivot components. The mechanism fixed pivots $\mathbf{a}_0$ and $\mathbf{b}_0$ and moving pivots $\mathbf{a}_1$ and $\mathbf{b}_1$ are illustrated in Figure 6.6. Equation (6.15) is a constant-length constraint. It constrains the crank and follower lengths to a constant value through each precision position.[11]

$$\left(\mathbf{a}_j-\mathbf{a}_0\right)^T\left(\mathbf{a}_j-\mathbf{a}_0\right)-\left(\mathbf{a}_1-\mathbf{a}_0\right)^T\left(\mathbf{a}_1-\mathbf{a}_0\right)=0 \tag{6.16}$$

In Equation (6.16), $\mathbf{a}_0=\left[a_{0x},a_{0y},1\right]^T$ and $\mathbf{a}_1=\left[a_{1x},a_{1y},1\right]^T$. Also, $\mathbf{a}_j=\left[D_j\right]\mathbf{a}_1$ and

$$\left[D_j\right]=\begin{bmatrix}\cos\alpha_j & -\sin\alpha_j & p_{jx}-p_{1x}\cos\alpha_j+p_{1y}\sin\alpha_j\\ \sin\alpha_j & \cos\alpha_j & p_{jy}-p_{1x}\sin\alpha_j-p_{1y}\cos\alpha_j\\ 0 & 0 & 1\end{bmatrix}\quad j=2,3,4 \tag{6.17}$$

Matrix (6.17) defines the displacement of the coupler by a point $\mathbf{p}$ and a displacement angle α. Given four precision positions, Equation (6.16) forms a set of three nonlinear simultaneous equations to solve for three of the four unknown variables for a single dyad with an arbitrary choice of one variable. The arbitrary choice of a single variable enables the user to calculate unique crank and follower dyad solutions. Although as presented, Equation (6.16) includes dyad variables $\mathbf{a}_0$ and $\mathbf{a}_1$, it can be used directly to calculate dyad variables $\mathbf{b}_0$ and $\mathbf{b}_1$. Appendix F.5 includes the MATLAB file user instructions for Equation (6.16) for approximate four-bar motion generation. Although this equation forms a set of three nonlinear simultaneous equations for four-position synthesis, in the Appendix F.5 MATLAB file, they are solved as an unconstrained nonlinear optimization model for N precision positions.[12]

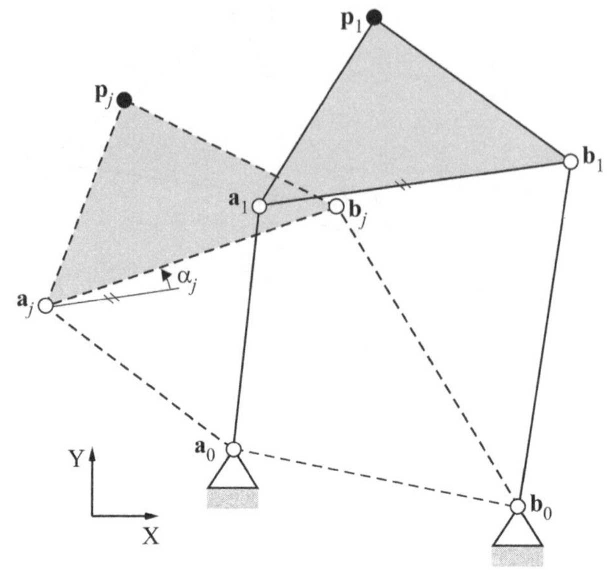

FIGURE 6.6 Four-bar mechanism in starting and displaced positions (alternate synthesis model).

Equation (6.16) can be an alternative to Equations (6.1) and (6.2)—particularly for four-position synthesis. Because Equation (6.16) does not include crank/follower rotation angle variables, the user must determine the crank link rotation angles required to approximate the precision positions *after* calculating a mechanism solution. This can be accomplished using the Appendix H.1 (or I.1) MATLAB file since it is the spatial form of the displacement model in Section 6.5.

Example 6.5

Problem Statement: Using the Appendix F.5 MATLAB files, synthesize a planar four-bar mechanism to approximate the landing gear precision positions (Figure E.6.17) in Table E.6.3.

Known Information: Table E.6.3 and Appendix F.5 MATLAB files

Solution Approach: In this example, planar four-bar dyad variables a_{0x} and b_{0x} were specified to $a_{0x} = 0$ and $b_{0x} = 5$, respectively.[13] Figure E.6.18 includes the input in bold type in the Appendix F.5 MATLAB file *a0_a1_quasi_newton.m* when $a_{0x} = 0$. Figure E.6.19 includes the output from the same file. Figure E.6.20 includes the output from the Appendix F.5 MATLAB file *a0_a1_quasi_newton.m* when $b_{0x} = a_{0x} = 1$. Figure E.6.21 illustrates the resulting planar four-bar mechanism. A kinematic analysis of this solution is presented in Example 6.13. The equation systems formulated from Equation (6.16) include no constraints to ensure that branch- and order defect-free solutions are calculated. With these equation systems, branch- and order defect-free solutions can be calculated through trial and error.

TABLE E.6.3
Landing Gear Precision Positions

Position	$\mathbf{p}_j$	α_j(deg)
1	0, 0	
2	0.203, 0.415	–25
3	0.397, 0.848	–40
4	0.385, 1.182	–70
5	0.182, 1.460	–90

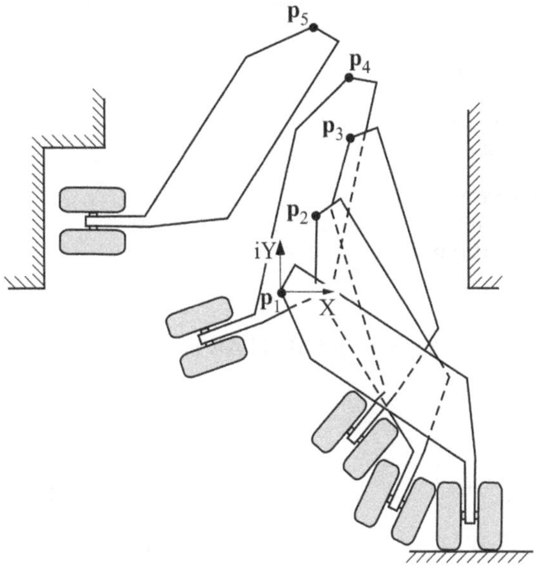

FIGURE E.6.17 Landing gear precision positions.

```
N = 5;

a0x = 0;

alpha2 = -25 *pi/180;
alpha3 = -40 *pi/180;
alpha4 = -70 *pi/180;
alpha5 = -90 *pi/180;

P1 = [0,0];
P2 = [0.203, 0.415];
P3 = [0.397, 0.848];
P4 = [0.385, 1.182];
P5 = [0.182, 1.460];

a0y = 1;
a1x = 1;
a1y = 1;
```

FIGURE E.6.18 Input in bold type in the Appendix F.5 MATLAB file *a0_a1_quasi_newton.m* (with $a_{0x} = 0$) for Example 6.5.

```
Iterations: 20
Func Evals: 21
Algorithm: quasi-newton
exit flag = 5
error = ( 3.814726e-03 )
a0y = ( 0.517066 )
a1x = ( 0.343369 )
a1y = ( -0.003546 )
Elapsed time is 0.209654 seconds.
Solution calculation is complete.
>>
```

FIGURE E.6.19 Output from the Appendix F.5 MATLAB file *a0_a1_quasi_newton.m* (with $a_{0x} = 0$) for Example 6.5.

```
Iterations: 31
Func Evals: 33
Algorithm: quasi-newton
exit flag = 5
error = ( 9.950957e-05 )
a0y = ( 1.045580 )
a1x = ( 0.838179 )
a1y = ( 0.570939 )
Elapsed time is 0.236547 seconds.
Solution calculation is complete.
>>
```

FIGURE E.6.20 Output from the Appendix F.5 MATLAB file *a0_a1_quasi_newton.m* (with $b_{0x} = a_{0x} = 1$) for Example 6.5.

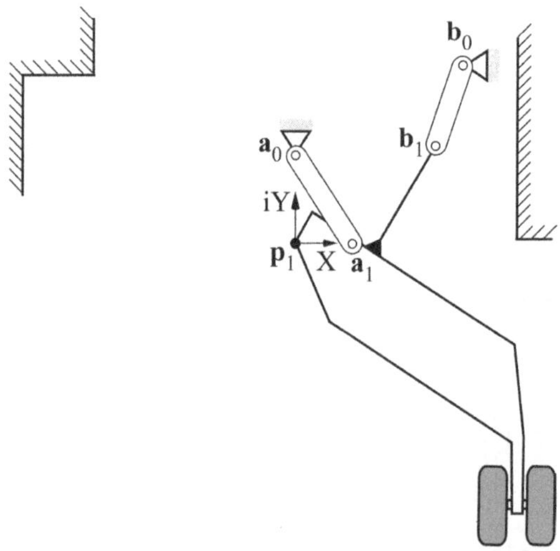

FIGURE E.6.21 Synthesized planar four-bar motion generator.

6.7 FOUR-BAR PATH AND MOTION GENERATION: CONSTRAINED OPTIMIZATION USING THE ALGEBRAIC DISPLACEMENT MODEL

The equation system in Section 6.6 has limited use for motion generation since it lacks a grounded link rotation angle variable, and subsequently, an order constraint. An alternate equation system, a constrained nonlinear optimization model, for planar four-bar path generation (for N precision points) includes the objective function

$$f(\mathbf{X}) = \sum_{j=2}^{N} \left\| \mathbf{p}_j - \mathbf{p} \right\|^2 {}^{14} \tag{6.18}$$

where $\mathbf{X} = (\mathbf{a}_0, \mathbf{a}_1, \mathbf{b}_0, \mathbf{b}_1, \theta_2 \ldots \theta_N, \alpha_2 \ldots \alpha_N)$. In Equation (6.18), the precision points are defined by $\mathbf{p}_j$, and the coupler points achieved by the synthesized mechanism (using Equation (6.14)) are defined by $\mathbf{p}$. In this objective function, the squared norms of the differences between $\mathbf{p}_j$ and $\mathbf{p}$ are to be minimized.

Along with the objective function, two types of equality constraint equations are used. To ensure the constant length of the follower link (link $\mathbf{b}_0 - \mathbf{b}_1$), Equation (6.5) is included for precision points

$2 \ldots N$. Likewise, to ensure the constant length of the crank link (link $\mathbf{a}_0 - \mathbf{a}_1$), Equation (6.19) is included for precision points $2 \ldots N$.

$$(\mathbf{a}-\mathbf{a}_0)^T(\mathbf{a}-\mathbf{a}_0)-(\mathbf{a}_1-\mathbf{a}_0)^T(\mathbf{a}_1-\mathbf{a}_0)=0 \tag{6.19}$$

Equations (6.5) and (6.19) include both Equations (6.6) and (6.7) from the algebraic planar four-bar displacement model in Section 6.5 to calculate the displaced moving pivots **a** and **b**.

To produce planar four-bar path generator solutions that are order defect-free, Inequality (6.20) is included for precision points $2 \ldots N$. The crank rotation angle in these equations is defined by θ, as shown in Figure 6.5.

$$\begin{cases} \theta_j > \theta_{j-1} \\ \theta_N < 2\pi \end{cases} \quad j = 2,3,\ldots,N \tag{6.20}$$

As written, Inequality (6.20) ensures counterclockwise crank link rotation. The inequality $\theta_j < \theta_{j-1}$ can be used instead of $\theta_j > \theta_{j-1}$ to ensure clockwise crank link rotation.

To produce planar four-bar path generator solutions that are branch defect-free, Inequality (6.21) is included for precision points $2 \ldots N$.

$$[(\mathbf{b}-\mathbf{b}_0)\times(\mathbf{a}-\mathbf{b}_0)]\cdot[(\mathbf{b}_1-\mathbf{b}_0)\times(\mathbf{a}_1-\mathbf{b}_0)]>0 \tag{6.21}$$

A change in the branch would result in a change in the sign in the product of the cross-products from positive to negative. By constraining the product of these cross-products to a positive value, Inequality (6.21) constrains the mechanism to a single branch.

Inequality (6.21) includes both Equations (6.6) and (6.7) from the algebraic planar four-bar displacement model in Section 6.5 to calculate displaced moving pivots $\mathbf{U}_1$ - $\mathbf{S}_1$ and $\mathbf{b}$. By ensuring a constant direction cross product between link $\mathbf{b}_0 - \mathbf{b}$ and distance $\mathbf{b}_0 - \mathbf{a}$, a single four-bar mechanism configuration is ensured throughout the precision points.

This combination of objective function, equality constraints, and inequality constraints allows for the direct minimization of the difference between the precision points and the achieved points while producing mechanism solutions that are free of order defects and branch defects. Because Equations (6.6), (6.7), and (6.14) are included in this constrained nonlinear optimization model (to calculate the displaced moving pivots **a**, **b**, and the displaced coupler point $\gamma_j = \gamma_j^*$), displacement angles θ and $\mathbf{p}_j$ are among the unknown variables to be calculated.[15] Appendix F.6 includes the MATLAB file user instructions for the constrained nonlinear optimization model presented in this section for planar four-bar path generation with order and branch defect elimination.

As presented, this constrained nonlinear optimization model is useful for planar four-bar path generation. By defining the orientation of the coupler link as a line having coupler point variables **p** and **q** as endpoints, Equation (6.22)—the objective function for planar four-bar motion generation—is formed for precision positions $2 \ldots N$.

$$f(\mathbf{X}) = \sum_{j=2}^{N} \left\{ \|\mathbf{p}_j - \mathbf{p}\|^2 + \|\mathbf{q}_j - \mathbf{q}\|^2 \right\} \tag{6.22}$$

where $\mathbf{X} = (\mathbf{a}_0, \mathbf{a}_1, \mathbf{b}_0, \mathbf{b}_1, \theta_2 \ldots \theta_N, \alpha_2 \ldots \alpha_N)$.[16] For planar four-bar motion generation with branch and order constraints, Equation (6.22) is substituted for Equation (6.18). Appendix F.7 includes the MATLAB file user instructions for the constrained nonlinear optimization model presented in this section for approximate planar four-bar motion generation with order and branch defect elimination.

Example 6.6

Problem Statement: Using the Appendix F.7 MATLAB files, synthesize a branch and order defect-free planar four-bar path generator to approximate the path points (Figure E.6.22) in Table E.6.4.

Known Information: Table E.6.4 and Appendix F.7 MATLAB files

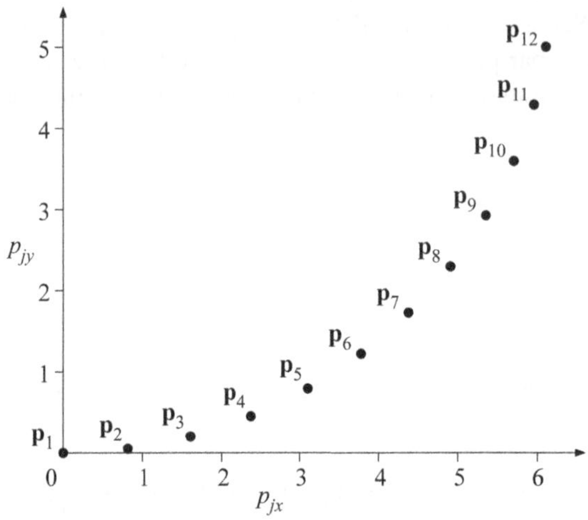

FIGURE E.6.22 Coupler curve precision points.

TABLE E.6.4
Coupler Curve Precision Points

Point	$\mathbf{p}_j$
1	0, 0
2	0.8116, 0.0509
3	1.6087, 0.2025
4	2.3771, 0.4518
5	3.1032, 0.7937
6	3.7743, 1.2213
7	4.3787, 1.7257
8	4.9063, 2.2968
9	5.3482, 2.9229
10	5.6975, 3.5913
11	5.9491, 4.2884
12	6.1, 5

Solution Approach: Figure E.6.23 includes the input in bold type in the Appendix F.7 MATLAB file *path_interior_point.m*. Figure E.6.24 includes the output from the same file. Figure E.6.25 illustrates the resulting planar four-bar mechanism. A kinematic analysis of this solution is presented in Example 6.14.

```
N = 12;

P1 = [0, 0];
P2 = [0.8116, 0.0509];
P3 = [1.6087, 0.2025];
P4 = [2.3771, 0.4518];
P5 = [3.1032, 0.7937];
P6 = [3.7743, 1.2213];
P7 = [4.3787, 1.7257];
P8 = [4.9063, 2.2968];
P9 = [5.3482, 2.9229];
P10 = [5.6975, 3.5913];
P11 = [5.9491, 4.2884];
P12 = [6.1, 5];

a0x = 1;
a0y = -1;

a1x = 1;
a1y = 1;

b0x = 3;
b0y = 3;

b1x = -3;
b1y = -3;

alpha2 = 1 *pi/180;
alpha3 = 2 *pi/180;
alpha4 = 3 *pi/180;
alpha5 = 4 *pi/180;
alpha6 = 5 *pi/180;
alpha7 = 6 *pi/180;
alpha8 = 7 *pi/180;
alpha9 = 8 *pi/180;
alpha10 = 9 *pi/180;
alpha11 = 10 *pi/180;
alpha12 = 11 *pi/180;

theta2 = 20 *pi/180;
theta3 = 35 *pi/180;
theta4 = 45 *pi/180;
theta5 = 55 *pi/180;
theta6 = 65 *pi/180;
theta7 = 75 *pi/180;
theta8 = 85 *pi/180;
theta9 = 95 *pi/180;
theta10 = 105 *pi/180;
theta11 = 115 *pi/180;
theta12 = 125 *pi/180;
```

FIGURE E.6.23 Input in bold type in the Appendix F.7 MATLAB file *path_interior_point.m* for Example 6.6 (counterclockwise crank rotation constraint applied).

```
Iterations: 53
Func Evals: 172
Algorithm: interior-point
exit flag = -2
error = ( 1.785273e-01)
a0x = ( 2.552682 )
a0y = ( 2.254125 )
a1x = ( -8.868663 )
a1y = ( 15.443918 )
b0x = ( 1.166834 )
b0y = ( 4.404037 )
b1x = ( -11.684031 )
b1y = ( -10.929525 )
alpha2 = ( 1.270241 )
alpha3 = ( 3.422608 )
alpha4 = ( 5.210378 )
alpha5 = ( 6.151612 )
alpha6 = ( 6.254538 )
alpha7 = ( 4.987040 )
alpha8 = ( 3.281339 )
alpha9 = ( 1.297585 )
alpha10 = ( -0.176938 )
alpha11 = ( -1.537176 )
alpha12 = ( -2.926034 )
theta2 = ( 7.813700 )
theta3 = ( 23.361184 )
theta4 = ( 41.199616 )
theta5 = ( 56.786206 )
theta6 = ( 86.623175 )
theta7 = ( 111.040508 )
theta8 = ( 130.837743 )
theta9 = ( 149.236267 )
theta10 = ( 161.455618 )
theta11 = ( 172.218314 )
theta12 = ( 183.004171 )
Elapsed time is 13.269310 seconds.
Solution calculation is complete.
>>
```

FIGURE E.6.24 Output from the Appendix F.7 MATLAB file *path_interior_point.m* for Example 6.6.

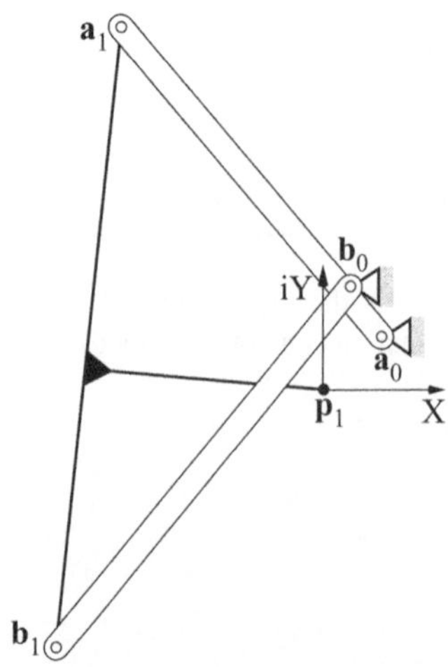

FIGURE E.6.25 Synthesized planar four-bar path generator.

Example 6.7

Problem Statement: Using the Appendix F.6 MATLAB files, synthesize a branch- and order defect-free planar four-bar motion generator to approximate the bucket precision positions in Figure E.6.26.

Known Information: Figure E.6.26 and Appendix F.6 MATLAB files

Solution Approach: Figure E.6.27 includes the input in bold type in the Appendix F.6 MATLAB file *motion_interior_point.m*. Figure E.6.28 includes the output from the same file. Figure E.6.29 illustrates the resulting planar four-bar mechanism. A kinematic analysis of this solution is presented in Example 6.15.

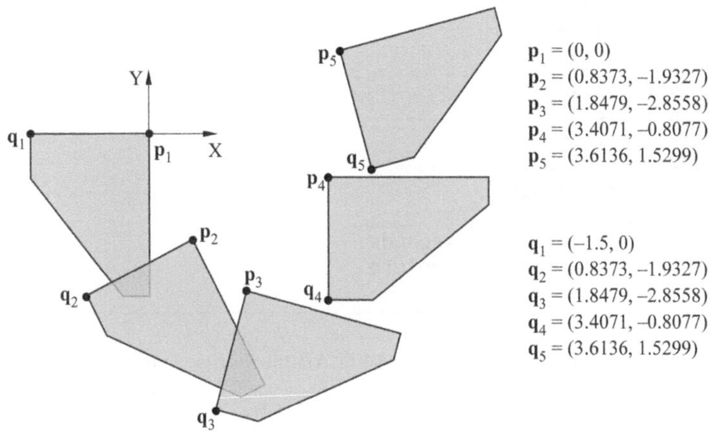

FIGURE E.6.26 Bucket precision positions.

```
N = 5;

P1 = [0, 0];
P2 = [0.8373, -1.9327];
P3 = [1.8479, -2.8558];
P4 = [3.4071, -0.8077];
P5 = [3.6136, 1.5299];

Q1 = [-1.5, 0];
Q2 = [-0.4992, -2.6137];
Q3 = [1.4597, -4.3047];
Q4 = [3.4071, -2.3077];
Q5 = [4.0018, 0.0810];

a0x = -1;
a0y = 1;

a1x = -1;
a1y = -1;

b0x = 1;
b0y = 1;

b1x = 1;
b1y = -1;

alpha2 = -30 *pi/180;
alpha3 = -30 *pi/180;
alpha4 = -30 *pi/180;
alpha5 = -30 *pi/180;

theta2 = 60 *pi/180;
theta3 = 120 *pi/180;
theta4 = 180 *pi/180;
theta5 = 240 *pi/180;
```

FIGURE E.6.27 Input in bold type in the Appendix F.6 MATLAB file *motion_interior_point.m* for Example 6.7 (counterclockwise crank rotation constraint applied).

```
Iterations: 38
Func Evals: 100
Algorithm: interior-point
exit flag = 2
error = ( 4.737472e-01 )
a0x = ( 0.298812 )
a0y = ( -0.753368 )
a1x = ( -4.146786 )
a1y = ( 1.032362 )
b0x = ( 0.942612 )
b0y = ( 1.761503 )
b1x = ( 1.063525 )
b1y = ( 3.480642 )
alpha2 = ( -23.880395 )
alpha3 = ( -40.847427 )
alpha4 = ( -50.144849 )
alpha5 = ( -56.128708 )
theta2 = ( 60.566902 )
theta3 = ( 94.087094 )
theta4 = ( 139.285950 )
theta5 = ( 178.375741 )
Elapsed time is 4.116399 seconds.
Solution calculation is complete.
>>
```

FIGURE E.6.28 Output from the Appendix F.6 MATLAB file *motion_interior_point.m* for Example 6.7.

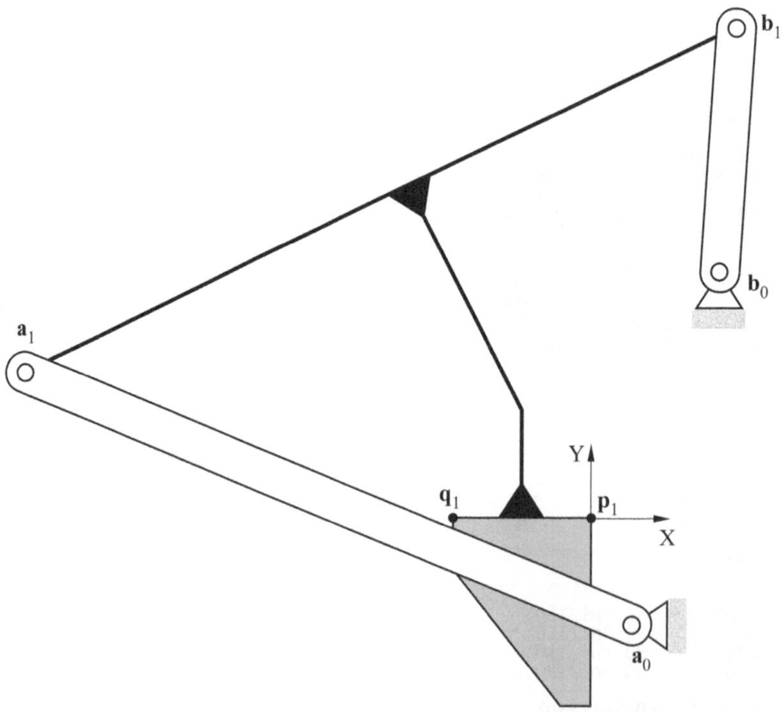

FIGURE E.6.29 Synthesized planar four-bar motion generator.

6.8 STEPHENSON III MOTION GENERATION

The planar four-bar motion generation models presented in Section 6.3 can be directly applied to Stephenson III motion generation for N precision positions. From the Appendix F.1 MATLAB file *W_Z_interior_point.m*, the $\mathbf{W}_1 - \mathbf{Z}_1$ dyad unknowns W_{1x}, W_{1y}, Z_{1x}, Z_{1y}, and $\beta_2 \ldots \beta_N$ are calculated. From the Appendix F.1 MATLAB file *U_S_quasi_newton.m*, the $\mathbf{U}_1 - \mathbf{S}_1$ dyad unknowns U_{1x}, U_{1y}, S_{1x}, S_{1y}, and $\gamma_2 \ldots \gamma_N$ are calculated. The Appendix F.1 MATLAB file *U_S_quasi_newton.m* can also be used to calculate the $\mathbf{U}_1^* - \mathbf{S}_1^*$ dyad unknowns knowing $\mathbf{U}_1 = \mathbf{U}_1^*$, $\mathbf{S}_1 = \mathbf{S}_1^*$, and $\gamma_j = \gamma_j^*$. Also, while $\mathbf{p}_j$ and α_j define the precision positions for dyads $\mathbf{W}_1 - \mathbf{Z}_1$ and $\mathbf{U}_1 - \mathbf{S}_1$, respectively, the precision positions for the $\mathbf{U}_1^* - \mathbf{S}_1^*$ dyad are defined by $\mathbf{p}_j$ and α_j^*, as illustrated in Figure 6.7.

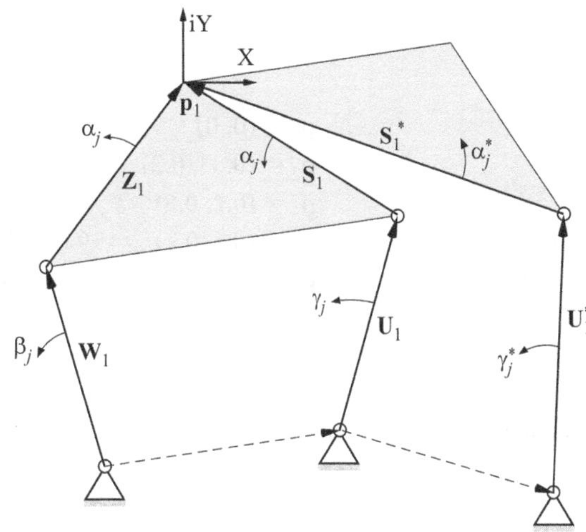

FIGURE 6.7 Stephenson III mechanism dyads.

Example 6.8

Problem Statement: Using the Appendix F.1 MATLAB files, synthesize a Stephenson III motion generator to approximate the gripper precision positions in Figure E.6.30. Dyads $\mathbf{W}_1 - \mathbf{Z}_1$ and $\mathbf{U}_1 - \mathbf{S}_1$ should approximate the upper jaw precision position data $\mathbf{p}_j$ and α_j, while the dyad $\mathbf{U}_1^* - \mathbf{S}_1^*$ should approximate the lower jaw precision position data $\mathbf{p}_j$ and α_j^*.

Known Information: Figure E.6.30 and Appendix F.1 MATLAB files

Solution Approach: Figure E.6.31 includes the input in bold type in the Appendix F.1 MATLAB file *W_Z_interior_point.m*. Figure E.6.32 includes the output from the same file.

Figure E.6.33 includes the input in bold type in the Appendix F.1 MATLAB file *U_S_quasi_newton.m*. Figure E.6.34 includes the output from the same file. Figure E.6.35 includes the input in bold type in the Appendix F.1 MATLAB file *U_S_quasi_newton.m* (for the $\mathbf{U}_1^* - \mathbf{S}_1^*$ dyad). Figure E.6.36 includes the output from the same file. Figure E.6.37 illustrates the resulting Stephenson III mechanism. A kinematic analysis of this solution is conducted in Example 6.16 in Section 6.9.

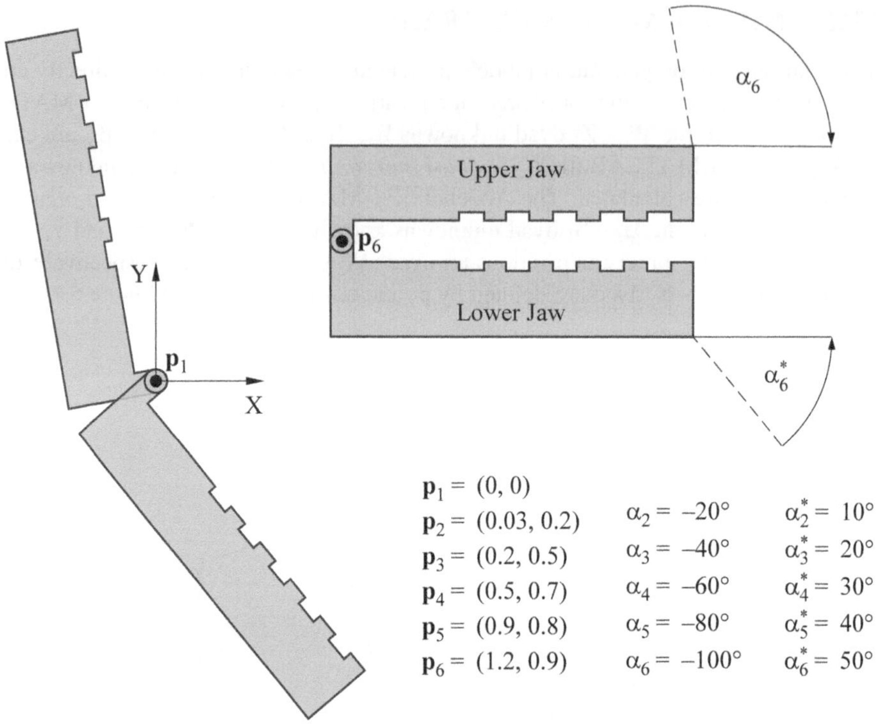

FIGURE E.6.30 Gripper precision positions.

```
N = 6;

alpha2 = -20 *pi/180;
alpha3 = -40 *pi/180;
alpha4 = -60 *pi/180;
alpha5 = -80 *pi/180;
alpha6 = -100 *pi/180;

P1 = [0, 0];
P2 = [0, 0.2];
P3 = [0.2, 0.5];
P4 = [0.5, 0.7];
P5 = [0.9, 0.8];
P6 = [1.2, 0.9];

W1x = 1;
W1y = 1;
Z1x = 1;
Z1y = 1;

beta2 = -5 *pi/180;
beta3 = -10 *pi/180;
beta4 = -15 *pi/180;
beta5 = -20 *pi/180;
beta6 = -25 *pi/180;
```

FIGURE E.6.31 Input in bold type in the Appendix F.1 MATLAB file *W_Z_interior_point.m* for Example 6.8.

```
Iterations: 157
Func Evals: 224
Algorithm: interior-point
exit flag = 2
error = ( 8.260737e-05 )
W1x = ( 2.282730 )
W1y = ( -1.833264 )
Z1x = ( -3.241189 )
Z1y = ( 2.198972 )
beta2 = ( -24.132462 )
beta3 = ( -45.116233 )
beta4 = ( -65.874985 )
beta5 = ( -85.831519 )
beta6 = ( -109.390770 )
Elapsed time is 0.578297 seconds.
Solution calculation is complete.
>>
```

FIGURE E.6.32 Output from the Appendix F.1 MATLAB file *W_Z_interior_point.m* for Example 6.8.

```
N = 6;

alpha2 = -20 *pi/180;
alpha3 = -40 *pi/180;
alpha4 = -60 *pi/180;
alpha5 = -80 *pi/180;
alpha6 = -100 *pi/180;

P1 = [0, 0];
P2 = [0, 0.2];
P3 = [0.2, 0.5];
P4 = [0.5, 0.7];
P5 = [0.9, 0.8];
P6 = [1.2, 0.9];

U1x = 1;
U1y = -1;
S1x = -1;
S1y = -1;

gamma2 = 10 *pi/180;
gamma3 = 10 *pi/180;
gamma4 = 10 *pi/180;
gamma5 = 10 *pi/180;
gamma6 = 10 *pi/180;
```

FIGURE E.6.33 Input in bold type in the Appendix F.1 MATLAB file *U_S_quasi_newton.m* for Example 6.8.

```
Iterations: 131
Func Evals: 143
Algorithm: quasi-newton
exit flag = 5
error = ( 1.145952e-04 )
U1x = ( 0.048428 )
U1y = ( 0.058515 )
S1x = ( -0.998365 )
S1y = ( 0.080917 )
gamma2 = ( -167.200663 )
gamma3 = ( -170.413865 )
gamma4 = ( -157.995131 )
gamma5 = ( -103.279254 )
gamma6 = ( 42.046210 )
Elapsed time is 0.208750 seconds.
Solution calculation is complete.
>>
```

FIGURE E.6.34 Output from the Appendix F.1 MATLAB file *U_S_quasi_newton.m* for Example 6.8.

```
N = 6;

alpha2 = 10 *pi/180;
alpha3 = 20 *pi/180;
alpha4 = 30 *pi/180;
alpha5 = 40 *pi/180;
alpha6 = 50 *pi/180;

P1 = [0, 0];
P2 = [0, 0.2];
P3 = [0.2, 0.5];
P4 = [0.5, 0.7];
P5 = [0.9, 0.8];
P6 = [1.2, 0.9];

U1x = 1;
U1y = 1;
S1x = 1;
S1y = 1;

gamma2 = 50 *pi/180;
gamma3 = 50 *pi/180;
gamma4 = 50 *pi/180;
gamma5 = 50 *pi/180;
gamma6 = 50 *pi/180;
```

FIGURE E.6.35 Input in bold type in the Appendix F.1 MATLAB file *U_S_quasi_newton.m* for Example 6.8 (for dyad $\mathbf{U}_1^* - \mathbf{S}_1^*$).

```
Iterations: 137
Func Evals: 158
Algorithm: quasi-newton
exit flag = 5
error = ( 1.190122e-03 )
U1x = ( -0.234668 )
U1y = ( 3.129947 )
S1x = ( 1.306212 )
S1y = ( -0.315486 )
gamma2 = ( 0.685440 )
gamma3 = ( -3.132686 )
gamma4 = ( -9.439138 )
gamma5 = ( -18.574099 )
gamma6 = ( -26.487632 )
Elapsed time is 0.217306 seconds.
Solution calculation is complete.
>>
```

FIGURE E.6.36 Output from the Appendix F.1 MATLAB file *U_S_quasi_newton.m* for Example 6.8 (for dyad $\mathbf{U}_1^* - \mathbf{S}_1^*$).

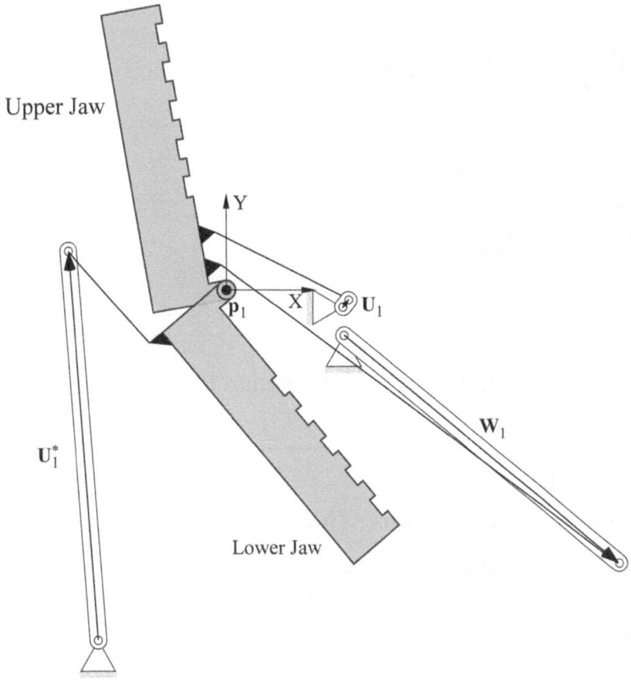

FIGURE E.6.37 Synthesized Stephenson III motion generator.

6.9 FROM DIMENSIONAL SYNTHESIS TO KINEMATIC DISPLACEMENT ANALYSIS

After synthesizing planar four-bar, geared five-bar, and Stephenson III motion and path generators, kinematic displacement analyses should be conducted to determine if the mechanisms achieve their precision positions or points, and do so without encountering any circuit, branch, or order defects. Either the Appendix B MATLAB files or the Appendix C MATLAB and Simscape Multibody files are available for the user to verify the accuracy and defect presence of synthesized mechanism solutions.

Example 6.9

Problem Statement: Using the Appendix B.1 or C.1 files, calculate the landing gear positions achieved by the motion generator solution given in Example 6.1.

Known Information: Example 6.1 solution and Appendix B.1 MATLAB file

Solution Approach: Figure E.6.38 includes the input specified (in bold text) in the Appendix B.1 MATLAB file. Table E.6.5 includes the landing gear positions achieved. While the values of α_j can be directly compared to the prescribed values in Example 6.1 (Table E.6.1), $\boldsymbol{p}_1$ must be subtracted from each $\boldsymbol{p}_j$ value to produce results that can also be compared to those in Table E.6.1. As shown by the order of the crank angle solutions and the corresponding order of the achieved landing gear positions in Table E.6.5, the planar four-bar motion generator solution is free of order defects.

```
W1 = 0.648748 - i*0.561522;
Z1 = -0.283731 + i*0.136138;
U1 = -0.178559 - i*0.450057;
S1 = -0.839784 - i*0.582417;

L1 = Z1;
V1 = Z1 - S1;
G1 = W1 + Z1 - S1 - U1;

start_ang    = 0;
step_ang     = 0.01;
stop_ang     = 102.19;

angular_vel  = 0 * ones(N+1,1);
angular_acc  = 0 * ones(N+1,1);
```

FIGURE E.6.38 Specified input (in bold text) in the Appendix B.1 MATLAB file for Example 6.9.

TABLE E.6.5
Landing Gear Positions Achieved

Position	β_j (deg)	$\mathbf{p}_j$	α_j (deg)	$\mathbf{p}_j - \mathbf{p}_1$
1	0	0.3650, –0.4254	0.0000	0, 0
2	22.59	0.6060, –0.0336	–22.8400	0.2410, 0.3918
3	49.54	0.7309, 0.4213	–42.4850	0.3659, 0.8467
4	72.54	0.7619, 0.7635	–70.1310	0.3969, 1.1889
5	102.19	0.5487, 1.0361	–90.1300	0.1837, 1.4615

Example 6.10

Problem Statement: Using the Appendix B.1 or C.1 files, calculate the dwell path points achieved by the path generator solution given in Example 6.2.

Known Information: Example 6.2 solution and Appendix B.1 MATLAB file

Solution Approach: Figure E.6.39 includes the input specified (in bold text) in the Appendix B.1 MATLAB file. Table E.6.6 includes the dwell path points achieved. The initial path point p_1 must be subtracted from each p_j value to produce results that can be compared to those in Table E.6.2. As shown by the order of the crank angle solutions and the corresponding order of the achieved dwell path points in Table E.6.6, the planar four-bar path generator solution is free of order defects.

```
W1 = 1.579630 + i*1.581584;
Z1 = -12.386988 - i*19.454246;
U1 = 1.582902 - i*9.010446;
S1 = 1.943278 - i*20.536207;

L1 = Z1;
V1 = Z1 - S1;
G1 = W1 + Z1 - S1 - U1;

start_ang    = 0;
step_ang     = 0.01;
stop_ang     = 193.68;

angular_vel  = 0 * ones(N+1,1);
angular_acc  = 0 * ones(N+1,1);
```

FIGURE E.6.39 Specified input (in bold text) in the Appendix B.1 MATLAB file for Example 6.10.

TABLE E.6.6
Dwell Path Points Achieved

Point	β_j (deg)	$\mathbf{p}_j$	$\mathbf{p}_j - \mathbf{p}_1$
1	0	−10.807, −17.873	0, 0
2	56.22	−11.809, −17.872	−1.002, 0.001
3	73.11	−12.806, −17.873	−1.999, 0.000
4	87.68	−13.806, −17.874	−2.999, −0.001
5	101.84	−14.808, −17.872	−4.001, 0.001
6	116.89	−15.810, −17.871	−5.003, 0.002
7	134.87	−16.806, −17.874	−5.999, −0.001
8	193.68	−17.808, −17.872	−7.001, 0.001

Example 6.11

Problem Statement: Using the Appendix B.3 or C.3 files, calculate the landing gear positions achieved by the motion generator solution given in Example 6.3.

Known Information: Example 6.3 solution and Appendix B.3 MATLAB file

Solution Approach: Figure E.6.40 includes the input specified (in bold text) in the Appendix B.3 MATLAB file. Table E.6.7 includes the landing gear positions achieved. While the values of α_j can be directly compared to the prescribed values in Example 6.1 (Table E.6.1), $\boldsymbol{p}_1$ must be subtracted from each $\boldsymbol{p}_j$ value to produce results that can also be compared to those in Table E.6.1. As shown by the order of the crank angle solutions and the corresponding order of the achieved landing gear positions in Table E.6.7, the geared five-bar motion generator solution is free of order defects.

```
W1 = 0.648748 - i*0.561522;
Z1 = -0.283731 + i*0.136138;
U1 = 0.657244 - i*0.546383;
S1 = 0.200861 - i*0.613504;
T1 = -0.38314 - i*0.153482;

L1 = Z1;
V1 = Z1 - T1;
G1 = W1 + Z1 - T1 - S1 - U1;
ratio = 3/2;

start_ang    = 0;
step_ang     = 0.01;
stop_ang     = 102.19;

angular_vel  = 0 * ones(N+1,1);
angular_acc  = 0 * ones(N+1,1);
```

FIGURE E.6.40 Specified input (in bold text) in the Appendix B.3 MATLAB file for Example 6.11.

TABLE E.6.7
Landing Gear Positions Achieved

Position	β_j (deg)	$\mathbf{p}_j$	α_j (deg)	$\mathbf{p}_j - \mathbf{p}_1$
1	0	0.3650, −0.4254	0.0000	0.0000, 0.0000
2	22.59	0.5856, −0.0535	−17.6470	0.2206, 0.3719
3	49.54	0.7393, 0.4245	−44.1180	0.3743, 0.8499
4	72.54	0.7722, 0.7623	−72.0170	0.4072, 1.1877
5	102.19	0.5498, 1.0356	−90.3520	0.1847, 1.4610

Example 6.12

Problem Statement: Using the Appendix B.2 or C.2 files, calculate the dwell path points achieved by the path generator solution given in Example 6.4.

Known Information: Example 6.4 solution and Appendix B.2 MATLAB file

Solution Approach: Figure E.6.41 includes the input specified (in bold text) in the Appendix B.2 MATLAB file. Table E.6.8 includes the dwell path points achieved. The initial path point $\boldsymbol{p}_1$ must be subtracted from each $\boldsymbol{p}_j$ value to produce results that can be compared to those in Table E.6.2. As shown by the order of the crank angle solutions and the corresponding order of the achieved dwell path points in Table E.6.8, the geared five-bar path generator solution is free of order defects.

```
W1 = 1.579630 + i*1.581584;
Z1 = -12.386988 - i*19.454246;
U1 = -2.794764 - i*2.116332;
S1 = -21.287198 + i*19.108909;
T1 = -9.360478 - i*27.544693;

L1 = Z1;
V1 = Z1 - T1;
G1 = W1 + Z1 - T1 - S1 - U1;
ratio = -5/4;

start_ang   = 0;
step_ang    = 0.01;
stop_ang    = 193.68;

angular_vel  = 0 * ones(N+1,1);
angular_acc  = 0 * ones(N+1,1);
```

FIGURE E.6.41 Specified input (in bold text) in the Appendix B.2 MATLAB file for Example 6.12.

TABLE E.6.8
Dwell Path Points Achieved

Point	β_j (deg)	$\mathbf{p}_j$	$\mathbf{p}_j - \mathbf{p}_1$
1	0	−10.807, −17.873	0.000, 0.000
2	56.22	−11.807, −17.873	−1.000, 0.000
3	73.11	−12.808, −17.872	−2.001, 0.001
4	87.68	−13.807, −17.873	−3.000, 0.000
5	101.84	−14.807, −17.873	−4.000, 0.000
6	116.89	−15.807, −17.872	−5.000, 0.001
7	134.87	−16.807, −17.873	−6.000, 0.000
8	193.68	−17.807, −17.873	−7.000, 0.000

Example 6.13

Problem Statement: Using the Appendix H.1 or I.1 files, calculate the landing gear positions achieved by the motion generator solution given in Example 6.5.

Known Information: Example 6.5 solution and Appendix H.1 MATLAB file

Solution Approach: Figure E.6.42 includes the input specified (in bold text) in the Appendix H.1 MATLAB file. Table E.6.9 includes the landing gear positions achieved.[17]

```
a0 = [0, 0.517066, 0]';
a1 = [0.343369, -0.003546, 0]';
ua0 = [0, 0, 1]';
ua1 = [0, 0, 1]';
b0 = [1, 1.045580, 0]';
b1 = [0.838179, 0.570939, 0]';
p1 = [0, 0, 0]';
q1 = p1;
r1 = p1;

start_ang   = 0;
step_ang    = 0.01;
stop_ang    = 115.12;

angular_vel  = 0 * ones(N+1,1);
angular_acc  = 0 * ones(N+1,1);
```

FIGURE E.6.42 Specified input (in bold text) in the Appendix H.1 MATLAB file for Example 6.13.

TABLE E.6.9
Landing Gear Positions Achieved

Position	θ_j (deg)	$\mathbf{p}_j$	α_j (deg)
1	0	0, 0	0
2	38.17	0.282, 0.468	–24.997
3	62.70	0.359, 0.807	–40.002
4	90.65	0.403, 1.190	–70.004
5	115.12	0.329, 1.392	–90.001

Example 6.14

Problem Statement: Using the Appendix H.1 or I.1 files, calculate the path points achieved by the path generator solution given in Example 6.6.

Known Information: Example 6.6 solution and Appendix H.1 MATLAB file

Solution Approach: Figure E.6.43 includes the input specified (in bold text) in the Appendix H.1 MATLAB file. Table E.6.10 includes the path points achieved. Because a single path generator solution achieves the results given in Table E.6.10, it is free of order and branch defects.

```
a0 = [2.552682, 2.254125, 0]';
a1 = [-8.868663, 15.443918, 0]';
ua0 = [0, 0, 1]';
ua1 = [0, 0, 1]';
b0 = [1.166834, 4.404037, 0]';
b1 = [-11.684031, -10.929525, 0]';
p1 = [0, 0, 0]';
q1 = p1;
r1 = p1;

start_ang   = 0;
step_ang    = 0.01;
stop_ang    = 183;

angular_vel  = 0 * ones(N+1,1);
angular_acc  = 0 * ones(N+1,1);
```

FIGURE E.6.43 Specified input (in bold text) in the Appendix H.1 MATLAB file for Example 6.14.

TABLE E.6.10
Path Points Achieved

Point	θ_j (deg)	$\mathbf{p}_j$
1	0	0, 0
2	7.81	0.6352, −0.0849
3	23.36	1.7143, 0.0436
4	41.20	2.5734, 0.4365
5	56.79	3.0496, 0.7924
6	86.62	3.6932, 1.2608
7	111.04	4.3288, 1.6049
8	130.84	4.9553, 2.1063
9	149.24	5.4973, 2.8873
10	161.46	5.7443, 3.5905
11	172.22	5.8321, 4.3157
12	183	5.7596, 5.1123

Example 6.15

Problem Statement: Using the Appendix H.1 or I.1 files, calculate the bucket positions achieved by the motion generator solution given in Example 6.7.

Known Information: Example 6.7 solution and Appendix H.1 MATLAB file

Solution Approach: Figure E.6.44 includes the input specified (in bold text) in the Appendix. H.1 MATLAB file. Table E.6.11 includes the bucket positions achieved. Because a single motion generator solution achieves the results given in Table E.6.11, it is free of order and branch defects.

```
a0 = [0.298812, -0.753368, 0]';
a1 = [-4.146786, 1.032362, 0]';
ua0 = [0, 0, 1]';
ua1 = [0, 0, 1]';
b0 = [0.942612, 1.761503, 0]';
b1 = [1.063525, 3.480642, 0]';
p1 = [0, 0, 0]';
q1 = [-1.5, 0, 0]';
r1 = q1;

start_ang    = 0;
step_ang     = 0.01;
stop_ang     = 178.38;

angular_vel  = 0 * ones(N+1,1);
angular_acc  = 0 * ones(N+1,1);
```

FIGURE E.6.44 Specified input (in bold text) in the Appendix H.1 MATLAB file for Example 6.15.

TABLE E.6.11
Bucket Positions Achieved

Position	θ_j (deg)	$\mathbf{p}_j$	$\mathbf{q}_j$
1	0	0, 0	−1.5, 0
2	60.57	0.5012, −2.0982	−0.7016, −2.9944
3	94.09	2.1434, −2.6106	1.2457, −3.8123
4	139.29	3.5981, −0.8756	3.5757, −2.3755
5	178.38	3.3524, 1.3939	4.1529, 0.1253

Example 6.16

Problem Statement: Using the Appendix B.4 or C.4 files, calculate the gripper positions achieved by the motion generator solution given in Example 6.8.

Known Information: Example 6.8 solution and Appendix B.4 MATLAB file

Solution Approach: Figure E.6.45 includes the input specified (in bold text) in the Appendix B.4 MATLAB file. Table E.6.12 includes the gripper positions achieved. Because a single motion generator solution achieves the results given in Table E.6.12, it is free of order and branch defects.

```
W1 = 2.282730 - i*1.833264;
Z1 = -3.241189 + i*2.198972;
U1 = 0.048428 + i*0.058515;
S1 = -0.998365 + i*0.080917;
U1s = -0.234668 + i*3.129947;
S1s = 1.306212 - i*0.315486;

L1 = Z1;
V1 = Z1 - S1;
G1 = W1 + Z1 - S1 - U1;

V1s = -S1s;
G1s = W1 + Z1 - S1s - U1s - G1;
L1s = 1;

start_ang   = 0;
step_ang    = -0.01;
stop_ang    = -109.39;

angular_vel  = 0 * ones(N+1,1);
angular_acc  = 0 * ones(N+1,1);
```

FIGURE E.6.45 Specified input (in bold text) in the Appendix B.4 MATLAB file for Example 6.16.

TABLE E.6.12
Gripper Positions Achieved

Position	β_j (deg)	$\mathbf{p}_j$	$\mathbf{p}_j - \mathbf{p}_1$	α_j (deg)	α_j^* (deg)
1	0	–0.9585, 0.3657	0, 0	0	0
2	–24.13	–0.8282, 0.6597	0.1303, 0.2940	–22.342	12.424
3	–45.12	–0.6652, 0.8818	0.2932, 0.5161	–41.401	21.795
4	–65.87	–0.3925, 1.0686	0.5660, 0.7029	–60.934	30.836
5	–85.83	–0.0564, 1.1623	0.9021, 0.7966	–80.053	38.408
6	–109.39	0.3526, 1.1529	1.3111, 0.7872	–102.320	46.096

6.10 SUMMARY

What distinguishes path generation from motion generation is that coupler path points (called *precision points*) are achieved in path generation while coupler positions (called *precision positions*) are achieved in motion generation. The analytical motion generation equations in Chapter 4 have limited use in path generation.

A *coupler curve atlas* includes a wide array of coupler curves and their corresponding four-bar mechanism link proportions. Such information can provide designers with judicious starting points for design applications. Two commonly used coupler curve features are straight lines and circular arcs. These features constitute *dwell regions* and result in a cam-like zero radial output motion for a nonzero crank supplied input motion.

While analytical motion and path generation equations enable one to calculate mechanism solutions that precisely achieve precision positions and precision points, respectively, they become more limited and ultimately inadequate as the number of precision positions and points increases. Under such a condition, it is more useful to apply *nonlinear optimization models* to calculate mechanism solutions to approximate prescribed parameters. This chapter includes two types of nonlinear models for motion and path generation: *unconstrained nonlinear models* and *constrained nonlinear optimization models*. The *Quasi-Newton* algorithm is implemented in MATLAB to calculate solutions for the unconstrained nonlinear optimization models, and the *Interior-Point* algorithm is employed in MATLAB to calculate solutions for the constrained nonlinear optimization models. The latter optimization models include constraints to eliminate branch defects and order defects. The optimization models presented in this chapter are solved using the Appendix F MATLAB files.

While the planar four-bar mechanism includes a single member (the coupler link) that undergoes complex motion, planar multi-loop mechanisms such as the *Stephenson III mechanism* and planar five-bar linkages such as the *geared five-bar mechanism* include at least two links that undergo complex motion. Like the planar four-bar mechanism, vector loop equations were formulated for these mechanisms for motion and path generation.

PROBLEMS

1. Using the Appendix F.1 MATLAB files, synthesize an order, branch, and circuit defect-free planar four-bar mechanism for the four digger bucket positions in Figure P.6.1. Branch defect elimination can be verified through *Filemon's Construction* or a displacement analysis (using the Appendix B.1 or C.1 files). Order, branch and circuit defect elimination can also be verified through a displacement analysis.

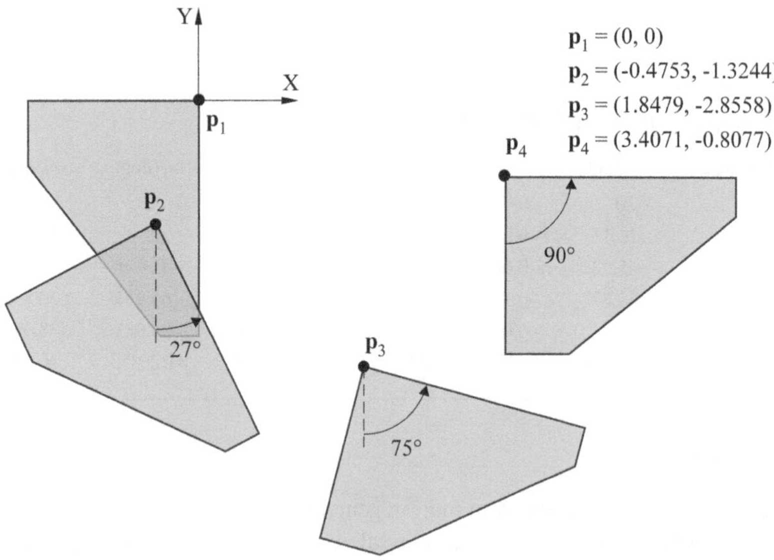

FIGURE P.6.1 Four digger bucket positions.

2. Using the Appendix F.1 MATLAB files, synthesize an order, branch, and circuit defect-free planar four-bar mechanism for Problem 10 in Chapter 5. Branch defect elimination can be verified through *Filemon's Construction* or a displacement analysis (using the Appendix B.1 or C.1 files). Order, branch and circuit defect elimination can also be verified through a displacement analysis.

3. Using the Appendix F.1 MATLAB files, synthesize an order, branch, and circuit defect-free planar four-bar mechanism for the five latch positions in Figure P.6.2. Branch defect elimination can be verified through *Filemon's Construction* or a displacement analysis (using the Appendix B.1 or C.1 files). Order, branch, and circuit defect elimination can also be verified through a displacement analysis.

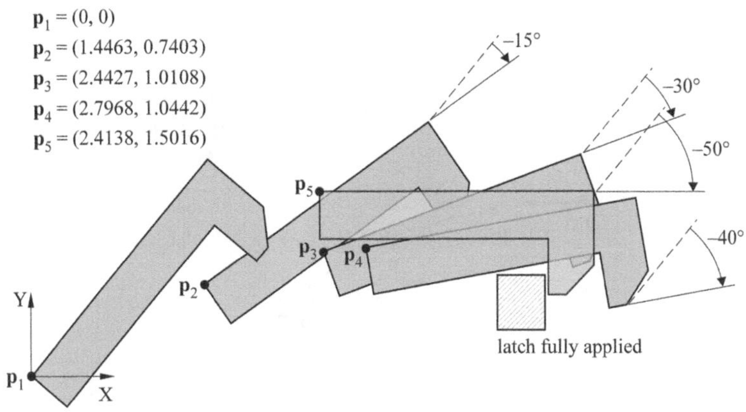

FIGURE P.6.2 Five latch positions.

4. Using the Appendix F.1 MATLAB files, synthesize an order, branch, and circuit defect-free planar four-bar mechanism for the seven assembly component positions in Figure P.6.3. Branch defect elimination can be verified through *Filemon's Construction* or a displacement analysis (using the Appendix B.1 or C.1 files). Order, branch, and circuit defect elimination can also be verified through a displacement analysis.

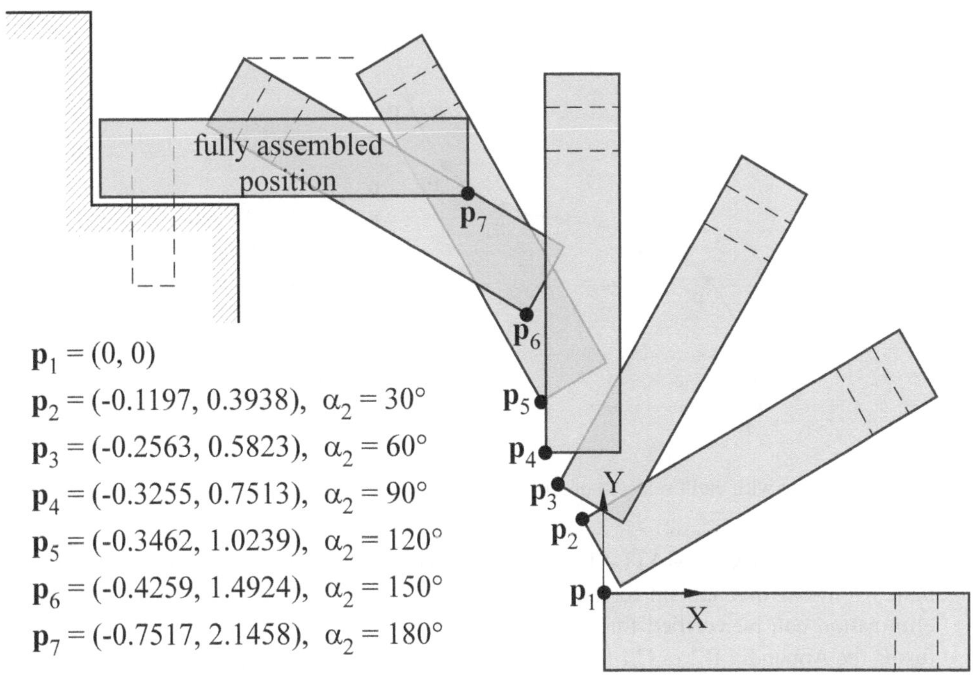

FIGURE P.6.3 Seven assembly component positions.

5. Using the Appendix F.2 MATLAB files, synthesize an order, branch, and circuit defect-free planar four-bar mechanism for the five precision points in Figure P.6.4. Branch defect elimination can be verified through *Filemon's Construction* or a displacement analysis (using the Appendix B.1 or C.1 files). Order, branch, and circuit defect elimination can also be verified through a displacement analysis.

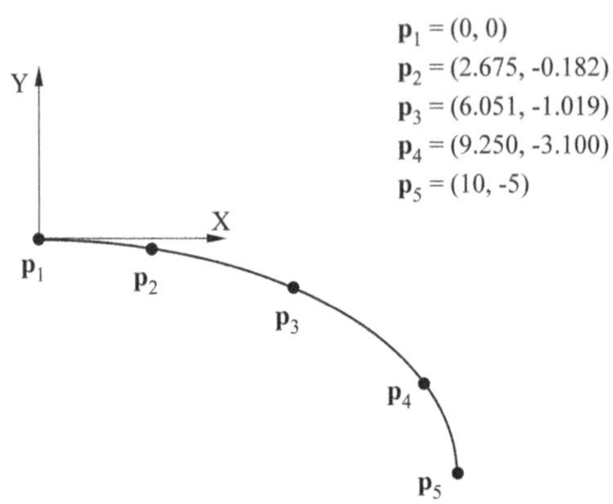

FIGURE P.6.4 Curve with five selected precision points.

6. Using the Appendix F.2 MATLAB files, synthesize an order, branch, and circuit defect-free planar four-bar mechanism for the eight precision points in Figure P.6.5. Branch defect elimination can be verified through *Filemon's Construction* or a displacement analysis (using the Appendix B.1 or C.1 files). Order, branch, and circuit defect elimination can also be verified through a displacement analysis.

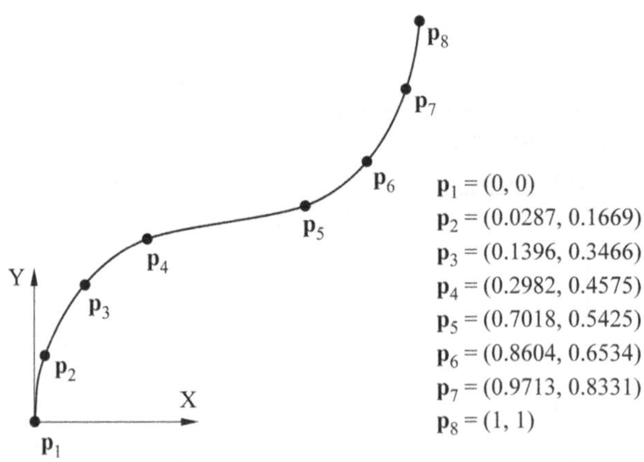

FIGURE P.6.5 Curve with eight selected precision points.

7. Using the Appendix F.2 MATLAB files, synthesize an order, branch, and circuit defect-free planar four-bar mechanism for the eight precision points in Figure P.6.6. Branch defect elimination can be verified through *Filemon's Construction* or a displacement analysis (using the Appendix B.1 or C.1 files). Order, branch, and circuit defect elimination can also be verified through a displacement analysis.

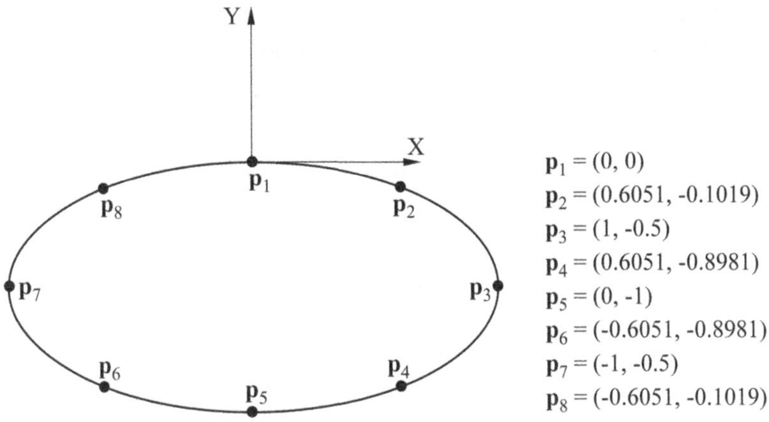

FIGURE P.6.6 Curve with eight selected precision points.

8. Using the Appendix F.2 MATLAB files, synthesize an order, branch, and circuit defect-free planar four-bar mechanism for the eight precision points in Figure P.6.7. Branch defect elimination can be verified through *Filemon's Construction* or a displacement analysis (using the Appendix B.1 or C.1 files). Order, branch, and circuit defect elimination can also be verified through a displacement analysis.

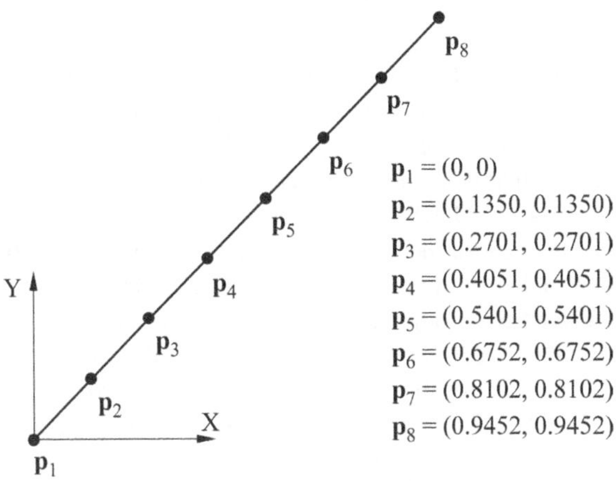

FIGURE P.6.7 Curve with eight selected precision points.

9. Using the Appendix F.3 MATLAB files, synthesize an order, branch, and circuit defect-free geared five-bar mechanism for the four precision positions in Problem 1. Use the $W_1 - Z_1$ and β_j solutions from Problem 1 along with $\gamma_j = 17\beta_j/20$. Order, branch, and circuit defect elimination can be verified through a displacement analysis.
10. Using the Appendix F.3 MATLAB files, synthesize an order, branch, and circuit defect-free geared five-bar mechanism for the five precision positions in Problem 3. Use the $W_1 - Z_1$ and β_j solutions from Problem 3 along with $\gamma_j = -16\beta_j/20$. Order, branch, and circuit defect elimination can be verified through displacement analysis.
11. Using the Appendix F.3 MATLAB files, synthesize an order, branch, and circuit defect-free geared five-bar mechanism for the seven precision positions in Problem 4. Use the $W_1 - Z_1$ and β_j solutions from Problem 4 along with $\gamma_j = 13\beta_j/20$. Order, branch, and circuit defect elimination can be verified through a displacement analysis.

12. Using the Appendix F.3 MATLAB files, synthesize an order, branch, and circuit defect-free geared five-bar mechanism for the four loading–unloading bucket positions in Figure P.6.8. In this problem, $\gamma_j = -11\beta_j/20$. Order, branch, and circuit defect elimination can be verified through a displacement analysis.

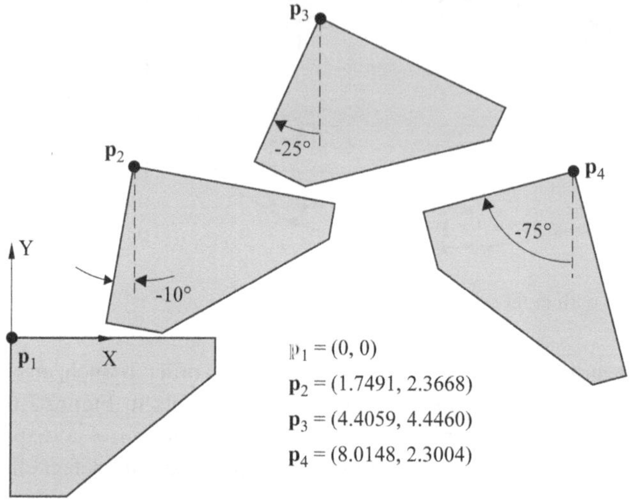

FIGURE P.6.8 Four loading–unloading bucket positions.

13. Using the Appendix F.4 MATLAB files, synthesize an order, branch, and circuit defect-free geared five-bar mechanism for the seven precision points in Problem 5. Use the $W_1 - Z_1$ and β_j solutions from Problem 5 along with $\gamma_j = -11\beta_j/20$. Order, branch, and circuit defect elimination can be verified through a displacement analysis.
14. Using the Appendix F.4 MATLAB files, synthesize an order, branch, and circuit defect-free geared five-bar mechanism for the seven precision points in Problem 6. Use the $W_1 - Z_1$ and β_j solutions from Problem 6 along with $\gamma_j = 13\beta_j/20$. Order, branch, and circuit defect elimination can be verified through a displacement analysis.
15. Using the Appendix F.4 MATLAB files, synthesize an order, branch, and circuit defect-free geared five-bar mechanism for the five precision points in Problem 7. Use the $W_1 - Z_1$ and β_j solutions from Problem 7 along with $\gamma_j = -16\beta_j/20$. Order, branch, and circuit defect elimination can be verified through a displacement analysis.
16. Using the Appendix F.4 MATLAB files, synthesize an order, branch, and circuit defect-free geared five-bar mechanism for the four precision points in Problem 8. Use the $W_1 - Z_1$ and β_j solutions from Problem 8 along with $\gamma_j = 17\beta_j/20$. Order, branch, and circuit defect elimination can be verified through a displacement analysis.
17. Using the Appendix F.5 MATLAB files, synthesize an order, branch, and circuit defect-free planar four-bar mechanism for the four loading–unloading bucket positions in Figure P.6.9. Branch defect elimination can be verified through *Filemon's Construction* or a displacement analysis (using the Appendix H.1 or I.1 files). Order, branch, and circuit defect elimination can also be verified through a displacement analysis.

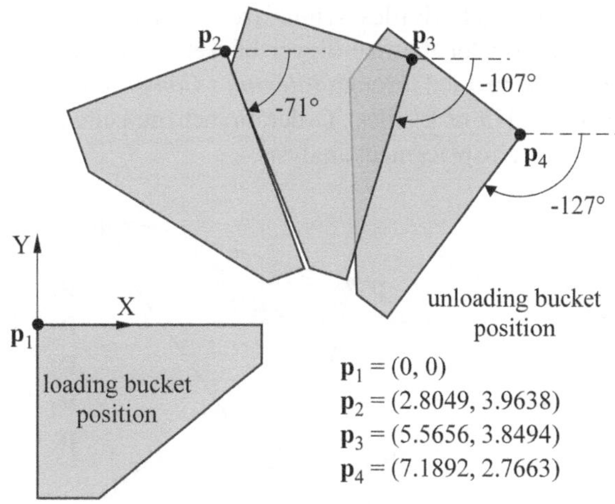

FIGURE P.6.9 Four loading–unloading bucket positions.

18. Using the Appendix F.5 MATLAB files, synthesize an order, branch, and circuit defect-free planar four-bar mechanism for the five assembly component positions in Figure P.6.10. Branch defect elimination can be verified through *Filemon's Construction* or a displacement analysis (using the Appendix H.1 or I.1 files). Order, branch, and circuit defect elimination can also be verified through a displacement analysis.

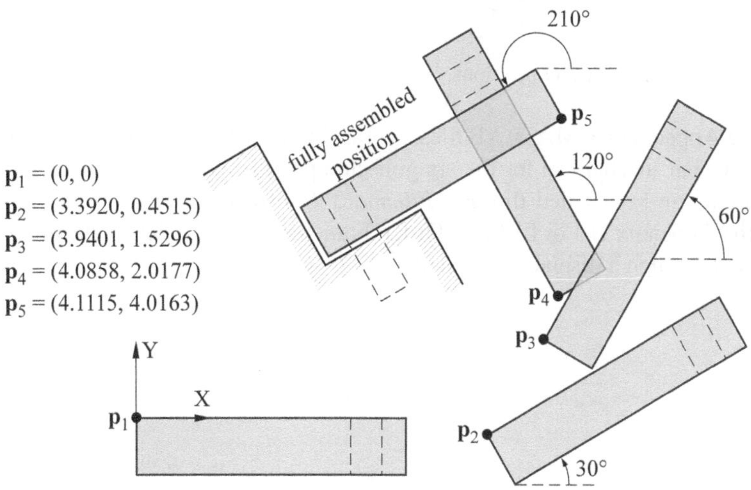

FIGURE P.6.10 Five assembly component positions.

19. Using the Appendix F.5 MATLAB files, synthesize an order, branch, and circuit defect-free planar four-bar mechanism for the planar four-bar mechanism for Problem 10 in Chapter 5. Branch defect elimination can be verified through *Filemon's Construction* or a displacement analysis (using the Appendix H.1 or I.1 files). Order, branch, and circuit defect elimination can also be verified through a displacement analysis.

20. Using the Appendix F.6 MATLAB files, synthesize an order, branch, and circuit defect-free planar four-bar mechanism for the five digger bucket positions in Figure P.6.11. Branch defect elimination can be verified through *Filemon's Construction* or a displacement analysis (using the Appendix H.1 or I.1 files). Order, branch, and circuit defect elimination can also be verified through a displacement analysis.

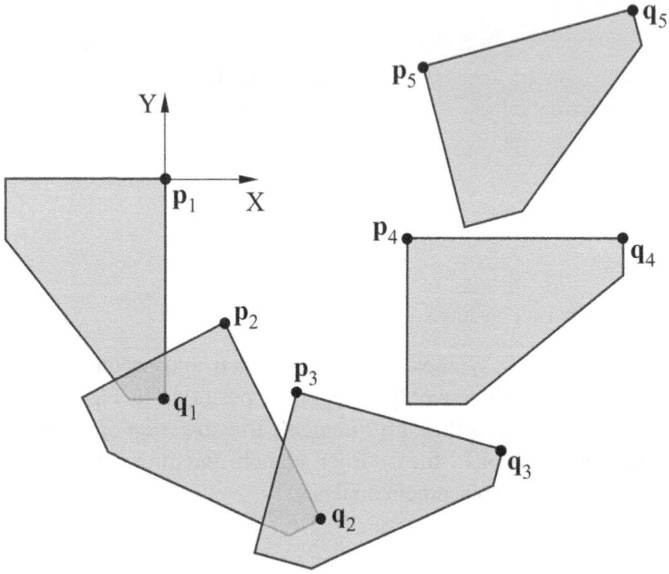

$\mathbf{p}_1 = (0, 0)$
$\mathbf{p}_2 = (0.8373, -1.9327)$
$\mathbf{p}_3 = (1.8479, -2.8558)$
$\mathbf{p}_4 = (3.4071, -0.8077)$
$\mathbf{p}_5 = (3.6136, 1.5299)$

$\mathbf{q}_1 = (0.0000, -3.2500)$
$\mathbf{q}_2 = (2.3128, -4.8285)$
$\mathbf{q}_3 = (4.9872, -3.6970)$
$\mathbf{q}_4 = (6.6571, -0.8077)$
$\mathbf{q}_5 = (6.7529, 2.3711)$

FIGURE P.6.11 Five digger bucket positions.

21. Using the Appendix F.6 MATLAB files, synthesize an order, branch, and circuit defect-free planar four-bar mechanism for the six golf club positions in Figure P.6.12. Branch defect elimination can be verified through *Filemon's Construction* or a displacement analysis (using the Appendix H.1 or I.1 files). Order, branch, and circuit defect elimination can also be verified through a displacement analysis.

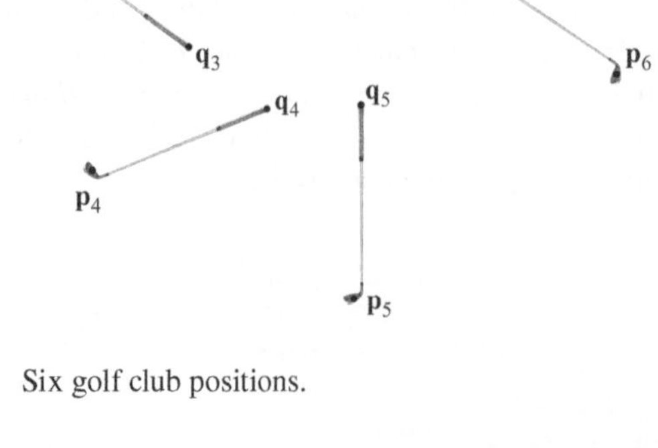

$\mathbf{p}_1 = (0, 0)$
$\mathbf{p}_2 = (-5.2712, -0.2125)$
$\mathbf{p}_3 = (-8.9671, -4.2108)$
$\mathbf{p}_4 = (-7.7905, -9.6713)$
$\mathbf{p}_5 = (-1.6841, -12.5390)$
$\mathbf{p}_6 = (4.4420, -7.5103)$

$\mathbf{q}_1 = (-3.8302, -3.2139)$
$\mathbf{q}_2 = (-4.8354, -5.1935)$
$\mathbf{q}_3 = (-4.8713, -7.0787)$
$\mathbf{q}_4 = (-3.2590, -7.5582)$
$\mathbf{q}_5 = (-1.6841, -7.5390)$
$\mathbf{q}_6 = (0.2486, -4.7871)$

FIGURE P.6.12 Six golf club positions.

22. Table P.6.1 includes the five landing gear precision positions in Example 6.1 expressed as points **p** and **q**. Using the Appendix F.6 MATLAB files, synthesize an order, branch, and circuit defect-free planar four-bar mechanism for these positions. Branch defect elimination can be verified through *Filemon's Construction* or a displacement analysis (using the Appendix H.1 or I.1 files). Order, branch, and circuit defect elimination can also be verified through a displacement analysis.

TABLE P.6.1
Landing Gear Precision Positions

Precision Position	$\mathbf{p}_j$	$\mathbf{q}_j$
1	0, 0	2, 0
2	0.203, 0.415	2.016, −0.430
3	0.397, 0.848	1.929, −0.438
4	0.385, 1.182	1.069, −0.697
5	0.182, 1.460	0.182, −0.540

23. Table P.6.2 includes the five gripper jaw precision positions in Example 5.3 expressed as points **p** and **q**. Using the Appendix F.6 MATLAB files, synthesize an order, branch, and circuit defect-free planar four-bar mechanism for these positions. Branch defect elimination can be verified through *Filemon's Construction* or a displacement analysis (using the Appendix H.1 or I.1 files). Order, branch, and circuit defect elimination can also be verified through a displacement analysis.

TABLE P.6.2
Gripper Jaw Precision Positions

Precision Position	$\mathbf{p}_j$	$\mathbf{q}_j$
1	0, 0	0.6470, 2.4148
2	0.1104, 0.1789	1.3604, 2.3440
3	0.1815, 0.4882	1.9493, 2.2560
4	0.3700, 0.9775	2.6880, 1.9140
5	0.6647, 1.4078	3.1647, 1.4078

24. Using the Appendix F.7 MATLAB files, synthesize a defect-free planar four-bar mechanism for the eight precision points in Problem 6. Defect elimination can be verified through a displacement analysis using the Appendix H.1 or I.1 files.
25. Using the Appendix F.7 MATLAB files, synthesize a defect-free planar four-bar mechanism for the eight precision points in Problem 7. Defect elimination can be verified through a displacement analysis using the Appendix H.1 or I.1 files.
26. Using the Appendix F.7 MATLAB files, synthesize a defect-free planar four-bar mechanism for the eight precision points in Problem 8. Defect elimination can be verified through a displacement analysis using the Appendix H.1 or I.1 files.
27. Using the Appendix F.7 MATLAB files, synthesize a defect-free planar four-bar mechanism for the eight precision points in Figure P.6.13. Defect elimination can be verified through a displacement analysis using the Appendix H.1 or I.1 files.

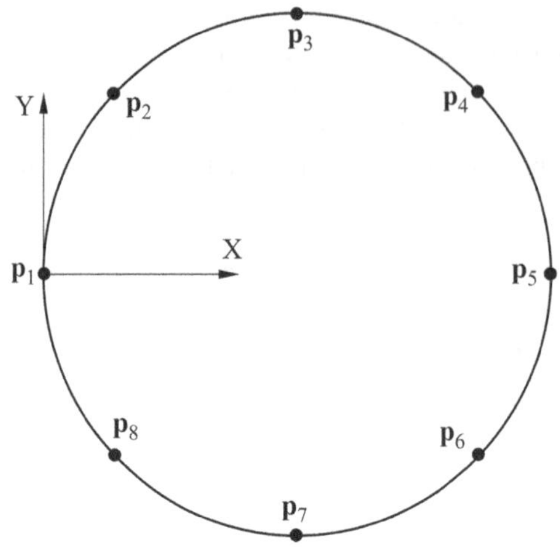

$\mathbf{p}_1 = (0, 0)$
$\mathbf{p}_2 = (0.2792, 0.6931)$
$\mathbf{p}_3 = (1, 1)$
$\mathbf{p}_4 = (1.7208, 0.6931)$
$\mathbf{p}_5 = (2, 0)$
$\mathbf{p}_6 = (1.7208, -0.6931)$
$\mathbf{p}_7 = (1, -1)$
$\mathbf{p}_8 = (0.2792, -0.6931)$

FIGURE P.6.13 Curve with eight selected precision points.

28. Using the Appendix F.1 MATLAB files, synthesize a defect-free Stephenson III mechanism for Problem 22 in Chapter 5. Defect elimination can be verified through a displacement analysis using the Appendix B.4 or C.4 files.
29. Using the Appendix F.1 MATLAB files, synthesize a defect-free Stephenson III mechanism for Problem 23 in Chapter 5. Defect elimination can be verified through a displacement analysis using the Appendix B.4 or C.4 files.
30. Using the Appendix F.1 MATLAB files, synthesize a defect-free Stephenson III mechanism for Problem 24 in Chapter 5. Defect elimination can be verified through a displacement analysis using the Appendix B.4 or C.4 files.

NOTES

1 In path generation, the prescribed rigid-body path points are also called *precision points*.
2 As derived by Wunderlich (1963), the maximum order of a coupler curve m for a mechanism with n links connected with revolute joints only is $m = 3 \cdot 2^{\left(n/2 - 1\right)}$.
3 Other straight-line tracing linkages include the Watt, Chebyschev, Evans, Hart, Peaucellier, and Hoeken straight-line linkages.
4 An arc has a constant finite radius, while a straight line has a constant radius of infinity.
5 As a result, the phrase *approximate synthesis* is sometimes used to describe dimensional synthesis via numerical optimization.
6 This algorithm was also used to solve the nonlinear simultaneous equation systems presented in this book.
7 The unknown objective function variables to be calculated are denoted by $\mathbf{X} = (\ldots)$.
8 A spatial form of these planar four-bar displacement equations is presented in Section 8.2.1 for the kinematic analysis of the RRSS and 4R spherical mechanisms.
9 Matrix I in Equation (6.10) is a 3×3 identity matrix.
10 Because the planar four-bar mechanism has two configurations or *branches* for a given crank position (the *open* and *crossed* configurations), two α solutions exist (angles α_1 and α_2 in Equations (5.12)) for every value of θ.
11 Unlike the vector-loop closure equations, Equation (6.16) does not include crank/follower link rotation angle variables and subsequently does not include crank rotation angle constraints.
12 This unconstrained equation system is the squared sum of Equation (6.16) for precision positions $2 \ldots N$.
13 Because the same Appendix F.5 MATLAB files used to calculate a_{1y} and $\mathbf{a}_1$ are also used to calculate b_{1y} and $\mathbf{b}_1$, when calculating the latter, $a_{0y} = b_{0y}$ and $\mathbf{a}_1 = \mathbf{b}_1$.

14 The length of a vector **u** is called the **norm** of **u**. The norm of a vector $\mathbf{u} = (u_x, u_y)$ in two-space is $\|\mathbf{u}\| = \sqrt{u_x^2 + u_y^2}$. The squared norm of a vector difference **u**–**v** is $\|\mathbf{u} - \mathbf{v}\|^2 = (u_x - v_x)^2 + (u_y - v_y)^2$.

15 It can be observed from Equation (6.13) that the coupler displacement angle α is a function of the crank displacement angle θ in the algebraic planar four-bar displacement model. In the constrained nonlinear optimization models, however, both angles are calculated simultaneously. Because α and θ are calculated, Equations (6.9) through (6.13) are not used in the optimization models. Equation (6.5) does constrain α and θ such that an "$\alpha(\theta)$" relationship is ensured.

16 In this motion generation equations, the coupler link rotation variable α is not prescribed because the precision position is defined by two points rather than a point and an angle.

17 The alpha values in Table E.6.9 were calculated using $\alpha = \text{acos}\left(\mathbf{V}_1 \cdot \mathbf{V}_j / \|\mathbf{V}_1\| \|\mathbf{V}_j\|\right)$, where $\mathbf{V}_1 = \mathbf{a}_1 - \mathbf{b}_1$ and $\mathbf{V} = \mathbf{a} - \mathbf{b}$. This is because the α values originally calculated in the Appendix H.1 file are the differences between angles $\angle \mathbf{b} - \mathbf{a} - \mathbf{a}_0$ and $\angle \mathbf{b}_1 - \mathbf{a}_1 - \mathbf{a}_0$.

REFERENCES

1. Wunderlich, Walter. 1963. Höhere Koppelkurven. *Österreichisches Ingenieur Archiv*, 17: 162–165.
2. Norton, R. L. 2008. *Design of Machinery*, 4th Edition. pages 140–147. New York: McGraw-Hill.
3. Hrones, J. A. and Nelson, G. L. 1951. *Analysis of the Fourbar Linkage.* Cambridge: MIT Technology Press.
4. Norton, R. L. 2008. *Design of Machinery*, 4th Edition. pages 147–152. New York: McGraw-Hill.
5. Norton, R. L. 2008. *Design of Machinery*, 4th Edition. chapter 8. pages 147–152. New York: McGraw-Hill.
6. Myszka, D. H. 2005. *Machines and Mechanisms: Applied Kinematic Analysis*, 3rd Edition. chapter 9. Saddle River: Prentice Hall.
7. Waldron, K. J. and Kinzel, G. L. 2004. *Kinematics, Dynamics and Design of Machinery*, 2nd Edition. chapter 8. Saddle River: Prentice-Hall.
8. Sandor, G. N. and Erdman, A. G. 1984. *Advanced Mechanism Design: Analysis and Synthesis*, Volume 2. pages 53–54, 62. Englewood Cliffs: Prentice-Hall.
9. Nocedal, J. and Wright, S. J. 2006. *Numerical Optimization.* pages 73–91. New York: Springer.
10. Powell, M. J. D. 1970. A new algorithm for unconstrained optimization. In *Nonlinear Programming*, ed. J. B. Rosen, O. L. Mangasarian, and K. Ritter. pages 31–66. London: Academic Press.
11. Nocedal, J. and Wright, S. J. 2006. *Numerical Optimization.* pages 578–593. New York: Springer.
12. Suh, C. H. and Radcliffe, C. W. 1978. *Kinematics and Mechanisms Design.* pages 83–85. New York: John Wiley & Sons.
13. Suh, C. H. and Radcliffe, C. W. 1978. *Kinematics and Mechanisms Design.* pages 51, 52, 129–132. New York: John Wiley & Sons.

7 Planar Four-Bar Function Generation

CONCEPT OVERVIEW

In this chapter, the reader will gain a central understanding regarding the following:

1. Mathematical function representation as link displacement angles for function generation
2. Analytical planar four-bar function generation with finitely separated and multiply separated positions
3. Planar four-bar function generation with finitely separated and multiply separated positions using simultaneous equations and unconstrained optimization models and constrained optimization models

7.1 INTRODUCTION TO FUNCTION GENERATION

In function generation, the mechanism dimensions required to achieve or approximate the prescribed link displacement angles are calculated.[1] The displacement angles are commonly defined according to a mathematical function. In four-bar function generation, the angular displacements of both the crank and follower links are typically specified, where the angular displacement of the follower link is a function of the crank link angular displacement (Figure 7.1a).[2] To further convey function generation, Figure 7.1b illustrates a four-bar lawn sprinkling mechanism (where the sprinkler is affixed to the follower link). For effective operation, the sprinkler should oscillate within a particular range to avoid over-sprinkling or under-sprinkling of the lawn area. The planar four-bar function generation models presented in this chapter are useful for calculating the planar four-bar mechanism dimensions required to achieve or approximate the prescribed link angular displacements, velocities, and accelerations.

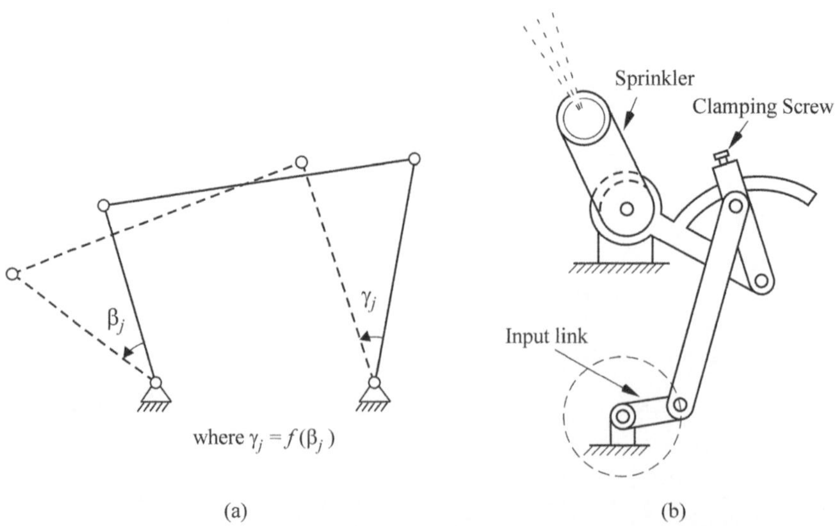

FIGURE 7.1 Function generation concepts.

DOI: 10.1201/9781003591375-7

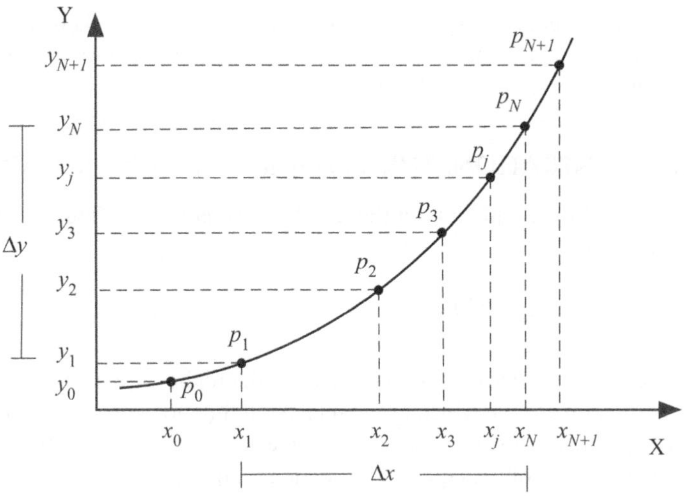

FIGURE 7.2 Function curve with precision points.

As explained earlier, the precision points in four-bar function generation are commonly crank and follower displacement angles, where the follower angles are defined as a function of crank angles. In Figure 7.1a, the crank displacement angles β_j correspond to the independent function variable "x" and the follower displacement angles γ_j correspond to a user-defined function $f(x)$ [1,2].

The curve in Figure 7.2 represents an example function to be achieved through function generation. From this function, precision points (p_j) are selected. Because only a finite number of precision points can be prescribed through analytical function generation, the resulting function generator will achieve $f(x)$ at p_j or simply $f(x_j)$, as opposed to achieving the function continuously.

The abscissa and ordinate ranges for the function are Δx and Δy, respectively, and the corresponding crank and follower displacement angle ranges are $\Delta\beta$ and $\Delta\gamma$, respectively (Figure 7.2). Linear relationships between the selected p_j values of a given function and the corresponding crank and follower precision points (β_j and γ_j, respectively) can be expressed as

$$\begin{aligned} \frac{\beta_j-\beta_1}{x_j-x_1} &= \frac{\Delta\beta}{\Delta x} \\ \frac{\gamma_j-\gamma_1}{y_j-y_1} &= \frac{\Delta\gamma}{\Delta y} \end{aligned} \tag{7.1}$$

Angles β_1 and γ_1 are 0 because they are the displacement angles at the initial mechanism position. Solving for β_j and γ_j, Equation (7.1) can be expressed as Equation (7.2), where β_j and γ_j are the crank and follower precision points corresponding to the coordinates x_j and y_j of the function precision points p_j, respectively. Crank and follower link displacement angles are among the precision points for the planar four-bar function generation equations in this chapter.[3]

$$\begin{aligned} \beta_j &= \frac{\Delta\beta}{\Delta x}\left(x_j-x_1\right) \\ \gamma_j &= \frac{\Delta\gamma}{\Delta y}\left(y_j-y_1\right) \end{aligned} \tag{7.2}$$

While Equation (7.2) is needed to specify β_j and γ_j according to a mathematical function, this is not the only option. One can also specify these displacement angles according to one's own preferences. This approach is demonstrated in Example 7.1.

7.2 FUNCTION GENERATION: THREE, FOUR, AND FIVE PRECISION POINTS

The general vector loop closure equation for the four-bar mechanism in Figure 7.3 in the *j*th position (the displaced position) is

$$W_1 e^{i(\theta+\beta_j)} + V_1 e^{i(\rho+\alpha_j)} - U_1 e^{i(\sigma+\gamma_j)} - \mathbf{G}_1 = 0 \tag{7.3}$$

Being always in-line with the *x*-axis, vector $\mathbf{G}_1$ only has the real component G_{1x}. Because link *proportions* affect link rotation angles (which are of interest in function generation) and not the *scale* of the mechanism, a single mechanism link length variable can be specified.[4] After specifying $G_{1x} = 1$ and moving it to the right side of the equation, Equation (7.3) becomes

$$W_1 e^{i(\theta+\beta_j)} + V_1 e^{i(\rho+\alpha_j)} - U_1 e^{i(\sigma+\gamma_j)} = 1 \tag{7.4}$$

Expanding Equation (7.4) and grouping the real and imaginary terms as separate equations produce

$$\begin{aligned} &W_1 \cos\theta\cos\beta_j - W_1 \sin\theta\sin\beta_j + V_1\cos\rho\cos\alpha_j - V_1\sin\rho\sin\alpha_j - U_1\cos\sigma\cos\gamma_j \\ &+U_1\sin\sigma\sin\gamma_j - 1 = 0 \\ &W_1\sin\theta\cos\beta_j + W_1\cos\theta\sin\beta_j + V_1\sin\rho\cos\alpha_j + V_1\cos\rho\sin\alpha_j - U_1\sin\sigma\cos\gamma_j \\ &-U_1\cos\sigma\sin\gamma_j = 0 \end{aligned} \tag{7.5}$$

After setting $W_1\cos\theta = W_{1x}$, $W_1\sin\theta = W_{1y}$, $V_1\cos\rho = V_{1x}$, $V_1\sin\rho = V_{1y}$, $U_1\cos\sigma = U_{1x}$, and $U_1\sin\sigma = U_{1y}$, Equation (7.5) can be rewritten as

$$\begin{aligned} &W_{1x}\cos\beta_j - W_{1y}\sin\beta_j + V_{1x}\cos\alpha_j - V_{1y}\sin\alpha_j - U_{1x}\cos\gamma_j + U_{1y}\sin\gamma_j - 1 = 0 \\ &W_{1y}\cos\beta_j + W_{1x}\sin\beta_j + V_{1y}\cos\alpha_j + V_{1x}\sin\alpha_j - U_{1y}\cos\gamma_j - U_{1x}\sin\gamma_j = 0 \end{aligned} \tag{7.6}$$

When expressed in matrix form for three precision points, Equation (7.6) becomes

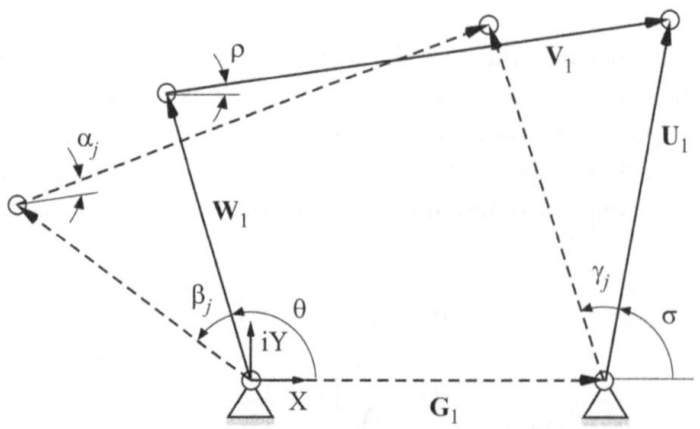

FIGURE 7.3 Four-bar mechanism in starting and (dashed) displaced positions.

$$\begin{bmatrix} 1 & 0 & 1 & 0 & -1 & 0 \\ 0 & 1 & 0 & 1 & 0 & -1 \\ \cos\beta_2 & -\sin\beta_2 & \cos\alpha_2 & -\sin\alpha_2 & -\cos\gamma_2 & \sin\gamma_2 \\ \sin\beta_2 & \cos\beta_2 & \sin\alpha_2 & \cos\alpha_2 & -\sin\gamma_2 & -\cos\gamma_2 \\ \cos\beta_3 & -\sin\beta_3 & \cos\alpha_3 & -\sin\alpha_3 & -\cos\gamma_3 & \sin\gamma_3 \\ \sin\beta_3 & \cos\beta_3 & \sin\alpha_3 & \cos\alpha_3 & -\sin\gamma_3 & -\cos\gamma_3 \end{bmatrix} \begin{Bmatrix} W_{1x} \\ W_{1y} \\ V_{1x} \\ V_{1y} \\ U_{1x} \\ U_{1y} \end{Bmatrix} = \begin{bmatrix} 1 \\ 0 \\ 1 \\ 0 \\ 1 \\ 0 \end{bmatrix} \quad (7.7)$$

The first two rows in Equation (7.7) are the result of displacement angles β_1, α_1, and γ_1 (corresponding to the initial mechanism position), all being 0 in Equation (7.6).

Therefore, with three precision points, Equation (7.6) forms Equation (7.7)—a linear simultaneous equation system to calculate the components of link vectors $\mathbf{W}_1$, $\mathbf{V}_1$, and $\mathbf{U}_1$. In addition to prescribing the crank and follower displacement angles, the coupler displacement angles α_2 and α_3 are also prescribed in Equation (7.7). This simultaneous equation system can be solved analytically using Cramer's rule.

Given four or five precision points, Equation (7.6) produces Equation (7.8)—a nonlinear simultaneous equation system (where $j = 2, 3, \ldots, N$). With four precision points, $N = 4$. The user can specify any one of the three coupler displacement angles (where the remaining coupler displacement angles and the components of $\mathbf{W}_1$, $\mathbf{V}_1$, and $\mathbf{U}_1$ are calculated). With five precision points, $N = 5$, and all of the four coupler displacement angles are calculated along with the components of $\mathbf{W}_1$, $\mathbf{V}_1$, and $\mathbf{U}_1$.

$$\begin{gathered} W_{1x} + V_{1x} - U_{1x} - 1 = 0 \\ W_{1y} + V_{1y} - U_{1y} = 0 \\ W_{1x}\cos\beta_j - W_{1y}\sin\beta_j + V_{1x}\cos\alpha_j - V_{1y}\sin\alpha_j - U_{1x}\cos\gamma_j + U_{1y}\sin\gamma_j - 1 = 0 \\ W_{1y}\cos\beta_j + W_{1x}\sin\beta_j + V_{1y}\cos\alpha_j + V_{1x}\sin\alpha_j - U_{1y}\cos\gamma_j - U_{1x}\sin\gamma_j = 0 \\ \vdots \\ W_{1x}\cos\beta_N - W_{1y}\sin\beta_N + V_{1x}\cos\alpha_N - V_{1y}\sin\alpha_N - U_{1x}\cos\gamma_N + U_{1y}\sin\gamma_N - 1 = 0 \\ W_{1y}\cos\beta_N + W_{1x}\sin\beta_N + V_{1y}\cos\alpha_N + V_{1x}\sin\alpha_N - U_{1y}\cos\gamma_N - U_{1x}\sin\gamma_N = 0 \end{gathered} \quad (7.8)$$

Table 7.1 includes the maximum number of equations, unknowns, free choices and solutions in planar four-bar function generation for 3.4 and 5 precisions points.

TABLE 7.1

Maximum Number of Solutions for Equation (7.8) for Three, Four, and Five Precision Points

Number of Precision Points	Number of Equations	Number of Unknowns	Number of Free Choices	Number of Solutions
3	6	(8) $W_{1x}, W_{1y}, V_{1x}, V_{1y}, U_{1x}, U_{1y}, \alpha_2, \alpha_3$	2	Infinite
4	8	(9) $W_{1x}, W_{1y}, V_{1x}, V_{1y}, U_{1x}, U_{1y}, \alpha_2, \alpha_3, \alpha_4$	1	Infinite
5	10	(10) $W_{1x}, W_{1y}, V_{1x}, V_{1y}, U_{1x}, U_{1y}, \alpha_2, \alpha_3, \alpha_4, \alpha_5$	0	Finite

Example 7.1

Problem Statement: Synthesize a planar four-bar sprinkler mechanism to achieve a follower rotation range of 60° for a corresponding crank rotation range of 180°.

Known Information: Equation (7.7)

Solution Approach: The crank and follower rotation ranges have been equally divided into the displacement angles given in Table E.7.1. This table also includes arbitrarily specified coupler link displacement angles.

Figure E.7.1 includes the calculation procedure in the MATLAB command window and Figure E.7.2 illustrates the resulting planar four-bar mechanism. Being an analytically calculated result, this mechanism achieves the precision points precisely.

TABLE E.7.1
Four-Bar Sprinkler Mechanism Precision Points

Precision Point	β_j (deg)	γ_j (deg)	α_j (deg)
1	0	0	0
2	90	30	–5
3	180	60	10

```
>> B2 = 90*pi/180; B3 = 180*pi/180;
>> A2 = -5*pi/180; A3 = 10*pi/180;
>> G2 = 30*pi/180; G3 = 60*pi/180;

>> CB2 = cos(B2); SB2 = sin(B2); CB3 = cos(B3); SB3 = sin(B3);
>> CG2 = cos(G2); SG2 = sin(G2); CG3 = cos(G3); SG3 = sin(G3);
>> CA2 = cos(A2); SA2 = sin(A2); CA3 = cos(A3); SA3 = sin(A3);

>> WVU = inv([1, 0, 1, 0, -1, 0
0, 1, 0, 1, 0, -1
CB2, -SB2, CA2, -SA2, -CG2, SG2
SB2, CB2, SA2, CA2, -SG2, -CG2
CB3, -SB3, CA3, -SA3, -CG3, SG3
SB3, CB3, SA3, CA3, -SG3, -CG3])*[1, 0, 1, 0, 1, 0]'

WVU =

    0.2670
    0.0816
    1.0598
    0.4564
    0.3268
    0.5380

>>
```

FIGURE E.7.1 Example 7.1 $\mathbf{W}_1 - \mathbf{V}_1 - \mathbf{U}_1$ calculation procedure in the MATLAB command window.

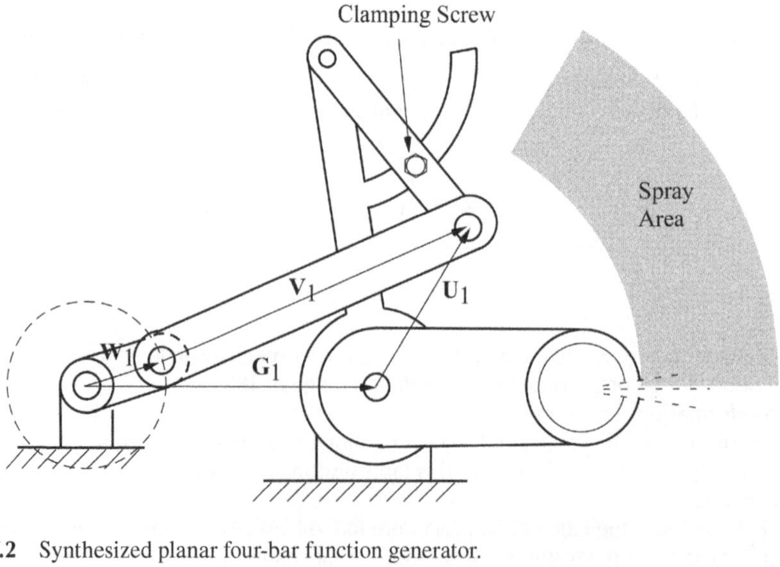

FIGURE E.7.2 Synthesized planar four-bar function generator.

Example 7.2

Problem Statement: Synthesize a planar four-bar mechanism to achieve a follower rotation range of $\Delta\gamma = 60°$ for a corresponding crank rotation range of $\Delta\beta = 90°$ according to the function $f(x) = y = \sin(x)$. Consider precision points at $x_1 = 20°$, $x_2 = 40°$, $x_3 = 60°$, and $x_4 = 80°$. Also, let $\alpha_4 = -9°$.

Known Information: Equation (7.8)

Solution Approach: Figure E.7.3 includes the calculation procedure in the MATLAB command Window. In this figure, the calculated solution appears in the following order: W_{1x}, W_{1y}, V_{1x}, V_{1y}, U_{1x}, U_{1y}, α_2, and α_3 (where α_4 is the specified free choice). In Figure E.7.3, α_2 and α_3 are given in radians. Figure E.7.4 illustrates the resulting planar four-bar mechanism. A kinematic analysis of this solution is presented in Example 7.8 in Section 7.7.

```
>> DELTA_B = 90*pi/180; DELTA_G = 60*pi/180;
>> x1 = 20*pi/180; x2 = 40*pi/180; x3 = 60*pi/180; x4 = 80*pi/180;
>> y1 = sin(x1); y2 = sin(x2); y3 = sin(x3); y4 = sin(x4);
>> DELTA_x = x4 - x1; DELTA_y = y4 - y1; A4 = -9*pi/20;

>> B2=(DELTA_B/DELTA_x)*(x2 - x1); G2=(DELTA_G/DELTA_y)*(y2 - y1);
>> B3=(DELTA_B/DELTA_x)*(x3 - x1); G3=(DELTA_G/DELTA_y)*(y3 - y1);
>> B4=(DELTA_B/DELTA_x)*(x4 - x1); G4=(DELTA_G/DELTA_y)*(y4 - y1);

>> CB2 = cos(B2); SB2 = sin(B2); CG2 = cos(G2); SG2 = sin(G2);
>> CB3 = cos(B3); SB3 = sin(B3); CG3 = cos(G3); SG3 = sin(G3);
>> CB4 = cos(B4); SB4 = sin(B4); CG4 = cos(G4); SG4 = sin(G4);

>> x0 = [1, 1, 1, 1, 1, 1, -pi/20, -pi/20];

>> fx=@(x)[x(1) + x(3) - x(5) -1;
x(2) + x(4) - x(6);
x(1)*CB2 - x(2)*SB2 + x(3)*cos(x(7)) - x(4)*sin(x(7)) - x(5)*CG2 +
x(6)*SG2 - 1;
x(2)*CB2 + x(1)*SB2 + x(4)*cos(x(7)) + x(3)*sin(x(7)) - x(6)*CG2 -
x(5)*SG2;
x(1)*CB3 - x(2)*SB3 + x(3)*cos(x(8)) - x(4)*sin(x(8)) - x(5)*CG3 +
x(6)*SG3 - 1;
x(2)*CB3 + x(1)*SB3 + x(4)*cos(x(8)) + x(3)*sin(x(8)) - x(6)*CG3 -
x(5)*SG3;
x(1)*CB4 - x(2)*SB4 + x(3)*cos(A4) - x(4)*sin(A4) - x(5)*CG4 +
x(6)*SG4 - 1;
x(2)*CB4 + x(1)*SB4 + x(4)*cos(A4) + x(3)*sin(A4) - x(6)*CG4 -
x(5)*SG4];

>> [x,fval] = fsolve(fx,x0);

>> x

x =

-0.0452 -0.1414 0.9157 0.0116 -0.1295 -0.1298 -0.0459 -0.0925
```

FIGURE E.7.3 Example 7.2 $\mathbf{W}_1$, $\mathbf{V}_1$, $\mathbf{U}_1$, α_2, and α_3 calculation procedure in the MATLAB command window.

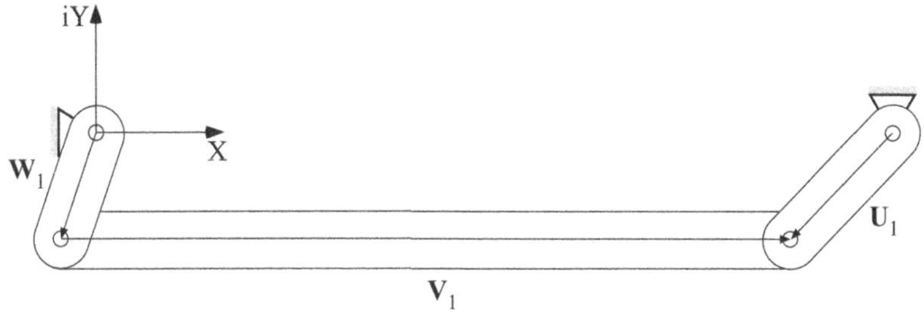

FIGURE E.7.4 Synthesized planar four-bar function generator.

Example 7.3

Problem Statement: Synthesize a planar four-bar mechanism to achieve a follower rotation range of $\Delta\gamma = 75°$ for a corresponding crank rotation range of $\Delta\beta = 120°$ according to the function $f(x) = y = x^{1.25} + 2$. Consider precision points at $x_1 = 2$, $x_2 = 4$, $x_3 = 6$, $x_4 = 8$, and $x_5 = 10$.

Known Information: Equation (7.8)

Solution Approach: Figure E.7.5 includes the calculation procedure in the MATLAB command Window. In this figure, the calculated solution appears in the following order: W_{1x}, W_{1y}, V_{1x}, V_{1y}, U_{1x}, U_{1y}, α_2, α_3, α_4, and α_5. In Figure E.7.5, α_2 through α_5 are given in radians. Figure E.7.6 illustrates the resulting planar four-bar mechanism.

```
>> DELTA_B = 90*pi/180; DELTA_G = 60*pi/180;
>> x1 = 2; x2 = 4; x3 = 6; x4 = 8; x5 = 10;
>> y1 = 2 + x1^1.25; y2 = 2 + x2^1.25; y3 = 2 + x3^1.25;
>> y4 = 2 + x4^1.25; y5 = 2 + x5^1.25;
>> DELTA_x = x4 - x1; DELTA_y = y4 - y1;

>> B2=(DELTA_B/DELTA_x)*(x2 - x1); G2=(DELTA_G/DELTA_y)*(y2 - y1);
>> B3=(DELTA_B/DELTA_x)*(x3 - x1); G3=(DELTA_G/DELTA_y)*(y3 - y1);
>> B4=(DELTA_B/DELTA_x)*(x4 - x1); G4=(DELTA_G/DELTA_y)*(y4 - y1);
>> B5=(DELTA_B/DELTA_x)*(x5 - x1); G5=(DELTA_G/DELTA_y)*(y5 - y1);

>> CB2 = cos(B2); SB2 = sin(B2); CG2 = cos(G2); SG2 = sin(G2);
>> CB3 = cos(B3); SB3 = sin(B3); CG3 = cos(G3); SG3 = sin(G3);
>> CB4 = cos(B4); SB4 = sin(B4); CG4 = cos(G4); SG4 = sin(G4);
  >> CB5 = cos(B5); SB5 = sin(B5); CG5 = cos(G5); SG5 = sin(G5);

>> x0 = [10, 10, 10, 10, 10, 10, -pi/20, -pi/20, -pi/20, -pi/20];

>> fx=@(x)[x(1) + x(3) - x(5) - 1;
x(2) + x(4) - x(6);
x(1)*CB2 - x(2)*SB2 + x(3)*cos(x(7)) - x(4)*sin(x(7)) - x(5)*CG2 +
x(6)*SG2 - 1;
x(2)*CB2 + x(1)*SB2 + x(4)*cos(x(7)) + x(3)*sin(x(7)) - x(6)*CG2 -
x(5)*SG2;
x(1)*CB3 - x(2)*SB3 + x(3)*cos(x(8)) - x(4)*sin(x(8)) - x(5)*CG3 +
x(6)*SG3 - 1;
x(2)*CB3 + x(1)*SB3 + x(4)*cos(x(8)) + x(3)*sin(x(8)) - x(6)*CG3 -
x(5)*SG3;
x(1)*CB4 - x(2)*SB4 + x(3)*cos(x(9)) - x(4)*sin(x(9)) - x(5)*CG4 +
x(6)*SG4 - 1;
x(2)*CB4 + x(1)*SB4 + x(4)*cos(x(9)) + x(3)*sin(x(9)) - x(6)*CG4 -
x(5)*SG4;
x(1)*CB5 - x(2)*SB5 + x(3)*cos(x(10)) - x(4)*sin(x(10)) - x(5)*CG5
+ x(6)*SG5 -1;
x(2)*CB5 + x(1)*SB5 + x(4)*cos(x(10)) + x(3)*sin(x(10)) - x(6)*CG5
- x(5)*SG5];

>> [x,fval] = fsolve(fx,x0);

>> x

x =

-0.7162 -0.3565 1.2802 -0.9664 -0.4360 -1.3230 0.1769 0.3214
0.4440 0.5579
```

FIGURE E.7.5 Example 7.3 $\mathbf{W}_1$, $\mathbf{V}_1$, $\mathbf{U}_1$, α_2, α_3, α_4, and α_5 calculation procedure in the MATLAB command window.

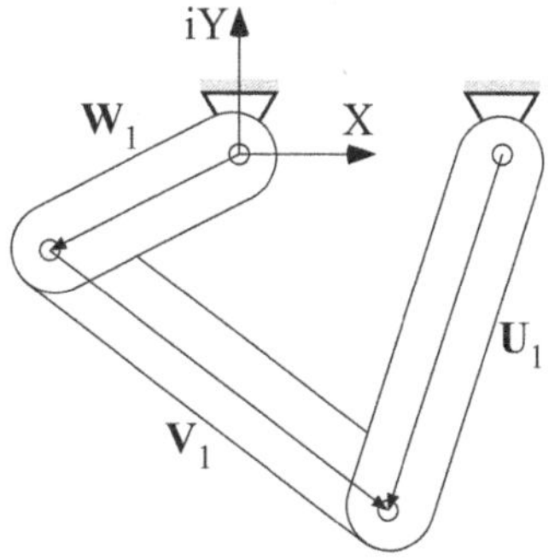

FIGURE E.7.6 Synthesized planar four-bar function generator.

7.3 FUNCTION GENERATION FOR INDEFINITE PRECISION POINTS

As described in Section 7.2, three-precision point function generation produces Equation (7.7)—a linear simultaneous equation system which can be solved analytically using Cramer's rule. Four- or five-precision point function generation produces Equation (7.8)—a nonlinear simultaneous equation system that can be solved numerically in MATLAB using the optimization algorithms in Section 2.7.1. Beyond five precision points, or for N points, the number of equations exceeds the number of unknowns in $\mathbf{W}_1$, $\mathbf{V}_1$, $\mathbf{U}_1$, and $\alpha_2 \ldots \alpha_N$. In this case, a single objective function (to be minimized) can be formulated by taking the squared sum of Equation (7.6) for $j = 1,2,3,\ldots,N$ precision points.[5] This objective function is an unconstrained nonlinear optimization model that can be solved numerically in MATLAB using the optimization algorithms in Section 2.7.1.

Equation (7.9) is an objective function formed from Equation (7.6) for N precision points, where the initial mechanism position displacement angles $\alpha_1 = \beta_1 = \gamma_1 = 0$. For Equation (7.9), $\mathbf{X} = (W_{1x}, W_{1y}, V_{1x}, V_{1y}, U_{1x}, U_{1y}, \alpha_2 \ldots \alpha_4)$ when four precision points are prescribed and $\mathbf{X} = (W_{1x}, W_{1y}, V_{1x}, V_{1y}, U_{1x}, U_{1y}, \alpha_2 \ldots \alpha_5)$ when five precision points are prescribed. For N precision points, where N is greater than 5, $\mathbf{X} = (W_{1x}, W_{1y}, V_{1x}, V_{1y}, U_{1x}, U_{1y}, \alpha_2 \ldots \alpha_N)$.

$$f(\mathbf{X}) = \sum_{j=1}^{N} \left\{ \begin{array}{l} \left(W_{1x}\cos\beta_j - W_{1y}\sin\beta_j + V_{1x}\cos\alpha_j - V_{1y}\sin\alpha_j - U_{1x}\cos\gamma_j + U_{1y}\sin\gamma_j - 1\right)^2 \\ +\left(W_{1y}\cos\beta_j + W_{1x}\sin\beta_j + V_{1y}\cos\alpha_j + V_{1x}\sin\alpha_j - U_{1y}\cos\gamma_j - U_{1x}\sin\gamma_j\right)^2 \end{array} \right\} \tag{7.9}$$

Appendix G.1 includes user instructions for the MATLAB files for Equation (7.9)—the unconstrained nonlinear optimization model for planar four-bar function generation. Using the Appendix G.1 MATLAB files (which are available for download at https://www.routledge.com/9781032956800), solutions for $\mathbf{W}_1$, $\mathbf{V}_1$, $\mathbf{U}_1$, and $\alpha_2 \ldots \alpha_N$ are calculated.

Example 7.4

Problem Statement: Synthesize a planar four-bar function generator to approximate the six flow rate (Q) and pump head (h) readings corresponding to the data points given in Figure E.7.7. Let $\Delta\beta = \Delta\gamma = 60°$.

Known Information: Figure E.7.7 and the Appendix G.1 MATLAB files

Solution Approach: Figure E.7.8 includes the calculation procedure in the MATLAB command Window for β and γ using Equation (7.2). Figure E.7.9 includes the input and output in bold type in the Appendix G.1 MATLAB file *UVW_quasi_newton.m*. In this file, $N = 6$, and initial values of $\mathbf{W}_1 = \mathbf{U}_1 = (-1,-1)$, $\mathbf{V}_1 = (-1,1)$, and 20°, 40°, ...,100° for $\alpha_2 \ldots \alpha_6$, respectively, were specified. Figure E.7.10 illustrates the resulting planar four-bar mechanism. A kinematic analysis of this solution is presented in Example 7.9.

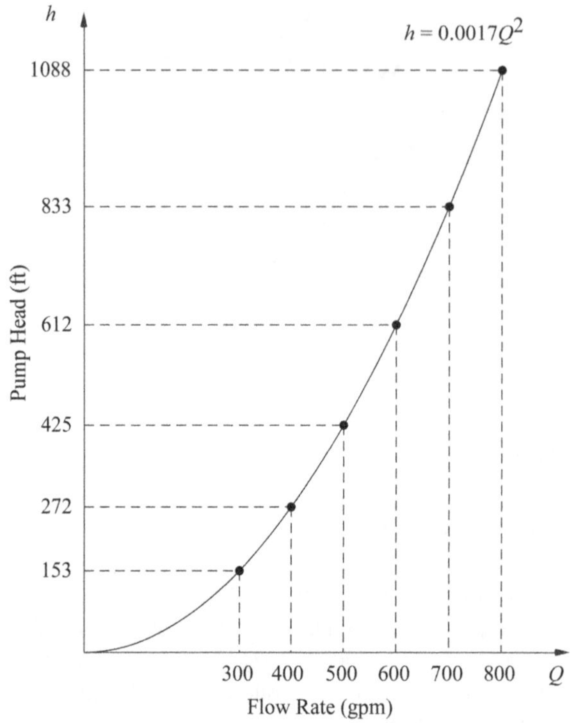

FIGURE E.7.7 Pump head vs. flow rate curve with selected data points.

```
>> DELTA_B = 60*pi/180; DELTA_G = 60*pi/180;
>> Q1 = 300; Q2 = 400; Q3 = 500; Q4 = 600; Q5 = 700; Q6 = 800;
>> h1 = 153; h2 = 272; h3 = 425; h4 = 612; h5 = 833; h6 = 1088;
>> DELTA_Q = Q6 - Q1; DELTA_h = h6 - h1;

>> B = [1:6]; G = [1:6];

>> B(1) = 0; G(1) = 0;
>> B(2) = (DELTA_B/DELTA_Q)*(Q2 - Q1);
>> B(3) = (DELTA_B/DELTA_Q)*(Q3 - Q1);
>> B(4) = (DELTA_B/DELTA_Q)*(Q4 - Q1);
>> B(5) = (DELTA_B/DELTA_Q)*(Q5 - Q1);
>> B(6) = (DELTA_B/DELTA_Q)*(Q6 - Q1);

>> G(3) = (DELTA_G/DELTA_h)*(h3 - h1);
>> G(4) = (DELTA_G/DELTA_h)*(h4 - h1);
>> G(5) = (DELTA_G/DELTA_h)*(h5 - h1);
>> G(6) = (DELTA_G/DELTA_h)*(h6 - h1);

>> BETA = B*180/pi

BETA =

         0   12.0000   24.0000   36.0000   48.0000   60.0000

>> GAMMA = G*180/pi

GAMMA =

         0    7.6364   17.4545   29.4545   43.6364   60.0000

>>
```

FIGURE E.7.8 Example 7.4 β and γ calculation procedure in the MATLAB command window.

```
N = 6;

beta2 = 12 *pi/180;
beta3 = 24 *pi/180;
beta4 = 36 *pi/180;
beta5 = 48 *pi/180;
beta6 = 60 *pi/180;

gamma2 = 7.6364 *pi/180;
gamma3 = 17.4545 *pi/180;
gamma4 = 29.4545 *pi/180;
gamma5 = 43.6364 *pi/180;
gamma6 = 60 *pi/180;

U1x = -1;
U1y = -1;
V1x = -1;
V1y = 1;
W1x = -1;
W1y = -1;

alpha2 = 20 *pi/180;
alpha3 = 20 *pi/180;
alpha4 = 20 *pi/180;
alpha5 = 20 *pi/180;
alpha6 = 20 *pi/180;

Iterations: 497
Func Evals: 621
Algorithm: quasi-newton
exit flag = 5
error = ( 2.233656e-09 )
U1x = ( -0.383422 )
U1y = ( -0.734592 )
V1x = ( 3.289653 )
V1y = ( -1.321894 )
W1x = ( -2.673080 )
W1y = ( 0.587305 )
alpha2 = ( 8.888312 )
alpha3 = ( 17.604281 )
alpha4 = ( 26.225223 )
alpha5 = ( 34.887564 )
alpha6 = ( 43.784938 )
Elapsed time is 0.271374 seconds.
Solution calculation is complete.
>>
```

FIGURE E.7.9 Input and output in bold type in the Appendix G.1 MATLAB file *UVW_quasi_newton.m* for Example 7.4.

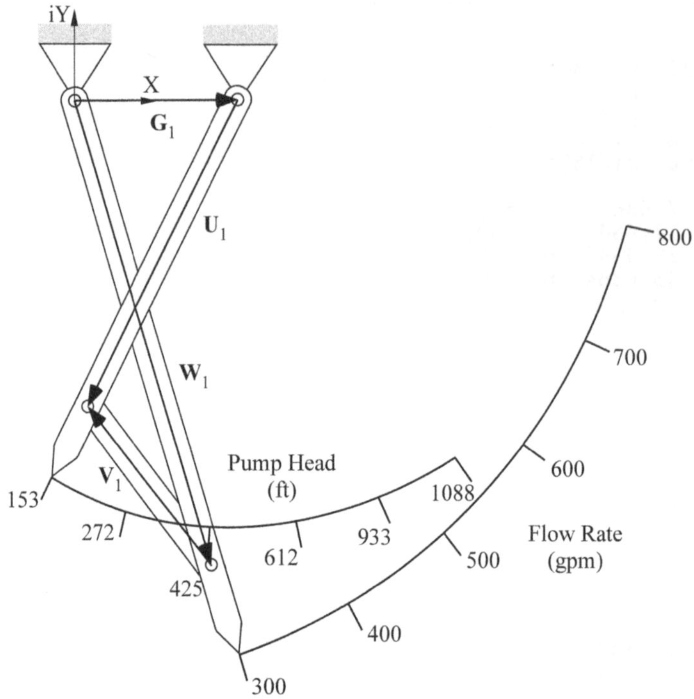

FIGURE E.7.10 Synthesized planar four-bar function generator.

7.4 VELOCITY AND ACCELERATION CONSTRAINTS FOR FUNCTION GENERATION

In addition to link displacement angles, it is also possible to synthesize a function generator to achieve or approximate prescribed link angular velocities and accelerations. Angular displacements (being discrete, finitely separated quantities) are called *finitely separated positions* (FSPs), and derivative quantities such as angular velocities and accelerations are *multiply separated positions* (MSPs). Figure 7.4 illustrates the function generator angular velocity and angular acceleration variables—the first and second derivatives of angular displacements β, α, and γ.

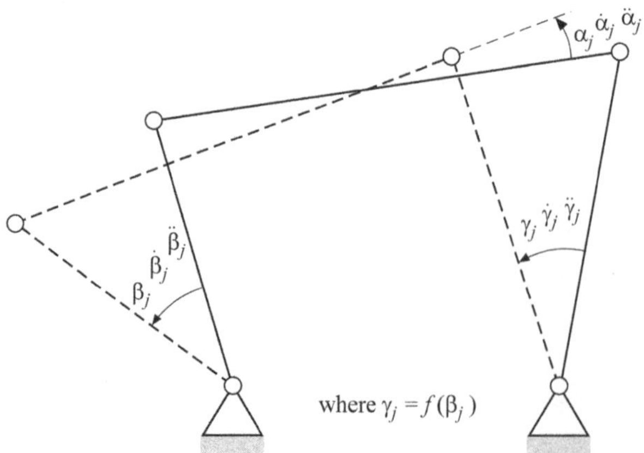

FIGURE 7.4 Planar four-bar function generator with angular velocity and acceleration variables.

By differentiating, expanding, and separating vector loop Equation (7.4), mechanism velocity and acceleration constraint equations are derived for the planar four-bar function generator. Differentiating Equation (7.4) yields

$$\dot{\beta}_j W_1 e^{i(\theta+\beta_j)} + \dot{\alpha}_j V_1 e^{i(\rho+\alpha_j)} - \dot{\gamma}_j U_1 e^{i(\sigma+\gamma_j)} = 0 \tag{7.10}$$

Expanding Equation (7.10) and grouping the real and imaginary terms into separate equations produce

$$\begin{aligned} &\dot{\beta}_j\left(W_{1x}\cos\beta_j - W_{1y}\sin\beta_j\right) + \dot{\alpha}_j\left(V_{1x}\cos\alpha_j - V_{1y}\sin\alpha_j\right) - \dot{\gamma}_j\left(U_{1x}\cos\gamma_j - U_{1y}\sin\gamma_j\right) = 0 \\ &\dot{\beta}_j\left(W_{1x}\sin\beta_j + W_{1y}\cos\beta_j\right) + \dot{\alpha}_j\left(V_{1x}\sin\alpha_j + V_{1y}\cos\alpha_j\right) - \dot{\gamma}_j\left(U_{1x}\sin\gamma_j + U_{1y}\cos\gamma_j\right) = 0 \end{aligned} \tag{7.11}$$

Equation (7.11) constitutes planar four-bar function generator velocity constraints (introducing link velocity terms $\dot{\alpha}$, $\dot{\beta}$, and $\dot{\gamma}$). Differentiating Equation (7.10) yields

$$i\ddot{\beta}_j W_1 e^{i(\theta+\beta_j)} - \dot{\beta}_j^2 W_1 e^{i(\theta+\beta_j)} + i\ddot{\alpha}_j V_1 e^{i(\rho+\alpha_j)} - \dot{\alpha}_j^2 V_1 e^{i(\rho+\alpha_j)} - i\ddot{\gamma}_j U_1 e^{i(\sigma+\gamma_j)} + \dot{\gamma}_j^2 U_1 e^{i(\sigma+\gamma_j)} = 0 \tag{7.12}$$

Expanding Equation (7.12) and grouping the real and imaginary terms into separate equations produce

$$\begin{aligned} &-\ddot{\beta}_j\left(W_{1x}\sin\beta_j + W_{1y}\cos\beta_j\right) - \dot{\beta}_j^2\left(W_{1x}\cos\beta_j - W_{1y}\sin\beta_j\right) - \ddot{\alpha}_j\left(V_{1x}\sin\alpha_j + V_{1y}\cos\alpha_j\right) \\ &-\dot{\alpha}_j^2\left(V_{1x}\cos\alpha_j - V_{1y}\sin\alpha_j\right) + \ddot{\gamma}_j\left(U_{1x}\sin\gamma_j + U_{1y}\cos\gamma_j\right) + \dot{\gamma}_j^2\left(U_{1x}\cos\gamma_j - U_{1y}\sin\gamma_j\right) = 0 \\ &\ddot{\beta}_j\left(W_{1x}\cos\beta_j - W_{1y}\sin\beta_j\right) - \dot{\beta}_j^2\left(W_{1x}\sin\beta_j + W_{1y}\cos\beta_j\right) + \ddot{\alpha}_j\left(V_{1x}\cos\alpha_j - V_{1y}\sin\alpha_j\right) \\ &-\dot{\alpha}_j^2\left(V_{1x}\sin\alpha_j + V_{1y}\cos\alpha_j\right) - \ddot{\gamma}_j\left(U_{1x}\cos\gamma_j - U_{1y}\sin\gamma_j\right) + \dot{\gamma}_j^2\left(U_{1x}\sin\gamma_j + U_{1y}\cos\gamma_j\right) = 0 \end{aligned} \tag{7.13}$$

Equation (7.13) constitutes planar four-bar function generator acceleration constraints—introducing link angular acceleration terms $\ddot{\alpha}$, $\ddot{\beta}$, and $\ddot{\gamma}$, in addition to the link angular velocity terms $\dot{\alpha}$, $\dot{\beta}$, and $\dot{\gamma}$ introduced in Equation (7.11).

7.5 FUNCTION GENERATION WITH FINITE AND MULTIPLY SEPARATED POSITIONS

As presented in Table 7.1, three precision points for planar four-bar function generation produce a linear simultaneous equation system to calculate the scalar components of link vectors $\mathbf{W}_1$, $\mathbf{V}_1$, and $\mathbf{U}_1$. A similar equation system to calculate the same vector components can also be formed from Equation (7.6) (with $\beta = \alpha = \gamma = 0$) and combinations of Equations (7.6), (7.11), and/or (7.13). With such an equation set, the user can specify not only link angular displacements (FSPs) but also corresponding link angular velocities and/or accelerations (MSPs). This equation set can be solved analytically using Cramer's rule. The resulting planar four-bar mechanism will precisely achieve the prescribed angular displacements, velocities, and/or accelerations.

There are two practical combinations of Equations (7.6), (7.11), and (7.13) to produce a 6×6 linear equation system to calculate $\mathbf{W}_1$, $\mathbf{V}_1$, and $\mathbf{U}_1$. One combination is Equation (7.6) with $\beta = \alpha = \gamma = 0$;

Equation (7.6) with nonzero β, α, and γ values; and Equation (7.11) with the same nonzero β, α, and γ values. Another combination is Equations (7.6), (7.11), and (7.13), all with $\beta = \alpha = \gamma = 0$.

Example 7.5

Problem Statement: Synthesize a planar four-bar function generator to achieve angular displacements of $\beta = 55°$, $\alpha = -15°$, and $\gamma = -37°$, along with angular velocities of $\dot{\beta} = 0.1$ rad/s, $\dot{\alpha} = -0.25$ rad/s, and $\dot{\gamma} = -0.2$ rad/s.

Known Information: Equations (7.6) and (7.11)

Solution Approach: A 6 × 6 linear simultaneous equation system to calculate the components of link vectors $\mathbf{W}_1$, $\mathbf{V}_1$, and $\mathbf{U}_1$ is formed from Equation (7.6), with $\beta = \alpha = \gamma = 0$; Equation (7.6) with the given nonzero β, α, and γ values; and Equation (7.11) with the given β, $\dot{\beta}$, α, $\dot{\alpha}$, γ, and $\dot{\gamma}$ values.[6] When expressed in matrix form, these equations become

$$\begin{bmatrix} 1 & 0 & 1 & 0 & -1 & 0 \\ 0 & 1 & 0 & 1 & 0 & -1 \\ \cos\beta_2 & -\sin\beta_2 & \cos\alpha_2 & -\sin\alpha_2 & -\cos\gamma_2 & \sin\gamma_2 \\ \sin\beta_2 & \cos\beta_2 & \sin\alpha_2 & \cos\alpha_2 & -\sin\gamma_2 & -\cos\gamma_2 \\ \dot{\beta}_2\cos\beta_2 & -\dot{\beta}_2\sin\beta_2 & \dot{\alpha}_2\cos\alpha_2 & -\dot{\alpha}_2\sin\alpha_2 & -\dot{\gamma}_2\cos\gamma_2 & \dot{\gamma}_2\sin\gamma_2 \\ \dot{\beta}_2\sin\beta_2 & \dot{\beta}_2\cos\beta_2 & \dot{\alpha}_2\sin\alpha_2 & \dot{\alpha}_2\cos\alpha_2 & -\dot{\gamma}_2\sin\gamma_2 & -\dot{\gamma}_2\cos\gamma_2 \end{bmatrix} \begin{Bmatrix} W_{1x} \\ W_{1y} \\ V_{1x} \\ V_{1y} \\ U_{1x} \\ U_{1y} \end{Bmatrix} = \begin{bmatrix} 1 \\ 0 \\ 1 \\ 0 \\ 0 \\ 0 \end{bmatrix} \tag{7.14}$$

Figure E.7.11 includes the calculation procedure for Equation (7.14) in the MATLAB command window, and Figure E.7.12 illustrates the resulting planar four-bar mechanism. Being an analytically calculated result, this mechanism achieves the precision points precisely. A kinematic analysis of this solution is presented in Example 7.10.

```
>> B = 55*pi/180; dB = 0.1;
>> A = -15*pi/180; dA = -0.25;
>> G = -37*pi/180; dG = -0.2;

>> CB = cos(B); SB = sin(B);
>> CA = cos(A); SA = sin(A);
>> CG = cos(G); SG = sin(G);

>> WVU = inv([1, 0, 1, 0, -1, 0
0, 1, 0, 1, 0, -1
CB, -SB, CA, -SA, -CG, SG
SB, CB, SA, CA, -SG, -CG
dB*CB, -dB*SB, dA*CA, -dA*SA, -dG*CG, dG*SG
dB*SB, dB*CB, dA*SA, dA*CA, -dG*SG, -dG*CG
])*[1, 0, 1, 0, 0, 0]'

WVU =

    0.4366
   -0.4190
   -0.6056
    0.5667
   -1.1689
    0.1478

>>
```

FIGURE E.7.11 Example 7.5 $\mathbf{W}_1 - \mathbf{V}_1 - \mathbf{U}_1$ calculation procedure in the MATLAB command window.

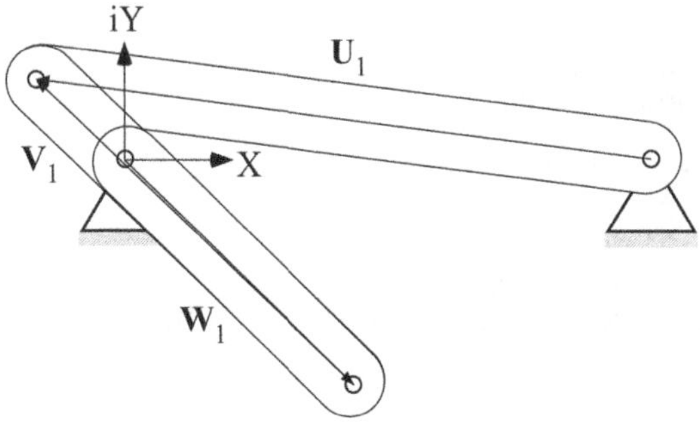

FIGURE E.7.12 Synthesized planar four-bar function generator.

Example 7.6

Problem Statement: Synthesize a planar four-bar function generator to achieve angular velocities of $\dot{\beta} = 1$ rad/s, $\dot{\alpha} = -1.5$ rad/s, and $\dot{\gamma} = -0.75$ rad/s and angular accelerations of $\ddot{\beta} = 0$ rad/s^2, $\ddot{\alpha} = -1$ rad/s^2, and $\ddot{\gamma} = -0.1$ rad/s^2, all at the initial mechanism position.

Known Information: Equations (7.6), (7.11), and (7.13)

Solution Approach: A 6 × 6 linear simultaneous equation system to calculate the components of link vectors $\mathbf{W}_1$, $\mathbf{V}_1$, and $\mathbf{U}_1$ is formed from Equation (7.6) with $\beta = \alpha = \gamma = 0$; Equation (7.11) with the given β, $\dot{\beta}$, α, $\dot{\alpha}$, γ, and $\dot{\gamma}$ values; and Equation (7.13) with the given β, $\dot{\beta}$, $\ddot{\beta}$, α, $\dot{\alpha}$, $\ddot{\alpha}$, γ, $\dot{\gamma}$, and $\ddot{\gamma}$ values. When expressed in matrix form, these equations become

$$\begin{bmatrix} 1 & 0 & 1 & 0 & -1 & 0 \\ 0 & 1 & 0 & 1 & 0 & -1 \\ \dot{\beta}c_\beta & -\dot{\beta}s_\beta & \dot{\alpha}c_\alpha & -\dot{\alpha}s_\alpha & -\dot{\gamma}c_\gamma & \dot{\gamma}s_\gamma \\ \dot{\beta}s_\beta & \dot{\beta}c_\beta & \dot{\alpha}s_\alpha & \dot{\alpha}c_\alpha & -\dot{\gamma}s_\gamma & -\dot{\gamma}c_\gamma \\ \dot{\beta}^2s_\beta+\ddot{\beta}c_\beta & -\dot{\beta}^2c_\beta-\ddot{\beta}s_\beta & \dot{\alpha}^2s_\alpha+\ddot{\alpha}c_\alpha & -\dot{\alpha}^2c_\alpha-\ddot{\alpha}s_\alpha & -\dot{\gamma}^2s_\gamma-\ddot{\gamma}c_\gamma & \dot{\gamma}^2c_\gamma+\ddot{\gamma}s_\gamma \\ \dot{\beta}^2c_\beta+\ddot{\beta}s_\beta & -\dot{\beta}^2s_\beta+\ddot{\beta}c_\beta & \dot{\alpha}^2c_\alpha+\ddot{\alpha}s_\alpha & -\dot{\alpha}^2s_\alpha+\ddot{\alpha}c_\alpha & -\dot{\gamma}^2c_\gamma-\ddot{\gamma}s_\gamma & \dot{\gamma}^2s_\gamma-\ddot{\gamma}c_\gamma \end{bmatrix} \begin{Bmatrix} W_{1x} \\ W_{1y} \\ V_{1x} \\ V_{1y} \\ U_{1x} \\ U_{1y} \end{Bmatrix} = \begin{bmatrix} 1 \\ 0 \\ 0 \\ 0 \\ 0 \\ 0 \end{bmatrix} \tag{7.15}$$

In Equation (7.15), $c_{\text{ang}} = \cos(\text{ang})$ and $s_{\text{ang}} = \sin(\text{ang})$.

Figure E.7.13 includes the calculation procedure for Equation (7.15) in the MATLAB command window, and Figure E.7.14 illustrates the resulting planar four-bar mechanism. Being an analytically calculated result, this mechanism achieves the precision points precisely. A kinematic analysis of this solution is conducted in Example 7.11.

```
>> dB = 1; ddB = 0;
>> dA = -1.5; ddA = -1;
>> dG = -0.75; ddG = -0.1;

>> WVU = inv([1, 0, 1, 0, -1, 0
0, 1, 0, 1, 0, -1
dB, 0, dA, 0, -dG, 0
0, dB, 0, dA, 0, -dG
ddB, -dB^2, ddA, -dA^2, -ddG, dG^2
dB^2, ddB, dA^2, ddA, -dG^2, -ddG])*[1, 0, 0, 0, 0, 0]'

WVU =

    0.2818
    0.0540
   -0.3424
    0.1260
   -1.0605
    0.1801

>>
```

FIGURE E.7.13 Example 7.6 $\mathbf{W}_1 - \mathbf{V}_1 - \mathbf{U}_1$ calculation procedure in the MATLAB command window.

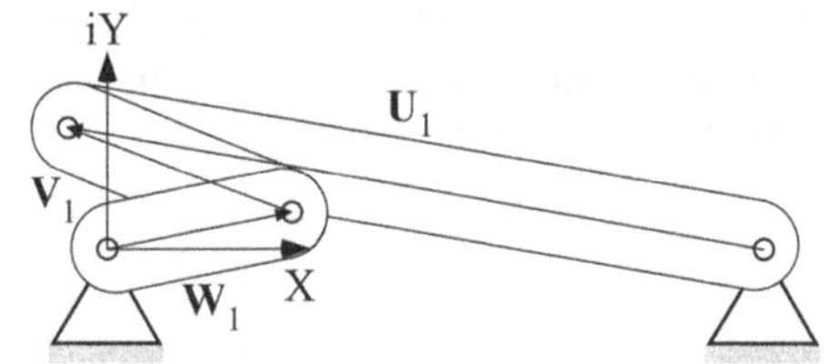

FIGURE E.7.14 Synthesized planar four-bar function generator.

7.6 FUNCTION GENERATION FOR INDEFINITE FINITE AND MULTIPLY SEPARATED POSITIONS

Using Equation (7.8) as an objective function and Equation (7.11) for each prescribed group of β_j - $\dot{\alpha}_j$ - $\dot{\gamma}_j$ values, a constrained nonlinear optimization model with equality constraints can be formulated to synthesize a planar four-bar function generator to approximate the prescribed FSPs and velocity MSPs.[7] From this equation model, $\mathbf{W}_1$, $\mathbf{V}_1$, $\mathbf{U}_1$, and $\alpha_2 \ldots \alpha_N$ are calculated that approximate the prescribed relationships $\dot{\alpha}(\beta)$ and $\dot{\gamma}(\beta)$ as well as the conventional prescribed $\alpha(\beta)$ and $\gamma(\beta)$ relationships. Appendix G.2 includes MATLAB file user instructions for a constrained nonlinear optimization model comprising Equations (7.8) and (7.11). From this model, solutions for $\mathbf{W}_1$, $\mathbf{V}_1$, $\mathbf{U}_1$, and $\alpha_2 \ldots \alpha_N$ are calculated to approximate the prescribed FSPs and angular velocity MSPs. The Appendix G.2 MATLAB files are available for download at https://www.routledge.com/9781032956800.

Example 7.7

Problem Statement: Synthesize a planar four-bar mechanism to drive an oscillating paddle that will be the follower link. The paddle should approximate an angular displacement and velocity of 35° and 0.7 rad/s, respectively, at a crank angular displacement of 90°. The paddle should also approximate an angular displacement and velocity and 70° and 0 rad/s, respectively, at a crank displacement of 180°. The angular velocity of the crank link is a constant 1 rad/s.

Known Information: $\beta_2 = 90°, \beta_3 = 180°, \dot{\beta} = 1$ rad/s, $\gamma_2 = 35°, \gamma_3 = 70°, \dot{\gamma}_2 = 0.7$ rad/s, $\dot{\gamma}_3 = 0$ rad/s, and Appendix G.2 MATLAB files

Solution Approach: Figure E.7.15 includes the input and output in bold type in the Appendix G.2 MATLAB file *UVW_interior-point.m*. In this file, $N = 3$, and initial values of $\mathbf{W}_1 = \mathbf{V}_1 = \mathbf{U}_1 = (1,1)$ and $\alpha_2 = \alpha_3 = 10°$ were specified. Also, arbitrary values of $\dot{\alpha}_2 = \dot{\alpha}_3 = -0.3$ rad/s were specified. Figure E.7.16 illustrates the resulting planar four-bar mechanism. A kinematic analysis of this solution is presented in Example 7.12.

```
N = 3;

beta2 = -90 *pi/180;
beta3 = -180 *pi/180;

gamma2 = 35 *pi/180;
gamma3 = 70 *pi/180;

betap2 = -1;
betap3 = -1;

gammap2 = 0.7;
gammap3 = 0.1;

alphap2 = 0.15;
alphap3 = -0.1;

U1x = 1;
U1y = 1;
V1x = 1;
V1y = 1;
W1x = 1;
W1y = 1;

alpha2 = 10 *pi/180;
alpha3 = 10 *pi/180;

Iterations: 11
Func Evals: 26
Algorithm: interior-point
exit flag = -2
error = ( 2.302268e-05 )
U1x = ( 0.252000 )
U1y = ( 0.101415 )
V1x = ( 1.143055 )
V1y = ( 0.041280 )
W1x = ( 0.108613 )
W1y = ( 0.060122 )
alpha2 = ( 15.125730 )
alpha3 = ( 14.686715 )
Elapsed time is 0.415467 seconds.
Solution calculation is complete.
>>
```

FIGURE E.7.15 Input and output in bold type in the Appendix G.2 MATLAB file *UVW_interior-point.m* for Example 7.7.

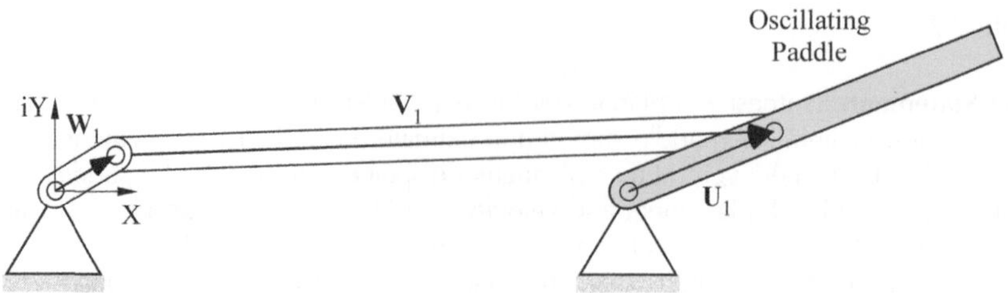

FIGURE E.7.16 Synthesized planar four-bar function generator.

7.7 FROM FUNCTION GENERATION TO KINEMATIC ANALYSIS

Figure 7.5a includes the dyad vectors calculated in planar four-bar function generation, and Figure 7.5b includes the mechanism vectors specified in planar four-bar kinematic analysis. But as shown in this figure, vectors $\mathbf{G}_1$ and $\mathbf{L}_1$ do not explicitly appear in a function generator solution. Because the ground link vector was assigned a value of unity when formulating Equation (7.4), $\mathbf{G}_1 = (1,0)$ in kinematic analysis. The same unity value for $\mathbf{G}_1$ can be calculated, as shown in Table 7.2. Vector $\mathbf{L}_1$ can be assigned a dummy value in kinematic analysis. This is because the purpose of vector $\mathbf{L}_1$ is to indicate a coupler location of interest, and coupler point motion is not of concern in function generation.

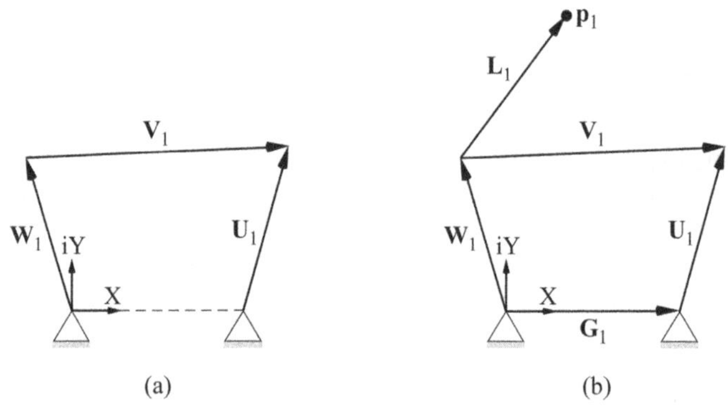

FIGURE 7.5 Planar four-bar mechanism vectors for (a) function generation and (b) kinematic analysis.

TABLE 7.2
Vectors for Planar Four-Bar Function Generation and Kinematic Analysis

Motion Generation		Kinematic Analysis	
Vector	Vector Expression	Vector	Vector Expression
$\mathbf{W}_1$	$\mathbf{W}_1 = W_{1x} + iW_{1y}$	$\mathbf{W}_1$	$\mathbf{W}_1 = W_{1x} + iW_{1y}$
$\mathbf{V}_1$	$\mathbf{V}_1 = V_{1x} + iV_{1y}$	$\mathbf{V}_1$	$\mathbf{V}_1 = V_{1x} + iV_{1y}$
$\mathbf{U}_1$	$\mathbf{U}_1 = U_{1x} + iU_{1y}$	$\mathbf{U}_1$	$\mathbf{U}_1 = U_{1x} + iU_{1y}$
		$\mathbf{G}_1$	$\mathbf{G}_1 = \mathbf{W}_1 + \mathbf{V}_1 - \mathbf{U}_1 = (1,0)$
		$\mathbf{L}_1$	Assign a dummy value here

Example 7.8

Problem Statement: Using the Appendix B.1 or C.1 files, calculate the precision points achieved by the function generator solution given in Example 7.2.

Known Information: Example 7.2 solution and Appendix B.1 MATLAB file

Solution Approach: Figure E.7.17 includes the input specified (in bold text) in the Appendix B.1 MATLAB file. Table E.7.2 includes the calculated mechanism output.

```
W1 = -0.0452 - i*0.1414;
V1 = 0.9157 + i*0.0116;
U1 = -0.1295 - i*0.1298;
G1 = W1 + V1 - U1;
L1 = 0;

start_ang    = 0;
step_ang     = 1;
stop_ang     = 90;

angular_vel  = 0 * ones(N+1,1);
angular_acc  = 0 * ones(N+1,1);
```

FIGURE E.7.17 Specified input (in bold text) in the Appendix B.1 MATLAB file for Example 7.8.

TABLE E.7.2
Calculated Precision Points

Precision Point	β_j (deg)	α_j (deg)	γ_j (deg)
1	0	0.0000	0.0000
2	30	−2.6291	28.0750
3	60	−5.3021	48.9090
4	90	−9.0043	59.9920

Example 7.9

Problem Statement: Using the Appendix B.1 or C.1 files, calculate the precision points achieved by the function generator solution given in Example 7.4.

Known Information: Example 7.4 solution and Appendix B.1 MATLAB file

Solution Approach: Figure E.7.18 includes the input specified (in bold text) in the Appendix B.1 MATLAB file. Table E.7.3 includes the calculated mechanism output.[8]

```
W1 = -2.673080 + i*0.587305;
V1 = 3.289653 - i*1.321894;
U1 = -0.383422 - i*0.734592;
G1 = W1 + V1 - U1;
L1 = 0;

start_ang    = 0;
step_ang     = 1;
stop_ang     = 60;

angular_vel  = 0 * ones(N+1,1);
angular_acc  = 0 * ones(N+1,1);
```

FIGURE E.7.18 Specified input (in bold text) in the Appendix B.1 MATLAB file for Example 7.9.

TABLE E.7.3
Calculated Precision Points

Precision Point	β_j (deg)	α_j (deg)	γ_j (deg)
1	0	0.0000	0.0000
2	12	8.8883	7.6384
3	24	17.6040	17.4520
4	36	26.2250	29.4580
5	48	34.8880	43.6350
6	60	43.7850	60.0000

Example 7.10

Problem Statement: Using the Appendix B.1 or C.1 files, calculate the precision points achieved by the function generator solution given in Example 7.5.

Known Information: Example 7.5 solution and Appendix B.1 MATLAB file

Solution Approach: Figure E.7.19 includes the input specified (in bold text) in the Appendix B.1 MATLAB file. Table E.7.4 includes the calculated mechanism output.

```
W1 = 0.4366 - i*0.4190;
V1 = -0.6056 + i*0.5667;
U1 = -1.1689 + i*0.1478;
G1 = W1 + V1 - U1;
L1 = 0;

start_ang    = 0;
step_ang     = 1;
stop_ang     = 55;

angular_vel  = 0.1 * ones(N+1,1);
angular_acc  = 0 * ones(N+1,1);
```

FIGURE E.7.19 Specified input (in bold text) in the Appendix B.1 MATLAB file for Example 7.10.

TABLE E.7.4
Calculated Precision Points

Precision Point	β_j (deg)	α_j (deg)	γ_j (deg)	$\dot{\alpha}_j$ (rad/s)	$\dot{\gamma}_j$ (rad/s)
1	0	0.0000	0.0000	0.0742	0.0011
2	55	–15.0020	–37.0040	–0.2501	–0.2001

Example 7.11

Problem Statement: Using the Appendix B.1 or C.1 files, calculate the precision points achieved by the function generator solution given in Example 7.6.

Known Information: Example 7.6 solution and Appendix B.1 MATLAB file

Solution Approach: Figure E.7.20 includes the input specified (in bold text) in the Appendix B.1 MATLAB file. Table E.7.5 includes the calculated mechanism output.

```
W1 = 0.2818 + i*0.0540;
V1 = -0.3424 + i*0.1260;
U1 = -1.0605 + i*0.1801;
G1 = W1 + V1 - U1;
L1 = 0;

start_ang   = 0;
step_ang    = 1;
stop_ang    = 0;

angular_vel  = 1 * ones(N+1,1);
angular_acc  = 0 * ones(N+1,1);
```

FIGURE E.7.20 Specified input (in bold text) in the Appendix B.1 MATLAB file for Example 7.11.

TABLE E.7.5
Calculated Precision Points

β_j (deg)	α_j (deg)	γ_j (deg)	$\dot{\alpha}_j$ (rad/s)	$\dot{\gamma}_j$ (rad/s)
0	0	0	−1.5012	−0.7504
$\ddot{\alpha}_j$ (rad/s²)	$\ddot{\gamma}_j$ (rad/s²)			
−0.9904	−0.0967			

Example 7.12

Problem Statement: Using the Appendix B.1 or C.1 files, calculate the precision points achieved by the function generator solution given in Example 7.7.

Known Information: Example 7.7 solution and Appendix B.1 MATLAB file

Solution Approach: Figure E.7.21 includes the input specified (in bold text) in the Appendix B.1 MATLAB file. Table E.7.6 includes the calculated mechanism output.

```
W1 = 0.108613 + i*0.060122;
V1 = 1.143055 + i*0.041280;
U1 = 0.252000 + i*0.101415;
G1 = W1 + V1 - U1;
L1 = 0;

start_ang   = 0;
step_ang    = -1;
stop_ang    = -180;

angular_vel  = -1 * ones(N+1,1);
angular_acc  = 0 * ones(N+1,1);
```

FIGURE E.7.21 Specified input (in bold text) in the Appendix B.1 MATLAB file for Example 7.12.

TABLE E.7.6
Calculated Precision Points

Precision Point	β_j (deg)	α_j (deg)	γ_j (deg)	$\dot{\alpha}_j$ (rad/s)	$\dot{\gamma}_j$ (rad/s)
1	0	0.0000	0.0000	−0.0392	−0.6088
2	−90	14.7830	33.2500	0.1571	0.7206
3	−180	14.7750	70.9360	−0.1004	0.0989

7.8 SUMMARY

The objective in *function generation* is to calculate the mechanism dimensions required to achieve or approximate prescribed link displacement angles. For the planar four-bar linkage, displacement angles are typically specified for the crank and follower links. Often, the displacement angles correspond to a mathematical function (where the crank and follower displacement angles are input and output function values, respectively).

Equations for analytical function generation are derived for the planar four-bar mechanism by taking the vector loop sum of the entire mechanism. The resulting equations can be used to calculate the dimensions of a planar four-bar mechanism to precisely achieve three prescribed crank and follower displacement angle pairs (also called *precision points*).

While analytical function generation is feasible for three precision points, beyond three, the resulting equations form nonlinear simultaneous equations. In this chapter, unconstrained nonlinear optimization models have also been formulated for four-, five-, and N precision point function generation.

In addition to link displacement angles (or *FSPs*), it is also possible to synthesize function generators to achieve or approximate prescribed link angular velocities and/or accelerations (or *MSPs*). Differentiating, expanding, and separating the planar four-bar vector loop equation produce velocity and acceleration equations (after the first and second derivatives, respectively). Two combinations of the velocity equation and/or the acceleration equation and the initial position vector loop equation produce two linear simultaneous equation systems to analytically calculate the planar four-bar mechanism dimensions required to precisely achieve the prescribed FSPs and/or MSPs.

Using the objective function from the unconstrained nonlinear optimization model for N precision point function generation along with the derived velocity equation as a constraint produces a constrained nonlinear optimization model to calculate the mechanism dimensions required to approximate N prescribed FSPs and velocity MSPs.

The constrained and unconstrained optimization models presented in this chapter for approximating N FSPs and N FSPs and velocity MSPs are solved using the Appendix G MATLAB files.

PROBLEMS

1. Using Equation (7.7), synthesize a planar four-bar mechanism for a 75° total wiper blade rotation range (Figure P.7.1) and a 45° total crank rotation range (where the wiper blade is affixed to the follower link). Verify the solution using the Appendix B.1 or C.1 files.

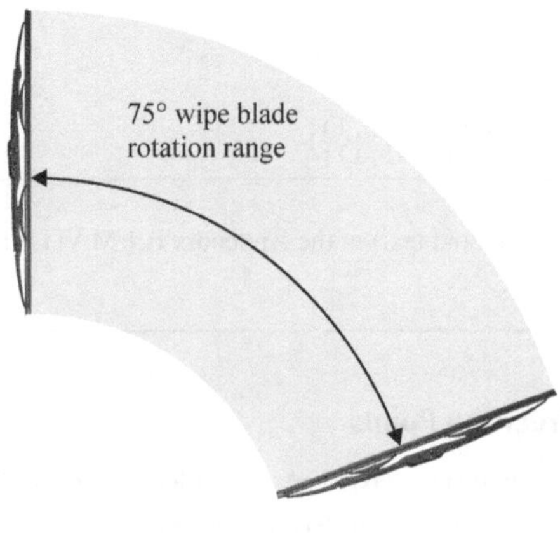

FIGURE P.7.1 Wiper blade rotation range.

2. Using Equation (7.7), synthesize a planar four-bar mechanism for a 75° total wheel rotation range (Figure P.7.2) and a 50° total crank rotation range (where the wheel is affixed to the follower link). Verify the solution using the Appendix B.1 or C.1 files.

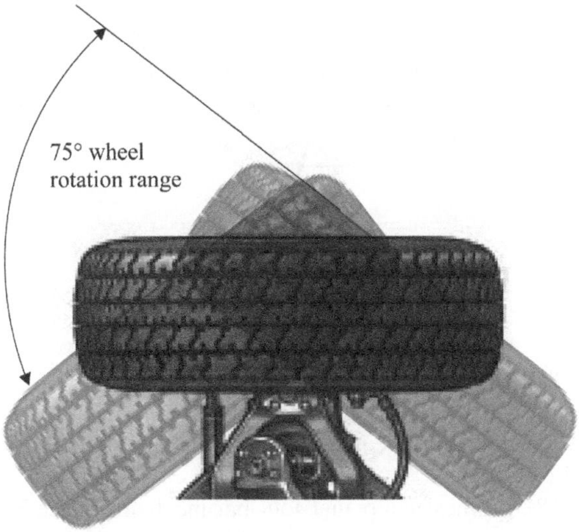

FIGURE P.7.2 Wheel rotation range.

3. Using Equation (7.7), synthesize a planar four-bar mechanism for a 90° valve rotation range (Figure P.7.3) and a 40° crank rotation range (where the valve is affixed to the follower link). Verify the solution using the Appendix B.1 or C.1 files.

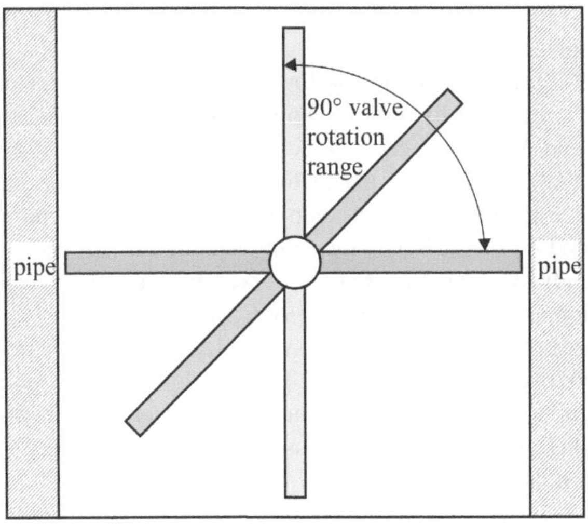

FIGURE P.7.3 Valve rotation range.

4. Using Equation (7.7), synthesize a planar four-bar mechanism for a 130° hatch rotation range (Figure P.7.4) and a 180° crank rotation range (where the hatch is affixed to the follower link). Verify the solution using the Appendix B.1 or C.1 files.

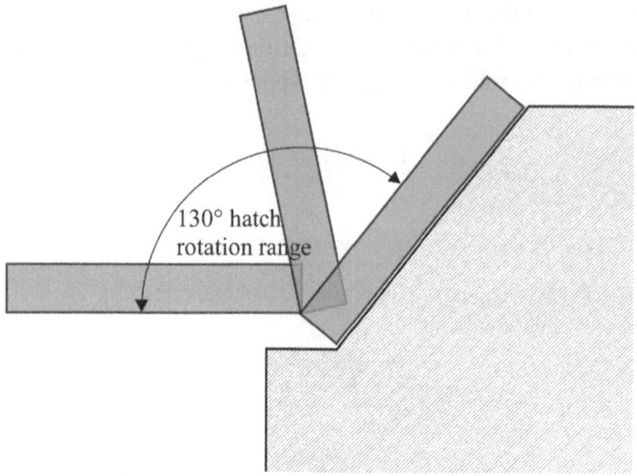

FIGURE P.7.4 Hatch rotation range.

5. Using Equation (7.8), repeat Problem 1 for four precision points. Verify the solution using the Appendix B.1 or C.1 files.
6. Using Equation (7.8), synthesize a planar four-bar mechanism (see Figure E.7.10) for four precision points to measure the spring energy with respect to spring deflection according to the equation SPRING ENERGY = $f(x) = kx^2/2$. The spring rate (k) is 1.75 lbf/in (218.91 N/m), and the spring deflection range is 1.25 in $\leq x \leq$ 4.75 in (0.032 m $\leq x \leq$ 0.121 m). To calculate β_j and γ_j use Equation (7.2) with $\Delta\beta = 90°$ and $\Delta\gamma = 135°$. Verify the solution using the Appendix B.1 or C.1 files.
7. Using Equation (7.8), synthesize a planar four-bar mechanism for the function $f(x) = \sin x$ over the range $0 \leq x \leq \pi/4$ for four precision points. To calculate β_j and γ_j, use Equation (7.2) with $\Delta\beta = 45°$ and $\Delta\gamma = 90°$. Verify the solution using the Appendix B.1 or C.1 files.
8. Using Equation (7.8), repeat Problem 2 for five precision points. Verify the solution using the using the Appendix B.1 or C.1 files.
9. Using Equation (7.8), synthesize a planar four-bar mechanism for the function $f(x) = \log x$ over the range $5 \leq x \leq 30$ for five precision points. To calculate β_j and γ_j, use Equation (7.2) with $\Delta\beta = -60°$ and $\Delta\gamma = 90°$. Verify the solution using the Appendix B.1 or C.1 files.
10. Using Equation (7.8), synthesize a planar four-bar mechanism for the function $f(x) = x^2$ over the range $5 \leq x \leq 10$ for five precision points. To calculate β_j and γ_j, use Equation (7.2) with $\Delta\beta = 180°$ and $\Delta\gamma = 90°$. Verify the solution using the Appendix B.1 or C.1 files.
11. Using the Appendix G.1 MATLAB files, repeat Problem 3 for eight precision points. Verify the solution using the Appendix B.1 or C.1 files.
12. Using the Appendix G.1 MATLAB files, synthesize a planar four-bar mechanism for the function $y = \tan x$ over the range $1° \leq x \leq 45°$ for eight precision points. Use Chebyshev spacing (see Section 7.1) to determine x_1 through x_7 (where $x_0 = 0$, $x_{N+1} = 46°$, $N = 8$, and $j = 2, 3, \ldots, N-1$). To calculate β_j and γ_j, use Equation (7.2) with $\Delta\beta = 60°$ and $\Delta\gamma = 90°$. Verify the solution using the Appendix B.1 or C.1 files.
13. Using Appendix G.1 MATLAB files, synthesize a planar four-bar mechanism for the function $y = \sqrt{x}$ over the range $1 \leq x \leq 100$ for eight precision points. Use Chebyshev spacing (see Section 7.1) to determine x_j (where $x_0 = 0$, $x_{N+1} = 101$, $N = 8$, and $j = 2, 3, \ldots, N-1$). To calculate β_j and γ_j, use Equation (7.2) with $\Delta\beta = 75°$ and $\Delta\gamma = 135°$. Verify the solution using the Appendix B.1 or C.1 files.
14. Using the Appendix G.1 MATLAB files, synthesize a planar four-bar mechanism to measure the vehicle speed with respect to the transmission gear according to the data in

TABLE P.7.1
Transmission Gears and Corresponding Vehicle Speeds

Transmission Gear	Vehicle Speed (km/h)
1	35
2	78
3	125
4	155
5	190
6	215

Table P.7.1. To calculate β_j and γ_j, use Equation (7.2) with $\Delta\beta = 60°$ and $\Delta\gamma = 120°$. Verify the solution using the Appendix B.1 or C.1 files.

15. Using the Appendix G.1 MATLAB files, synthesize a planar four-bar mechanism to measure fluid velocity with respect to fluid pressure at six precision points according to the equation VELOCITY $= f(x) = \sqrt{2x/\rho}$. The fluid density (ρ) is 1,000 kg/m^3 (62.4 lbm/ft^3), and the pressure range is 1,000 Pa $\le x \le$ 6,000 Pa (0.145 psi $\le x \le$ 0.870 psi). To calculate β_j and γ_j, use Equation (7.2) with $\Delta\beta = 50°$ and $\Delta\gamma = 110°$. Verify the solution using the Appendix B.1 or C.1 files.
16. Using the Appendix G.1 MATLAB files, synthesize a planar four-bar mechanism for the function $y = \cos x$ over the range $1° \le x \le 90°$ at six precision points. Use Chebyshev spacing (see Section 7.1) to determine x_j (where $x_0 = 0, x_{N+1} = 91°, N = 6$, and $j = 2,3,\ldots,N-1$). To calculate β_j and γ_j, use Equation (7.2) with $\Delta\beta = 60°$ and $\Delta\gamma = 145°$. Verify the solution using the Appendix B.1 or C.1 files.
17. Using Equation (7.14), synthesize a planar four-bar mechanism to achieve a follower angular displacement and velocity of $\gamma_2 = \pi/4$ rad and $\dot{\gamma}_2 = 1.15$ rad/s and a coupler angular displacement and velocity of $\alpha_2 = \pi/18$ rad and $\dot{\alpha}_2 = 0.25$ rad/s, respectively. The angular displacement and constant angular velocity for the crank are $\beta_2 = \pi/6$ rad and $\dot{\beta} = 1$ rad/s, respectively. Verify the solution using the Appendix B.1 or C.1 files.
18. Using Equation (7.14), synthesize a planar four-bar mechanism to achieve a follower angular displacement and velocity of $\gamma_2 = -55°$ and $\dot{\gamma}_2 = -3.95$ rad/s and a coupler angular displacement and velocity of $\alpha_2 = -15°$ and $\dot{\alpha}_2 = -0.45$ rad/s, respectively. The angular displacement and constant angular velocity for the crank are $\beta_2 = -25°$ and $\dot{\beta} = -0.75$ rad/s, respectively. Verify the solution using the Appendix B.1 or C.1 files.
19. Using Equation (7.15), synthesize a planar four-bar mechanism to achieve a follower angular velocity and acceleration of $\dot{\gamma} = 1.15$ rad/s and $\ddot{\gamma} = -0.15$ rad/s^2 and a coupler angular velocity and acceleration of $\dot{\alpha} = 0.65$ rad/s and $\ddot{\alpha} = 0.10$ rad/s^2 for a crank angular velocity and acceleration of $\dot{\beta} = 2$ rad/s and $\ddot{\beta} = 0.5$ rad/s^2, respectively. Verify the solution using the Appendix B.1 or C.1 files.
20. Using Equation (7.15), repeat Problem 20 using a crank angular velocity and angular acceleration of $\dot{\beta} = -2$ rad/s and $\ddot{\beta} = -0.5$ rad/s^2, respectively. Verify the solution using the Appendix B.1 or C.1 files.
21. Using Equation (7.15), synthesize a planar four-bar mechanism to achieve a follower angular velocity and acceleration of $\dot{\gamma} = -3.5$ rad/s and $\ddot{\gamma} = -0.35$ rad/s^2 and a coupler angular velocity and acceleration of $\dot{\alpha} = 1.65$ rad/s and $\ddot{\alpha} = 0.75$ rad/s^2 for a crank angular velocity and acceleration of $\dot{\beta} = -1$ rad/s and $\ddot{\beta} = 0.5$ rad/s^2, respectively. Verify the solution using the Appendix B.1 or C.1 files.

22. Using the Appendix G.2 MATLAB files, synthesize a planar four-bar mechanism for the valve rotation range in Problem 3 for three precision points. The valve angular velocities are $\dot{\gamma}_2 = 2$ and $\dot{\gamma}_3 = 0.2$ rad/s for a constant crank angular velocity of $\dot{\beta} = 1.5$ rad/s. Verify the solution using the Appendix B.1 or C.1 files.
23. Using the Appendix G.2 MATLAB files, synthesize a planar four-bar mechanism for the finite and multiply separated positions given in Table P.7.2. Consider a constant crank angular velocity of $\dot{\beta} = 1.5$ rad/s. Verify the solution using the Appendix B.1 or C.1 files.

TABLE P.7.2
Precision Point Data for Finite and Multiply Separated Positions

Point	β_j (deg)	γ_j (deg)	$\dot{\gamma}_j$ (rad/s)
2	10	3.4260	0.25
3	20	13.3688	0.5
4	30	28.8550	0.75
5	40	48.3688	1
6	50	70	1.25

24. Using the Appendix G.2 MATLAB files, synthesize a planar four-bar mechanism for precision points 2 through 4 in Problem 12, the MSPs in Table P.7.3, and a constant crank angular velocity of $\dot{\beta} = 1.75$ rad/s. Verify the solution using the Appendix B.1 or C.1 files.

TABLE P.7.3
Multiply Separated Positions

Point	$\dot{\gamma}_j$ (rad/s)
2	2.75
3	2.50
4	2.25

NOTES

1 In addition to link angular displacements, derivative quantities such as angular velocities and angular accelerations are among the prescribed parameters that are considered in function generation.
2 Prescribed displacement angles are also called precision points in function generation.
3 *Chebyshev Spacing* is a classic and commonly employed technique for selecting precision points. For N precision points (where $x_0 \le x_j \le x_{N+1}$), the Chebyshev spacing equation to calculate the function's independent variables is $x_j = \frac{1}{2}(x_{N+1} + x_0) - \frac{1}{2}(x_{N+1} - x_0)\cos\frac{(2j-1)\pi}{2N}$ $j = 1,2,\ldots,N$.
4 Because the motion of a function generator is not determined by its scale, a function generator can be uniformly scaled to any size.
5 While this objective function is applicable beyond five precision points, it is also applicable for four- and five-point function generation.
6 To ensure the function generator solution forms a closed loop, Equation (6.6) (with $\beta = \alpha = \gamma = 0$) must be included in the linear simultaneous equation system.
7 For each $\dot{\beta}_j - \dot{\alpha}_j - \dot{\gamma}_j$ group specified, Equation (7.11) is included as an equality constraint in the equation system.

8 While a positive β range was specified and the γ values in this table were achieved with a positive β range, it may be necessary to reverse the sign of the specified β range when conducting a kinematic analysis.

REFERENCES

1. Sandor, G. N. and Erdman, A. G. 1984. *Advanced Mechanism Design: Analysis and Synthesis*, Volume 2. pages 52–53. Englewood Cliffs: Prentice-Hall.
2. Waldron, K. J. and Kinzel, G. L. 2004. *Kinematics, Dynamics and Design of Machinery*, 2nd Edition. pages 287–288. Saddle River: Prentice-Hall.

ADDITIONAL READING

Norton, R. L. 2008. *Design of Machinery*, 4th Edition. pages 256–259. New York: McGraw-Hill.
Sandor, G. N. and Erdman, A. G. 1984. *Advanced Mechanism Design: Analysis and Synthesis*, Volume 2. pages 127–133. Englewood Cliffs: Prentice-Hall.
Waldron, K. J. and Kinzel, G. L. 2004. *Kinematics, Dynamics and Design of Machinery*, 2nd Edition. pages 283–294. Saddle River: Prentice-Hall.

8 Spatial Mechanism Kinematics and Dimensional Synthesis

CONCEPT OVERVIEW

In this chapter, the reader will gain a central understanding regarding the following:

1. Advantages and disadvantages of spatial mechanisms and the ongoing aim in spatial mechanism modeling
2. Intermediate and total spatial kinematics of the *RRSS*, *RSSR*, *4R Spherical,* and *RSSR-SS* mechanisms
3. Constrained and unconstrained nonlinear optimization for R-R, R-S, and Spherical R-R dyad dimensional synthesis
4. Constrained nonlinear optimization for RRSS, 4R Spherical, and RSSR mechanism loop dimensional synthesis

8.1 INTRODUCTION

Planar mechanisms are restricted to motion in two-dimensional or planar space. *Spatial* mechanisms can exhibit three-dimensional or spatial motion. Spatial mechanism motion is predominantly determined by the degrees of freedom of the mechanism joints used and spatial orientation of the joints.

The cylindrical joint (Figure 3.7) can enable spatial mechanism motion because this joint has a translational degree of freedom (DOF) along the z-axis in addition to the planar rotational DOF about the z-axis. The spherical joint (Figure 3.7) can enable spatial mechanism motion because this joint has rotational DOFs about the x- and y-axes, in addition to the planar rotational DOF about the z-axis. Even revolute and prismatic joints, though possessing only a single DOF, will enable spatial mechanism motion when positioned to have joint axes (or lines of action in the case of prismatic joints) that are skewed (rather than orthogonal) to the x–y plane.

Because spatial mechanisms have the capacity to exhibit spatial motion, they offer a greater variety of possible motions and are structurally more general than planar mechanisms. Because the equations for spatial mechanism analysis and synthesis are often much larger in scale and greater in complexity than those for planar mechanisms however (in addition to the actual design of the former compared to the latter), their applications in practice are often limited. In practice, it is not uncommon to find complicated planar mechanism solutions when in fact a simpler spatial mechanism solution is possible. It is, therefore, an ongoing task to devise simple methods of calculation, to produce design aids with diagrams, and to set design standards for spatial mechanisms [1]. Several technological developments have greatly facilitated the design and analysis of spatial mechanisms. These developments include commercial mathematical analysis and mechanical simulation software such as MATLAB and Simscape Multibody and the high-speed computing hardware that they operate on. While modern-day computing hardware and mathematical modeling and mechanical simulation software don't simplify spatial mechanism models, they greatly reduce the time it takes to calculate results from them (thus making spatial mechanisms more practical for everyday use).

Like planar mechanisms, there is also a variety of spatial mechanisms. This chapter considers three types of four-bar spatial mechanisms: the *RRSS*, *4R Spherical,* and *RSSR* mechanisms (Figures 8.1a–c, respectively). The RRSS, 4R Spherical, and RSSR are among the more basic four-bar spatial mechanisms in terms of the types of joints used and the required linkage assembly conditions for motion.

DOI: 10.1201/9781003591375-8

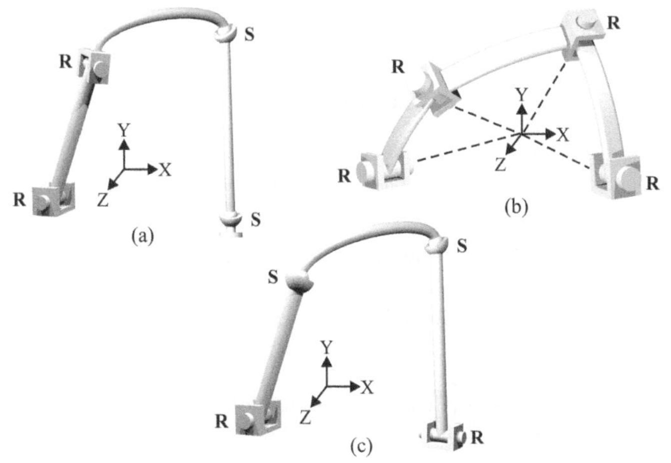

FIGURE 8.1 (a) RRSS; (b) 4R Spherical; and (c) RSSR mechanisms.

8.2 RRSS AND 4R SPHERICAL MECHANISM KINEMATIC ANALYSIS

8.2.1 Displacement Equations

The *Revolute–Revolute–Spherical–Spherical* mechanism or RRSS mechanism (Figure 8.1a) is a spatial, four-link mechanism with a mobility of 2 (with $L = 4$ and $J_1 = J_3 = 4$). Although one DOF of the RRSS mechanism enables it to be crank-driven, the other DOF is produced by the follower link. The follower link (link $\mathbf{b}_0 - \mathbf{b}_1$ in Figure 8.2) is bounded by spherical joints. As a result, this link can rotate about its own axis of symmetry. This DOF is called a *passive degree of freedom*. Passive DOFs generally do not contribute to the overall kinematic motion of a mechanism.[1]

Suh and Radcliffe presented displacement equations for the RRSS mechanism [2]. Equation (8.1), the follower-link constant-length equation, forms the basis for RRSS displacement equations.

$$(\mathbf{b}-\mathbf{b}_0)^T(\mathbf{b}-\mathbf{b}_0)-(\mathbf{b}_1-\mathbf{b}_0)^T(\mathbf{b}_1-\mathbf{b}_0)=0 \tag{8.1}$$

The rotation of the crank and coupler links (links $\mathbf{a}_0 - \mathbf{a}_1$ and $\mathbf{a}_1 - \mathbf{b}_1$, respectively, in Figure 8.2) and the rotation of the crank moving-pivot joint axis $\mathbf{ua}_1$ about the crank fixed-pivot joint axis $\mathbf{ua}_0$ (by displacement angle θ) are given by

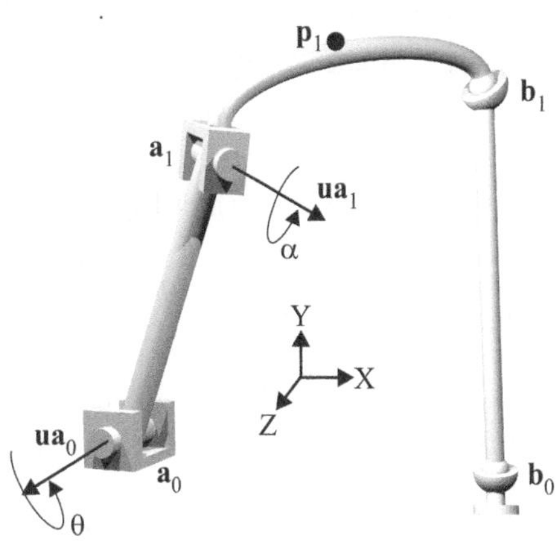

FIGURE 8.2 RRSS mechanism with displacement variables.

$$\mathbf{a} = \left[R_{\theta,\mathbf{ua}_0}\right](\mathbf{a}_1 - \mathbf{a}_0) + \mathbf{a}_0 \tag{8.2}$$

$$\mathbf{ua} = \left[R_{\theta,\mathbf{ua}_0}\right]\mathbf{ua}_1 \tag{8.3}$$

$$\mathbf{b}_1' = \left[R_{\theta,\mathbf{ua}_0}\right](\mathbf{b}_1 - \mathbf{a}_0) + \mathbf{a}_0 \tag{8.4}$$

The global displacement of the follower moving pivot $\mathbf{b}_1$ is expressed as

$$\mathbf{b} = \left[R_{\alpha,\mathbf{ua}}\right](\mathbf{b}_1' - \mathbf{a}) + \mathbf{a} \tag{8.5}$$

The rotational displacement matrices $\left[R_{\theta,\mathbf{ua}_0}\right]$ and $\left[R_{\alpha,\mathbf{ua}}\right]$ are identical in form to Matrix (2.30).

Substituting Equation (8.5) into Equation (8.1) produces

$$E\cos(\alpha) + F\sin(\alpha) + G = 0 \tag{8.6}$$

In Equation (8.6),

$$E = (\mathbf{a} - \mathbf{b}_0)^T \{[I - Q_{\mathbf{ua}}](\mathbf{b}_1' - \mathbf{a})\}^2 \tag{8.7}$$

$$F = (\mathbf{a} - \mathbf{b}_0)^T \{[P_{\mathbf{ua}}](\mathbf{b}_1' - \mathbf{a})\} \tag{8.8}$$

$$G = (\mathbf{a} - \mathbf{b}_0)^T \{[Q_{\mathbf{ua}}](\mathbf{b}_1' - \mathbf{a})\} + \frac{1}{2}\begin{Bmatrix} (\mathbf{b}_1' - \mathbf{a})^T(\mathbf{b}_1' - \mathbf{a}) + (\mathbf{a} - \mathbf{b}_0)^T(\mathbf{a} - \mathbf{b}_0) \\ -(\mathbf{b}_1 - \mathbf{b}_0)^T(\mathbf{b}_1 - \mathbf{b}_0) \end{Bmatrix} \tag{8.9}$$

$$[P_{\mathbf{ua}}] = \begin{bmatrix} 0 & -ua_z & ua_y \\ ua_z & 0 & -ua_x \\ -ua_y & ua_x & 0 \end{bmatrix} \tag{8.10}$$

and

$$[Q_{\mathbf{ua}}] = \begin{bmatrix} ua_x^2 & ua_x ua_y & ua_x ua_z \\ ua_x ua_y & ua_y^2 & ua_y ua_z \\ ua_x ua_z & ua_y ua_z & ua_z^2 \end{bmatrix} \tag{8.11}$$

The two coupler link displacement angle solutions (α) for Equation (8.6) are

$$\alpha_{1,2} = 2\tan^{-1}\frac{-F \pm \sqrt{E^2 + F^2 - G^2}}{G - E}{}_3 \tag{8.12}$$

This angle is the difference between the angles $\angle \mathbf{b} - \mathbf{a} - \mathbf{a}_0$ and $\angle \mathbf{b}_1 - \mathbf{a}_1 - \mathbf{a}_0$. With α known, the displaced RRSS moving pivot $\mathbf{b}$ from Equation (8.5) and also the displaced RRSS coupler point $\mathbf{p}$ from Equation (8.13) can be calculated.

Two sets of link displacement angles are calculated for a given crank displacement angle in algebraic four-bar mechanism displacement equations. The $\pm$ term in Equation (8.12) corresponds to two coupler displacement angle solutions—one solution for each mechanism assembly configuration.

The global displacement of an arbitrary RRSS coupler point $\mathbf{p}_1$ is expressed as

$$\mathbf{p} = [R_{\alpha,\mathbf{ua}}](\mathbf{p}_1' - \mathbf{a}) + \mathbf{a} \quad (8.13),$$

where

$$\mathbf{p}_1' = [R_{\theta,\mathbf{ua}_0}](\mathbf{p}_1 - \mathbf{a}_0) + \mathbf{a}_0 \quad (8.14)$$

RRSS variables $\mathbf{a}_0$, $\mathbf{a}_1$, $\mathbf{ua}_0$, $\mathbf{ua}_1$, $\mathbf{b}_0$, $\mathbf{b}_1$, and $\mathbf{p}_1$ are 3×1 vectors containing x-, y-, and z-components.

8.2.2 Velocity Equations

Figure 8.3 includes the velocity variables for the RRSS mechanism. Equation (8.15), the derivative of the follower-link constant-length equation, forms the basis for the RRSS velocity equations.

$$(\dot{\mathbf{b}})^T(\mathbf{b} - \mathbf{b}_0) = 0 \quad (8.15)$$

The global velocity of the RRSS moving pivot $\mathbf{b}_1$ (variable $\dot{\mathbf{b}}$ in Equation (8.15)) is expressed as

$$\dot{\mathbf{b}} = \dot{\alpha}[P_{\mathbf{ua}}](\mathbf{b} - \mathbf{a}) + \dot{\mathbf{b}}' = [V_{\dot{\alpha},\mathbf{ua}}](\mathbf{b} - \mathbf{a}) + \dot{\mathbf{b}}' \quad (8.16),$$

where

$$\dot{\mathbf{b}}' = \dot{\theta}[P_{\mathbf{ua}_0}](\mathbf{b} - \mathbf{a}_0) = [V_{\dot{\theta},\mathbf{ua}_0}](\mathbf{b} - \mathbf{a}_0) \quad (8.17)$$

The rotational velocity matrices $[V_{\dot{\theta},\mathbf{ua}_0}]$ and $[V_{\dot{\alpha},\mathbf{ua}}]$ are identical in form to Matrix (2.35).

In an identical form, the global velocity of an arbitrary RRSS coupler point $\mathbf{p}_1$ is expressed as

$$\dot{\mathbf{p}} = [V_{\dot{\alpha},\mathbf{ua}}](\mathbf{p} - \mathbf{a}) + \dot{\mathbf{p}}' \quad (8.18)$$

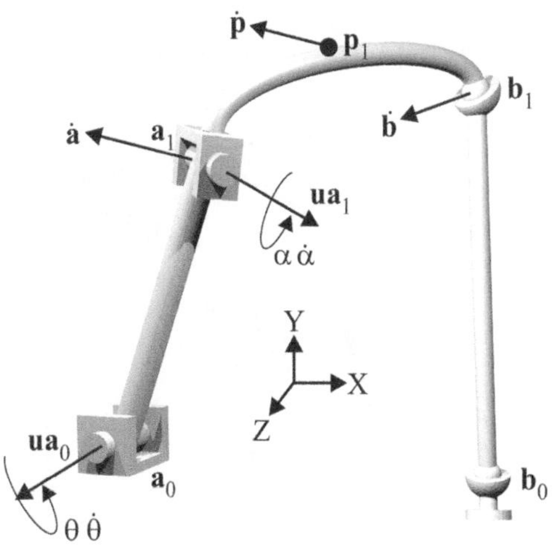

FIGURE 8.3 RRSS mechanism with velocity variables.

where

$$\dot{\mathbf{p}}' = \left[V_{\dot{\theta},\mathbf{ua}_0}\right](\mathbf{p} - \mathbf{a}_0) \tag{8.19}$$

Substituting Equations (8.16) and (8.17) into Equation (8.15) and solving for the coupler angular velocity $\dot{\alpha}$ produce the coupler angular velocity equation

$$\dot{\alpha} = -\dot{\theta}\frac{(\mathbf{b} - \mathbf{b}_0)^T\left\{[P_{\mathbf{ua}_0}](\mathbf{b} - \mathbf{a}_0)\right\}}{(\mathbf{b} - \mathbf{b}_0)^T\left\{[P_{\mathbf{ua}}](\mathbf{b} - \mathbf{a})\right\}} \tag{8.20}$$

With $\dot{\alpha}$ known, the displaced RRSS moving-pivot velocity $\dot{\mathbf{b}}$ from Equation (8.16) and also the displaced RRSS coupler point velocity $\dot{\mathbf{p}}$ from Equation (10.18) can be calculated.

Lastly, the global velocity of the RRSS moving pivot $\mathbf{a}_1$ is expressed as

$$\dot{\mathbf{a}} = \dot{\theta}\left[P_{\mathbf{ua}_0}\right](\mathbf{a} - \mathbf{a}_0) = \left[V_{\dot{\theta},\mathbf{ua}_0}\right](\mathbf{a} - \mathbf{a}_0) \tag{8.21}$$

which is identical in form to Equations (8.17) and (8.19).

8.2.3 Acceleration Equations

Figure 8.4 includes the acceleration variables for the RRSS mechanism. Equation (8.22), the second derivative of the follower-link constant-length equation, forms the basis for the RRSS acceleration equations.

$$\left(\ddot{\mathbf{b}}\right)^T(\mathbf{b} - \mathbf{b}_0) + \left(\dot{\mathbf{b}}\right)^T\left(\dot{\mathbf{b}}\right) = 0 \tag{8.22}$$

The global acceleration of the RRSS moving pivot $\mathbf{b}_1$ (variable $\ddot{\mathbf{b}}$ in Equation (8.22)) is expressed as

$$\begin{aligned}\ddot{\mathbf{b}} &= \ddot{\mathbf{b}}' + \left\{\ddot{\alpha}[P_{\mathbf{ua}}] + \dot{\alpha}\left[\cancel{\dot{P}_{\mathbf{ua}}}\right] + \dot{\alpha}^2[P_{\mathbf{ua}}][P_{\mathbf{ua}}]\right\}(\mathbf{b} - \mathbf{a}) + 2\dot{\theta}[P_{\mathbf{ua}_0}]\left\{\dot{\alpha}[P_{\mathbf{ua}}](\mathbf{b} - \mathbf{a})\right\} \\ &= \ddot{\mathbf{b}}' + \left[A_{\dot{\alpha},\ddot{\alpha},\mathbf{ua}}\right](\mathbf{b} - \mathbf{a}) + 2\left[V_{\dot{\theta},\mathbf{ua}_0}\right]\left\{[V_{\dot{\alpha},\mathbf{ua}}](\mathbf{b} - \mathbf{a})\right\}\end{aligned} \tag{8.23}$$

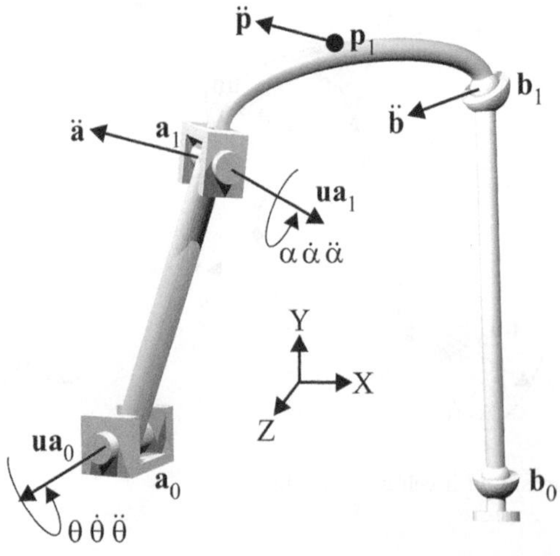

FIGURE 8.4 RRSS mechanism with acceleration variables.

where

$$\ddot{\mathbf{b}}' = \left\{\ddot{\theta}\left[P_{\mathbf{ua}_0}\right] + \dot{\theta}\left[\cancel{\dot{P}_{\mathbf{ua}_0}}\right] + \dot{\theta}^2\left[P_{\mathbf{ua}_0}\right]\left[P_{\mathbf{ua}_0}\right]\right\}(\mathbf{b} - \mathbf{a}_0) = \left[A_{\dot{\theta},\ddot{\theta},\mathbf{ua}_0}\right](\mathbf{b} - \mathbf{a}_0) \tag{8.24}$$

The matrices $\left[\dot{P}_{\mathbf{ua}}\right]$ and $\left[\dot{P}_{\mathbf{ua}_0}\right]$ that appear in Equations (10.23) and (10.24) become 0 since $\mathbf{ua}_0$ and $\mathbf{ub}_0$, being grounded, do no exhibit motion [3]. The rotational acceleration matrices $\left[A_{\dot{\theta},\ddot{\theta},\mathbf{ua}_0}\right]$ and $\left[A_{\dot{\alpha},\ddot{\alpha},\mathbf{ua}}\right]$ are identical in form to Matrix (2.39).

In an identical form, the global acceleration of an arbitrary RRSS coupler point $\mathbf{p}_1$ is expressed as

$$\ddot{\mathbf{p}} = \ddot{\mathbf{p}}' + \left[A_{\dot{\alpha},\ddot{\alpha},\mathbf{ua}}\right](\mathbf{p} - \mathbf{a}) + 2\left[V_{\dot{\theta},\mathbf{ua}_0}\right]\left\{\left[V_{\dot{\alpha},\mathbf{ua}}\right](\mathbf{p} - \mathbf{a})\right\} \tag{8.25}$$

where

$$\ddot{\mathbf{p}}' = \left[A_{\dot{\theta},\ddot{\theta},\mathbf{ua}_0}\right](\mathbf{p} - \mathbf{a}_0) \tag{8.26}$$

Substituting Equations (8.23) and (8.24) into Equation (8.22) and solving for the coupler angular acceleration $\ddot{\alpha}$ produce the coupler angular acceleration equation

$$\ddot{\alpha} = -\frac{(\mathbf{b} - \mathbf{b}_0)^T\left(\ddot{\mathbf{b}}' + \left\{\dot{\alpha}^2\left[P_{\mathbf{ua}}\right]\left[P_{\mathbf{ua}}\right]\right\}(\mathbf{b} - \mathbf{a}) + 2\left[V_{\dot{\theta},\mathbf{ua}_0}\right]\left\{\left[V_{\dot{\alpha},\mathbf{ua}}\right](\mathbf{b} - \mathbf{a})\right\}\right) + \left(\dot{\mathbf{b}}\right)^T\left(\dot{\mathbf{b}}\right)}{(\mathbf{b} - \mathbf{b}_0)^T\left\{\left[P_{\mathbf{ua}}\right](\mathbf{b} - \mathbf{a})\right\}} \tag{8.27}$$

With $\ddot{\alpha}$ known, the displaced RRSS moving point acceleration $\ddot{\mathbf{b}}$ from Equation (8.23) and also the displaced RRSS coupler point acceleration $\ddot{\mathbf{p}}$ from Equation (8.25) can be calculated.

Lastly, the global acceleration of the RRSS moving pivot $\mathbf{a}_1$ is expressed as

$$\ddot{\mathbf{a}} = \left[A_{\dot{\theta},\ddot{\theta},\mathbf{ua}_0}\right](\mathbf{a} - \mathbf{a}_0) \tag{8.28}$$

which is identical in form to Equations (8.24) and (8.26).

Appendix H.1 includes the MATLAB file user instructions for RRSS displacement, velocity, and acceleration analysis. This MATLAB file (which is available for download at https://www.routledge.com/9781032956800) utilizes the RRSS displacement, velocity, and acceleration equations presented in Sections 8.2.1 through 8.2.3, respectively.

Although the equations presented in Sections 8.2.1 through 8.2.3 are presented for the spatial RRSR mechanism, the same equations can also be used for the displacement, velocity, and acceleration analysis of a planar four-bar mechanism (specifically a planar RRSS mechanism). To accomplish this, the user should specify planar mechanism values for the RRSS mechanism variables (e.g. $\mathbf{ua}_0 = \mathbf{ua}_1 = (0,0,1)$ and $a_{0z} = a_{1z} = b_{0z} = b_{1z} = p_{1z} = 0$).[4]

Example 8.1

Problem Statement: Figure E.8.1 illustrates an RRSS mechanism used to guide a solar panel. Using the Appendix H.1 MATLAB file, calculate the solar panel positions (the displaced values of points $\mathbf{p}_1$, $\mathbf{q}_1$, and $\mathbf{r}_1$) achieved by the RRSS mechanism in Table E.8.1 at crank displacement angles of 50°, 100°, and 200°.

Known Information: Table E.8.1 and Appendix H.1 MATLAB file

Solution Approach: Figure E.8.2 includes the input specified (in bold text) in the Appendix F.1 MATLAB file. Table E.8.2 includes the spatial coordinates of **p**, **q**, and **r** calculated using the Appendix F.1 MATLAB file for the three crank displacement angles. The dimensions in Table E.8.1 and the results in Table E.8.2 are for the initial RRSS mechanism configuration (which corresponds to the first branch in the Appendix H.1 MATLAB file).[5]

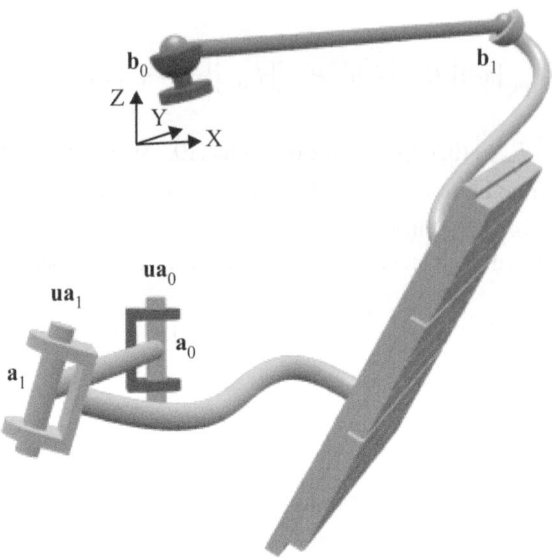

FIGURE E.8.1 RRSS mechanism used to guide a solar panel.

TABLE E.8.1
RRSS Mechanism Dimensions

Variable	Value
$\mathbf{a}_0$	–0.0576, 0.289, –1.4112
$\mathbf{a}_1$	0.1452, –2.5421, –1.18
$\mathbf{ua}_0$	–0.0003, 0.0814, 0.9967
$\mathbf{ua}_1$	0.304, 0.0992, 0.9475
$\mathbf{b}_0$	0.0851, 0.457, 0.5096
$\mathbf{b}_1$	1.7725, 5.1566, 0.6499
$\mathbf{p}_1$	1.7321, 0, –1
$\mathbf{q}_1$	1.2321, 0, –1.866
$\mathbf{r}_1$	1.9486, 0, –1.125

```
a0 = [-0.0576, 0.289, -1.4112]';
a1 = [0.1452, -2.5421, -1.18]';
ua0 = [-0.0003, 0.0814, 0.9967]';
ua1 = [0.304, 0.0992, 0.9475]';
b0 = [0.0851, 0.457, 0.5096]';
b1 = [1.7725, 5.1566, 0.6499]';
p1 = [1.7321, 0, -1]';
q1 = [1.2321, 0, -1.866]';
r1 = [1.9486, 0, -1.125]';

start_ang   = 0;
step_ang    = 50;
stop_ang    = 200;

angular_vel  = 0 * ones(N+1,1);
angular_acc  = 0 * ones(N+1,1);
```

FIGURE E.8.2 Specified input (in bold text) in the Appendix H.1 MATLAB file for Example 8.1.

TABLE E.8.2
Solar Panel Positions Achieved by RRSS Mechanism

θ	p	q	r
50°	1.0926, 1.3848, −1.0402	0.7287, 0.9589, −1.8685	1.2045, 1.5598, −1.1793
100°	−0.1671, 1.7348, −1.0237	−0.1141, 1.1570, −1.8382	−0.2507, 1.9244, −1.1637
200°	−1.7270, −0.0051, −0.9672	−1.2325, 0.0396, −1.8352	−1.9224, −0.1086, −1.0839

Example 8.2

Problem Statement: Using the Appendix H.1 MATLAB file, repeat Example 4.2 using the displacement equations for the spatial RRSS mechanism. When expressed as *x*–*y* coordinates, the mechanism dimensions in Table E.4.2 become $\mathbf{a}_0 = (0,0,0)$, $\mathbf{a}_1 = (1.3266, 1.4428, 0)$, $\mathbf{b}_0 = (0.6075, -0.6909, 0)$, $\mathbf{b}_1 = (1.5312, 1.1839, 0)$, and $\mathbf{p}_1 = (2.0642, 0.509, 0)$.

Known Information: Example 4.2 and Appendix H.1 MATLAB file

Solution Approach: Figure E.8.3 includes the input specified (in bold text) in the Appendix H.1 MATLAB file. Figure E.8.4 includes the level-luffing crane and coupler curve calculated. The crane dimensions and results given in this figure are for the initial RRSS mechanism configuration (which corresponds to the first branch in the Appendix H.1 MATLAB file).

```
a0 = [0, 0, 0]';
a1 = [1.3266, 1.4428, 0]';
ua0 = [0, 0, 1]';
ua1 = [0, 0, 1]';
b0 = [0.6075, -0.6909, 0]';
b1 = [1.5312, 1.1839, 0]';
p1 = [2.0642, 0.509, 0]';
q1 = p1;
r1 = p1;

start_ang  = 10;
step_ang   = -1;
stop_ang   = -32;

angular_vel  = 0 * ones(N+1,1);
angular_acc  = 0 * ones(N+1,1);
```

FIGURE E.8.3 Specified input (in bold text) in the Appendix H.1 MATLAB file for Example 8.2.

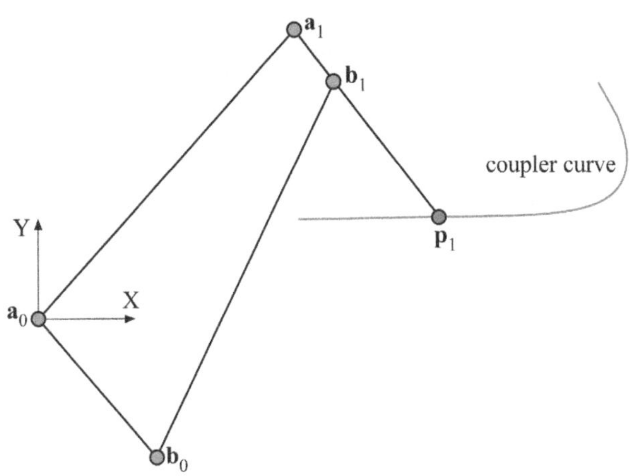

FIGURE E.8.4 Level-luffing crane mechanism with the calculated coupler curve.

8.2.4 RRSS Mechanism as a 4R Spherical Mechanism

The *Revolute–Revolute–Revolute–Revolute* spherical mechanism or four revolute spherical mechanism or simply 4R Spherical mechanism (Figure 8.5) is a single-DOF four-bar mechanism that exhibits *spherical motion*—a unique type of spatial motion [4].[6] Contrary to Gruebler's Equation, from which a mobility of −2 is calculated for the 4R Spherical mechanism (with $L = J_1 = 4$), it actually has a mobility of 1.[7] Because the links of the 4R Spherical mechanism are circular arcs (with all links having a common center) and the mechanism joint axes all intersect at that common center, the links of this mechanism have spherical surface workspaces.

The RRSS kinematic equations introduced in Sections 8.2.1 through 8.2.3 can be applied directly to the 4R Spherical mechanism (Figure 8.6) [5]. This application is particularly useful for calculating 4R Spherical mechanism link positions, velocities, and accelerations. To apply the RRSS kinematic equations, the user should specify the appropriate 4R Spherical mechanism values for $\mathbf{a}_0$, $\mathbf{a}_1$, $\mathbf{b}_0$, $\mathbf{b}_1$, $\mathbf{p}_1$, $\mathbf{ua}_0$, and $\mathbf{ua}_1$. Variables $\mathbf{a}_0$, $\mathbf{a}_1$, $\mathbf{b}_0$, and $\mathbf{b}_1$ should all lie on the surface of a sphere, and variables $\mathbf{ua}_0$ and $\mathbf{ua}_1$ should intersect at the center of the sphere. If it is assumed that the 4R Spherical mechanism should lie on a sphere of unit radius ($r_{\text{sphere}} = 1$), then $\mathbf{a}_0 = \mathbf{ua}_0$ and $\mathbf{a}_1 = \mathbf{ua}_1$.

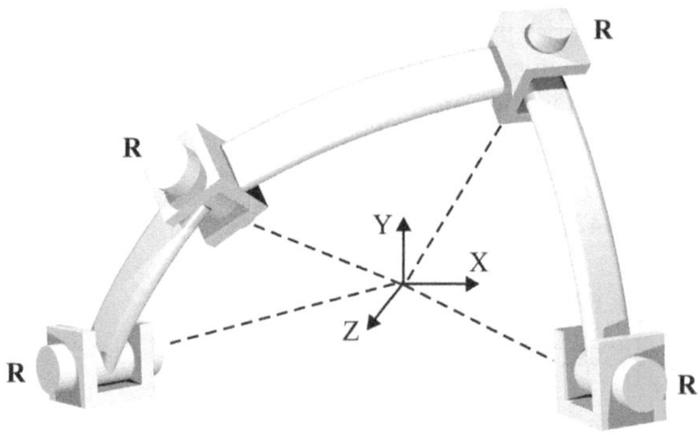

FIGURE 8.5 4R Spherical mechanism.

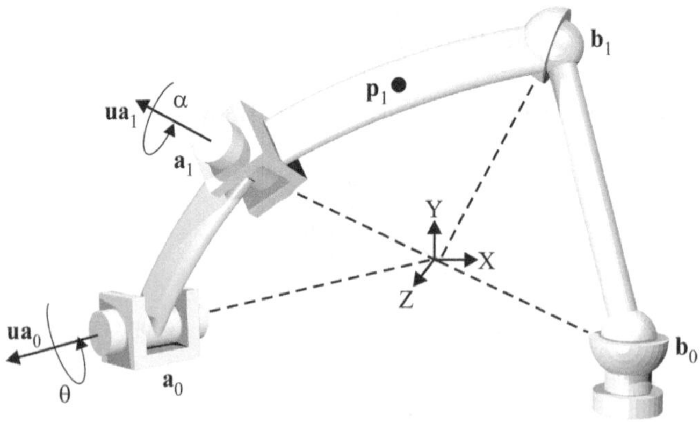

FIGURE 8.6 RRSS mechanism configured as a 4R Spherical mechanism.

Example 8.3

Problem Statement: Figure E.8.5 illustrates a 4R Spherical mechanism used in a folding wing system. Using the Appendix H.1 MATLAB file, calculate the location and velocity of point $\mathbf{p}_1$ achieved by the 4R Spherical mechanism mentioned in Table E.8.3 over a crank displacement angle range of 168° (at 28° increments). The rotational speed of the crank is 0.5 rpm.

Known Information: Table E.8.3 and Appendix H.1 MATLAB file

Solution Approach: Figure E.8.6 includes the input specified (in bold text) in the Appendix H.1 MATLAB file. Table E.8.4 includes the $\mathbf{p}_1$ location and velocity values calculated using the Appendix H.1 MATLAB file. The dimensions in Table E.8.3 and the results in Table E.8.4 are for the initial RRSS mechanism configuration (which corresponds to the first branch in the Appendix H.1 MATLAB file).

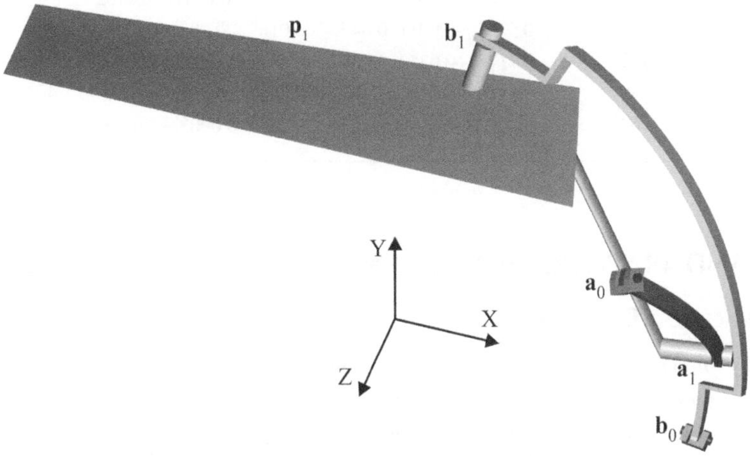

FIGURE E.8.5 4R Spherical mechanism used in a folding wing system.

TABLE E.8.3
4R Spherical Mechanism Dimensions (with Link Lengths in Meters)

Variable	Value
$\mathbf{a}_0$	0.2612, 0.7274, 0.6346
$\mathbf{a}_1$	−0.0151, 0.9879, 0.1552
$\mathbf{b}_0$	0.1793, 0.9837, 0.0172
$\mathbf{b}_1$	−0.1761, 0.2162, 0.9603
$\mathbf{p}_1$	−0.2081, 0.1675, 0.8227

```
a0 = [0.2612, 0.7274, 0.6346]';
a1 = [-0.0151, 0.9879, 0.1552]';
ua0 = a0;
ua1 = a1;
b0 = [0.1793, 0.9837, 0.0172]';
b1 = [-0.1761, 0.2162, 0.9603]';
p1 = [-0.2081, 0.1675, 0.8227]';
q1 = p1;
r1 = p1;

start_ang   = 0;
step_ang    = 28;
stop_ang    = 168;

angular_vel  = (0.5*pi/30) * ones(N+1,1);
angular_acc  = 0 * ones(N+1,1);
```

FIGURE E.8.6 Specified input (in bold text) in the Appendix H.1 MATLAB file for Example 8.3.

TABLE E.8.4
Wing Point Locations and Velocities Achieved by 4R Spherical Mechanism

θ (deg)	p (m)	ṗ (m/s)
28	0.1998, 0.0772, 0.838	0.0261, −0.0059, −0.0057
56	0.4058, 0.0319, 0.7632	0.0195, −0.0041, −0.0102
84	0.5745, −0.0011, 0.6466	0.0168, −0.003, −0.0149
112	0.7174, −0.0252, 0.4827	0.0136, −0.0021, −0.0203
140	0.8209, −0.04, 0.2697	0.0082, −0.001, −0.0251
168	0.8637, −0.0434, 0.023	0.0007, 0.0003, −0.0271

8.3 RSSR AND 4R SPHERICAL MECHANISM KINEMATIC ANALYSIS

8.3.1 Displacement Equations

The *Revolute–Spherical–Spherical–Revolute* mechanism or RSSR mechanism (Figure 8.7) is a spatial, four-bar mechanism having two degrees of freedom (with $L = 4$ and $J_1 = J_3 = 2$)—the rotation of the coupler link (link $\mathbf{a}_1 - \mathbf{b}_1$) about its own axis of symmetry being a *passive DOF.*

Suh and Radcliffe presented displacement equations for the RSSR mechanism [6]. Equation (8.29), the coupler link constant-length equation, forms the basis for the RSSR displacement equations.

$$(\mathbf{a}-\mathbf{b})^T(\mathbf{a}-\mathbf{b})-(\mathbf{a}_1-\mathbf{b}_1)^T(\mathbf{a}_1-\mathbf{b}_1)=0 \tag{8.29}$$

The rotation of the crank and follower links (link $\mathbf{a}_0 - \mathbf{a}_1$ and $\mathbf{b}_0 - \mathbf{b}_1$, respectively, in Figure 8.7) about their fixed-pivot joint axes by crank and follower displacement angles θ and ϕ, respectively, are given by

$$\mathbf{a}=\left[R_{\theta,\mathbf{ua}_0}\right](\mathbf{a}_1-\mathbf{a}_0)+\mathbf{a}_0 \tag{8.30}$$

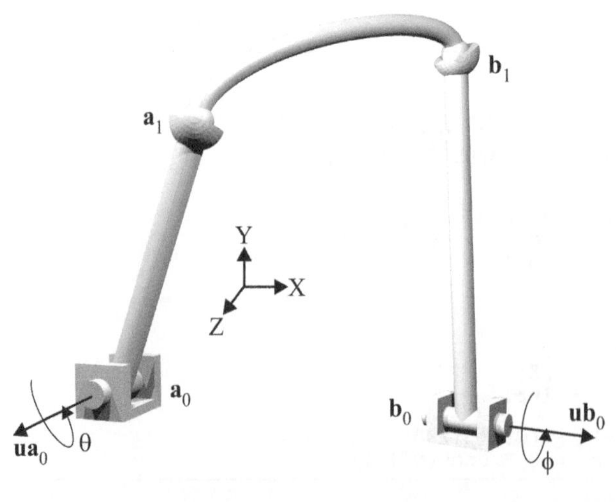

FIGURE 8.7 RSSR mechanism with displacement variables.

and

$$\mathbf{b} = \left[R_{\phi,\mathbf{ub}_0}\right](\mathbf{b}_1 - \mathbf{b}_0) + \mathbf{b}_0 \tag{8.31}$$

The rotational displacement matrices $\left[R_{\theta,\mathbf{ua}_0}\right]$ and $\left[R_{\phi,\mathbf{ub}_0}\right]$ are identical in form to Matrix (2.30).

Substituting Equations (8.30) and (8.31) into Equation (8.29) produces

$$E\cos(\phi) + F\sin(\phi) + G = 0 \tag{8.32}$$

In Equation (8.32),

$$E = (\mathbf{a} - \mathbf{b}_0)^T \left\{\left[I - Q_{\mathbf{ub}_0}\right](\mathbf{b}_1 - \mathbf{b}_0)\right\} \tag{8.33}$$

$$F = (\mathbf{a} - \mathbf{b}_0)^T \left\{\left[P_{\mathbf{ub}_0}\right](\mathbf{b}_1 - \mathbf{b}_0)\right\} \tag{8.34}$$

$$\begin{aligned} G &= (\mathbf{a} - \mathbf{b}_0)^T \left\{\left[Q_{\mathbf{ub}_0}\right](\mathbf{b}_1 - \mathbf{b}_0)\right\} \\ &\quad + \frac{1}{2}\left\{(\mathbf{a}_1 - \mathbf{b}_1)^T(\mathbf{a}_1 - \mathbf{b}_1) - (\mathbf{a} - \mathbf{b}_0)^T(\mathbf{a} - \mathbf{b}_0) - (\mathbf{b}_1 - \mathbf{b}_0)^T(\mathbf{b}_1 - \mathbf{b}_0)\right\} \end{aligned} \tag{8.35}$$

In Equations (8.33) through (8.35), matrices $\left[Q_{\mathbf{ub}_0}\right]$ and $\left[P_{\mathbf{ub}_0}\right]$ are identical in form to Equations (8.10) and (8.11). The two follower-link displacement angle solutions (ϕ) for Equation (8.32) are

$$\phi_{1,2} = 2\tan^{-1}\frac{-F \pm \sqrt{E^2 + F^2 - G^2}}{G - E}{}_8 \tag{8.36}$$

With ϕ known, the displaced RSSR moving pivot **b** from Equation (8.32) can be calculated. Variables $\mathbf{a}_0$, $\mathbf{a}_1$, $\mathbf{ua}_0$, $\mathbf{b}_0$, $\mathbf{b}_1$, and $\mathbf{ub}_0$ are 3 × 1 vectors containing *x*-, *y*-, and *z*-components.

8.3.2 Velocity Equations

Figure 8.8 includes the velocity variables for the RSSR mechanism. Equation (8.37), the derivative of the coupler link constant-length equation, forms the basis for RSSR velocity equations.

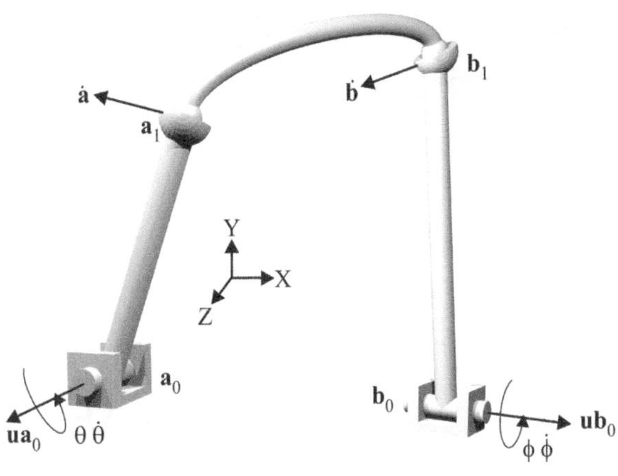

FIGURE 8.8 RSSR mechanism with velocity variables.

$$\left(\dot{\mathbf{a}}-\dot{\mathbf{b}}\right)^T(\mathbf{a}-\mathbf{b})=0 \tag{8.37}$$

The global velocity of the RSSR moving pivot $\mathbf{b}_1$ (variable $\dot{\mathbf{b}}$ in Equation (8.37)) is expressed as

$$\dot{\mathbf{b}}=\dot{\phi}\left[P_{\mathbf{ub}_0}\right](\mathbf{b}-\mathbf{b}_0)=\left[V_{\dot{\phi},\mathbf{ub}_0}\right](\mathbf{b}-\mathbf{b}_0) \tag{8.38}$$,

and the global velocity of the RSSR moving pivot $\mathbf{a}_1$ (variable $\dot{\mathbf{a}}$ in Equation (8.38)) is expressed as

$$\dot{\mathbf{a}}=\dot{\theta}\left[P_{\mathbf{ua}_0}\right](\mathbf{a}-\mathbf{a}_0)=\left[V_{\dot{\theta},\mathbf{ua}_0}\right](\mathbf{a}-\mathbf{a}_0) \tag{8.39}$$

The rotational velocity matrices $\left[V_{\dot{\theta},\mathbf{ua}_0}\right]$ and $\left[V_{\dot{\phi},\mathbf{ub}_0}\right]$ are identical in form to Matrix (2.35).

Substituting Equations (8.38) and (8.39) into Equation (8.37) and solving for the follower angular velocity $\dot{\phi}$ produce the follower angular velocity equation

$$\dot{\phi}=\frac{(\dot{\mathbf{a}})^T(\mathbf{a}-\mathbf{b})}{(\mathbf{a}-\mathbf{b})^T\left\{\left[P_{\mathbf{ub}_0}\right](\mathbf{b}-\mathbf{b}_0)\right\}} \tag{8.40}$$

With $\dot{\phi}$ known, the displaced RSSR moving-pivot velocity $\dot{\mathbf{b}}$ from Equation (8.38) can be calculated.

8.3.3 Acceleration Equations

Figure 8.9 includes the acceleration variables for the RSSR mechanism. Equation (8.41), the second derivative of the coupler link constant-length equation, forms the basis for RSSR acceleration equations.

$$\left(\ddot{\mathbf{a}}-\ddot{\mathbf{b}}\right)^T(\mathbf{a}-\mathbf{b})+\left(\dot{\mathbf{a}}-\dot{\mathbf{b}}\right)^T\left(\dot{\mathbf{a}}-\dot{\mathbf{b}}\right)=0 \tag{8.41}$$

The global acceleration of the RSSR moving pivot $\mathbf{b}_1$ (variable $\ddot{\mathbf{b}}$ in Equation (10.41)) is expressed as

$$\ddot{\mathbf{b}}=\left\{\ddot{\phi}\left[P_{\mathbf{ub}_0}\right]+\dot{\phi}^2\left[P_{\mathbf{ub}_0}\right]\left[P_{\mathbf{ub}_0}\right]\right\}(\mathbf{b}-\mathbf{b}_0)=\left[A_{\dot{\phi},\ddot{\phi},\mathbf{ub}_0}\right](\mathbf{b}-\mathbf{b}_0) \tag{8.42}$$,

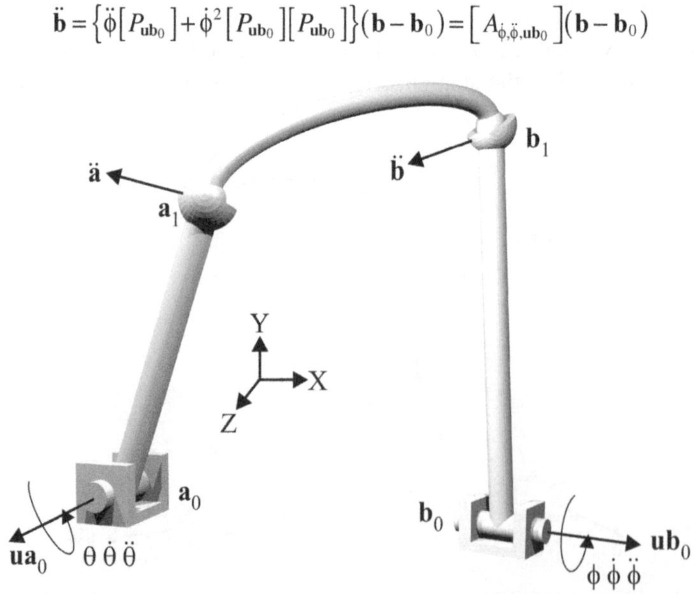

FIGURE 8.9 RSSR mechanism with acceleration variables.

and the global acceleration of the RSSR moving pivot $\mathbf{a}_1$ (variable $\ddot{\mathbf{a}}$ in Equation (8.41)) is expressed as

$$\ddot{\mathbf{a}} = \left\{\ddot{\theta}\left[P_{\mathbf{ua}_0}\right] + \dot{\theta}^2\left[P_{\mathbf{ua}_0}\right]\left[P_{\mathbf{ua}_0}\right]\right\}(\mathbf{a} - \mathbf{a}_0) = \left[A_{\dot{\theta},\ddot{\theta},\mathbf{ua}_0}\right](\mathbf{a} - \mathbf{a}_0)^9 \quad (8.43)$$

The rotational acceleration matrices $\left[A_{\dot{\theta},\ddot{\theta},\mathbf{ua}_0}\right]$ and $\left[A_{\dot{\phi},\ddot{\phi},\mathbf{ub}_0}\right]$ are identical in form to Matrix (2.39).

Substituting Equations (8.42) and (8.43) into Equation (8.41) and solving for the follower angular acceleration $\ddot{\phi}$ produce the follower angular acceleration equation

$$\ddot{\phi} = \frac{(\mathbf{a} - \mathbf{b})^T\left\{\ddot{\mathbf{a}} - \dot{\phi}^2\left[P_{\mathbf{ub}_0}\right]\left[P_{\mathbf{ub}_0}\right](\mathbf{b} - \mathbf{b}_0)\right\} + \left(\dot{\mathbf{a}} - \dot{\mathbf{b}}\right)^T\left(\dot{\mathbf{a}} - \dot{\mathbf{b}}\right)}{(\mathbf{a} - \mathbf{b})^T\left\{\left[P_{\mathbf{ub}_0}\right](\mathbf{b} - \mathbf{b}_0)\right\}} \quad (8.44)$$

With $\ddot{\phi}$ known, the displaced RSSR moving-pivot acceleration $\ddot{\mathbf{b}}$ from Equation (8.42) can be calculated.

Appendix H.2 includes the MATLAB file user instructions for RSSR displacement, velocity, and acceleration analyses. This MATLAB file (which is available for download at https://www.routledge.com/9781032956800), utilizes the RSSR displacement, velocity, and acceleration equations presented in Sections 8.3.1 through 8.3.3.

Like the RRSS mechanism equations in Section 8.2, the RSSR equations can also be used for the displacement, velocity, and acceleration analyses of a planar four-bar mechanism (specifically a planar RSSR mechanism). To accomplish this, the user should specify planar mechanism values for the RSSR mechanism variables (e.g., $\mathbf{ua}_0 = \mathbf{ub}_0 = (0,0,1)$ and $a_{0z} = a_{1z} = b_{0z} = b_{1z} = 0$).[10]

Example 8.4

Problem Statement: Figure E.8.7 illustrates an RSSR mechanism used in a steering system. All of the gear pairs shown have 1:1 gear ratios. Using the Appendix H.2 MATLAB file, calculate the wheel displacement angles achieved by the RSSR mechanism in Table E.8.5 over a steering wheel displacement angle range of 50° (at 10° increments).

Known Information: Table E.8.5 and Appendix H.2 MATLAB file

Solution Approach: Figure E.8.8 includes the input specified (in bold text) in the Appendix H.2 MATLAB file. Table E.8.6 includes the wheel displacement angles calculated using the Appendix H.2 MATLAB file over the steering wheel displacement angles. The dimensions in Table E.8.5 and the results in Table E.8.6 are for the initial RSSR mechanism configuration (which corresponds to the first branch in the Appendix H.2 MATLAB file).[11]

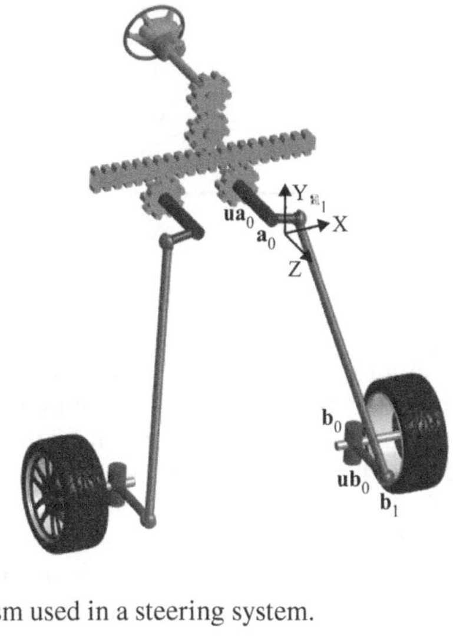

FIGURE E.8.7 RSSR mechanism used in a steering system.

TABLE E.8.5
RSSR Mechanism Dimensions

Variable	Value
$\mathbf{a}_0$	0, 0, –0.4023
$\mathbf{a}_1$	0.3356, –0.0708, –0.4023
$\mathbf{ua}_0$	0, 0, 1
$\mathbf{b}_0$	1, –2.3885, 0
$\mathbf{b}_1$	1, –2.3885, 0.73
$\mathbf{ub}_0$	0, –1, 0

```
a0 = [0, 0, -0.4023]';
a1 = [0.3356, -0.0708, -0.4023]';
ua0 = [0, 0, 1]';
b0 = [1, -2.3885, 0]';
b1 = [1, -2.3885, 0.73]';
ub0 = [0, -1, 0]';

start_ang   = 0;
step_ang    = 10;
stop_ang    = 50;

angular_vel  = 0 * ones(N+1,1);
angular_acc  = 0 * ones(N+1,1);
```

FIGURE E.8.8 Specified input (in bold text) in the Appendix H.2 MATLAB file for Example 8.4.

TABLE E.8.6
Wheel Displacement Angles Achieved by the RSSR Mechanism

θ (deg)	ϕ (deg)
10	15.041
20	29.998
30	44.907
40	60.009
50	76.05

Example 8.5

Problem Statement: Using the Appendix H.2 MATLAB file, plot the follower angular displacement versus the crank angular displacement (for both mechanism branches) for the mechanism in Example 4.2. When expressed as *x*–*y* coordinates, the mechanism dimensions in Table E.4.2 become $\mathbf{a}_0 = (0,0,0)$, $\mathbf{a}_1 = (1.3266,1.4428,0)$, $\mathbf{b}_0 = (0.6075,-0.6909,0)$, and $\mathbf{b}_1 = (1.5312,1.1839,0)$.

Known Information: Example 4.2 and Appendix H.2 MATLAB file

Solution Approach: Figure E.8.9 includes the input specified (in bold text) in the Appendix H.2 MATLAB file. Figure E.8.10 includes the follower versus crank angular displacement plots for the level-luffing crane mechanism calculated using the Appendix H.2 MATLAB file.

```
a0 = [0, 0, 0]';
a1 = [1.3266, 1.4428, 0]';
ua0 = [0, 0, 1]';
b0 = [0.6075, -0.6909, 0]';
b1 = [1.5312, 1.1839, 0]';
ub0 = [0, 0, 1]';

start_ang    = 10;
step_ang     = -1;
stop_ang     = -32;

angular_vel  = 0 * ones(N+1,1);
angular_acc  = 0 * ones(N+1,1);
```

FIGURE E.8.9 Specified input (in bold text) in the Appendix F.2 MATLAB file for Example 8.5.

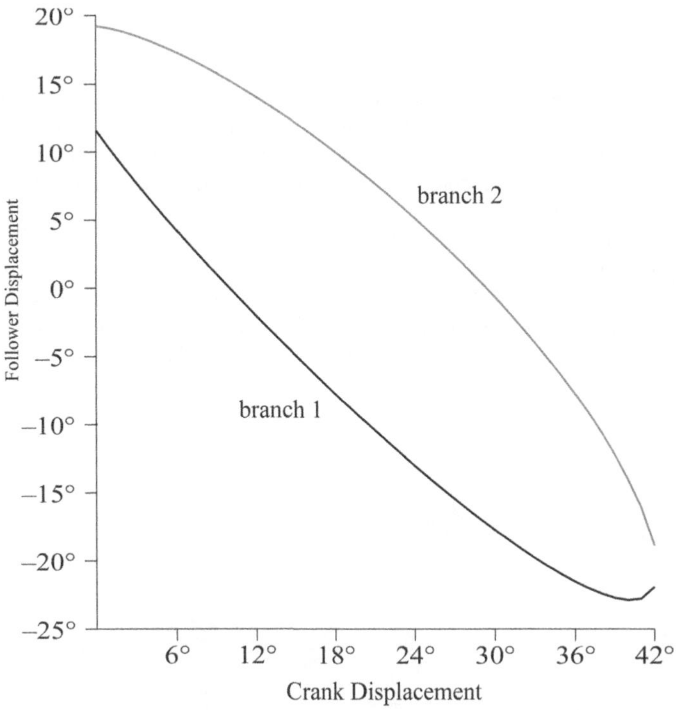

FIGURE E.8.10 Follower versus crank angular displacement plots for the level-luffing crane mechanism.

8.3.4 RSSR Mechanism as a 4R Spherical Mechanism

The RSSR kinematic equations introduced in Sections 8.3.1 through 8.3.3 can be applied directly to the 4R Spherical mechanism (Figure 8.10). This application is particularly useful for calculating 4R Spherical mechanism link angular positions, velocities, and accelerations. To apply the RSSR kinematic equations, the user should specify the appropriate 4R Spherical mechanism values for $\mathbf{a}_0$, $\mathbf{a}_1$, $\mathbf{b}_0$, $\mathbf{b}_1$, $\mathbf{ua}_0$, and $\mathbf{ub}_0$. Variables $\mathbf{a}_0$, $\mathbf{a}_1$, $\mathbf{b}_0$, and $\mathbf{b}_1$ should all lie on the surface of a sphere, and variables $\mathbf{ua}_0$ and $\mathbf{ub}_0$ should intersect at the center of the sphere.

If it is assumed that the 4R Spherical mechanism should lie on a sphere of unit radius ($r_{sphere} = 1$), then $\mathbf{a}_0 = \mathbf{ua}_0$ and $\mathbf{b}_0 = \mathbf{ub}_0$.

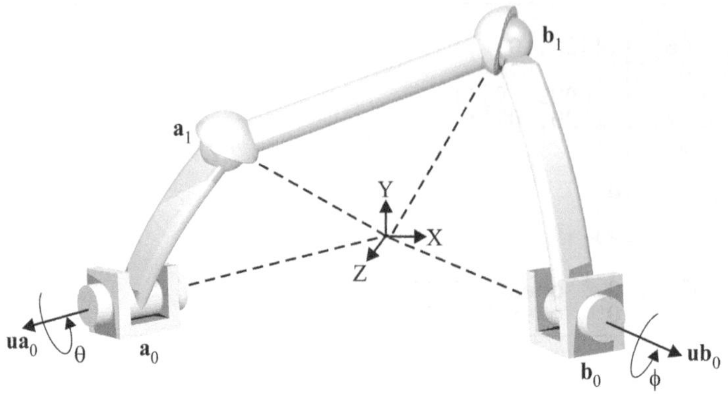

FIGURE 8.10 RSSR mechanism configured as a 4R Spherical mechanism.

Example 8.6

Problem Statement: Figure E.8.11 illustrates the initial and final positions of a 4R Spherical mechanism used to displace a camera (affixed to the follower link). The mechanism dimensions are $\mathbf{a}_0 = (1,0,0)$, $\mathbf{a}_1 = (0.9083,-0.3824,0.1695)$, $\mathbf{b}_0 = (\sin 45°, \cos 45°, 0)$, and $\mathbf{b}_1 = (-0.0166,-0.1928,0.9811)$. Given a 180° crank displacement, calculate the resulting follower angular displacement using the Appendix H.2 MATLAB file.

Known Information: Appendix H.2 MATLAB file

Solution Approach: Figure E.8.12 includes the input specified (in bold text) in the Appendix H.2 MATLAB file. A follower displacement of 80.255° was calculated. The 4R Spherical dimensions and results given in this example problem are for the initial RSSR mechanism configuration (which corresponds to the first branch in the Appendix H.2 MATLAB file).

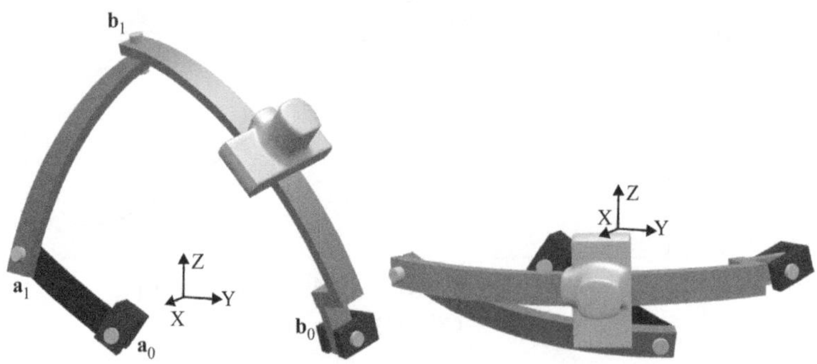

FIGURE E.8.11 4R Spherical camera rotation mechanism at (left) initial and (right) final positions.

```
a0 = [1, 0, 0]';
a1 = [0.9083, -0.3824, 0.1695]';
ua0 = a0;
b0 = [sin(45*pi/180), cos(45*pi/180), 0]';
b1 = [-0.0166, -0.1928, 0.9811]';
ub0 = b0;

start_ang   = 0;
step_ang    = 180;
stop_ang    = 180;

angular_vel  = 0 * ones(N+1,1);
angular_acc  = 0 * ones(N+1,1);
```

FIGURE E.8.12 Specified input (in bold text) in the Appendix H.2 MATLAB file for Example 8.6.

8.4 RRSS AND 4R SPHERICAL MOTION AND PATH GENERATION

The RRSS mechanism includes a revolute–revolute crank link and a spherical–spherical follower link. Suh and Radcliffe introduced motion generation equations to calculate the R-R and S-S dyad variables to approximate spatial precision positions.

Equations (8.45) through (8.50) encompass the constraint equation set for the R-R dyad (Figure 8.11) [7]. Equations (8.45) and (8.46) are orthogonality constraints between the R-R link and its revolute joint axes (axes $\mathbf{ua}_0$ and $\mathbf{ua}_1$, respectively). Equations (8.47) and (8.48) are unit vector constraints for $\mathbf{ua}_0$ and $\mathbf{ua}_1$, respectively. Equations (8.49) and (8.50) constrain the scalar length of the R-R link to a constant magnitude throughout each precision position.[12]

$$(\mathbf{ua}_0)^T(\mathbf{a}_j-\mathbf{a}_0)=0 \quad j=1,2,3,\ldots \tag{8.45}$$

$$(\mathbf{ua}_j)^T(\mathbf{a}_j-\mathbf{a}_0)=0 \quad j=1,2,3,\ldots \tag{8.46}$$

$$(\mathbf{ua}_0)^T(\mathbf{ua}_0)-1=0 \tag{8.47}$$

$$(\mathbf{ua}_1)^T(\mathbf{ua}_1)-1=0 \tag{8.48}$$

$$(\mathbf{a}_j-\mathbf{a}_0)^T(\mathbf{a}_j-\mathbf{a}_0)-(\mathbf{a}_1-\mathbf{a}_0)^T(\mathbf{a}_1-\mathbf{a}_0)=0 \quad j=2,3,\ldots \tag{8.49}$$

$$\begin{aligned}&\left[(\mathbf{a}_j+\mathbf{ua}_j)-(\mathbf{a}_0+\mathbf{ua}_0)\right]^T\left[(\mathbf{a}_j+\mathbf{ua}_j)-(\mathbf{a}_0+\mathbf{ua}_0)\right]\\&-\left[(\mathbf{a}_1+\mathbf{ua}_1)-(\mathbf{a}_0+\mathbf{ua}_0)\right]^T\left[(\mathbf{a}_1+\mathbf{ua}_1)-(\mathbf{a}_0+\mathbf{ua}_0)\right]=0 \; j=2,3\ldots\end{aligned} \tag{8.50}$$

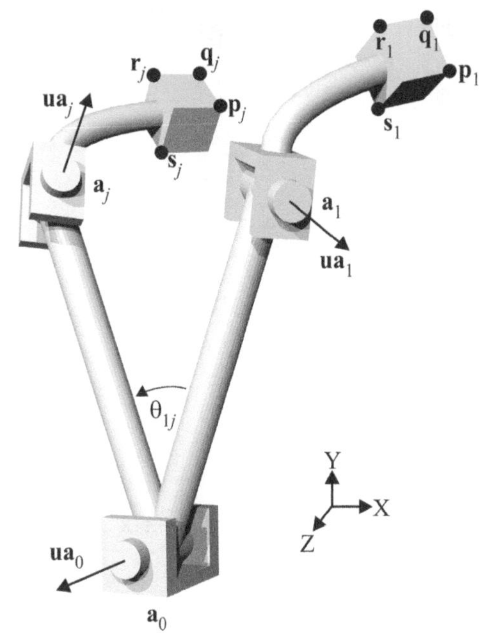

FIGURE 8.11 Revolute–Revolute (R-R) dyad at rigid-body positions 1 and *j*.

In Equations (8.45)–(8.50),

$$\mathbf{a}_0 = \left[a_{0x}, a_{0y}, a_{0z}, 1\right]^T, \quad \mathbf{a}_1 = \left[a_{1x}, a_{1y}, a_{1z}, 1\right]^T, \quad \mathbf{ua}_0 = \left[ua_{0x}, ua_{0y}, ua_{0z}, 0\right]^T,$$

$$\mathbf{ua}_1 = \left[ua_{1x}, ua_{1y}, ua_{1z}, 0\right]^T, \quad \mathbf{a}_j = \left[D_{1j}\right]\mathbf{a}_1 \quad and \quad \mathbf{ua}_j = \left[R_{\mathbf{ua}_0,\theta_{1j}}\right]\mathbf{ua}_1$$

Matrix $\left[D_{1j}\right]$ defines the coupler displacement from precision position 1 to precision position j. It is expressed as

$$\left[D_{1j}\right] = \begin{bmatrix} p_{xj} & q_{xj} & r_{xj} & s_{xj} \\ p_{yj} & q_{yj} & r_{yj} & s_{yj} \\ p_{zj} & q_{zj} & r_{zj} & s_{zj} \\ 1 & 1 & 1 & 1 \end{bmatrix} \begin{bmatrix} p_{1x} & q_{1x} & r_{1x} & s_{1x} \\ p_{1y} & q_{1y} & r_{1y} & s_{1y} \\ p_{1z} & q_{1z} & r_{1z} & s_{1z} \\ 1 & 1 & 1 & 1 \end{bmatrix}^{-1}_{13} \tag{8.51}$$

where the precision position is defined by the x-, y-, and z-coordinates of $\mathbf{p}$, $\mathbf{q}$, $\mathbf{r}$, and $\mathbf{s}$. Matrix $\left[R_{\mathbf{ua}_0,\theta_{1j}}\right]$ defines the rotation of $\mathbf{ua}_1$ about $\mathbf{ua}_0$ by R-R displacement angle θ and is identical in form to Matrix (2.30). Equations (8.45)–(8.50) form a 12-equation nonlinear simultaneous equation system to solve for the 12 unknown R-R dyad variables in $\mathbf{a}_0$, $\mathbf{a}_1$, $\mathbf{b}_0$, $\mathbf{b}_1$, $\mathbf{ua}_0$, and $\mathbf{ub}_0$ given three precision positions.

Only a constant-length constraint equation is required to synthesize the S-S link (Figure 8.12) [8]. This constraint is expressed as

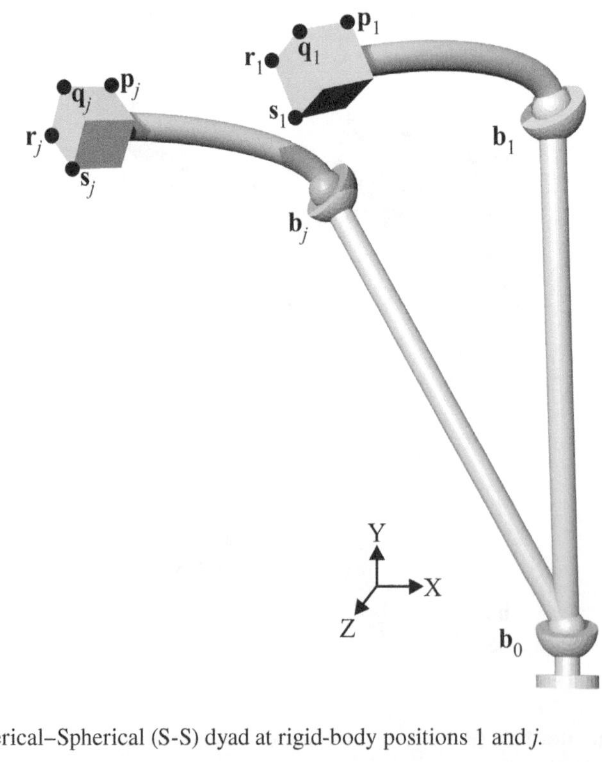

FIGURE 8.12 Spherical–Spherical (S-S) dyad at rigid-body positions 1 and j.

$$\left(\mathbf{b}_j-\mathbf{b}_0\right)^T\left(\mathbf{b}_j-\mathbf{b}_0\right)-\left(\mathbf{b}_1-\mathbf{b}_0\right)^T\left(\mathbf{b}_1-\mathbf{b}_0\right)=0 \quad j=2,3,\ldots \tag{8.52}$$

where $\mathbf{b}_j=\left[D_{1j}\right]\mathbf{b}_1$ and matrix $\left[D_{1j}\right]$ is identical to Matrix (8.51). In Equation (8.52), $\mathbf{b}_0=\left[b_{0x},b_{0y},b_{0z},1\right]^T\mathbf{b}_0=\left[b_{0x},b_{0y},b_{0z},1\right]^T$ and $\mathbf{b}_1=\left[b_{1x},b_{1y},b_{1z},1\right]^T$. Equation (7.32) forms a six-equation nonlinear simultaneous equation system to solve for the six unknown S-S dyad variables in $\mathbf{b}_0$ and $\mathbf{b}_1$ given seven precision positions. Appendices J.1 and J.2 include user instructions for the MATLAB files to solve the R-R and S-S dyad motion generation equations presented in this section, respectively, although the R-R and S-S dyad equations, as presented, accommodate three and six precision positions, respectively, in the Appendices J.1 and J.2. MATLAB files are solved as constrained and unconstrained nonlinear optimization models for an indefinite number of precision positions.

The 4R Spherical mechanism includes revolute–revolute crank and follower links configured so that the fixed- and moving-pivot joint axes intersect at a common point—the center of a sphere. Suh and Radcliffe introduced motion generation equations to calculate the spherical R-R dyad variables to approximate spatial precision positions.

Equations (8.53) through (8.55) encompass the constraint equation set for the spherical R-R dyad (Figure 8.13) [9]. Equation (8.53) constrains the chord length of the spherical R-R link to a constant magnitude throughout each precision positions. Equations (8.54) and (8.55) constrain $\mathbf{a}_0$ and $\mathbf{a}_1$ to a sphere of unit radius, respectively.

$$\left(\mathbf{a}_j-\mathbf{a}_0\right)^T\left(\mathbf{a}_j-\mathbf{a}_0\right)-\left(\mathbf{a}_1-\mathbf{a}_0\right)^T\left(\mathbf{a}_1-\mathbf{a}_0\right)=0 \quad j=2,3,\ldots \tag{8.53}$$

$$\left(\mathbf{a}_0\right)^T\left(\mathbf{a}_0\right)-1=0 \tag{8.54}$$

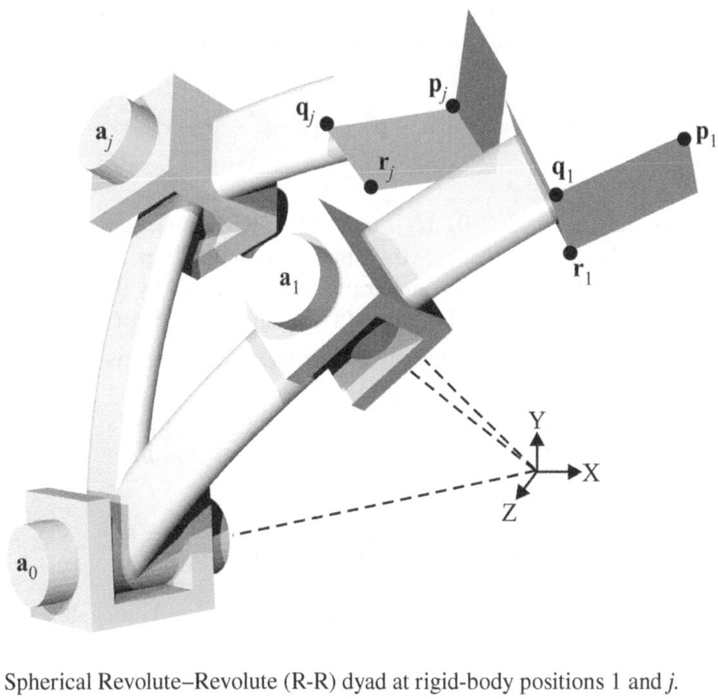

FIGURE 8.13 Spherical Revolute–Revolute (R-R) dyad at rigid-body positions 1 and *j*.

$$(\mathbf{a}_1)^T(\mathbf{a}_1)-1=0 \tag{8.55}$$

In Equations (8.53) through (8.55), $\mathbf{a}_0=\left[a_{0x},a_{0y},a_{0z},1\right]^T$, $\mathbf{a}_1=\left[a_{1x},a_{1y},a_{1z},1\right]^T$, and $\mathbf{a}_j=\left[D_{1j}\right]\mathbf{a}_1$. Matrix $\left[D_{1j}\right]$ defines the coupler displacement from precision position 1 to precision position j. It is expressed as

$$\left[D_{1j}\right]=\begin{bmatrix} p_{x_j} & q_{x_j} & r_{x_j} & 0 \\ p_{y_j} & q_{y_j} & r_{y_j} & 0 \\ p_{z_j} & q_{z_j} & r_{z_j} & 0 \\ 1 & 1 & 1 & 1 \end{bmatrix}\begin{bmatrix} p_{1x} & q_{1x} & r_{1x} & 0 \\ p_{1y} & q_{1y} & r_{1y} & 0 \\ p_{1z} & q_{1z} & r_{1z} & 0 \\ 1 & 1 & 1 & 1 \end{bmatrix}^{-1} \quad 14 \tag{8.56}$$

where the precision position is defined by the x-, y-, and z-coordinates of $\mathbf{p}$, $\mathbf{q}$, and $\mathbf{r}$. Equations (8.53) through (8.55) form a five-equation nonlinear simultaneous equation system to solve for five of the six unknown spherical R-R dyad variables in $\mathbf{a}_0$ and $\mathbf{a}_1$ (with one arbitrary variable choice) given four precision positions. Appendix J.3 includes user instructions for the MATLAB files to solve the spherical R-R equations presented in this section for 4R Spherical motion generation. Although the spherical R-R dyad equations, as presented, accommodate four precision positions, in the Appendix J.3 MATLAB files, they are solved as constrained nonlinear optimization models for an indefinite number of precision positions.

Planar four-bar motion and path generation equation systems with branch and order constraints are presented in Section 6.6. Similar equation systems can also be formulated for spatial RRSS and 4R Spherical motion and path generation. By defining the spatial position of the coupler link by coupler point variables $\mathbf{p}$, $\mathbf{q}$, and $\mathbf{r}$, an objective function to be minimized is formed by taking the sum of the squared norms of the differences between the prescribed $\mathbf{p}$, $\mathbf{q}$, and $\mathbf{r}$ values and those achieved by the synthesized mechanism. This objective function is expressed as

$$f(\mathbf{X})=\sum_{j=2}^{N}\left\{\left\|\mathbf{p}_j-\mathbf{p}\right\|^2+\left\|\mathbf{q}_j-\mathbf{q}\right\|^2+\left\|\mathbf{r}_j-\mathbf{r}\right\|^2\right\} \tag{8.57}$$

where $\mathbf{X}=\left(\mathbf{a}_0,\mathbf{a}_1,\mathbf{ua}_0,\mathbf{ua}_1,\mathbf{b}_0,\mathbf{b}_1,\theta_2\ldots\theta_N,\alpha_2\ldots\alpha_N\right)$ for the RRSS mechanism. For the 4R Spherical mechanism, $\mathbf{X}=\left(\mathbf{a}_0,\mathbf{a}_1,\mathbf{b}_0,\mathbf{b}_1,\theta_2\ldots\theta_N,\alpha_2\ldots\alpha_N\right)$. In Equation (8.57), $\mathbf{p}_j$, $\mathbf{q}_j$, and $\mathbf{r}_j$ define the precision positions, and $\mathbf{p}$, $\mathbf{q}$, and $\mathbf{r}$ define the coupler positions achieved by the synthesized mechanism (using Equation (8.13)). As presented, Equation (8.57) can be used for RRSS and 4R Spherical motion generation. For RRSS and 4R Spherical path generation, the terms $\left\|\mathbf{q}_j-\mathbf{q}\right\|^2$ and $\left\|\mathbf{r}_j-\mathbf{r}\right\|^2$ are removed from Equation (8.57).

Along with the objective function, several types of equality constraints are required for the RRSS mechanism. To ensure the constant length of the crank and follower links, Equation (8.1) along with the equation

$$(\mathbf{a}-\mathbf{a}_0)^T(\mathbf{a}-\mathbf{a}_0)-(\mathbf{a}_1-\mathbf{a}_0)^T(\mathbf{a}_1-\mathbf{a}_0)=0 \tag{8.58}$$

are included for precision positions $2\ldots N$. Equations (8.58) and (8.1) include Equations (8.2) and (8.5), respectively, to calculate the displaced moving pivots $\mathbf{a}$ and $\mathbf{b}$. Equations (8.59) and (8.60) ensure that the crank link is orthogonal to the fixed-pivot and moving-pivot joint axes and are included for precision positions $2\ldots N$. Unit vector Equations (8.61) and (8.62) are included for the R-R link fixed- and moving-pivot joint axes. Equation (8.63) ensures a constant moving-pivot joint axis orientation with respect to the crank link and is included for precision positions $2 \ldots N$. Equations (8.59), (8.60), and (8.63) include Equations (8.2) and/or (8.3) to calculate the displaced moving pivot **a** and the displaced moving-pivot joint axis **ua**.

$$(\mathbf{ua}_0)^T(\mathbf{a}-\mathbf{a}_0)=0 \tag{8.59}$$

$$(\mathbf{ua})^T(\mathbf{a}-\mathbf{a}_0)=0 \tag{8.60}$$

$$(\mathbf{ua}_0)^T(\mathbf{ua}_0)-1=0 \tag{8.61}$$

$$(\mathbf{ua}_1)^T(\mathbf{ua}_1)-1=0 \tag{8.62}$$

$$\begin{aligned}&\left[(\mathbf{a}+\mathbf{ua})-(\mathbf{a}_0+\mathbf{ua}_0)\right]^T\left[(\mathbf{a}+\mathbf{ua})-(\mathbf{a}_0+\mathbf{ua}_0)\right]\\&-\left[(\mathbf{a}_1+\mathbf{ua}_1)-(\mathbf{a}_0+\mathbf{ua}_0)\right]^T\left[(\mathbf{a}_1+\mathbf{ua}_1)-(\mathbf{a}_0+\mathbf{ua}_0)\right]=0\end{aligned} \tag{8.63}$$

To produce RRSS linkage solutions that are order defect-free, Inequality (8.64) is included for precision positions $2 \ldots N$.

$$\begin{cases}\theta_j > \theta_{j-1} \\ \theta_N < 2\pi\end{cases} \quad j=2,3,\ldots \tag{8.64}$$

As written, Inequality (8.64) ensures counterclockwise crank link rotation. The inequality $\theta_j < \theta_{j-1}$ can be used instead in Inequality (8.64) to ensure clockwise crank link rotation.

To produce RRSS linkage solutions that are branch defect-free, Inequality (8.65) is included for each of precision positions $2 \ldots N$.

$$\left[(\mathbf{b}-\mathbf{b}_0)\times(\mathbf{a}-\mathbf{b}_0)\right]\cdot\left[(\mathbf{b}_1-\mathbf{b}_0)\times(\mathbf{a}_1-\mathbf{b}_0)\right]>0 \tag{8.65}$$

A change in branch would result in a change in sign in the product of the cross products from positive to negative. By constraining the product of these cross products to a positive value, Inequality (8.65) constrains the mechanism to a single branch.

Inequality (8.65) includes Equations (8.2) and (8.5) to calculate the displaced moving pivots **a** and **b**, respectively. This combination of objective function, equality constraints, and inequality constraints allows for the direct minimization of the difference between the prescribed and achieved RRSS precision positions or precision points while producing RRSS solutions that are order defect-free and branch defect-free.

Like the planar four-bar constrained nonlinear equation systems in Section 6.6, RRSS displacement angles θ and α are among the unknown variables calculated.[15] Appendices J.4 and J.5 include user instructions for the MATLAB files to solve the constrained nonlinear optimization model presented in this section for RRSS motion and path generation.

As presented, this constrained nonlinear optimization model is for RRSS motion and path generation with branching and order constraints. For 4R Spherical motion and path generation with branching and order constraints, Equations (8.59) through (8.62) are replaced with

Equations (8.66)–(8.69), respectively. These equations constrain the 4R Spherical mechanism's fixed and moving pivots to a sphere of unit radius.

$$(\mathbf{a}_0)^T(\mathbf{a}_0)-1=0 \tag{8.66}$$

$$(\mathbf{a}_1)^T(\mathbf{a}_1)-1=0 \tag{8.67}$$

$$(\mathbf{b}_0)^T(\mathbf{b}_0)-1=0 \tag{8.68}$$

$$(\mathbf{b}_1)^T(\mathbf{b}_1)-1=0 \tag{8.69}$$

Appendices J.6 and J.7 include user instructions for the MATLAB files to solve the constrained nonlinear optimization model presented in this section for 4R Spherical motion and path generation.

Example 8.7

Problem Statement: Using the Appendix J.4 MATLAB files, synthesize a branch- and order defect-free RRSS mechanism for the five solar panel positions given in Table E.8.7 (Figure E.8.13).

Known Information: Table E.8.7 and Appendix J.4 MATLAB files

Solution Approach: Figure E.8.14 includes the input in bold type in the Appendix J.4 MATLAB file *RRSS_interior_point.m*. Figure E.8.15 includes the output from the same file.

Figure E.8.16 illustrates the resulting RRSS mechanism. A kinematic analysis of this solution is conducted in Example 8.14 in Section 8.6.

TABLE E.8.7
Solar Panel Precision Positions

Precision Position	p	q	r
1	1.7321, 0, −1	1.2321, 0, −1.8660	1.9486, 0, −1.1250
2	1.2247, 1.2247, −1	0.8712, 0.8712, −1.8660	1.3778, 1.3778, −1.1250
3	0, 1.7321, −1	0, 1.2321, −1.8660	−0.0631, 1.9416, −1.1210
4	−1.2247, 1.2247, −1	−0.8712, 0.8712, −1.8660	−1.4308, 1.3048, −1.1168
5	−1.7321, 0, −1	−1.2321, 0, −1.8660	−1.9258, −0.1117, −1.1118

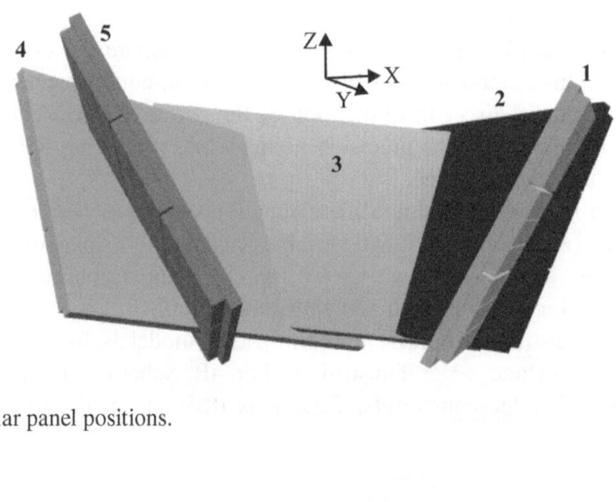

FIGURE E.8.13 Solar panel positions.

```
N = 5;

P1 = [1.7321, 0, -1];
P2 = [1.2247, 1.2247, -1];
P3 = [0, 1.7321, -1];
P4 = [-1.2247, 1.2247, -1];
P5 = [-1.7321, 0, -1];

Q1 = [1.2321, 0, -1.8660];
Q2 = [0.8712, 0.8712, -1.8660];
Q3 = [0, 1.2321, -1.8660];
Q4 = [-0.8712, 0.8712, -1.8660];
Q5 = [-1.2321, 0, -1.8660];

R1 = [1.9486, 0, -1.1250];
R2 = [1.3778, 1.3778, -1.1250];
R3 = [-0.0631, 1.9416, -1.1210];
R4 = [-1.4308, 1.3048, -1.1168];
R5 = [-1.9258, -0.1117, -1.1118];

a0x = 0;
a0y = 1;
a0z = 0;

a1x = 5;
a1y = 1;
a1z = 0;

ua0x = 0;
ua0y = 0;
ua0z = 1;

ua1x = 0;
ua1y = 0;
ua1z = 1;

b0x = 0;
b0y = 0;
b0z = 0;

b1x = 2;
b1y = -1;
b1z = 1;

alpha2 = 25 *pi/180;
alpha3 = 50 *pi/180;
alpha4 = 75 *pi/180;
alpha5 = 100 *pi/180;

theta2 = 45 *pi/180;
theta3 = 90 *pi/180;
theta4 = 135 *pi/180;
theta5 = 180 *pi/180;
```

FIGURE E.8.14 Input in bold type in the Appendix J.4 MATLAB file *RRSS_interior_point.m* for Example 8.7 (counterclockwise crank rotation constraint applied).

```
Iterations: 31
Func Evals: 93
Algorithm: interior-point
exit flag = 2
error = ( 5.320619e-02 )
a0x = ( 0.036138 )
a0y = ( 0.095884 )
a0z = ( 0.347070 )
a1x = ( 5.773299 )
a1y = ( 0.609854 )
a1z = ( 0.337957 )
ua0x = ( -0.005795 )
ua0y = ( 0.082352 )
ua0z = ( 0.996589 )
ua1x = ( -0.025801 )
ua1y = ( 0.304883 )
ua1z = ( 0.952047 )
b0x = ( 0.051871 )
b0y = ( 0.435930 )
b0z = ( 0.450254 )
b1x = ( 3.365637 )
b1y = ( -2.116441 )
b1z = ( 1.058884 )
alpha2 = ( 4.798059 )
alpha3 = ( 7.794214 )
alpha4 = ( 7.111336 )
alpha5 = ( 4.508328 )
theta2 = ( 56.581151 )
theta3 = ( 110.286397 )
theta4 = ( 152.665877 )
theta5 = ( 189.944655 )
Elapsed time is 1041.249666 seconds.
Solution calculation is complete.
>>
```

FIGURE E.8.15 Output from the Appendix J.4 MATLAB file *RRSS_interior_point.m* for Example 8.7.

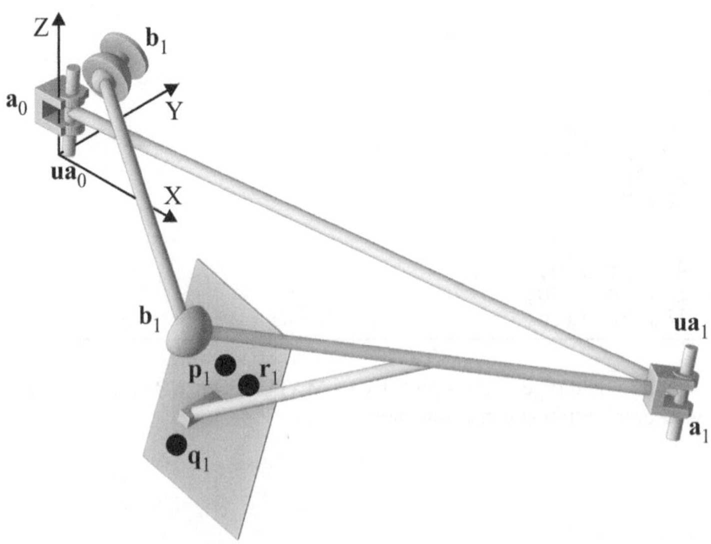

FIGURE E.8.16 Synthesized RRSS motion generator.

Example 8.8

Problem Statement: Using the Appendix J.6 MATLAB files, synthesize a branch- and order defect-free 4R Spherical mechanism for the five folding aircraft wing positions given in Table E.8.8 (Figure E.8.17).

Known Information: Table E.8.8 and Appendix J.6 MATLAB files

Solution Approach: Figure E.8.18 includes the input in bold type in the Appendix J.6 MATLAB file *Sphere_interior_point.m*. Figure E.8.19 includes the output from the same file.

Figure E.8.20 illustrates the resulting 4R Spherical mechanism. A kinematic analysis of this solution is presented in Example 8.15 in Section 8.6.

TABLE E.8.8
Folding Wing Precision Positions

Precision Position	p	q	r
1	−0.2081, 0.1675, 0.8227	0.2272, 0.1665, 0.8179	0, 0.3087, 0.8080
2	0.1442, 0.0859, 0.8486	0.5013, 0.2482, 0.6598	0.2613, 0.2946, 0.7702
3	0.5009, 0.0166, 0.7050	0.6727, 0.3175, 0.4415	0.5104, 0.2616, 0.6475
4	0.7676, −0.0298, 0.3977	0.7554, 0.3639, 0.2124	0.7179, 0.2147, 0.4321
5	0.8637, −0.0462, −0.0095	0.7769, 0.3803, −0.0095	0.8041, 0.1737, 0.1319

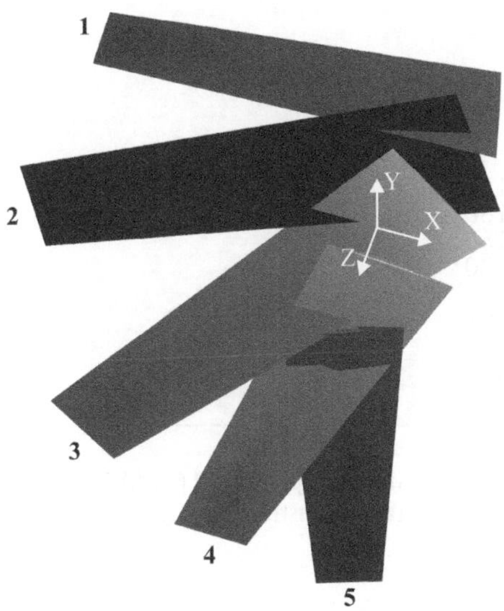

FIGURE E.8.17 Folding wing positions.

```
N = 5;

P1 = [-0.2081, 0.1675, 0.8227];
P2 = [0.1442, 0.0859, 0.8486];
P3 = [0.5009, 0.0166, 0.7050];
P4 = [0.7676, -0.0298, 0.3977];
P5 = [0.8637, -0.0462, -0.0095];

Q1 = [0.2272, 0.1665, 0.8179];
Q2 = [0.5013, 0.2482, 0.6598];
Q3 = [0.6727, 0.3175, 0.4415];
Q4 = [0.7554, 0.3639, 0.2124];
Q5 = [0.7769, 0.3803, -0.0095];

R1 = [0, 0.3087, 0.8080];
R2 = [0.2613, 0.2946, 0.7702];
R3 = [0.5104, 0.2616, 0.6475];
R4 = [0.7179, 0.2147, 0.4321];
R5 = [0.8041, 0.1737, 0.1319];

a0x = 1;
a0y = 1;
a0z = 1;

a1x = 1;
a1y = 1;
a1z = 1;

b0x = 0;
b0y = 1;
b0z = 0;

b1x = 0;
b1y = 0;
b1z = 1;

alpha2 = -20 *pi/180;
alpha3 = -40 *pi/180;
alpha4 = -60 *pi/180;
alpha5 = -80 *pi/180;

theta2 = 25 *pi/180;
theta3 = 35 *pi/180;
theta4 = 45 *pi/180;
theta5 = 55 *pi/180;
```

FIGURE E.8.18 Input in bold type in the Appendix J.6 MATLAB file *Sphere_interior_point.m* for Example 8.8 (counterclockwise crank rotation constraint applied).

```
Iterations: 86
Func Evals: 257
Algorithm: interior-point
exit flag = 2
error = ( 2.113997e-03 )
a0x = ( -0.643678 )
a0y = ( -0.393975 )
a0z = ( -0.656096 )
a1x = ( -0.358079 )
a1y = ( -0.529870 )
a1z = ( -0.768776 )
b0x = ( -0.802561 )
b0y = ( 0.399208 )
b0z = ( 0.443317 )
b1x = ( 0.621598 )
b1y = ( 0.444330 )
b1z = ( 0.645125 )
alpha2 = ( -68.277277 )
alpha3 = ( -123.483996 )
alpha4 = ( -180.148154 )
alpha5 = ( -250.948120 )
theta2 = ( 32.834864 )
theta3 = ( 60.211378 )
theta4 = ( 88.737030 )
theta5 = ( 121.986587 )
Elapsed time is 3528.466885 seconds.
Solution calculation is complete.
>>
```

FIGURE E.8.19 Output from the Appendix J.6 MATLAB file *Sphere_interior_point.m* for Example 8.8.

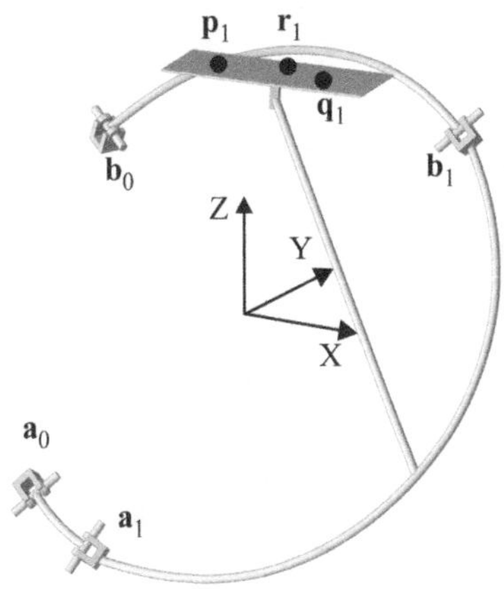

FIGURE E.8.20 Synthesized 4R Spherical motion generator.

Example 8.9

Problem Statement: Using the Appendix J.7 MATLAB files, synthesize a branch- and order defect-free 4R Spherical mechanism for the curve points given in Table E.8.9 (Figure E.8.21).

Known Information: Table E.8.9 and Appendix J.7 MATLAB files

Solution Approach: Figure E.8.22 includes the input in bold type in Appendix J.7 MATLAB file *Sphere_interior_point.m*. Figure E.8.23 includes the output from the same file.

Figure E.8.24 illustrates the resulting 4R Spherical mechanism. A kinematic analysis of this solution is presented in Example 8.16 in Section 8.6.

TABLE E.8.9
Prescribed Spherical Curve Points

Point	$\mathbf{p}_j$
1	0.7071, 0.0000, 0.7071
2	0.7057, 0.0186, 0.7083
3	0.6999, 0.0414, 0.7130
4	0.6852, 0.0718, 0.7248
5	0.6515, 0.1129, 0.7502
6	0.5652, 0.1746, 0.8063
7	0.4509, 0.2238, 0.8641
8	0.3340, 0.2561, 0.9071
9	0.2181, 0.2764, 0.9360
10	0.1432, 0.2845, 0.9479
11	0.0609, 0.2894, 0.9553
12	0.0000, 0.2905, 0.9569

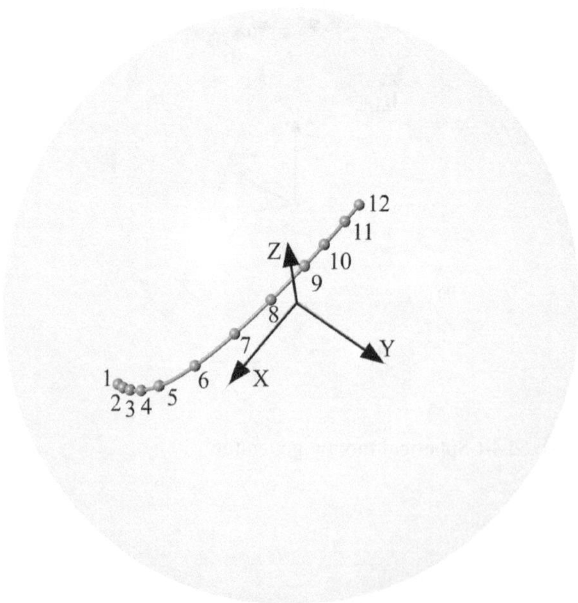

FIGURE E.8.21 Prescribed spherical curve points.

```
N = 12;

P1 = [0.7071, 0.0000, 0.7071];
P2 = [0.7057, 0.0186, 0.7083];
P3 = [0.6999, 0.0414, 0.7130];
P4 = [0.6852, 0.0718, 0.7248];
P5 = [0.6515, 0.1129, 0.7502];
P6 = [0.5652, 0.1746, 0.8063];
P7 = [0.4509, 0.2238, 0.8641];
P8 = [0.3340, 0.2561, 0.9071];
P9 = [0.2181, 0.2764, 0.9360];
P10 = [0.1432, 0.2845, 0.9479];
P11 = [0.0609, 0.2894, 0.9553];
P12 = [0.0000, 0.2905, 0.9569];

a0x = 0;
a0y = -0.7;
a0z = 0.7;

a1x = 0;
a1y = 0.7;
a1z = 0.7;

b0x = 0;
b0y = 0;
b0z = -1;

b1x = 0;
b1y = -1;
b1z = 0;

alpha2 = -10 *pi/180;
alpha3 = -20 *pi/180;
alpha4 = -30 *pi/180;
alpha5 = -40 *pi/180;
alpha6 = -50 *pi/180;
alpha7 = -60 *pi/180;
alpha8 = -70 *pi/180;
alpha9 = -80 *pi/180;
alpha10 = -90 *pi/180;
alpha11 = -100 *pi/180;
alpha12 = -110 *pi/180;

theta2 = 10 *pi/180;
theta3 = 20 *pi/180;
theta4 = 30 *pi/180;
theta5 = 40 *pi/180;
theta6 = 50 *pi/180;
theta7 = 60 *pi/180;
theta8 = 70 *pi/180;
theta9 = 80 *pi/180;
theta10 = 90 *pi/180;
theta11 = 100 *pi/180;
theta12 = 110 *pi/180;
```

FIGURE E.8.22 Input in bold type in the Appendix J.7 MATLAB file *Sphere_interior_point.m* for Example 8.9 (counterclockwise crank rotation constraint applied).

```
Iterations: 29
Func Evals: 114
Algorithm: interior-point
exit flag = 1
error = ( 5.636644e-04 )
a0x = ( -0.221639 )
a0y = ( -0.118504 )
a0z = ( 0.967901 )
a1x = ( 0.045033 )
a1y = ( -0.188244 )
a1z = ( 0.981089 )
b0x = ( 0.226497 )
b0y = ( 0.860384 )
b0z = ( -0.456552 )
b1x = ( -0.856003 )
b1y = ( -0.301045 )
b1z = ( -0.420275 )
alpha2 = ( -24.934414 )
alpha3 = ( -44.886020 )
alpha4 = ( -59.541362 )
alpha5 = ( -74.259195 )
alpha6 = ( -95.397850 )
alpha7 = ( -117.895748 )
alpha8 = ( -138.217373 )
alpha9 = ( -157.377718 )
alpha10 = ( -172.987692 )
alpha11 = ( -190.165962 )
alpha12 = ( -200.841673 )
theta2 = ( 20.559047 )
theta3 = ( 38.800133 )
theta4 = ( 53.164483 )
theta5 = ( 68.380963 )
theta6 = ( 91.647403 )
theta7 = ( 118.502044 )
theta8 = ( 145.148299 )
theta9 = ( 173.007877 )
theta10 = ( 198.140309 )
theta11 = ( 228.936695 )
theta12 = ( 251.659957 )
Elapsed time is 12562.001365 seconds.
Solution calculation is complete.
>>
```

FIGURE E.8.23 Output from the Appendix J.7 MATLAB file *Sphere_interior_point.m* for Example 8.9.

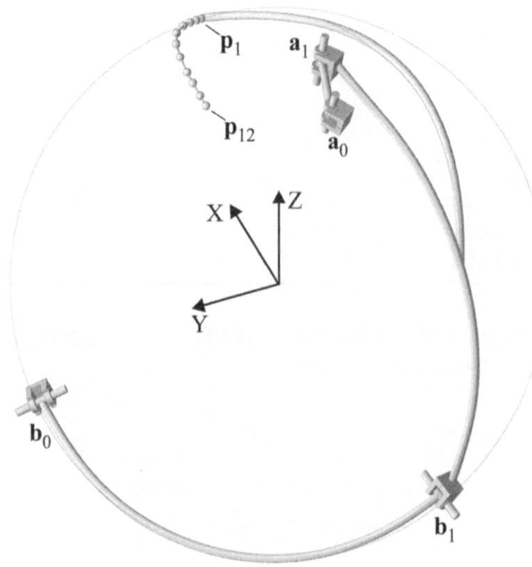

FIGURE E.8.24 Synthesized 4R Spherical path generator.

Example 8.10

Problem Statement: Using the Appendix J.1 and J.2 MATLAB files, synthesize an RRSS mechanism for the first three precision positions given in Example 8.1 (Table E.8.10).

Known Information: Table E.8.10 and Appendix J.1 and J.2 MATLAB files

Solution Approach: In this example, S-S dyad variable $\mathbf{b}_1$ was specified to $\mathbf{b}_1 = (1,-1,0)$, and the R-R dyad displacement angle variables θ_2 and θ_3 were specified to 50° and 90°, respectively. Figure E.8.25 includes the input in bold type in the Appendix J.1 MATLAB file *RR_interior_point.m*. Figure E.8.26 includes the output from the same file. Figure E.8.27 includes the input in bold type in the Appendix J.2 MATLAB file *SS_quasi_newton.m*. Figure E.8.28 includes the output from the same file. Figure E.8.29 illustrates the resulting RRSS mechanism. A kinematic analysis of this solution is presented in Example 8.17 in Section 8.6. The equation systems formulated from Equations (8.45) through (8.50) include no constraints to ensure that branch defect-free solutions are calculated.[16] With these equation sets, branch defect-free solutions can be calculated through trial and error.

TABLE E.8.10
Solar Panel Positions (with Point s Included for Matrix (8.51))

Precision Position	p	q	r	s
1	1.7321, 0, −1	1.2321, 0, −1.8660	1.9486, 0, −1.1250	1.7321, 2, −1
2	1.2247, 1.2247, −1	0.8712, 0.8712, −1.8660	1.3778, 1.3778, −1.1250	−0.1895, 2.6390, −1
3	0, 1.7321, −1	0, 1.2321, −1.8660	−0.0631, 1.9416, −1.1210	−1.9353, 1.2951, −0.7477

```
N = 3;

theta2 = 50 *pi/180;
theta3 = 90 *pi/180;

P1 = [1.7321, 0, -1];
P2 = [1.2247, 1.2247, -1];
P3 = [0, 1.7321, -1];

Q1 = [1.2321, 0, -1.8660];
Q2 = [0.8712, 0.8712, -1.8660];
Q3 = [0, 1.2321, -1.8660];

R1 = [1.9486, 0, -1.1250];
R2 = [1.3778, 1.3778, -1.1250];
R3 = [-0.0631, 1.9416, -1.1210];

S1 = [1.7321, 2, -1];
S2 = [-0.1895, 2.6390, -1];
S3 = [-1.9353, 1.2951, -0.7477];

a0x = 0;
a0y = 0;
a0z = -1;

a1x = 0;
a1y = -2;
a1z = 1;

ua0x = 0;
ua0y = 0;
ua0z = 1;

ua1x = 0.7;
ua1y = 0;
ua1z = 0.7;
```

FIGURE E.8.25 Input in bold type in the Appendix J.1 MATLAB file *RR_interior_point.m* for Example 8.10.

```
Iterations: 8
Func Evals: 17
Algorithm: interior-point
exit flag = 1
error = ( 6.201193e-09 )
a0x = ( 0.171116 )
a0y = ( 0.120056 )
a0z = ( -0.420923 )
a1x = ( -2.209759 )
a1y = ( -0.496450 )
a1z = ( -0.287085 )
ua0x = ( 0.047468 )
ua0y = ( 0.033423 )
ua0z = ( 0.998314 )
ua1x = ( 0.047468 )
ua1y = ( 0.033423 )
ua1z = ( 0.998314 )
Elapsed time is 306.362616 seconds.
Solution calculation is complete.
>>
```

FIGURE E.8.26 Output from the Appendix J.1 MATLAB file *RR_interior_point.m* for Example 8.10.

```
N = 3;

b1x = 1;
b1y = -1;
b1z = 0;

b0x = 1;
b0y = 0;
b0z = 0;
```

FIGURE E.8.27 Input in bold type in the Appendix J.2 MATLAB file *SS_quasi_newton.m* for Example 8.10.

```
Iterations: 17
Func Evals: 19
Algorithm: quasi-newton
exit flag = 5
error = ( 6.933348e-31 )
b0x = ( -0.767385 )
b0y = ( 0.317566 )
b0z = ( -0.404121 )
Elapsed time is 2.420734 seconds.
Solution calculation is complete.
>>
```

FIGURE E.8.28 Output from the Appendix J.2 MATLAB file *SS_quasi_newton* for Example 8.10.

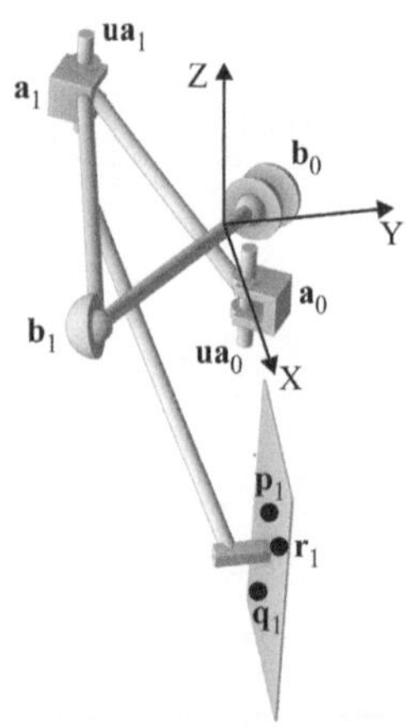

FIGURE E.8.29 Synthesized RRSS motion generator.

Example 8.11

Problem Statement: Using the Appendix J.3 MATLAB files, synthesize a 4R Spherical mechanism for the first four precision positions given in Example 5.2 (Table E.8.11).

Known Information: Table E.8.11 and Appendix J.3 MATLAB files

Solution Approach: In this example, Spherical R-R dyad variables a_{0x} and b_{0x} were specified to $a_{0x} = 0$ and $b_{0x} = 0.5$, respectively. Figure E.8.30 includes the input in bold type in the Appendix J.3 MATLAB file *RR_Sph_interior_point.m* when $a_{0x} = 0$. Figure E.8.31 includes the output from the same file. Figure E.8.32 includes the output from the Appendix J.2 MATLAB file *RR_Sph_interior_point.m* when $b_{0x} = a_{0x} = 0.5$. Figure E.8.33 illustrates the resulting 4R Spherical mechanism. A kinematic analysis of this solution is presented in Example 8.18 in Section 8.6. The equation systems formulated from Equations (8.53) through (8.55) include no constraints to ensure that branch- and order defect-free solutions are calculated. With these equation sets, branch- and order defect-free solutions can be calculated through trial and error.

TABLE E.8.11
Folding Wing Precision Positions

Precision Position	p	q	r
1	–0.2081, 0.1675, 0.8227	0.2272, 0.1665, 0.8179	0, 0.3087, 0.8080
2	0.1442, 0.0859, 0.8486	0.5013, 0.2482, 0.6598	0.2613, 0.2946, 0.7702
3	0.5009, 0.0166, 0.7050	0.6727, 0.3175, 0.4415	0.5104, 0.2616, 0.6475
4	0.7676, –0.0298, 0.3977	0.7554, 0.3639, 0.2124	0.7179, 0.2147, 0.4321

```
N = 4;

P1 = [-0.2081, 0.1675, 0.8227];
P2 = [0.1442, 0.0859, 0.8486];
P3 = [0.5009, 0.0166, 0.7050];
P4 = [0.7676, -0.0298, 0.3977];

Q1 = [0.2272, 0.1665, 0.8179];
Q2 = [0.5013, 0.2482, 0.6598];
Q3 = [0.6727, 0.3175, 0.4415];
Q4 = [0.7554, 0.3639, 0.2124];

R1 = [0, 0.3087, 0.8080];
R2 = [0.2613, 0.2946, 0.7702];
R3 = [0.5104, 0.2616, 0.6475];
R4 = [0.7179, 0.2147, 0.4321];

a0x = 0;
a0y = 1;
a0z = 1;

a1x = 1;
a1y = 1;
a1z = 1;
```

FIGURE E.8.30 Input in bold type in the Appendix J.3 MATLAB file *RR_Sph_interior_point.m* (with $a_{0x} = 0$) for Example 8.11.

```
Iterations: 8
Func Evals: 24
Algorithm: interior-point
exit flag = 1
error = ( 7.826523e-07 )
a0y = ( 0.800383 )
a0z = ( 0.599490 )
a1x = ( 0.569034 )
a1y = ( 0.377513 )
a1z = ( 0.730538 )
Elapsed time is 9.721568 seconds.
Solution calculation is complete.
>>
```

FIGURE E.8.31 Output from the Appendix J.3 MATLAB file *RR_Sph_interior_point.m* (with $a_{0x} = 0$) for Example 8.11.

```
Iterations: 20
Func Evals: 82
Algorithm: interior-point
exit flag = 1
error = ( 2.465959e-06 )
a0y = ( 0.754061 )
a0z = ( 0.425900 )
a1x = ( 0.188570 )
a1y = ( 0.765270 )
a1z = ( 0.615470 )
Elapsed time is 9.774350 seconds.
Solution calculation is complete.
>>
```

FIGURE E.8.32 Output from the Appendix J.3 MATLAB file *RR_Sph_interior_point.m* (with $b_{0x} = a_{0x} = 0.5$) for Example 8.11.

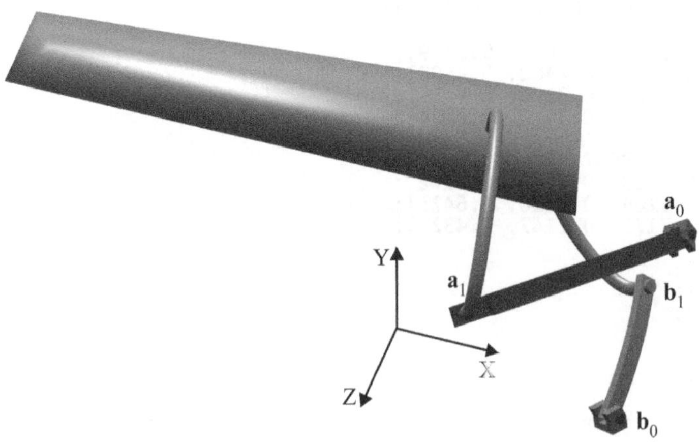

FIGURE E.8.33 Synthesized 4R Spherical motion generator.

8.5 RSSR AND 4R SPHERICAL FUNCTION GENERATION

Because the crank and follower links of the RSSR mechanism include revolute joints as fixed pivots (and subsequently, controlled follower rotation for supplied crank rotation), this mechanism is useful for spatial function generation applications. Suh and Radcliffe introduced function generation equations to calculate the RSSR moving-pivot variables to approximate the precision points.

Equation (8.70) encompasses the constraint equation set for the RSSR function generator [10]. As illustrated in Figure 8.14, the fixed-pivot joint axis $\mathbf{ua}_0$ lies on the z-axis, and the fixed-pivot joint axis $\mathbf{ub}_0$ intersects the x-axis at $x = 1$. The revolute joint unit vectors are defined as $\mathbf{ua}_0 = (0,0,1)$ and $\mathbf{ub}_0 = (0, -\sin\alpha, \cos\alpha)$. Once moving pivots $\mathbf{a}_1$ and $\mathbf{b}_1$ are known, fixed pivots $\mathbf{a}_0$ and $\mathbf{b}_0$ can be determined by analytic geometry in accordance with Figure 8.14. As illustrated in this figure, the fixed pivot $\mathbf{a}_0$ is the point of intersection of $\mathbf{ua}_0$ and an orthogonal line that includes $\mathbf{a}_1$ or $\mathbf{a}_0 = \left[0,0,a_{1z}\right]$. Likewise, the fixed pivot $\mathbf{b}_0$ is the point of intersection of $\mathbf{ub}_0$ and an orthogonal line that includes $\mathbf{b}_1$ or $\mathbf{b}_0 = \left[1, b_{1y}, 0\right]$.

$$(\mathbf{a}_j - \mathbf{b}_1)^T(\mathbf{a}_j - \mathbf{b}_1) - (\mathbf{a}_1 - \mathbf{b}_1)^T(\mathbf{a}_1 - \mathbf{b}_1) = 0 \quad j = 2,3,\ldots \tag{8.70}$$

In this RSSR function generation model, $\mathbf{a}_1 = \left[a_{1x}, a_{1y}, a_{1z}, 1\right]^T$, $\mathbf{b}_1 = \left[b_{1x}, b_{1y}, b_{1z}, 1\right]^T$, and $\mathbf{a}_j = \left[D_{1j}\right]\mathbf{a}_1$ $\mathbf{a}_j = \left[D_{1j}\right]\mathbf{a}_1$. Matrix $\left[D_{1j}\right]$ defines the relative rotations of the moving pivot $\mathbf{a}_1$ by the crank displacement angle θ_{1j} and the moving pivot $\mathbf{b}_1$ by the follower displacement angle ϕ_{1j} from position 1 to position j. This matrix is expressed as

$$\left[D_{1j}\right] = \left[D_{-\phi_{1j}}\right]\left[D_{\theta_{1j}}\right] = \begin{bmatrix} d_{11j} & d_{12j} & d_{13j} & d_{14j} \\ d_{21j} & d_{22j} & d_{23j} & d_{24j} \\ d_{31j} & d_{32j} & d_{33j} & d_{34j} \\ 0 & 0 & 0 & 1 \end{bmatrix} \tag{8.71}$$

where

$$d_{11j} = \cos\theta_{1j}\cos\phi_{1j} + \sin\theta_{1j}\cos\alpha\sin\phi_{1j}$$

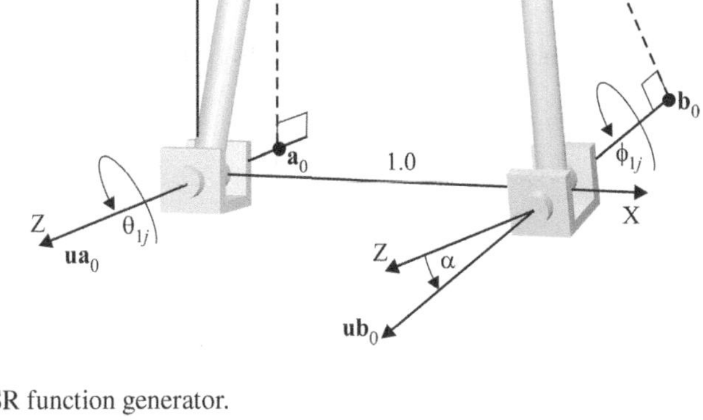

FIGURE 8.14 RSSR function generator.

$$d_{12j} = -\sin\theta_{1j}\cos\phi_{1j} + \cos\theta_{1j}\cos\alpha\sin\phi_{1j}$$

$$d_{13j} = \sin\alpha\sin\phi_{1j}$$

$$d_{14j} = 1 - \cos\phi_{1j}$$

$$d_{21j} = -\cos\theta_{1j}\cos\alpha\sin\phi_{1j} + \sin\theta_{1j}\left(\sin^2\alpha\left(1-\cos\phi_{1j}\right)+\cos\phi_{1j}\right)$$

$$d_{22j} = \sin\theta_{1j}\cos\alpha\sin\phi_{1j} + \cos\theta_{1j}\left(\sin^2\alpha\left(1-\cos\phi_{1j}\right)+\cos\phi_{1j}\right)$$

$$d_{23j} = -\sin\alpha\cos\alpha\left(1-\cos\phi_{1j}\right)$$

$$d_{24j} = \cos\alpha\sin\phi_{1j}$$

$$d_{31j} = -\cos\theta_{1j}\sin\alpha\sin\phi_{1j} - \sin\theta_{1j}\sin\alpha\cos\alpha\left(1-\cos\phi_{1j}\right)$$

$$d_{32j} = \sin\theta_{1j}\sin\alpha\sin\phi_{1j} - \cos\theta_{1j}\sin\alpha\cos\alpha\left(1-\cos\phi_{1j}\right)$$

$$d_{33j} = \cos^2\alpha\left(1-\cos\phi_{1j}\right)+\cos\phi_{1j}$$

$$d_{34j} = \sin\alpha\sin\phi_{1j}$$

Matrices $\left[D_{-\phi_{1j}}\right]$ and $\left[D_{\theta_{1j}}\right]$ are identical in form to Matrix (2.30). Equation (8.70) forms a six-equation nonlinear simultaneous equation system to solve for the six unknown variables in $\mathbf{a}_1$ and $\mathbf{b}_1$ given six precision points. Appendices J.8 and J.9 include user instructions for the MATLAB files to solve the equations presented in this section for RSSR function generation.

Although the RSSR function generation equations, as presented, accommodate six precision points, in the Appendix J.8 and J.9 MATLAB files, they are solved as unconstrained nonlinear optimization models for an indefinite number of precision points.

Suh and Radcliffe also introduced function generation equations to calculate the 4R Spherical mechanism moving-pivot variables to approximate precision points. Equations (8.72) through (8.74) encompass the constraint equation set for the 4R Spherical function generator [11]. As illustrated in Figure 8.15, the fixed-pivot $\mathbf{a}_0$ lies on the x-axis at $\mathbf{a}_0 = (1,0,0)$, and the fixed-pivot $\mathbf{b}_0$ is offset from $\mathbf{a}_0$ by the clockwise rotation α in the x–y plane.

$$\left(\mathbf{a}_j - \mathbf{b}_1\right)^T\left(\mathbf{a}_j - \mathbf{b}_1\right) - \left(\mathbf{a}_1 - \mathbf{b}_1\right)^T\left(\mathbf{a}_1 - \mathbf{b}_1\right) = 0 \quad j = 2,3,\ldots \tag{8.72}$$

$$\left(\mathbf{a}_1\right)^T\left(\mathbf{a}_1\right) - 1 = 0 \tag{8.73}$$

$$\left(\mathbf{b}_1\right)^T\left(\mathbf{b}_1\right) - 1 = 0 \tag{8.74}$$

In this 4R Spherical function generation model, $\mathbf{a}_1 = \left[a_{1x}, a_{1y}, a_{1z}, 1\right]^T$, $\mathbf{b}_1 = \left[b_{1x}, b_{1y}, b_{1z}, 1\right]^T$, and $\mathbf{a}_j = \left[D_{1j}\right]\mathbf{a}_1$. Matrix $\left[D_{1j}\right]$ defines the relative rotations of the moving pivot $\mathbf{a}_1$ by the crank

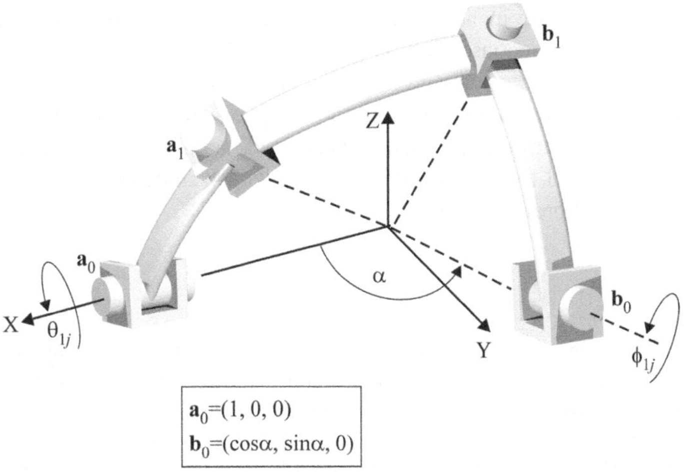

FIGURE 8.15 4R Spherical function generator.

displacement angle θ_{1j} and the moving pivot $\mathbf{b}_1$ by the follower displacement angle ϕ_{1j} from position 1 to position j. It is expressed as

$$[D_{1j}]=[D_{-\phi_{1j}}][D_{\theta_{1j}}]=\begin{bmatrix} d_{11j} & d_{12j} & d_{13j} & d_{14j} \\ d_{21j} & d_{22j} & d_{23j} & d_{24j} \\ d_{31j} & d_{32j} & d_{33j} & d_{34j} \\ 0 & 0 & 0 & 1 \end{bmatrix} \tag{8.75}$$

where

$$d_{11j} = \cos^2\alpha + \sin^2\alpha\cos\phi_{1j}$$

$$d_{12j} = \sin\alpha\cos\alpha\cos\theta_{1j}(1-\cos\phi_{1j}) - \sin\alpha\sin\theta_{1j}\sin\phi_{1j}$$

$$d_{13j} = -\sin\alpha\cos\alpha\cos\theta_{1j}(1-\cos\phi_{1j}) - \sin\alpha\sin\theta_{1j}\sin\phi_{1j}$$

$$d_{14j} = 0$$

$$d_{21j} = \sin\alpha\cos\alpha(1-\cos\phi_{1j})$$

$$d_{22j} = \sin^2\alpha\cos\phi_{1j} + \cos^2\alpha\cos\phi_{1j}\cos\theta_{1j} + \cos\alpha\sin\theta_{1j}\sin\phi_{1j}$$

$$d_{23j} = -\sin^2\alpha\cos\theta_{1j} - \cos^2\alpha\sin\theta_{1j}\cos\phi_{1j} + \cos\alpha\cos\theta_{1j}\sin\phi_{1j}$$

$$d_{24j} = 0$$

$$d_{31j} = \sin\alpha\sin\phi_{1j}$$

$$d_{32j} = \sin\theta_{1j}\cos\phi_{1j} - \cos\alpha\cos\theta_{1j}\sin\phi_{1j}$$

$$d_{33j} = \cos\theta_{1j}\cos\phi_{1j} + \cos\alpha\cos\theta_{1j}\sin\phi_{1j}$$

$$d_{34j} = 0$$

Matrices $\left[D_{-\phi_{1j}}\right]$ and $\left[D_{\theta_{1j}}\right]$ are identical in form to Matrix (2.30). Equations (8.72)–(8.74) form a six-equation nonlinear simultaneous equation system to solve for the six unknown variables in $\mathbf{a}_1$ and $\mathbf{b}_1$ given four precision points. Appendix J.10 includes user instructions for the MATLAB files to solve the equations presented in this section for 4R Spherical function generation. Although the 4R Spherical function generation equations, as presented, accommodate four precision points, in the Appendix J.10 MATLAB files, they are solved as a constrained nonlinear optimization model for an indefinite number of precision points.

Example 8.12

Problem Statement: Using the Appendix J.9 files, synthesize an RSSR steering mechanism for the steering wheel (θ) and steering direction (ϕ) displacement ranges given in Table E.8.12. The angular offset between the driver and wheel rotation axes (α) is 90°.

Known Information: Table E.8.12 and Appendix J.9 MATLAB files

Solution Approach: In this example, RSSR variables b_{1x} and α were specified to $b_{1x}=1$ and $\alpha=90°$, respectively. Figure E.8.34 includes the input in bold type in the Appendix J.9 MATLAB file *RSSR_quasi_newton.m*. Figure E.8.35 includes the output from the same file. Figure E.8.36 illustrates the resulting RSSR steering mechanism.

Calculating the point of intersection of $\mathbf{ua}_0$ and an orthogonal line that includes $\mathbf{a}_1$ produces $\mathbf{a}_0=\left[0,0,a_{1z}\right]=[0,0,-0.5050]$. Calculating the point of intersection of $\mathbf{ub}_0$ and an orthogonal line that includes $\mathbf{b}_1$ produces $\mathbf{b}_0=\left[1,b_{1y},0\right]=[1,-2.1435,0]$. A kinematic analysis of this solution is presented in Example 8.19 in Section 8.6.

TABLE E.8.12
Precision Points for the RSSR Steering Mechanism

Precision Point	θ_j (deg)	ϕ_j (deg)
2	–40	–60
3	–20	–30
4	20	30
5	40	60

```
N = 5;

theta2 = -40 *pi/180;
theta3 = -20 *pi/180;
theta4 = 20 *pi/180;
theta5 = 40 *pi/180;

phi2 = -60 *pi/180;
phi3 = -30 *pi/180;
phi4 = 30 *pi/180;
phi5 = 60 *pi/180;

b1x = 1;
ALPHA = 90 *pi/180;

a1x = 0.5;
a1y = 0;
a1z = -0.5;

b1y = -2;
b1z = 1;
```

FIGURE E.8.34 Input in bold type in the Appendix J.9 MATLAB file *RSSR_quasi_newton.m* for Example 8.12.

```
Iterations: 55
Func Evals: 64
Algorithm: quasi-newton
exit flag = 5
error = ( 1.479114e-30 )
a1x = ( 0.335448 )
a1y = ( -0.093379 )
a1z = ( -0.504968 )
b1y = ( -2.143466 )
b1z = ( 0.624966 )
Elapsed time is 0.882067 seconds.
Solution calculation is complete.
>>
```

FIGURE E.8.35 Output from the Appendix J.9 MATLAB file *RSSR_quasi_newton.m* for Example 8.12.

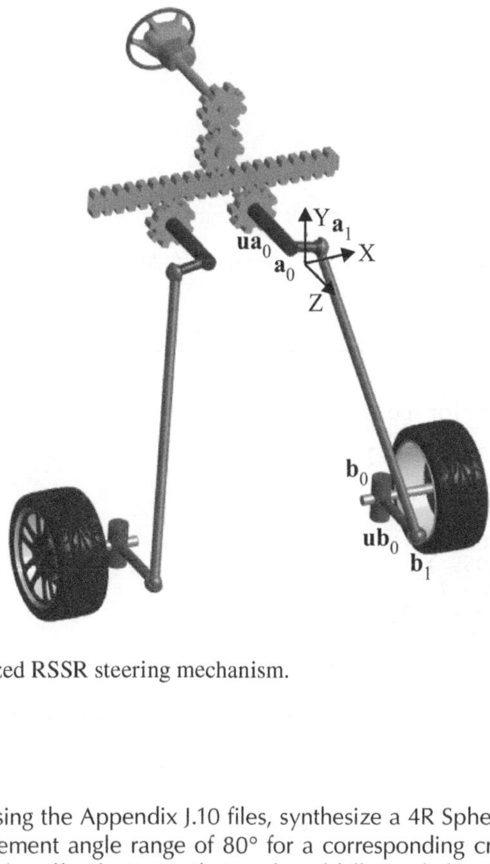

FIGURE E.8.36 Synthesized RSSR steering mechanism.

Example 8.13

Problem Statement: Using the Appendix J.10 files, synthesize a 4R Spherical mechanism to produce a camera displacement angle range of 80° for a corresponding crank displacement angle range of 180°. The angular offset between the crank and follower-link rotation axes (α) is 45°.

Known Information: Displacement angle ranges and increments (precision points) and Appendix J.9 MATLAB files

Solution Approach: Dividing the given displacement angle ranges into five precisions points produces $\theta_{1j} = 45°, 90°, 135°, 180°$ and $\phi_{1j} = 20°, 40°, 60°, 80°$.[17] Figure E.8.37 includes the input in bold type in the Appendix J.10 MATLAB file *Sphere_interior_point.m*. Figure E.8.38 includes the output from the same file. Figure E.8.39 illustrates the resulting 4R Spherical mechanism. A kinematic analysis of this solution is presented in Example 8.20 in Section 8.6.

```
N = 5;

theta2 = 45 *pi/180;
theta3 = 90 *pi/180;
theta4 = 135 *pi/180;
theta5 = 180 *pi/180;

phi2 = 20 *pi/180;
phi3 = 40 *pi/180;
phi4 = 60 *pi/180;
phi5 = 80 *pi/180;

a0x = 1;
a0y = 0;
a0z = 0;

ALPHA = 45;

b0x = cos(ALPHA *pi/180);
b0y = sin(ALPHA *pi/180);
b0z = 0;

a1x = 0.7;
a1y = 0;
a1z = 0.7;

b1x = 0;
b1y = 0.7;
b1z = 1;
```

FIGURE E.8.37 Input in bold type in the Appendix J.10 MATLAB file *Sphere_interior_point.m* for Example 8.13.

```
Iterations: 22
Func Evals: 82
Algorithm: interior-point
exit flag = 1
error = ( 5.727935e-05 )
a1x = ( 0.925534 )
a1y = ( -0.378007 )
a1z = ( 0.022261 )
b1x = ( -0.108603 )
b1y = ( -0.364055 )
b1z = ( 0.925019 )
Elapsed time is 10.781789 seconds.
Solution calculation is complete.
>>
```

FIGURE E.8.38 Output from the Appendix J.10 MATLAB file *Sphere_interior_point.m* for Example 8.13.

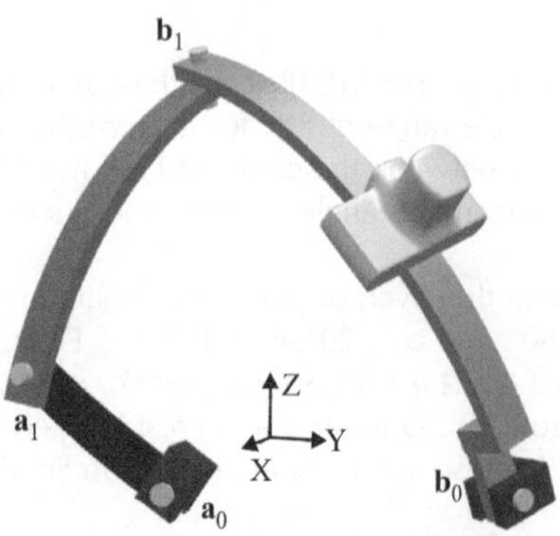

FIGURE E.8.39 Synthesized 4R Spherical camera rotation mechanism.

8.6 SPATIAL MECHANISM KINEMATIC ANALYSIS AND MODELING IN SIMSCAPE MULTIBODY™

It has been noted in Sections 8.2 and 8.3 that Appendices H.1 and H.2 include user instructions for the RRSS and RSSR mechanisms in MATLAB, respectively. In these files, the displacement, velocity, and acceleration equations formulated in Sections 8.2 and 8.3 are solved. These MATLAB files provide a means for the user to efficiently conduct RRSS and RSSR kinematic analyses by solving their displacement, velocity, and acceleration equations.

This book also utilizes Simscape Multibody as an alternate approach for simulation-based kinematic analysis. Simscape Multibody (formerly SimMechanics) is the latest MATLAB toolbox that provides a physical modeling environment for the mechanical modeling and simulation of rigid, multibody systems. The MATLAB and Simscape Multibody files that accompany this book use distinct approaches for kinematic analysis. In the Appendix H MATLAB files, for example, solutions are calculated from the spatial vector point-based equations in Sections 8.2 and 8.3. In Simscape Multibody, however, *Newtonian equations* (produced by building mechanism links and joints in a physical modeling environment) are solved using *ordinary differential equation* (ODE) solvers [6].

A library of Simscape Multibody files is available for download at https://www.routledge.com/9781032956800 to conduct RRSS and RSSR displacement, velocity, and acceleration analyses.* With these files, the user specifies the mechanism link dimensions and driving link parameters (e.g., crank displacements, velocities, and/or accelerations), and the displacements, velocities, and/or accelerations of the mechanism locations of interest are measured. Additionally, the motion of the mechanism itself is simulated. The Simscape Multibody file user instructions for the RRSS[18] and RSSR mechanisms are given in Appendices I.1 and I.2, respectively. The Appendix I files can also be used for 4R Spherical mechanism kinematic analysis. Because the link lengths are illustrated as straight lines (not circular arcs) in the Simscape Multibody simulation window in Appendix I files, the 4R Spherical mechanism will appear graphically as an RRSS mechanism. The mechanism, however, will be kinematically identical to a 4R Spherical mechanism when the joint axes and joint coordinates specified are for an actual 4R Spherical mechanism.

Example 8.14

Problem Statement: Using the Appendix I.1 Simscape Multibody files, calculate the coupler positions achieved by the RRSS mechanism solution in Example 8.7. Also confirm that this solution is order- and branch defect-free.

Known Information: Example 8.7 solution and Appendix I.1 Simscape Multibody files.

Solution Approach: Figure E.8.40 includes the input specified (in bold text) in the Appendix I.1 Simscape Multibody file. Table E.8.13 includes the achieved RRSS coupler positions measured

```
a0 = [0.036138, 0.095884, 0.347070]';
a1 = [5.773299, 0.609854, 0.337957]';
ua0 = [-0.005795, 0.082352, 0.996589]';
ua1 = [-0.025801, 0.304883, 0.952047]';
b0 = [0.051871, 0.435930, 0.450254]';
b1 = [3.365637, -2.116441, 1.058884]';
p1 = [1.7321, 0, -1]';
q1 = [1.2321, 0, -1.866]';
r1 = [1.9486, 0, -1.125]';

start_ang   = 0;
step_ang    = 0.01;
stop_ang    = 189.94;

angular_vel  = 0;
angular_acc  = 0;
```

FIGURE E.8.40 Specified input (in bold text) in the Appendix I.1 MATLAB and Simscape Multibody files for Example 8.14.

from the Appendix I.1 Simscape Multibody files. Figure E.8.41 is an initial position snapshot of the mechanism animation generated from the Appendix I.1 MATLAB and Simscape Multibody files. As indicated by the continually ascending order of the crank angle solutions given in Table E.8.13, the order of the achieved coupler positions, and the accuracy of these positions, the RRSS motion generator solution is order- and branch defect-free.

TABLE E.8.13
Coupler Positions Achieved by the Example 8.7 RRSS Mechanism Solution

θ	p	q	r
0°	1.7321, 0, −1	1.2321, 0, −1.8660	1.9486, 0, −1.1250
56.58°	1.2533, 1.2457, −1.0281	0.9457, 0.7540, −1.8427	1.3477, 1.4264, −1.1728
110.26°	−0.0737, 1.8101, −1.0341	0.1169, 1.2397, −1.8330	−0.1809, 1.9811, −1.1817
152.66°	−1.2856, 1.2100, −1.0026	−0.8096, 0.8908, −1.8220	−1.4872, 1.2633, −1.1405
189.94°	−1.6965, −0.0152, −0.9458	−1.1701, −0.0248, −1.7959	−1.9001, −0.0885, −1.0710

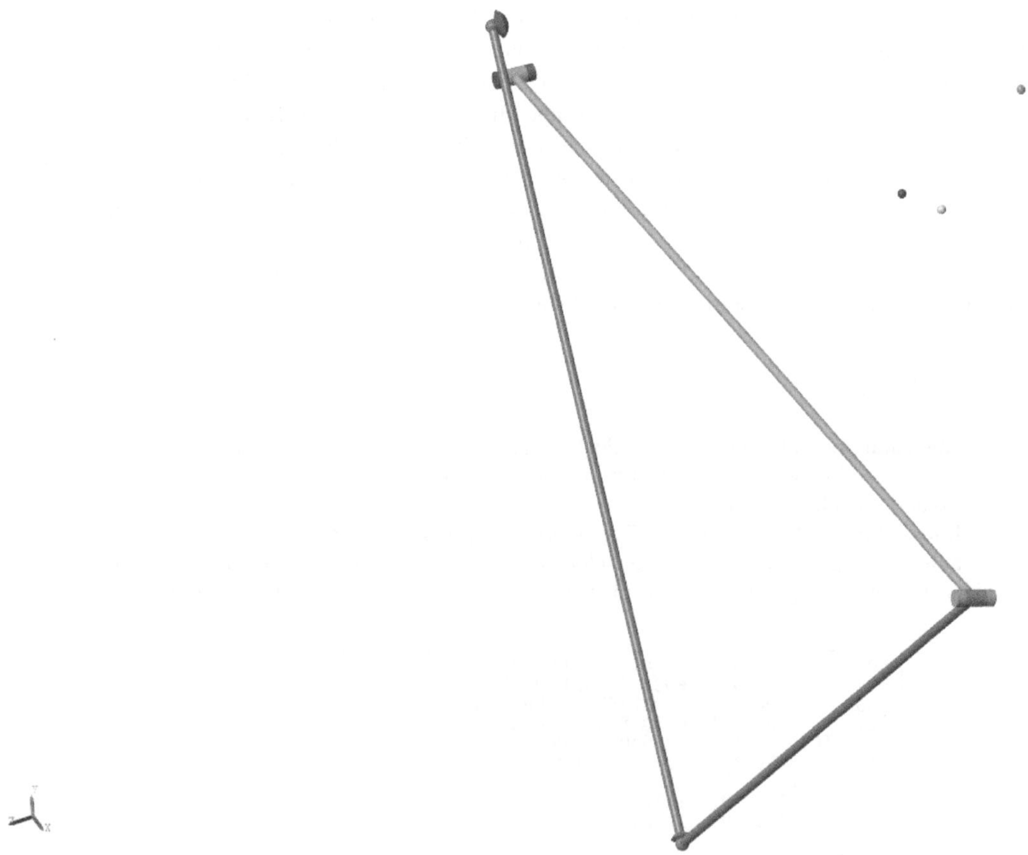

FIGURE E.8.41 Initial position of the RRSS mechanism in the Appendix I.1 Simscape Multibody animation for Example 8.14.

Example 8.15

Problem Statement: Using the Appendix I.1 Simscape Multibody files, calculate the coupler positions achieved by the 4R Spherical mechanism solution in Example 8.8. Also confirm that this solution is order- and branch defect-free.

Known Information: Example 8.8 solution and Appendix I.1 Simscape Multibody files.

Solution Approach: Figure E.8.42 includes the input specified (in bold text) in the Appendix I.1 Simscape Multibody file. Table E.8.14 includes the achieved 4R Spherical coupler positions measured from the Appendix I.1 Simscape Multibody files. Figure E.8.43 is an initial position snapshot of the mechanism animation generated from the Appendix I.1 MATLAB and Simscape Multibody files. As indicated by the continually ascending order of the crank angle solutions in Table E.8.14, the order of the achieved coupler positions, and the accuracy of these positions, the 4R Spherical motion generator solution is free of order and branch defects.

```
a0 = [-0.643678, -0.393975, -0.656096]';
a1 = [-0.358079, -0.529870, -0.768776]';
ua0 = a0;
ua1 = a1;
b0 = [-0.802561, 0.399208, 0.443317]';
b1 = [0.621598, 0.444330, 0.645125]';
p1 = [-0.2081, 0.1675, 0.8227]';
q1 = [0.2272, 0.1665, 0.8179]';
r1 = [0, 0.3087, 0.8080]';

start_ang   = 0;
step_ang    = 0.01;
stop_ang    = 121.99;

angular_vel  = 0;
angular_acc  = 0;
```

FIGURE E.8.42 Specified input (in bold text) in the Appendix I.1 MATLAB and Simscape Multibody files for Example 8.15.

TABLE E.8.14
Coupler Positions Achieved by the Example 8.8 4R Spherical Mechanism Solution

θ	p	q	r
0°	−0.2081, 0.1675, 0.8227	0.2272, 0.1665, 0.8179	0, 0.3087, 0.8080
32.84°	0.1505, 0.0662, 0.8492	0.4906, 0.2631, 0.6621	0.2494, 0.2863, 0.7771
60.23°	0.4977, 0.0284, 0.7069	0.6707, 0.3250, 0.4392	0.5073, 0.2725, 0.6454
88.76°	0.7590, −0.0299, 0.4138	0.7693, 0.3453, 0.1931	0.7210, 0.2189, 0.4247
121.99°	0.8640, −0.0402, −0.0122	0.7744, 0.3856, 0.0008	0.8387, 0.1625, 0.1351

FIGURE E.8.43 Initial position of the 4R Spherical mechanism in the Appendix I.1 Simscape Multibody animation for Example 8.15.

Example 8.16

Problem Statement: Using the Appendix I.1 Simscape Multibody files, calculate the precision points achieved by the 4R Spherical mechanism solution in Example 8.9. Also confirm that this solution is order- and branch defect-free.

Known Information: Example 8.9 solution and Appendix I.1 Simscape Multibody files.

Solution Approach: Figure E.8.44 includes the input specified (in bold text) in the Appendix I.1 Simscape Multibody file. Table E.8.15 includes the achieved 4R Spherical coupler points measured from the Appendix I.1 Simscape Multibody files. Figure E.8.45 is an initial position snapshot of the mechanism animation generated from the Appendix I.1 MATLAB and Simscape Multibody files. As indicated by the continually ascending order of the crank angle solutions in Tables E.8.14 and E.8.15, the order of the achieved coupler points, and the accuracy of these points, the 4R Spherical path generator solution is order- and branch defect-free.

```
a0 = [-0.221639, -0.118504, 0.967901]';
a1 = [0.045033, -0.188244, 0.981089]';
ua0 = a0;
ua1 = a1;
b0 = [0.226497, 0.860384, -0.456552]';
b1 = [-0.856003, -0.301045, -0.420275]';
p1 = [0.7071, 0, 0.7071]';
q1 = p1;
r1 = p1;

start_ang   = 0;
step_ang    = 0.01;
stop_ang    = 251.66;

angular_vel  = 0;
angular_acc  = 0;
```

FIGURE E.8.44 Specified input (in bold text) in the Appendix I.1 MATLAB and Simscape Multibody files for Example 8.16.

TABLE E.8.15
Coupler Points Achieved by the Example 8.9 4R Spherical Mechanism Solution

θ	p
0°	0.7071, 0, 0.7071
20.57°	0.7206, 0.0117, 0.6932
38.79°	0.7074, 0.0433, 0.7055
53.17°	0.6829, 0.0764, 0.7265
68.36°	0.6449, 0.1149, 0.7556
91.68°	0.5657, 0.1729, 0.8063
118.49°	0.4534, 0.2267, 0.8620
145.13°	0.3352, 0.2594, 0.9057
173.04°	0.2218, 0.2732, 0.9360
198.13°	0.1386, 0.2760, 0.9511
228.96°	0.0605, 0.2827, 0.9573
251.66°	0.0044, 0.3030, 0.9530

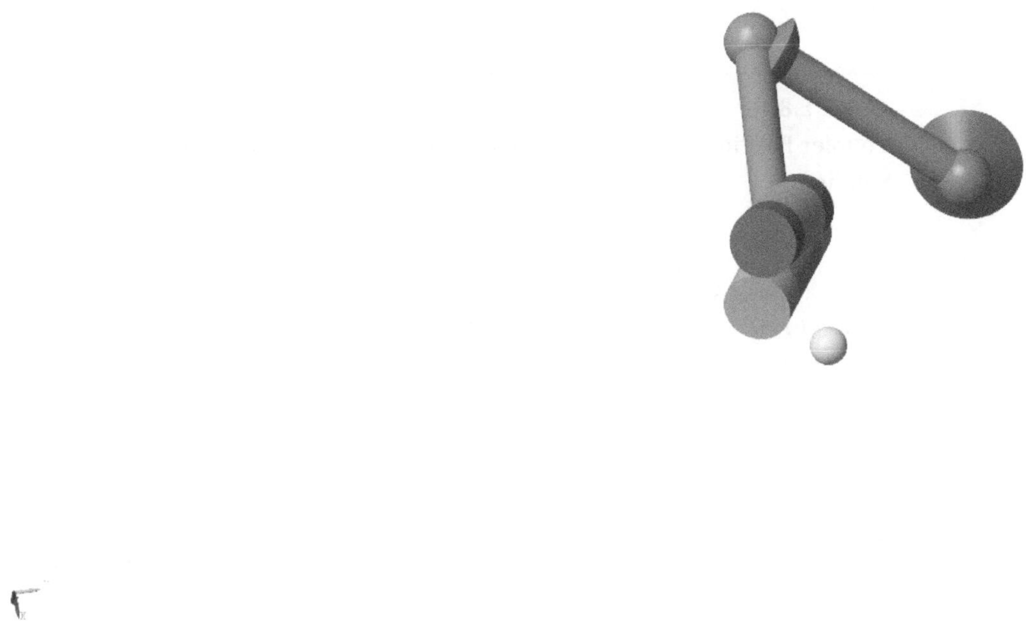

FIGURE E.8.45 Initial position of the 4R Spherical mechanism in the Appendix I.1 Simscape Multibody animation for Example 8.16.

Example 8.17

Problem Statement: Using the Appendix I.1 Simscape Multibody files, calculate the coupler positions achieved by the 4R Spherical mechanism solution in Example 8.10. Also confirm that this solution is order- and branch defect-free.

Known Information: Example 8.10 solution and Appendix I.1 Simscape Multibody files.

Solution Approach: Figure E.8.46 includes the input specified (in bold text) in the Appendix I.1 Simscape Multibody file. Table E.8.16 includes the achieved RRSS coupler positions measured from the Appendix I.1 Simscape Multibody files. Figure E.8.47 is an initial position snapshot of the mechanism animation generated from the Appendix I.1 MATLAB and Simscape Multibody files. As indicated by the continually ascending order of the crank angle solutions in Table E.8.16, the order of the achieved coupler positions, and the accuracy of these positions, the RRSS motion generator solution is order- and branch defect-free.

```
a0 = [0.171116, 0.120056, -0.420923]';
a1 = [-2.209759, -0.496450, -0.287085]';
ua0 = [0.047468, 0.033423, 0.998314]';
ua1 = [0.047468, 0.033423, 0.998314]';
b0 = [-0.767385, 0.317566, -0.404121]';
b1 = [1, -1, 0]';
p1 = [1.7321, 0, -1]';
q1 = [1.2321, 0, -1.8660]';
r1 = [1.9486, 0, -1.1250]';

start_ang   = 0;
step_ang    = 0.01;
stop_ang    = 90;

angular_vel  = 0;
angular_acc  = 0;
```

FIGURE E.8.46 Specified input (in bold text) in the Appendix I.1 MATLAB and Simscape Multibody files for Example 8.17.

TABLE E.8.16
Coupler Positions Achieved by the Example 8.10 RRSS Mechanism Solution

θ	p	q	r
0°	1.7321, 0, −1	1.2321, 0, −1.8660	1.9486, 0, −1.1250
50.02°	1.0289, 1.3794, −1.0127	0.6917, 0.9988, −1.8737	1.1519, 1.5566, −1.1392
90.00°	0.2967, 1.6950, −0.9885	0.2199, 1.2076, −1.8583	0.2898, 1.9132, −1.1102

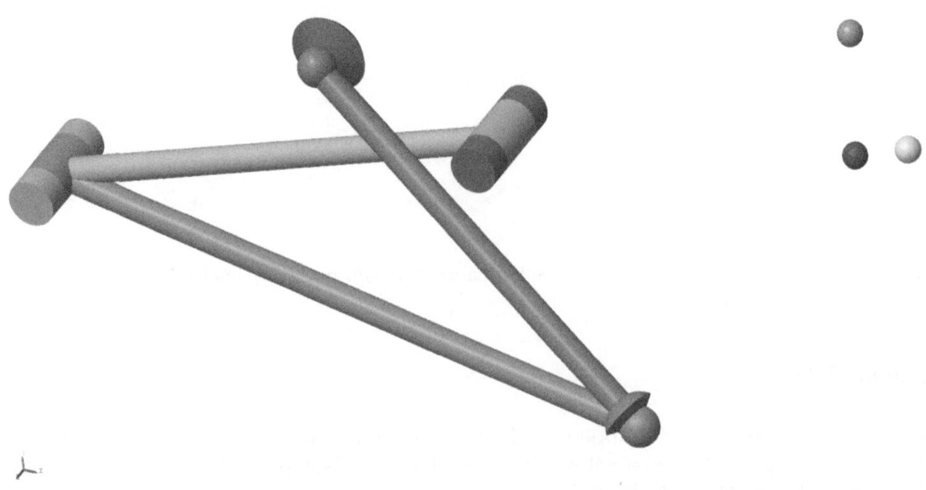

FIGURE E.8.47 Initial position of the RRSS mechanism in the Appendix I.1 Simscape Multibody animation for Example 8.17.

Example 8.18

Problem Statement: Using the Appendix I.1 Simscape Multibody files, calculate the coupler positions achieved by the 4R Spherical mechanism solution in Example 8.11. Also confirm that this solution is order- and branch defect-free.

Known Information: Example 8.11 solution and Appendix I.1 Simscape Multibody files.

Solution Approach: Figure E.8.48 includes the input specified (in bold text) in the Appendix I.1 Simscape Multibody file. Table E.8.17 includes the achieved 4R Spherical coupler positions and corresponding crank link displacement angles measured from the Appendix I.1 Simscape Multibody files. Figure E.8.49 is an initial position snapshot of the mechanism animation generated from the Appendix I.1 MATLAB and Simscape Multibody files. As indicated by the continually ascending order of the crank angle solutions in Table E.8.17, the order of the achieved coupler positions, and the accuracy of these positions, the 4R Spherical motion generator solution is order- and branch defect-free.

```
a0 = [0, 0.800383, 0.599490]';
a1 = [0.569034, 0.377513, 0.730538]';
ua0 = a0;
ua1 = a1;
b0 = [0.5, 0.754061, 0.425900]';
b1 = [0.188570, 0.765270, 0.615470]';
p1 = [-0.2081, 0.1675, 0.8227]';
q1 = [0.2272, 0.1665, 0.8179]';
r1 = [0, 0.3087, 0.8080]';

start_ang    = 0;
step_ang     = 0.01;
stop_ang     = 150;

angular_vel  = 0;
angular_acc  = 0;
```

FIGURE E.8.48 Specified input (in bold text) in the Appendix I.1 MATLAB and Simscape Multibody files for Example 8.18.

TABLE E.8.17
Coupler Positions Achieved by the Example 8.11 4R Spherical Mechanism Solution

θ	p	q	r
0°	−0.2081, 0.1675, 0.8227	0.2272, 0.1665, 0.8179	0, 0.3087, 0.8080
29.613°	0.1427, 0.0924, 0.8481	0.5003, 0.2529, 0.6588	0.2601, 0.3005, 0.7683
47.456°	0.5003, 0.0116, 0.7056	0.6733, 0.3132, 0.4437	0.5105, 0.2570, 0.6492
54.749°	0.7674, −0.0337, 0.3978	0.7575, 0.3599, 0.2121	0.7190, 0.2112, 0.4320

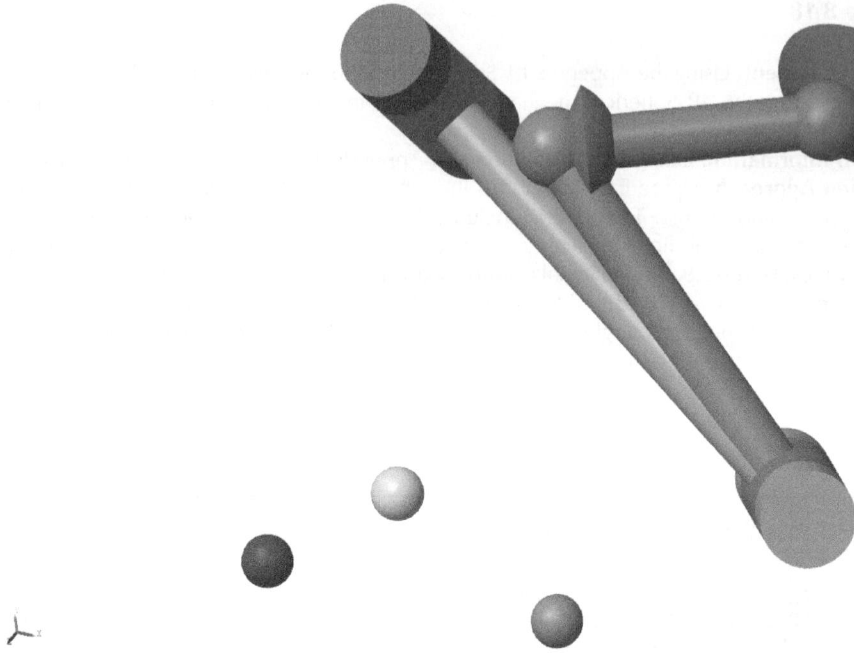

FIGURE E.8.49 Initial position of the 4R Spherical mechanism in the Appendix I.1 Simscape Multibody animation for Example 8.18.

Example 8.19

Problem Statement: Using the Appendix I.2 Simscape Multibody files, calculate the precision points achieved by the RSSR mechanism solution in Example 8.12.

Known Information: Example 8.12 solution and Appendix I.2 Simscape Multibody files.

Solution Approach: Figure E.8.50 includes the input specified (in bold text) in the Appendix I.2 Simscape Multibody file. Table E.8.18 includes the achieved RSSR precision points measured from the Appendix I.2 Simscape Multibody files. Figure E.8.51 is an initial position snapshot of the mechanism animation generated from the Appendix I.2 MATLAB and Simscape Multibody files.

```
a0 = [0, 0, -0.5050]';
a1 = [0.335448, -0.093379, -0.504968]';
ua0 = [0, 0, 1]';
b0 = [1, -2.1435, 0]';
b1 = [1, -2.143466, 0.624966]';
ub0 = [0, -1, 0]';

start_ang   = 0;
step_ang    = 20;
stop_ang    = 40;

angular_vel  = 0;
angular_acc  = 0;
```

FIGURE E.8.50 Specified input (in bold text) in the Appendix I.2 MATLAB and Simscape Multibody files for Example 8.19. (Here, the Appendix I.2 files were run twice. Once with the illustrated input and again with step_ang = –20 and stop_ang = –40.)

TABLE E.8.18
Crank and Follower Displacement Angles Achieved by the Example 8.12 RSSR Mechanism Solution

θ	ϕ
–40°	–55.674°
–19.989°	–29.984°
0°	0°
20.017°	30.026°
40.000°	60.000°

FIGURE E.8.51 Initial position of the RSSR mechanism in the Appendix I.2 Simscape Multibody animation for Example 8.19.

Example 8.20

Problem Statement: Using the Appendix I.2 Simscape Multibody files, calculate the precision points achieved by the 4R Spherical mechanism solution in Example 8.13.

Known Information: Example 8.13 solution and Appendix I.2 Simscape Multibody files.

Solution Approach: Figure E.8.52 includes the input specified (in bold text) in the Appendix I.2 Simscape Multibody file. Table E.8.19 includes the achieved 4R Spherical precision points measured from the Appendix I.2 Simscape Multibody files. Figure E.8.53 is an initial position snapshot of the mechanism animation generated from the Appendix I.2 MATLAB and Simscape Multibody files.

```
a0 = [1, 0, 0]';
a1 = [0.925534, -0.378007, 0.022261]';
ua0 = a0;
b0 = [cos(pi/4), sin(pi/4), 0]';
b1 = [-0.108603, -0.364055, 0.925019]';
ub0 = b0;

start_ang   = 0;
step_ang    = 45;
stop_ang    = 180;

angular_vel  = 0;
angular_acc  = 0;
```

FIGURE E.8.52 Specified input (in bold text) in the Appendix I.2 MATLAB and Simscape Multibody files for Example 8.20.

TABLE E.8.19
Crank and Follower Displacement Angles Achieved by Example 8.13 4R Spherical Mechanism Solution

θ	ϕ
0°	0°
45.004°	19.806°
89.982°	40.324°
135.02°	59.673°
179.65°	82.579°

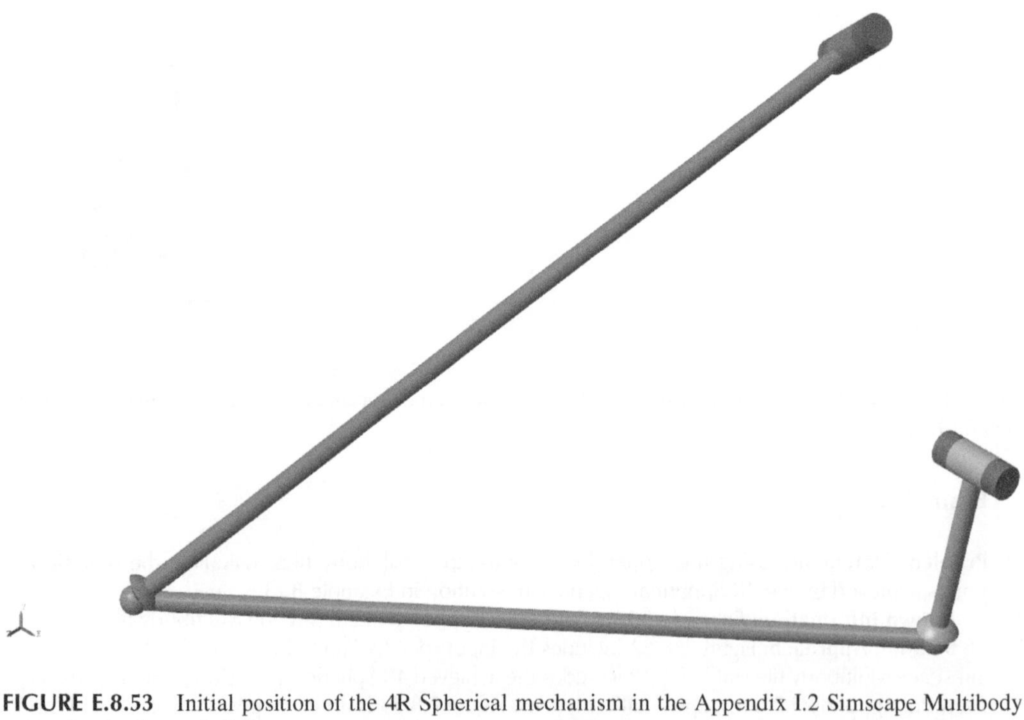

FIGURE E.8.53 Initial position of the 4R Spherical mechanism in the Appendix I.2 Simscape Multibody animation for Example 8.20.

8.7 SUMMARY

Just as planar mechanisms are configured so that mechanism motion is restricted to two-dimensional space, spatial mechanisms are configured to exhibit motion in three-dimensional space. Spatial mechanism motion is determined in part by the mobility of the mechanism joints used and the spatial orientation of the joints. The four-bar spatial mechanisms considered in this chapter are the RRSS, 4R Spherical, and RSSR mechanisms.

Suh and Radcliffe presented displacement, velocity, and acceleration equations for the RRSS and RSSR mechanisms. These equations can be directly applied to the 4R Spherical mechanism. To apply these models to the latter mechanism, the user should specify the appropriate 4R Spherical mechanism values to their corresponding RRSS and RSSR variables. The Appendix H.1 and H.2 MATLAB files utilize the Suh and Radcliffe equations for the kinematic analysis of RRSS and RSSR mechanisms, respectively. In the Appendix I.1 and I.2 MATLAB and Simscape Multibody files, simulation-based RRSS and RSSR kinematic analyses are conducted.

Suh and Radcliffe introduced nonlinear equation systems to synthesize R-R, S-S, and spherical R-R dyads for motion generation. As formulated, these equation systems can be used to synthesize RRSS motion generators to approximate three precision positions and 4R Spherical motion generators to approximate four precision positions. In this chapter, unconstrained and constrained nonlinear optimization models are used to synthesize R-R, S-S, and spherical R-R dyads or motion generation. Unconstrained and constrained nonlinear optimization models are also used to synthesize RRSS and 4R Spherical motion and path generators with branch and order defect elimination to approximate *N* precision positions and precision points.

Suh and Radcliffe introduced nonlinear equation systems to synthesize RSSR and 4R Spherical mechanisms to approximate six and four precision points, respectively, for function generation. In this chapter, unconstrained and constrained nonlinear optimization models are also used to synthesize RSSR and 4R Spherical function generators to approximate *N* precision points. The motion, path, and function generation equation systems presented in this chapter are solved using the Appendix J MATLAB files.

PROBLEMS

1. Using the Appendix J.1 and J.2 MATLAB files, synthesize an order-, branch-, and circuit defect-free RRSS mechanism for the component precision positions given in Table P.8.1 (see Figure P.8.1). Verify the solution using the Appendix H.1 or I.1 files.

TABLE P.8.1
Component Position Coordinates

Precision Position	p	q	r	s
1	0, 0, 0	1, 0, 0	0, 1, 0	0, 0, 1
2	−0.5165, −0.4387, 0.0212	0.4818, −0.4965, 0.0337	−0.4588, 0.5596, 0.0301	−0.5295, −0.4468, 1.0211
3	−0.7131, −0.9651, 0.1087	0.2856, −0.9139, 0.0972	−0.7643, 0.0337, 0.1015	−0.7020, −0.9573, 1.1086

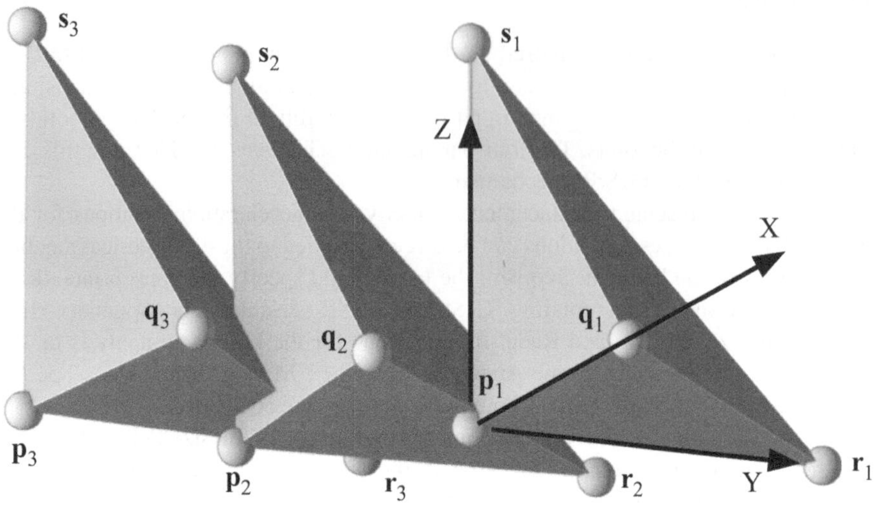

FIGURE P.8.1 Component positions.

2. Using the Appendix J.5 MATLAB files, synthesize an order-, branch-, and circuit defect-free RRSS mechanism for the tool path point coordinates given in Figure P.8.2. Verify the solution using the Appendix H.1 or I.1 files.

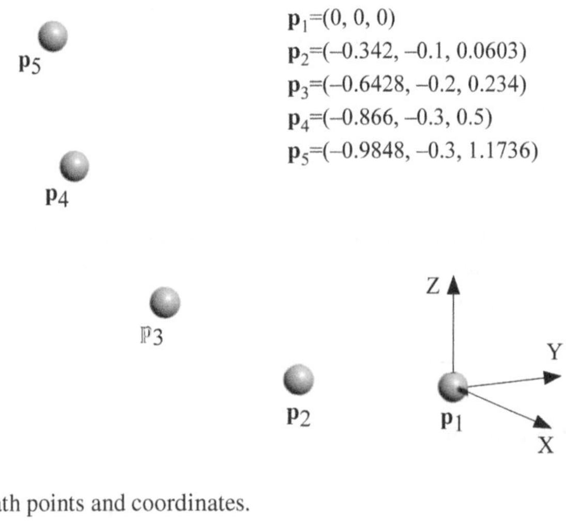

FIGURE P.8.2 Tool path points and coordinates.

3. Using the Appendix J.4 MATLAB files, synthesize an order-, branch-, and circuit defect-free RRSS mechanism for the antenna position coordinates given in Table P.8.2 (see Figure P.8.3). Verify the solution using the Appendix H.1 or I.1 files.

TABLE P.8.2
Antenna Position Coordinates

Precision Position	p	q	r
1	2, 0, 0	3, 0, 0	2, 0, 1
2	1.9021, 0.6180, 0	2.8387, 0.9224, −0.1736	2.0673, 0.6717, 0.9848
3	1.6180, 1.1756, 0	2.3783, 1.7279, −0.3420	1.8947, 1.3766, 0.9397
4	1.1756, 1.6180, 0	1.6846, 2.3187, −0.5	1.4695, 2.0225, 0.8660
5	0.6180, 1.9021, 0	0.8548, 2.6307, −0.6428	0.8167, 2.5134, 0.7660
6	0, 2, 0	0, 2.6428, −0.7660	0, 2.7660, 0.6428

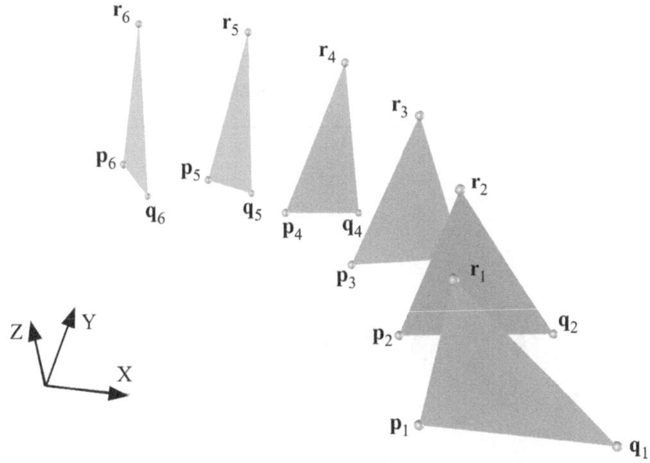

FIGURE P.8.3 Antenna positions.

4. Using the Appendix J.5 MATLAB files, synthesize an order-, branch-, and circuit defect-free RRSS mechanism for the tool path point coordinates given in Figure P.8.4. Verify the solution using the Appendix H.1 or I.1 files.

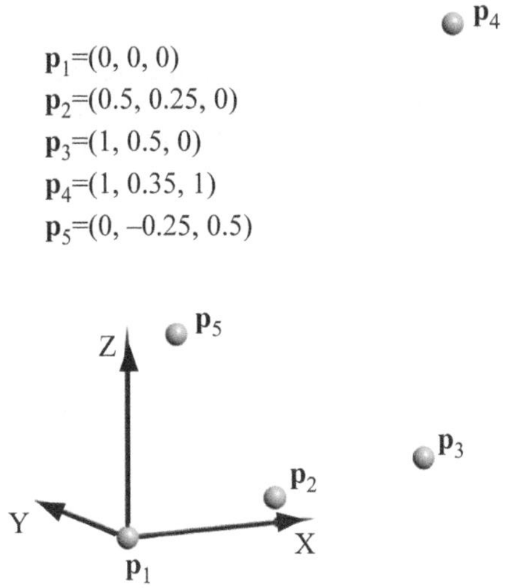

FIGURE P.8.4 Tool path points and coordinates.

5. Using the Appendix J.4 MATLAB files, synthesize an order-, branch-, and circuit defect-free RRSS mechanism for the seven component position coordinates given in Table P.8.3 (see Figure P.8.5). Verify the solution using the Appendix H.1 or I.1 files.

TABLE P.8.3
Component Position Coordinates

Precision Position	p	q	r
1	−0.75, −0.75, −0.5	0.2159,−1.0088, −0.5	−0.5154, 0.1254, −0.0774
2	−0.375, −0.375, −0.25	0.6164, −0.5055, −0.25	−0.2476, 0.5929, −0.0336
3	0, 0, 0	1, 0, 0	0, 1, 0
4	0.375, 0.375, 0.25	1.3664, 0.5055, 0.25	0.2476, 1.3429, 0.0336
5	0.75, 0.75, 0.5	1.7159, 1.0088, 0.5	0.5154, 1.6254, 0.0774
6	1.1250, 1.1250, 0.75	2.0489, 1.5077, 0.75	0.8318, 1.8327, 0.1072
7	1.5, 1.5, 1	2.3660, 2, 1	1.2132, 1.9967, 0.1808

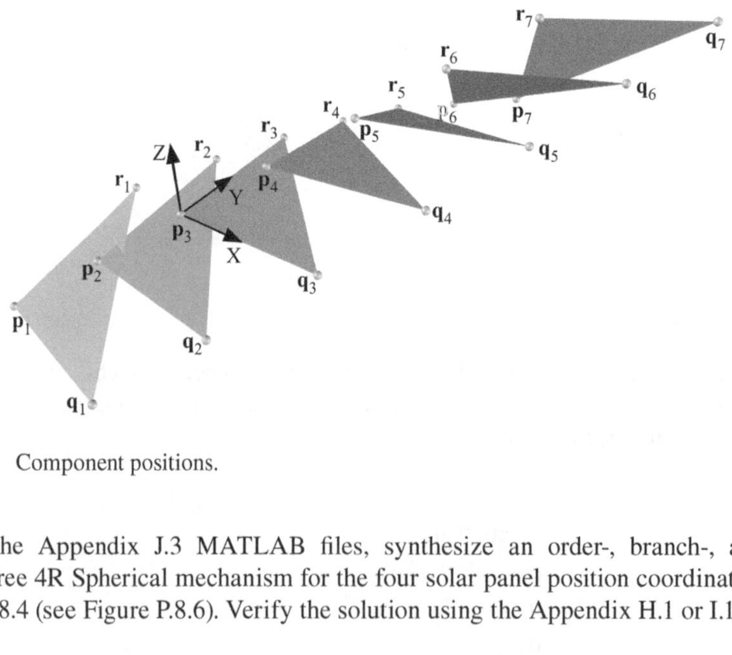

FIGURE P.8.5 Component positions.

6. Using the Appendix J.3 MATLAB files, synthesize an order-, branch-, and circuit defect-free 4R Spherical mechanism for the four solar panel position coordinates given in Table P.8.4 (see Figure P.8.6). Verify the solution using the Appendix H.1 or I.1 files.

TABLE P.8.4
Solar Panel Position Coordinates

Precision Position	p	q	r
1	1.4489, 0, 0.3882	2.4148, 0, 0.6470	1.4489, 1, 0.3882
2	1.2990, 0.7386, 0.1302	2.1651, 1.2310, 0.2171	0.8660, 1.3904, 0.7529
3	0.7500, 1.2941, 0.1132	1.2500, 2.1568, 0.1887	0.3170, 1.4677, 0.9977
4	0, 1.5, 0	0, 2.5, 0	0, 1.5, 1

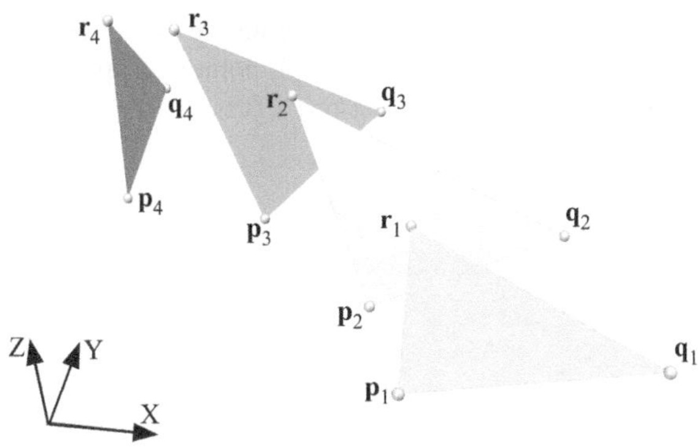

FIGURE P.8.6 Solar panel positions.

7. Repeat Problem 6 using the Appendix J.6 MATLAB files. Verify the solution using the Appendix H.1 or I.1 files.
8. Using the Appendix J.6 MATLAB files, synthesize an order-, branch-, and circuit defect-free 4R Spherical mechanism for the eight landing gear position coordinates given in Table P.8.5 (see Figure P.8.7). Verify the solution using the Appendix H.1 or I.1 files.

TABLE P.8.5
Landing Gear Precision Position Coordinates

Precision Position	p	q	r
1	1.5, 0, 0	2.5, 0, 0	1.5, 1, 0
2	1.4624, 0.3338, 0	2.4373, 0.5563, 0	1.2456, 1.2837, 0.2250
3	1.3515, 0.6508, 0	2.2524, 1.0847, 0	0.9615, 1.4606, 0.4384
4	1.1727, 0.9352, 0	1.9546, 1.5587, 0	0.6882, 1.5428, 0.6293
5	0.9352, 1.1727, 0	1.5587, 1.9546, 0	0.4539, 1.5566, 0.7880
6	0.6508, 1.3515, 0	1.0847, 2.2524, 0	0.2701, 1.5348, 0.9063
7	0.3338, 1.4624, 0	0.5563, 2.4373, 0	0.1311, 1.5087, 0.9781
8	0, 1.5, 0	0, 2.5, 0	0, 1.5, 1

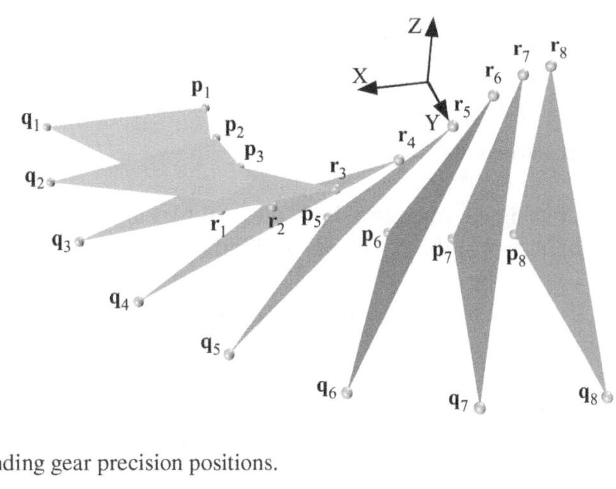

FIGURE P.8.7 Landing gear precision positions.

9. Using the Appendix J.7 MATLAB files, synthesize an order-, branch-, and circuit defect-free 4R Spherical mechanism for the path point coordinates given in Figure P.8.8. Verify the solution using the Appendix H.1 or I.1 files.

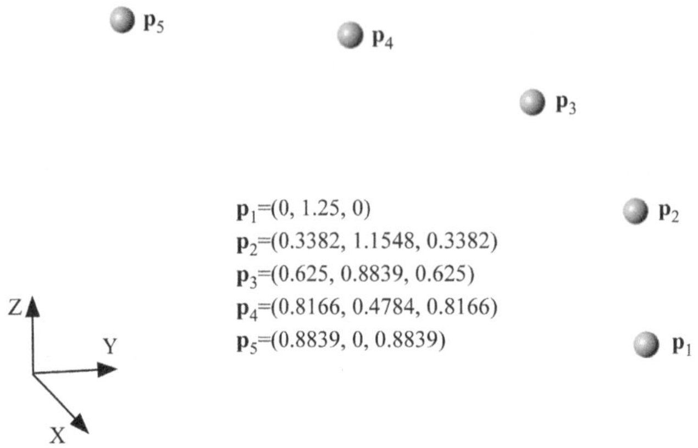

FIGURE P.8.8 Path points and coordinates.

10. Using the Appendix J.7 MATLAB files, synthesize an order-, branch-, and circuit defect-free 4R Spherical mechanism for the path coordinates given in Figure P.8.9. Verify the solution using the Appendix H.1 or I.1 files.

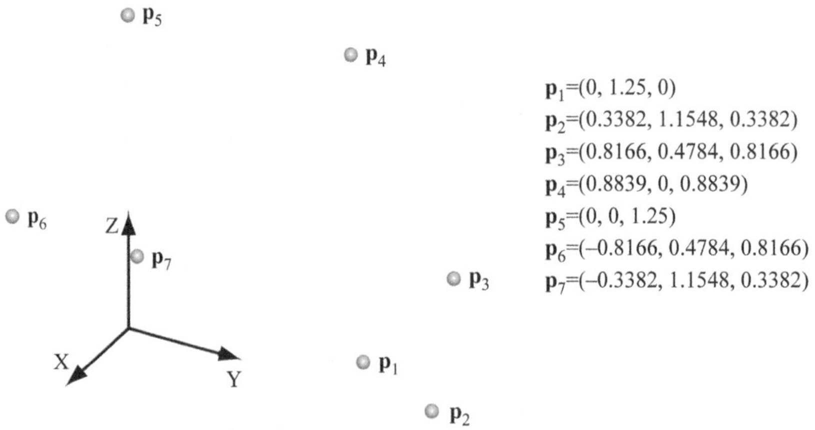

FIGURE P.8.9 Path points and coordinates.

11. Repeat Example 7.4 using the Appendix J.8 MATLAB files to synthesize an RSSR mechanism constrained to motion in the x–y plane. Verify the solution using the Appendix H.2 or I.2 files.
12. Repeat Problem 15 from Chapter 7 using the Appendix J.8 MATLAB files to synthesize an RSSR mechanism constrained to motion in the x–y plane. Verify the solution using the Appendix H.2 or I.2 files.
13. Using the Appendix J.8 MATLAB files, synthesize an RSSR mechanism (where $\alpha = 30°$) for the function $y = \sqrt{x}$ over the range $1 \le x \le 20$ for eight precision points. Use Chebyshev spacing (see Section 7.1) to determine x_j (where $x_0 = 0, x_{N+1} = 21, N = 8,$ and $j = 2,3,\ldots,N-1$).

To calculate θ_j and ϕ_j, use Equation (7.2) with $\Delta\beta = \Delta\theta = 45°$ and $\Delta\gamma = \Delta\phi = 90°$. Verify the solution using the Appendix H.2 or I.2 files.

14. Using the Appendix J.8 MATLAB files, synthesize an RSSR mechanism (where $\alpha = 30°$) for the function $y = x^2$ over the range $1 \le x \le 20$ for ten precision points. Use Chebyshev spacing (see Section 7.1) to determine x_j (where $x_0 = 0$, $x_{N+1} = 21$, $N = 10$, and $j = 2,3,\ldots,N-1$). To calculate θ_j and ϕ_j, use Equation (7.2) with $\Delta\beta = \Delta\theta = 60°$ and $\Delta\gamma = \Delta\phi = 30°$. Verify the solution using the Appendix H.2 or I.2 files.
15. Using the Appendix J.9 MATLAB files, synthesize an RSSR mechanism for the crank and rudder displacement angles given in Table P.8.6 (let $b_{0x} = b_{1x}$ and $\alpha = 90°$). Verify the solution using the Appendix H.2 or I.2 files.

TABLE P.8.6
Prescribed Crank and Rudder Displacement Angles

θ_j (deg)	ϕ_j (deg)
10	15
20	30
30	45
40	60

16. Using the Appendix J.9 MATLAB files, synthesize an RSSR mechanism for the crank and wheel displacement angles given in Table P.8.7 (let $b_{0x} = b_{1x}$ and $\alpha = 90°$). Verify the solution using the Appendix H.2 or I.2 files.

TABLE P.8.7
Prescribed Crank and Wheel Displacement Angles

θ_j (deg)	ϕ_j (deg)
−35	12
−30	18
−25	24
−20	30
−15	36
−10	42
−5	48
0	54

17. Repeat Problem 13 from Chapter 8 using the Appendix J.10 MATLAB files to synthesize a 4R Spherical mechanism (where $\alpha = 45°$). Verify the solution using the Appendix H.2 or I.2 files.
18. Repeat Problem 14 from Chapter 8 using the Appendix J.10 MATLAB files to synthesize a 4R Spherical mechanism (where $\alpha = 55°$). Verify the solution using the Appendix H.2 or I.2 files.

19. Using the Appendix J.10 MATLAB files, synthesize a 4R Spherical mechanism (where $\alpha = 40°$) for the crank and follower displacement angles given in Table P.8.8. Verify the solution using the Appendix H.2 or I.2 files.

TABLE P.8.8
Prescribed Crank and Follower Displacement Angles

θ_j (deg)	ϕ_j (deg)
5	−3
10	−6
15	−9
20	−12
25	−15
30	−18
35	−21
40	−24

20. Using the Appendix J.10 MATLAB files, synthesize a 4R Spherical mechanism (where $\alpha = 35°$) for the crank and follower displacement angles given in Table P.8.9. Verify the solution using the Appendix H.2 or I.2 files.

TABLE P.8.9
Prescribed Crank and Follower Displacement Angles

θ_j (deg)	ϕ_j (deg)
5	10
10	12
15	14
20	16
25	18
30	15
35	12
40	9

NOTES

1 The passive DOFs associated with the RSSR and RRSS mechanisms generally do not contribute to mechanism motion because the effects of these DOFs are typically highly localized.
2 Variable I in Equation (7.8) and Equation (7.20) is a 3×3 identity matrix.
3 Because four-link mechanisms have two assembly configurations (the *open* and *crossed* configurations), there are two α solutions (or two branches) for every θ value.
4 Due to the passive DOF of the follower link, rotation angles for this link cannot be directly calculated in the RRSS mechanism equations-even when restricted to planar motion (unlike the planar four-bar equations in Section 4.3).

5 The Appendix H.1 MATLAB file produces results for both the initial (branch 1) and second (branch 2) RRSS mechanism configurations.

6 While planar mechanisms have *planar-surface* workspaces and most spatial mechanisms have *volume* workspaces, spherical mechanisms have *spatial-surface* workspaces.

7 Mechanisms like the 4R Spherical mechanism, having true DOFs that are contrary to Gruebler's equation, are called *paradoxes* or *maverick mechanisms.*

8 Like Equation (8.12), there are two ϕ solutions corresponding the two RSSR assembly configurations (the open and crossed configurations) for every θ value.

9 In Equation (8.42) (and similarly in Equation (8.43)), the term $\ddot{\phi}\left[P_{\mathbf{ub}_0}\right](\mathbf{b}-\mathbf{b}_0)$ corresponds to the tangential acceleration, and the term $\dot{\phi}^2\left[P_{\mathbf{ub}_0}\right]\left[P_{\mathbf{ub}_0}\right](\mathbf{b}-\mathbf{b}_0)$ corresponds to the normal acceleration.

10 Due to the passive DOF of the coupler link, rotation angles for this link cannot be directly calculated in the RSSR mechanism equations—even when restricted to planar motion (unlike the planar four-bar equations in Section 4.3).

11 The Appendix H.2 MATLAB file produces results for both the initial (branch 1) and second (branch 2) RSSR mechanism configurations.

12 The constant twist constraint $(\mathbf{ua}_j)^T(\mathbf{ua}_0)=(\mathbf{ua}_1)^T(\mathbf{ua}_0)$ $j=2,3$ and constant moment constraint $(\mathbf{ua}_0)^T((\mathbf{a}_j-\mathbf{a}_0)\times\mathbf{ua}_j)=(\mathbf{ua}_0)^T((\mathbf{a}_1-\mathbf{a}_0)\times\mathbf{ua}_1)$ $j=2,3$ can also be used in place of Equations (7.28) and (7.29).

13 Precision position variables **p**, **q**, **r**, and **s** should not be *proportional* (i.e., chosen from the same plane).

14 To avoid producing proportional matrices, precision position variables **p**, **q**, and **r** should not all lie on a common line and $\mathbf{p}_1$, $\mathbf{q}_1$, and $\mathbf{r}_1$ should not all be selected from a common coordinate system plane..

15 It can be observed from Equation (8.12) that the coupler displacement angle α is a function of the crank displacement angle θ in the RRSS and 4R Spherical displacement equations. In the constrained nonlinear equation systems, however, both α and θ are calculated simultaneously. Because α and θ are calculated, Equations (8.6) through (8.12) are not used in the constrained equation systems. Equation (8.1) does constrain α and θ such that an $\alpha(\theta)$ relationship is ensured.

16 Order defect-free solutions are ensured because the R-R dyad rotation angles are prescribed (*prescribed timing*).

17 As illustrated in Figure 8.15, $\mathbf{ua}_0=(1,0,0)$ and $\mathbf{ub}_0=(\cos\alpha,\sin\alpha,0)$.

18 This library can also be downloaded from the authors' server http://www.softalink.com/kruss/mech_des_vis_prog_app_2nd_ed/MATLAB_SIMMULT_FILES.zip.

REFERENCES

1. Capellen, W. M. 1966. Kinematics-A Survey in Retrospect and Prospect. *Mechanism and Machine Theory*. 1: 211–228.
2. Suh, C. H. and Radcliffe, C. W. 1978. *Kinematics and Mechanisms Design*. pages 83–85. New York: John Wiley & Sons.
3. Suh, C. H. and Radcliffe, C. W. 1978. *Kinematics and Mechanisms Design*. page 85. New York: John Wiley & Sons.
4. Waldron, K. J. and Kinzel, G. L. 2004. *Kinematics, Dynamics and Design of Machinery*, 2nd Edition. pages 340–347. Saddle River: Prentice-Hall.
5. Russell, K., Shen, J. Q., and Sodhi, R. S. 2019. *Kinematic and Dynamics of Mechanical Systems: Implementation in MATLAB and Simscape Multibody: Third Edition*. Section 10.4. Boca Raton: CRC Press.
6. Suh, C. H. and Radcliffe, C. W. 1978. *Kinematics and Mechanisms Design*. pages 79–80. New York: John Wiley & Sons.
7. Suh, C. H. and Radcliffe, C. W. 1978. *Kinematics and Mechanisms Design*. pages 155–157. New York: John Wiley & Sons.
8. Suh, C. H. and Radcliffe, C. W. 1978. *Kinematics and Mechanisms Design*. pages 152–154. New York: John Wiley & Sons.
9. Suh, C. H. and Radcliffe, C. W. 1978. *Kinematics and Mechanisms Design*. page 199. New York: John Wiley & Sons.
10. Suh, C. H. and Radcliffe, C. W. 1978. *Kinematics and Mechanisms Design*. pages 181–182. New York: John Wiley & Sons.
11. Suh, C. H. and Radcliffe, C. W. 1978. *Kinematics and Mechanisms Design*. pages 179–181. New York: John Wiley & Sons.

9 Adjustable Planar and Spherical Four-Bar Mechanism Dimensional Synthesis

CONCEPT OVERVIEW

In this chapter, the reader will gain a central understanding regarding the following:

1. Advantages of incorporating adjustability in mechanism dyads
2. The adjustable options for mechanism dyads
3. Constrained and unconstrained nonlinear optimization for planar four-bar and 4R Spherical mechanism dyad dimensional synthesis
4. Constrained nonlinear optimization for planar four-bar and 4R Spherical mechanism loop dimensional synthesis

9.1 INTRODUCTION

The kinematic synthesis equations presented in Chapters 5–8 are useful for calculating the dimensions of a *single* mechanism configuration. As demonstrated in these chapters, this single mechanism configuration will precisely achieve or approximate either precision positions for motion generation or precision points for path generation or function generation.

In comparison to *classical* kinematic synthesis (where the dimensions of a single mechanism configuration are calculated), a substantial array of methodologies and mathematical models have been developed in recent decades for the kinematic synthesis of *adjustable* planar and spatial mechanisms [1–8]. An adjustable mechanism includes fixed and/or moving pivots that can be adjusted to achieve multiple mechanism configurations—each configuration achieving or approximating a particular group of precision positions or precision points.

To further convey the objective and advantage of adjustable mechanism kinematic synthesis, Figure 9.1 illustrates two groups of precision positions. For this illustration, let's assume a planar four-bar mechanism is required to approximate the position sequence in group 1 followed by the position sequence in group 2. Due to the prescribed order and arrangement of the precision positions in Figure 9.1, it is unlikely that the ideal mechanism solution to this problem (a mechanism solution that produces the closest precision position approximations) is a single planar four-bar mechanism configuration. One solution option for the motion generation problem is to synthesize a single mechanism configuration for the precision positions in group 1 and another single mechanism configuration for the precision positions in group 2 using the motion generation equations in Chapters 5 or 6. Although this solution is mathematically feasible and will likely produce acceptable precision point approximations, it will also produce two individual and independent motion generators.

Another solution option for this motion generation problem is to utilize an equation system for the kinematic synthesis of adjustable motion generators to calculate dimensions for a single-motion generator that can be adjusted to two distinct configurations—each configuration approximating a particular group of precision positions. A planar four-bar mechanism with adjustable moving pivots is illustrated in Figure 9.2. This mechanism can produce up to four distinct configurations—including configurations $\mathbf{a}_0 - \mathbf{a}_1 - \mathbf{b}_1 - \mathbf{b}_0$ and $\mathbf{a}_0 - \mathbf{a}_1^* - \mathbf{b}_1^* - \mathbf{b}_0$.[1] If this mechanism represented an adjustable

DOI: 10.1201/9781003591375-9

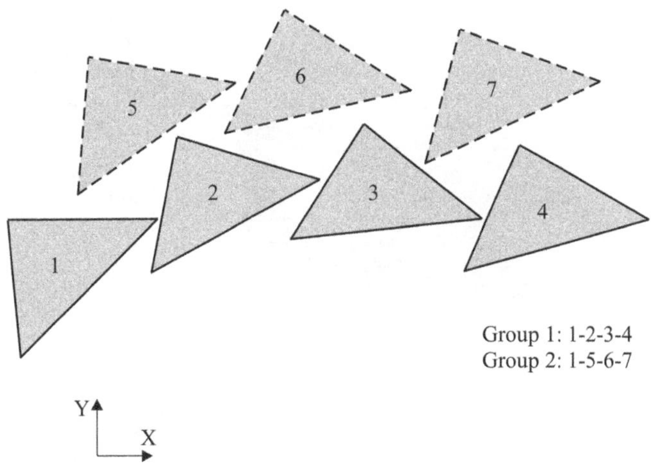

FIGURE 9.1 Two groups of precision positions.

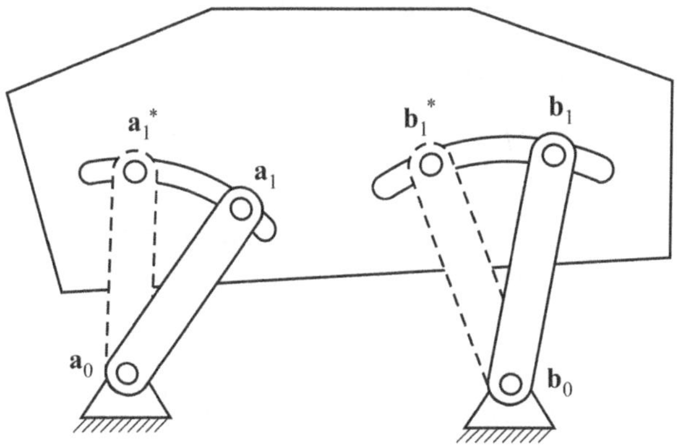

FIGURE 9.2 Planar four-bar motion generator with adjustable moving pivots.

mechanism solution for the precision positions in Figure 9.1, one configuration ($\mathbf{a}_0 - \mathbf{a}_1 - \mathbf{b}_1 - \mathbf{b}_0$ for example) will achieve the precision positions in group 1 and another configuration ($\mathbf{a}_0 - \mathbf{a}_1^* - \mathbf{b}_1^* - \mathbf{b}_0$ for example) will achieve the precision positions in group 2. It can be observed in Figure 9.2 that all adjustments employ the same mechanism hardware but produce kinematically distinct mechanism configurations. The primary advantage of adjustable kinematic synthesis is that it introduces a greater degree of flexibility in application to a single set of mechanism hardware.

Figure 9.3 illustrates a pair of locking pliers—an example of a commonly used adjustable planar four-bar mechanism. By turning the adjustment knob on the locking pliers, the effective link length L_1 is either increased or decreased. By adjusting the effective link length, alternate positions are achieved by the links of the locking pliers, which result in alternate degrees of tool locking strength.

Figure 9.4 illustrates the four adjustment options possible for a planar four-bar mechanism dyad. These options are *constant link length* moving- and fixed-pivot adjustments (Figures 9.4a and b, respectively) and *adjustable link length* moving- and fixed-pivot adjustments (Figures 9.4c and d, respectively). Links having adjustable lengths require extendable–collapsible links (e.g., telescoping links), while links having constant lengths enable the use of single-piece links. Moving-pivot adjustments maintain a single fixed-pivot placement throughout the mechanism adjustments, which can be advantageous if the fixed pivot is to be coupled to a driver that is in a fixed location. The four adjustment options in Figure 9.4 produce up to ten mechanism configurations (when considering every combination of adjustments).

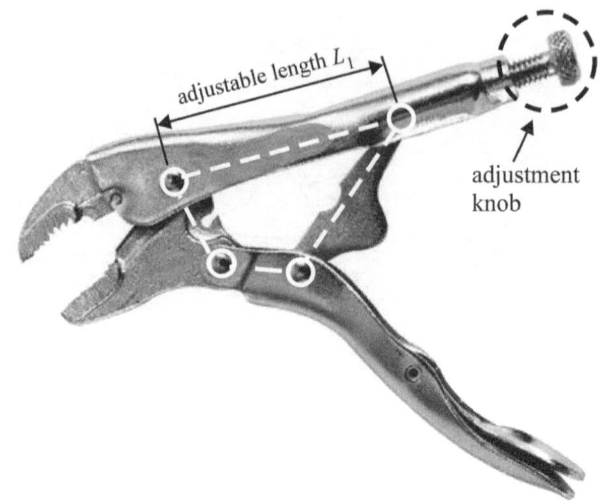

FIGURE 9.3 Adjustable planar four-bar lock pliers mechanism.

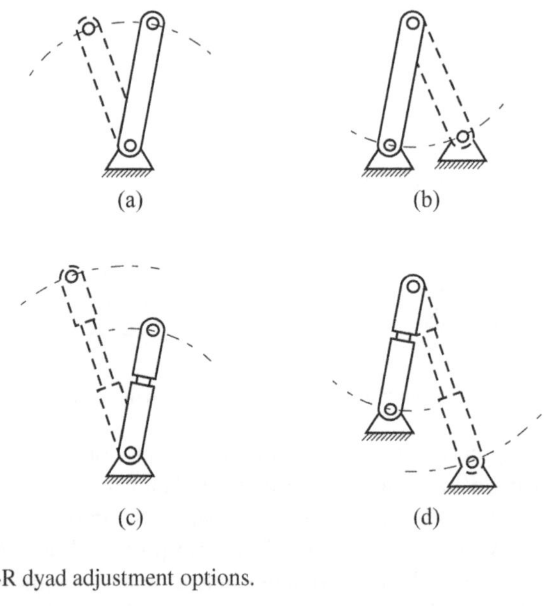

FIGURE 9.4 Planar R-R dyad adjustment options.

9.2 ADJUSTABLE PLANAR FOUR-BAR MOTION GENERATION

In Sections 6.6 and 6.7, unconstrained and constrained nonlinear equation systems for motion and path generation are presented to calculate the fixed and moving pivots of planar four-bar mechanism dyads. One equation system is an unconstrained nonlinear equation system for planar four-bar motion generation (Equation 6.16) [9]. Another equation system is a constrained nonlinear equation system for planar four-bar motion and path generation (Equations 6.18 through 6.22).

To synthesize an adjustable planar four-bar mechanism dyad to approximate two groups of precision positions (where one group includes positions 1–2–3 and the other group includes 1–4–5) using constant link length moving-pivot adjustments, Equation (6.16) becomes

$$\left(\mathbf{a}_j-\mathbf{a}_0\right)^T\left(\mathbf{a}_j-\mathbf{a}_0\right)=\left(\mathbf{a}_1-\mathbf{a}_0\right)^T\left(\mathbf{a}_1-\mathbf{a}_0\right) \quad j=2,3 \tag{9.1}$$

$$\left(\mathbf{a}_k^* - \mathbf{a}_0\right)^T \left(\mathbf{a}_k^* - \mathbf{a}_0\right) = \left(\mathbf{a}_1^* - \mathbf{a}_0\right)^T \left(\mathbf{a}_1^* - \mathbf{a}_0\right) \quad k = 4,5 \tag{9.2}$$

$$\left(\mathbf{a}_1 - \mathbf{a}_0\right)^T \left(\mathbf{a}_1 - \mathbf{a}_0\right) = \left(\mathbf{a}_1^* - \mathbf{a}_0\right)^T \left(\mathbf{a}_1^* - \mathbf{a}_0\right) \tag{9.3}$$

where $\mathbf{a}_j = \left[D_{1j}\right]\mathbf{a}_1$ and $\mathbf{a}_k^* = \left[D_{1k}\right]\mathbf{a}_1^*$. Equations (9.1) through (9.3) form a five-equation nonlinear simultaneous equation system to calculate five of the six dyad variables a_{0x}, a_{0y}, a_{1x}, a_{1y}, a_{1x}^*, and a_{1y}^* with a free choice of one variable.

To synthesize an adjustable planar four-bar mechanism dyad to approximate two groups of precision positions (where one group includes positions 1–2–3 and the other group includes 1–4–5) using constant link length fixed-pivot adjustments, Equation (6.16) becomes

$$\left(\mathbf{a}_j - \mathbf{a}_0\right)^T \left(\mathbf{a}_j - \mathbf{a}_0\right) = \left(\mathbf{a}_1 - \mathbf{a}_0\right)^T \left(\mathbf{a}_1 - \mathbf{a}_0\right) \quad j = 2,3 \tag{9.4}$$

$$\left(\mathbf{a}_k - \mathbf{a}_0^*\right)^T \left(\mathbf{a}_k - \mathbf{a}_0^*\right) = \left(\mathbf{a}_1 - \mathbf{a}_0^*\right)^T \left(\mathbf{a}_1 - \mathbf{a}_0^*\right) \quad k = 4,5 \tag{9.5}$$

$$\left(\mathbf{a}_1 - \mathbf{a}_0\right)^T \left(\mathbf{a}_1 - \mathbf{a}_0\right) = \left(\mathbf{a}_1 - \mathbf{a}_0^*\right)^T \left(\mathbf{a}_1 - \mathbf{a}_0^*\right) \tag{9.6}$$

where $\mathbf{a}_j = \left[D_{1j}\right]\mathbf{a}_1$ and $\mathbf{a}_k = \left[D_{1k}\right]\mathbf{a}_1$. Equations (9.4) through (9.6) form a five-equation nonlinear simultaneous equation system to calculate five of the six dyad variables a_{0x}, a_{0y}, a_{0x}^*, a_{0y}^*, a_{1x}, and a_{1y} with a free choice of one variable.

Equations (9.3) and (9.6) ensure constant crank and follower link lengths between mechanism adjustments, respectively. To synthesize mechanisms with adjustable crank and follower link lengths, Equations (9.3) and (9.6) are replaced by adding another precision position to one of the two groups of precision positions (e.g., k=4, 5, 6 in Equations (9.2) and (9.5)).

Appendices K.1 through K.4 include MATLAB file user instructions for the adjustable four-bar motion generation equations presented in this section. Although Equations (9.1) through (9.3) and Equations (9.4) through (9.6) form nonlinear simultaneous equation systems for two three-position synthesis problems, in Appendices K.1 through K.4, they are solved as unconstrained nonlinear optimization models in MATLAB for an indefinite number of precision positions.

Example 9.1

Problem Statement: Using the Appendix K.1 MATLAB files, synthesize an adjustable planar four-bar mechanism for the two groups of container positions given in Table E.9.1 (Figure E.9.1).

Known Information: Table E.9.1 and Appendix K.1 MATLAB files

Solution Approach: In this example, adjustable planar four-bar dyad variables a_{0x} and b_{0x} were specified to $a_{0x} = 0$ and $b_{0x} = 5$, respectively. Figure E.9.2 includes the input in bold type in the Appendix K.1 MATLAB file *a0_a1_a1s_quasi_newton.m* when $a_{0x} = 0$. Figure E.9.3 includes the output from the same file. Figure E.9.4 includes the output from the Appendix K.1 MATLAB file *a0_a1_a1s_quasi_newton.m* when $b_{0x} = a_{0x} = 5$. Figure E.9.5 illustrates the resulting adjustable planar four-bar mechanism. A kinematic analysis of this solution is presented in Example 9.10 in Section 9.7. The equation systems formulated from Equations (9.1) through (9.6) include no constraints to ensure that branch- and order defect-free solutions are calculated. With these equation systems, branch- and order defect-free solutions can be calculated through trial and error.

TABLE E.9.1
Precision Positions for Adjustable Four-Bar Mechanism

Adjustment 1			Adjustment 2		
Precision Position	$\mathbf{p}_j$	α_j **(deg)**	**Precision Position**	$\mathbf{p}_k$	α_k **(deg)**
1	6.5450, 20.1547		1	6.5450, 20.1547	
2	12.4409, 17.7173	–38.2965	4	18.4724, 15.4830	–74.4814
3	15.6261, 11.6238	–74.4814	5	17.1848, 8.5771	–98.0521

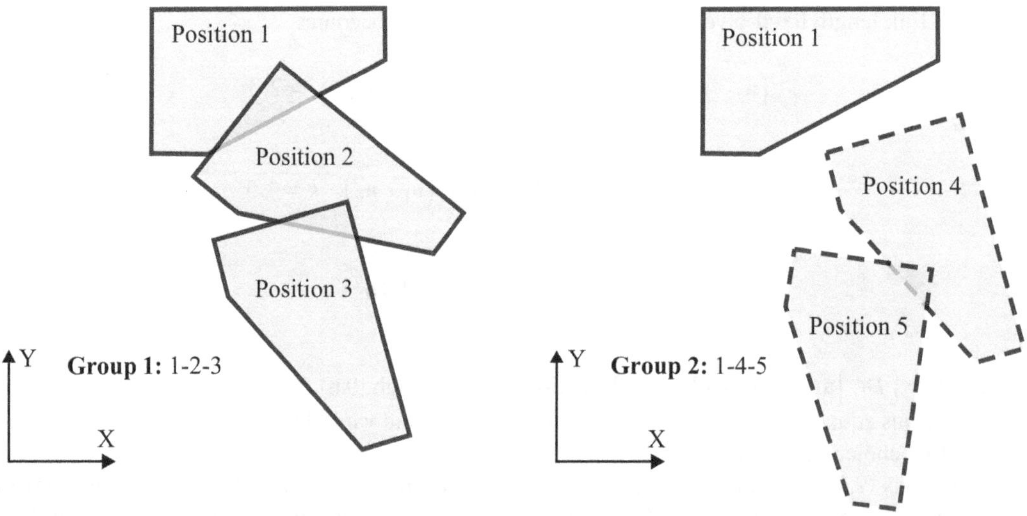

FIGURE E.9.1 Two groups of container positions.

```
M = 3;
N = 3;

a0x = 0;

alpha2 = -38.2965 *pi/180;
alpha3 = -74.4814 *pi/180;
alpha4 = -74.4814 *pi/180;
alpha5 = -98.0521 *pi/180;

P1 = [6.5450,20.1547];
P2 = [12.4409, 17.7173];
P3 = [15.6261, 11.6238];
P4 = [18.4724, 15.4830];
P5 = [17.1848, 8.5771];

a0y = 10;
a1x = 10;
a1y = 10;
a1sx = 10;
a1sy = 10;
```

FIGURE E.9.2 Input in bold type in the Appendix K.1 MATLAB file *a0_a1_a1s_quasi_newton.m* (with $a_{0x} = 0$) for Example 9.1.

```
Iterations: 35
Func Evals: 37
Algorithm: quasi-newton
exit flag = 5
error = ( 4.291403e-28 )
a0y = ( 16.280226 )
a1x = ( 8.482601 )
a1y = ( 9.273511 )
a1sx = ( 8.996460 )
a1sy = ( 9.946799 )
Elapsed time is 0.227861 seconds.
Solution calculation is complete.
>>
```

FIGURE E.9.3 Output from the Appendix K.1 MATLAB file *a0_a1_a1s_quasi_newton.m* (with $a_{0x} = 0$) for Example 9.1.

```
Iterations: 57
Func Evals: 58
Algorithm: quasi-newton
exit flag = 5
error = ( 2.966117e-28 )
a0y = ( 4.237433 )
a1x = ( 6.314356 )
a1y = ( 8.595783 )
a1sx = ( 9.398057 )
a1sy = ( 5.412097 )
Elapsed time is 0.413930 seconds.
Solution calculation is complete.
>>
```

FIGURE E.9.4 Output from the Appendix K.1 MATLAB file *a0_a1_a1s_quasi_newton.m* (with $b_{0x} = a_{0x} = 5$ and $a_{0y} = -10$) for Example 9.1.

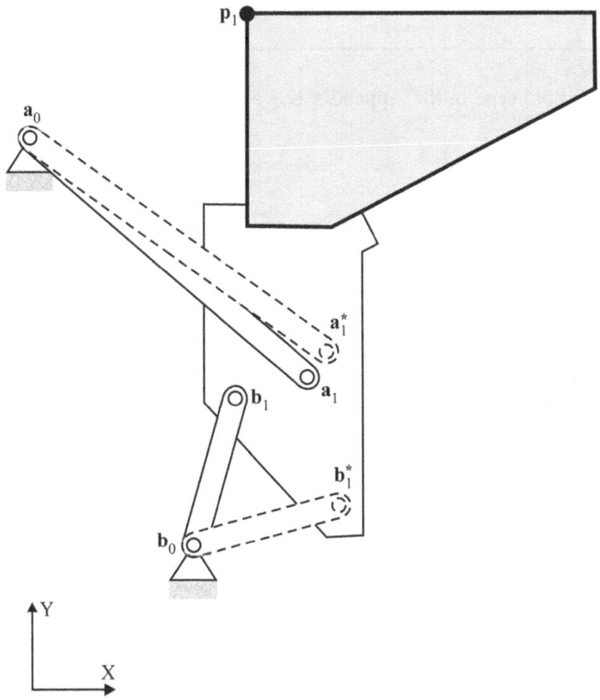

FIGURE E.9.5 Synthesized adjustable four-bar mechanism.

Example 9.2

Problem Statement: Using the Appendix K.3 MATLAB files, synthesize adjustable-length $\mathbf{a}_0 - \mathbf{a}_1$ and $\mathbf{a}_0 - \mathbf{a}_1^*$ solutions for Example 9.1.

Known Information: Table E.9.1 and Appendix K.3 MATLAB files

Solution Approach: In this example, adjustable planar four-bar dyad variable a_{0x} was specified to $a_{0x} = 0$. Figure E.9.6 includes the input in bold type in the Appendix K.3 MATLAB file *a0_a1_a1s_quasi_newton.m* when $a_{0x} = 0$. Figure E.9.7 includes the output from the same file. Figure E.9.8 illustrates the resulting adjustable planar four-bar mechanism. The equation systems formulated from Equations (9.1) through (9.6) include no constraints to ensure that branch- and order defect-free solutions are calculated. With these equation systems, branch- and order defect-free solutions can be calculated through trial and error.

```
M = 3;
N = 3;

a0x = 0;

alpha2 = -38.2965 *pi/180;
alpha3 = -74.4814 *pi/180;
alpha4 = -74.4814 *pi/180;
alpha5 = -98.0521 *pi/180;

P1 = [6.5450,20.1547];
P2 = [12.4409, 17.7173];
P3 = [15.6261, 11.6238];
P4 = [18.4724, 15.4830];
P5 = [17.1848, 8.5771];

a0y = 10;
a1x = 10;
a1y = 10;
a1sx = 10;
a1sy = 10;
```

FIGURE E.9.6 Input in bold type in the Appendix K.3 MATLAB file *a0_a1_a1s_quasi_newton.m* (with $a_{0x} = 0$) for Example 9.2.

```
Iterations: 30
Func Evals: 32
Algorithm: quasi-newton
exit flag = 5
error = ( 1.767048e-28 )
a0y = ( 16.000617 )
a1x = ( 8.499400 )
a1y = ( 9.340074 )
a1sx = ( 7.301120 )
a1sy = ( 9.799061 )
Elapsed time is 0.227430 seconds.
Solution calculation is complete.
>>
```

FIGURE E.9.7 Output from the Appendix K.3 MATLAB file *a0_a1_a1s_quasi_newton.m* (with $a_{0x} = 0$) for Example 9.2.

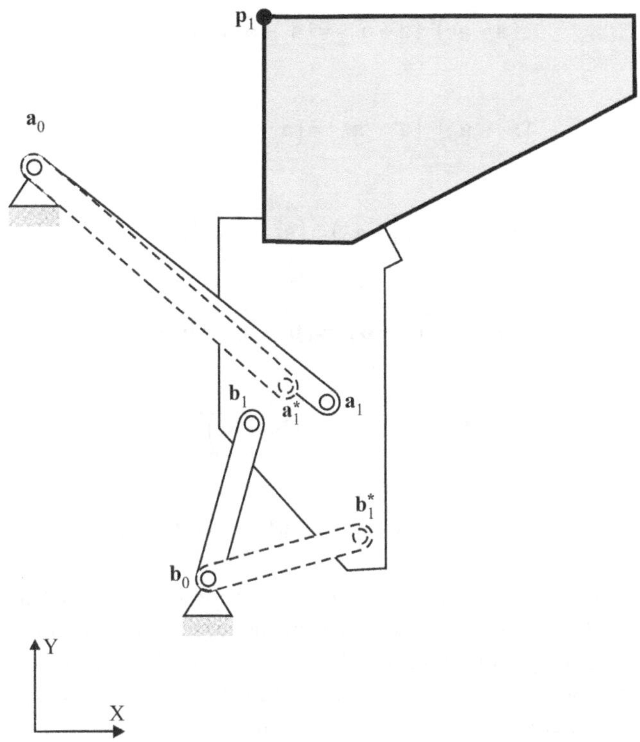

FIGURE E.9.8 Synthesized adjustable four-bar mechanism.

9.3 ADJUSTABLE PLANAR FOUR-BAR MOTION AND PATH GENERATION: CONSTRAINED OPTIMIZATION

A constrained nonlinear optimization model similar in form to the model presented in Section 6.7 can be formulated for adjustable planar four-bar motion or path generation. To synthesize adjustable planar four-bar motion generators to approximate an indefinite number of precision positions (two groups of positions where $j = 2,3,\ldots,M$ in one group and $k = M+1, M+2,\ldots,N$ in the other group), the objective function in Equation (9.7) is used:

$$f(\mathbf{X}) = \sum_{j=2}^{M}\left\{\|\mathbf{p}_j - \mathbf{p}\|^2 + \|\mathbf{q}_j - \mathbf{q}\|^2\right\} + \sum_{k=M+1}^{N}\left\{\|\mathbf{p}_k^* - \mathbf{p}^*\|^2 + \|\mathbf{q}_k^* - \mathbf{q}^*\|^2\right\} \tag{9.7}$$

where $\mathbf{X} = \left(\mathbf{a}_0, \mathbf{a}_1, \mathbf{a}_1^*, \mathbf{b}_0, \mathbf{b}_1, \mathbf{b}_1^*, \theta_2 \ldots \theta_M, \theta_{M+1} \ldots \theta_N, \alpha_2 \ldots \alpha_M, \alpha_{M+1} \ldots \alpha_N\right)$ for moving-pivot adjustments. In Equation (9.7), $\mathbf{p}_j$ and $\mathbf{q}_j$ are the precision position variables in the first group, while $\mathbf{p}$ and $\mathbf{q}$ are the coupler points achieved by the synthesized mechanism (using Equation (6.14)). In Equation (9.7), $\mathbf{p}_k^*$ and $\mathbf{q}_k^*$ are the precision position variables in the second group, while $\mathbf{p}^*$ and $\mathbf{q}^*$ are the coupler points achieved by the synthesized mechanism (also using Equation (6.14)). Angles θ_j and α_j are the crank and coupler link displacement angles for the mechanism configuration associated with the first group of precision positions, respectively, while angles θ_k and α_k are for the mechanism configuration associated with the second group of precision positions.

Along with Equation (9.7) as an objective function, Equations (9.8) through (9.13) are included (for precision positions $2\ldots M$ and $M+1\ldots N$) as constraints for constant link length moving-pivot adjustments. Equations (9.10) and (9.13) ensure constant crank and follower link lengths between mechanism adjustments.

$$(\mathbf{a}-\mathbf{a}_0)^T(\mathbf{a}-\mathbf{a}_0)=(\mathbf{a}_1-\mathbf{a}_0)^T(\mathbf{a}_1-\mathbf{a}_0) \tag{9.8}$$

$$(\mathbf{a}^*-\mathbf{a}_0)^T(\mathbf{a}^*-\mathbf{a}_0)=(\mathbf{a}_1^*-\mathbf{a}_0)^T(\mathbf{a}_1^*-\mathbf{a}_0) \tag{9.9}$$

$$(\mathbf{a}_1-\mathbf{a}_0)^T(\mathbf{a}_1-\mathbf{a}_0)=(\mathbf{a}_1^*-\mathbf{a}_0)^T(\mathbf{a}_1^*-\mathbf{a}_0) \tag{9.10}$$

$$(\mathbf{b}-\mathbf{b}_0)^T(\mathbf{b}-\mathbf{b}_0)=(\mathbf{b}_1-\mathbf{b}_0)^T(\mathbf{b}_1-\mathbf{b}_0) \tag{9.11}$$

$$(\mathbf{b}^*-\mathbf{b}_0)^T(\mathbf{b}^*-\mathbf{b}_0)=(\mathbf{b}_1^*-\mathbf{b}_0)^T(\mathbf{b}_1^*-\mathbf{b}_0 \tag{9.12}$$

$$(\mathbf{b}_1-\mathbf{b}_0)^T(\mathbf{b}_1-\mathbf{b}_0)=(\mathbf{b}_1^*-\mathbf{b}_0)^T(\mathbf{b}_1^*-\mathbf{b}_0 \tag{9.13}$$

To produce adjustable planar four-bar motion generator solutions that are order defect-free, Inequality (6.20) is included for the crank displacement angles corresponding to the first group of precision positions, and Inequality (9.14) is included for the crank displacement angles corresponding to the second group of precision positions. As with Inequality (6.20), the sign in Inequality (9.14) can be reversed to ensure clockwise crank link rotation.

$$\begin{cases} \theta_k > \theta_{k-1} \\ \theta_M < 2\pi \end{cases} \quad k = M+1, M+2, \ldots, N \tag{9.14}$$

To produce adjustable planar four-bar motion generator solutions that are branch defect-free, Inequality (6.21) is included for the first group of precision positions $(2\ldots M)$ along with Inequality (9.15) for the second group of precision positions $(M+1\ldots N)$.

$$\left[(\mathbf{b}^*-\mathbf{b}_0)\times(\mathbf{a}^*-\mathbf{b}_0)\right]\cdot\left[(\mathbf{b}_1^*-\mathbf{b}_0)\times(\mathbf{a}_1^*-\mathbf{b}_0)\right]>0 \tag{9.15}$$

Table 9.1 includes the constrained nonlinear optimization model for constant-length, moving-pivot-adjustable planar four-bar motion and path generation. Appendices K.5 and K.6 include MATLAB file user instructions for the adjustable four-bar motion and path generation optimization models presented in this section.

TABLE 9.1
Optimization Model: Constant-Length, Moving-Pivot Adjustable Planar Four-Bar Motion and Path Generation

Equation Type	Equation Number
Objective function	9.7[a]
Equality constraint	9.8–9.13
Inequality constraint	6.20, 6.21, 9.14, and 9.15

[a] For the synthesis of constant-length moving-pivot adjustable planar four-bar path generators, only the squared norms $\|\mathbf{p}_j-\mathbf{p}\|^2$ and $\|\mathbf{p}_k^*-\mathbf{p}^*\|^2$ are kept in Equation (9.7).

Example 9.3

Problem Statement: Using the Appendix K.5 MATLAB files, synthesize an adjustable planar four-bar mechanism for Example 9.1. In place of the α_j and α_k values included in Table E 9.1, the values for the additional rigid-body point **q** are $\mathbf{q}_1 = (17.3726, 20.1547)$, $\mathbf{q}_2 = (20.9386, 11.0071)$, $\mathbf{q}_3 = (18.5230, 1.1910)$, $\mathbf{q}_4 = (21.3693, 5.0502)$, and $\mathbf{q}_5 = (15.6681, -2.1437)$.

Known Information: Table E 9.1 and Appendix K.5 MATLAB files

Solution Approach: Figure E.9.9 includes the input in bold type in the Appendix K.1 MATLAB file *motion_interior_point*. Figure E.9.10 includes the output from the same file. Figure E.9.11 illustrates the resulting adjustable planar four-bar mechanism. A kinematic analysis of this solution is presented in Example 9.11 in Section 9.7.

```
M = 3;
N = 3;

P1 = [6.5450,20.1547];
P2 = [12.4409, 17.7173];
P3 = [15.6261, 11.6238];
P4 = [18.4724, 15.4830];
P5 = [17.1848, 8.5771];

Q1 = [17.3726, 20.1547];
Q2 = [20.9386, 11.0071];
Q3 = [18.5230, 1.1910];
Q4 = [21.3693, 5.0502];
Q5 = [15.6681, -2.1437];

a0x = -5;
a0y = 5;

a1x = 5;
a1y = 5;

a1sx = 5;
a1sy = 5;

b0x = 10;
b0y = 0;

b1x = 10;
b1y = 10;

b1sx = 10;
b1sy = 10;

alpha2 = -50 *pi/180;
alpha3 = -100 *pi/180;
alpha4 = -100 *pi/180;
alpha5 = -150 *pi/180;

theta2 = 50 *pi/180;
theta3 = 50 *pi/180;
theta4 = 50 *pi/180;
theta5 = 50 *pi/180;
```

FIGURE E.9.9 Input in bold type in the Appendix K.5 MATLAB file *motion_interior_point.m* for Example 9.3 (counterclockwise crank rotation constraints applied).

```
Iterations: 53
Func Evals: 141
Algorithm: interior-point
exit flag = 2
error = ( 2.235207e+00 )
a0x = ( -5.162407 )
a0y = ( 4.682209 )
a1x = ( 4.649614 )
a1y = ( 7.131542 )
a1sx = ( 4.885667 )
a1sy = ( 3.537139 )
b0x = ( 11.712962 )
b0y = ( -1.874513 )
b1x = ( 8.900739 )
b1y = ( 9.061422 )
b1sx = ( 9.579078 )
b1sy = ( 9.213761 )
alpha2 = ( -61.389008 )
alpha3 = ( -102.562426 )
alpha4 = ( -123.418626 )
alpha5 = ( -161.840575 )
theta2 = ( 18.802018 )
theta3 = ( 26.096293 )
theta4 = ( 51.170662 )
theta5 = ( 62.168229 )
Elapsed time is 2.697628 seconds.
Solution calculation is complete.
>>
```

FIGURE E.9.10 Output from the Appendix K.5 MATLAB file *motion_interior_point.m* for Example 9.3.

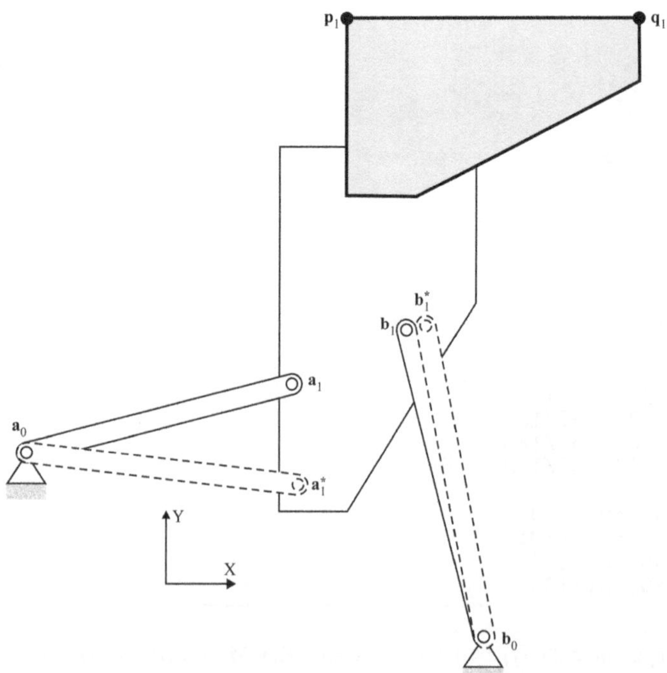

FIGURE E.9.11 Synthesized adjustable four-bar mechanism.

Example 9.4

Problem Statement: Using the Appendix K.6 MATLAB files, synthesize an adjustable planar four-bar mechanism for the coupler point coordinates given in Table E.9.2.

Known Information: Table E.9.2 and Appendix K.6 MATLAB files

Solution Approach: Figure E.9.12 includes the input in bold type in the Appendix K.6 MATLAB file *path_interior_point*. Figure E.9.13 includes the output from the same file. Figure E.9.14 illustrates the resulting adjustable planar four-bar mechanism. A kinematic analysis of this solution is presented in Example 9.12 in Section 9.7.

TABLE E.9.2
Coupler Point Coordinates for Adjustable Four-Bar Mechanism

Adjustment 1		Adjustment 2	
Precision Point	$\mathbf{p}_j$	Precision Point	$\mathbf{p}_k$
1	0, 0	1	0, 0
2	0.8, 0.1	5	1.1, 0.16
3	1.6, 0.7	6	2.3, 0.8
4	2, 1.5	7	2.7, 1.7

```
M = 4;
N = 4;

P1 = [0, 0];
P2 = [0.8, 0.1];
P3 = [1.6, 0.7];
P4 = [2, 1.5];
P5 = [1.1, 0.16];
P6 = [2.3, 0.8];
P7 = [2.7, 1.7];

a0x = 0;
a0y = 0;

a1x = 0;
a1y = -10;

a1sx = 0;
a1sy = -10;

b0x = 10;
b0y = 0;

b1x = 10;
b1y = -10;

b1sx = 10;
b1sy = -10;

alpha2 = 10 *pi/180;
alpha3 = 10 *pi/180;
alpha4 = 10 *pi/180;
alpha5 = 10 *pi/180;
alpha6 = 10 *pi/180;
alpha7 = 10 *pi/180;

theta2 = 10 *pi/180;
theta3 = 20 *pi/180;
theta4 = 30 *pi/180;
theta5 = 10 *pi/180;
theta6 = 20 *pi/180;
theta7 = 30 *pi/180;
```

FIGURE E.9.12 Input in bold type in the Appendix K.6 MATLAB file *path_interior_point* for Example 9.4 (counterclockwise crank rotation constraints applied).

```
Iterations: 19
Func Evals: 93
Algorithm: interior-point
exit flag = -2
error = ( 1.925454e-02 )
a0x = ( -0.250142 )
a0y = ( -1.967835 )
a1x = ( -2.181462 )
a1y = ( -12.647853 )
a1sx = ( -3.484017 )
a1sy = ( -12.328292 )
b0x = ( 3.253501 )
b0y = ( 2.761649 )
b1x = ( 17.438341 )
b1y = ( -8.098895 )
b1sx = ( 14.569997 )
b1sy = ( -11.067933 )
alpha2 = ( -6.416771 )
alpha3 = ( -12.920069 )
alpha4 = ( -18.139471 )
alpha5 = ( -8.420310 )
alpha6 = ( -17.175449 )
alpha7 = ( -22.462367 )
theta2 = ( 16.768006 )
theta3 = ( 33.881639 )
theta4 = ( 48.344345 )
theta5 = ( 21.027814 )
theta6 = ( 42.046893 )
theta7 = ( 54.884131 )
Elapsed time is 6.341098 seconds.
Solution calculation is complete.
>>
```

FIGURE E.9.13 Output from the Appendix K.6 MATLAB file *path_interior_point* for Example 9.4.

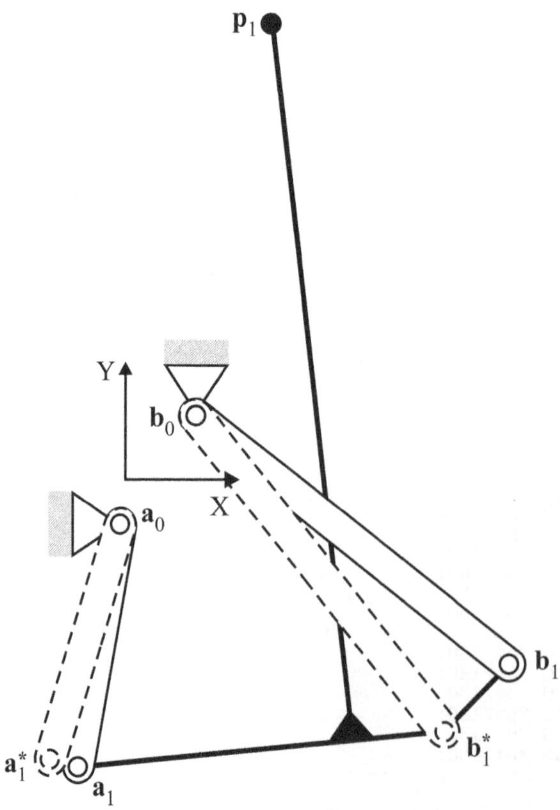

FIGURE E.9.14 Synthesized adjustable four-bar mechanism.

9.4 ADJUSTABLE PLANAR FOUR-BAR FUNCTION GENERATION

Equation (9.16) encompasses a planar four-bar function generation equation system introduced by Suh and Radcliffe to approximate four precision points [10]. This equation (a constant-length coupler link constraint) forms a four-equation nonlinear simultaneous equation system to calculate the components of moving pivots $\mathbf{a}_1$ and $\mathbf{b}_1$ (a_{1x}, a_{1y}, b_{1x} and b_{1y}). In this equation system, the fixed and moving pivots are $\mathbf{a}_0 = [0,0,1]^T$, $\mathbf{b}_0 = [1,0,1]^T$, $\mathbf{a}_1 = [a_{1x}, a_{1y}, 1]^T$ and $\mathbf{b}_1 = [b_{1x}, b_{1y}, 1]^T$. Matrix (9.17) defines the rotational displacements of the crank and follower links from position 1 to position j by displacement angles θ and ϕ (Figure 9.5).

$$(\mathbf{a}_j - \mathbf{b}_1)^T (\mathbf{a}_j - \mathbf{b}_1) = (\mathbf{a}_1 - \mathbf{b}_1)^T (\mathbf{a}_1 - \mathbf{b}_1) \tag{9.16}$$

where $\mathbf{a}_j = [D_{1j}]\mathbf{a}_1$ and

$$[D_{1j}] = \begin{bmatrix} \cos(\theta_{1j} - \phi_{1j}) & -\sin(\theta_{1j} - \phi_{1j}) & 1 - \cos\phi_{1j} \\ \sin(\theta_{1j} - \phi_{1j}) & \cos(\theta_{1j} - \phi_{1j}) & \sin\phi_{1j} \\ 0 & 0 & 1 \end{bmatrix} \quad j = 2,3,\ldots \tag{9.17}$$

To synthesize a planar four-bar mechanism having an adjustable follower link moving pivot (from $\mathbf{b}_1$ to $\mathbf{b}_1^*$ in Figure 9.6) to approximate two groups of three precision points (where one group includes angular displacements 1–2, 1–3, and 1–4 and the other group includes 1–5, 1–6, and 1–7), Equation (9.16) becomes

$$(\mathbf{a}_j - \mathbf{b}_1)^T (\mathbf{a}_j - \mathbf{b}_1) = (\mathbf{a}_1 - \mathbf{b}_1)^T (\mathbf{a}_1 - \mathbf{b}_1) \quad j = 2,3,4 \tag{9.18}$$

$$(\mathbf{a}_k - \mathbf{b}_1^*)^T (\mathbf{a}_k - \mathbf{b}_1^*) = (\mathbf{a}_1 - \mathbf{b}_1^*)^T (\mathbf{a}_1 - \mathbf{b}_1^*) \quad k = 5,6,7 \tag{9.19}$$

where $\mathbf{a}_j = [D_{1j}]\mathbf{a}_1$ and $\mathbf{a}_k = [D_{1k}]\mathbf{a}_1$. These equations form a six-equation nonlinear simultaneous equation system to calculate the components of moving pivots $\mathbf{a}_1$, $\mathbf{b}_1$, and $\mathbf{b}_1^*$ (a_{1x}, a_{1y}, b_{1x}, b_{1y}, b_{1x}^*, and b_{1y}^*).

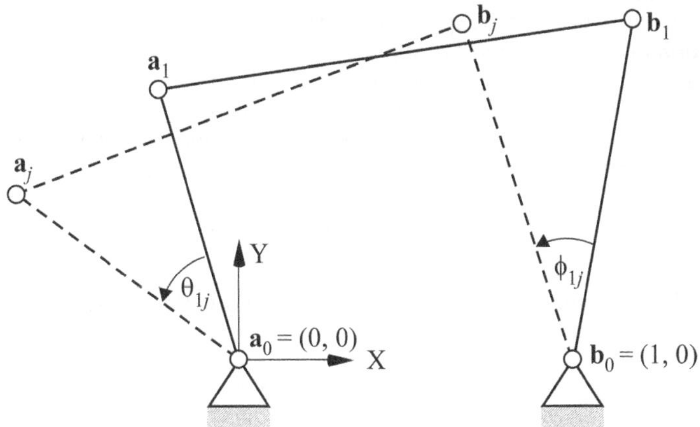

FIGURE 9.5 Planar four-bar function generator.

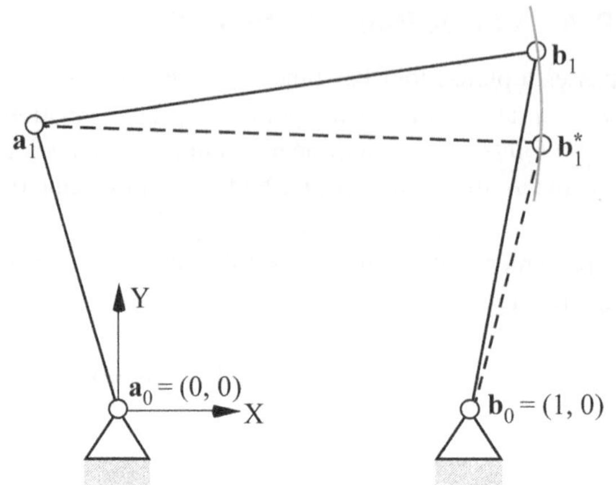

FIGURE 9.6 Adjustable planar four-bar function generator.

While the motion and path generation equation systems introduced by Suh and Radcliffe comprise crank and follower constant-length constraints, their function generation equations comprise constant-length coupler constraints. The equation system formed from Equations (9.18) and (9.19) does not include a coupler link length constraint between adjustments (giving the mechanism an adjustable effective coupler link length). To synthesize adjustable function generators having constant coupler link lengths between adjustments, one of the six simultaneous equations can be replaced with Equation (9.20) at a loss of one precision point.

$$(\mathbf{a}_1-\mathbf{b}_1)^T(\mathbf{a}_1-\mathbf{b}_1)=(\mathbf{a}_1-\mathbf{b}_1^*)^T(\mathbf{a}_1-\mathbf{b}_1^*) \tag{9.20}$$

Appendix K.7 includes MATLAB file user instructions for the approximate adjustable four-bar function generation equation set presented in this section. Although Equations (9.18) through (9.20) form a nonlinear simultaneous equation system for five precision points, in Appendix K.7, they are solved as an unconstrained nonlinear optimization model in MATLAB for an indefinite number of precision points ($j = 2,3,\ldots,M$ and $k = M+1, M+2,\ldots,N$ in Equations (9.18) and (9.19), respectively).

Example 9.5

Problem Statement: Using the Appendix K.7 MATLAB files, synthesize an adjustable planar four-bar sprinkler mechanism for the sprinkler precision points given in Table E.9.3.

Known Information: Table E.9.3 and Appendix K.7 MATLAB files

Solution Approach: Figure E.9.15 includes the input in bold type in the Appendix K.6 MATLAB file *func_quasi_newton.m*. Figure E.9.16 includes the output from the same file. Figure E.9.17 illustrates the resulting adjustable planar four-bar mechanism. A kinematic analysis of this solution is presented in Example 9.13 in Section 9.7.

TABLE E.9.3
Prescribed Displacements for Adjustable Planar Four-Bar Sprinkler Mechanism

Adjustment 1			Adjustment 2		
Precision Point	θ_j (deg)	ϕ_j (deg)	Precision Point	θ_j (deg)	ϕ_j (deg)
2	90	50	4	90	60
3	180	100	5	180	120

```
M = 3;
N = 3;

theta2 = 90 *pi/180;
theta3 = 180 *pi/180;
theta4 = 90 *pi/180;
theta5 = 180 *pi/180;

phi2 = 50 *pi/180;
phi3 = 100 *pi/180;
phi4 = 60 *pi/180;
phi5 = 120 *pi/180;

a1x = -1;
a1y = 1;

b1x = 1;
b1y = -1;

b1sx = 1;
b1sy = -1;
```

FIGURE E.9.15 Input in bold type in the Appendix K.7 MATLAB file *func_quasi_newton.m* for Example 9.5.

```
Iterations: 42
Func Evals: 44
Algorithm: quasi-newton
exit flag = 5
error = ( 7.509923e-32 )
a1 = [-0.262848,0.051665]
b1 = [0.686373,-0.154325]
b1s = [0.699519,-0.079874]
Elapsed time is 0.335178 seconds.
Solution calculation is complete.
>>
```

FIGURE E.9.16 Output from the Appendix K.7 MATLAB file *func_quasi_newton.m* for Example 9.5.

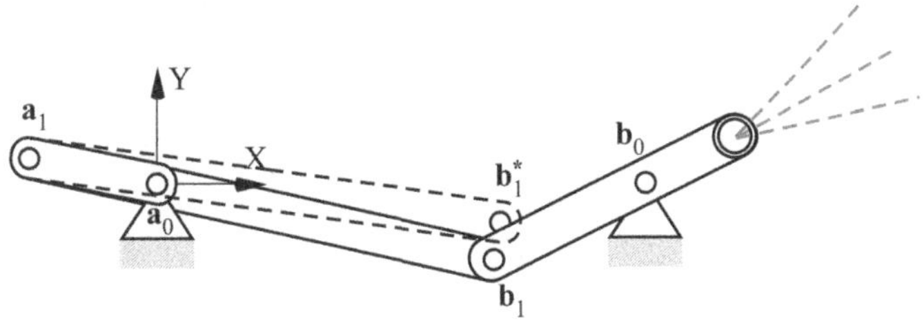

FIGURE E.9.17 Synthesized adjustable planar four-bar mechanism.

9.5 ADJUSTABLE 4R SPHERICAL MOTION GENERATION

A nonlinear equation system can be formulated from Equations (8.53) through (8.55) to synthesize an adjustable spherical R-R dyad to approximate two groups of precision positions (where one group includes positions 1–2–3 and the other group includes 1–4–5) using constant-length moving-pivot adjustments. To synthesize the dyads for an adjustable 4R Spherical motion generator, Equations (8.53) through (8.55) become

$$\left(\mathbf{a}_j - \mathbf{a}_0\right)^T \left(\mathbf{a}_j - \mathbf{a}_0\right) = \left(\mathbf{a}_1 - \mathbf{a}_0\right)^T \left(\mathbf{a}_1 - \mathbf{a}_0\right) \quad j = 2,3 \tag{9.21}$$

$$\left(\mathbf{a}_k^* - \mathbf{a}_0\right)^T \left(\mathbf{a}_k^* - \mathbf{a}_0\right) = \left(\mathbf{a}_1^* - \mathbf{a}_0\right)^T \left(\mathbf{a}_1^* - \mathbf{a}_0\right) \quad k = 4,5 \tag{9.22}$$

$$\left(\mathbf{a}_1 - \mathbf{a}_0\right)^T \left(\mathbf{a}_1 - \mathbf{a}_0\right) = \left(\mathbf{a}_1^* - \mathbf{a}_0\right)^T \left(\mathbf{a}_1^* - \mathbf{a}_0\right) \tag{9.23}$$

$$\left(\mathbf{a}_1\right)^T \left(\mathbf{a}_1\right) - 1 = 0 \tag{9.24}$$

$$\left(\mathbf{a}_0\right)^T \left(\mathbf{a}_0\right) - 1 = 0 \tag{9.25}$$

$$\left(\mathbf{a}_1^*\right)^T \left(\mathbf{a}_1^*\right) - 1 = 0 \tag{9.26}$$

where $\mathbf{a}_j = \left[D_{1j}\right]\mathbf{a}_1$ and $\mathbf{a}_k^* = \left[D_{1k}\right]\mathbf{a}_1^*$. Equations (9.21) through (9.26) form an eight-equation nonlinear simultaneous equation system to calculate eight of the nine spherical R-R dyad variables a_{0x}, a_{0y}, a_{0z}, a_{1x}, a_{1y}, a_{1z}, a_{1x}^*, a_{1y}^*, and a_{1z}^* with a free choice of one variable.

To synthesize an adjustable spherical R-R dyad to approximate two groups of precision positions (where one group includes positions 1–2–3 and the other group includes 1–4–5) having constant-length fixed-pivot adjustments, Equations (8.53) through (8.55) become

$$\left(\mathbf{a}_j - \mathbf{a}_0\right)^T \left(\mathbf{a}_j - \mathbf{a}_0\right) = \left(\mathbf{a}_1 - \mathbf{a}_0\right)^T \left(\mathbf{a}_1 - \mathbf{a}_0\right) \quad j = 2,3 \tag{9.27}$$

$$\left(\mathbf{a}_k - \mathbf{a}_0^*\right)^T \left(\mathbf{a}_k - \mathbf{a}_0^*\right) = \left(\mathbf{a}_1 - \mathbf{a}_0^*\right)^T \left(\mathbf{a}_1 - \mathbf{a}_0^*\right) \quad k = 4,5 \tag{9.28}$$

$$\left(\mathbf{a}_1 - \mathbf{a}_0\right)^T \left(\mathbf{a}_1 - \mathbf{a}_0\right) = \left(\mathbf{a}_1 - \mathbf{a}_0^*\right)^T \left(\mathbf{a}_1 - \mathbf{a}_0^*\right) \tag{9.29}$$

$$\left(\mathbf{a}_1\right)^T \left(\mathbf{a}_1\right) - 1 = 0 \tag{9.30}$$

$$\left(\mathbf{a}_0\right)^T \left(\mathbf{a}_0\right) - 1 = 0 \tag{9.31}$$

$$\left(\mathbf{a}_0^*\right)^T \left(\mathbf{a}_0^*\right) - 1 = 0 \tag{9.32}$$

where $\mathbf{a}_j = \left[D_{1j}\right]\mathbf{a}_1$ and $\mathbf{a}_k = \left[D_{1k}\right]\mathbf{a}_1$. Equations (9.27) through (9.32) form an eight-equation nonlinear simultaneous equation system to calculate eight of the nine spherical R-R dyad variables a_{0x}, a_{0z}, a_{1x}, a_{1y}, a_{1z}, a_{0x}^*, a_{0y}^*, and a_{0z}^* with a free choice of one variable.

Equations (9.23) and (9.29) ensure constant crank and follower link lengths between mechanism adjustments. To synthesize mechanisms with adjustable crank and follower link lengths, Equations (9.23) and (9.29) are removed, and another precision position is added to one of the two groups of precision positions (e.g., k=4, 5, 6 in Equations (9.22) and (9.28)).

Appendices K.8 through K.11 include MATLAB file user instructions for the adjustable 4R Spherical motion generation equations presented in this section. Although Equations (9.21) through (9.26) and Equations (9.27) through (9.32) form nonlinear simultaneous equation systems for two three-position synthesis problems, in Appendices K.8 through K.11, they are solved as constrained nonlinear optimization models in MATLAB for an indefinite number of precision positions.

Example 9.6

Problem Statement: Using the Appendix K.10 MATLAB files, synthesize an adjustable-length moving-pivot adjustable 4R Spherical mechanism for the two groups of solar panel position coordinates given in Table E.9.4 (group 1 includes positions 1–2–3, and group 2 includes positions 1–4–5–6).

Known Information: Table E.9.4 and Appendix K.10 MATLAB files

Solution Approach: In this example, adjustable planar four-bar dyad variables a_{0x} and b_{0x} were specified to $a_{0x} = 0$ and $b_{0x} = -0.5$. Figure E.9.18 includes the input in bold type in the Appendix

TABLE E.9.4
Solar Panel Position Coordinates

Precision Position	p	q	r
1	1.5, 0, 0	2.5, 0, 0.0175	1.5, 1, 0
2	1.2990, 0.75, 0	2.1651, 1.25, 0	0.8660, 1.5, 0.5
3	0.75, 1.2990, 0	1.25, 2.1651, 0	0.3170, 1.5490, 0.8660
4	1.2990, 0.6797, –0.3170	2.0235, 1.3481, –0.4858	0.6566, 1.4231, –0.1307
5	0.7500, 1.1773, –0.5490	1.0202, 2.1391, –0.5936	0.1462, 1.3827, 0.2213
6	0, 1.3595, –0.6339	0, 2.3595, –0.6339	0, 1.3595, 0.3661

```
M = 3;
N = 4;

P1 = [1.5, 0, 0];
P2 = [1.299, 0.75, 0];
P3 = [0.75, 1.299, 0];
P4 = [1.299, 0.6797, -0.317];
P5 = [0.75, 1.1773, -0.549];
P6 = [0, 1.3595, -0.6339];

Q1 = [2.5, 0, 0.0175];
Q2 = [2.1651, 1.25, 0];
Q3 = [1.25, 2.1651, 0];
Q4 = [2.0235, 1.3481, -0.4858];
Q5 = [1.0202, 2.1391, -0.5936];
Q6 = [0, 2.3595, -0.6339];

R1 = [1.5, 1, 0];
R2 = [0.8660, 1.5, 0.5];
R3 = [0.3170, 1.549, 0.866];
R4 = [0.6566, 1.4231, -0.1307];
R5 = [0.1462, 1.3827, 0.2213];
R6 = [0, 1.3595, 0.3661];

a0x = 0;

a0y = 0;
a0z = 1;
a1x = 1;
a1y = 1;
a1z = 1;
a1sx = 1;
a1sy = 1;
a1sz = 1;
```

FIGURE E.9.18 Input in bold type in the Appendix K.10 MATLAB file *a0_a1_a1s_interior_point.m* (with $a_{0x} = 0$) for Example 9.6.

K.10 MATLAB file *a0_a1_a1s_interior_point.m* when $a_{0x} = 0$. Figure E.9.19 includes the output from the same file. Figure E.9.20 includes the output from the Appendix K.10 MATLAB file *a0_a1_a1s_interior_point.m* when $b_{0x} = a_{0x} = -0.5$. Figure E.9.21 illustrates the resulting adjustable 4R Spherical mechanism. The equation systems formulated from Equations (9.21) through (9.32) include no constraints to ensure that branch- and order defect-free solutions are calculated. With these equation systems, branch- and order defect-free solutions can be calculated through trial and error.

```
Iterations: 22
Func Evals: 37
Algorithm: interior-point
exit flag = 2
error = ( 8.025862e-10 )
a0y = ( 0.786393 )
a0z = ( 0.617727 )
a1x = ( 0.293932 )
a1y = ( 0.518188 )
a1z = ( 0.803172 )
a1sx = ( 0.351886 )
a1sy = ( 0.935868 )
a1sz = ( 0.018120 )
Elapsed time is 23.184514 seconds.
The solution is complete.
>>
```

FIGURE E.919 Output from the Appendix K.10 MATLAB file *a0_a1_a1s_interior_point.m* (with $a_{0x} = 0$) for Example 9.6.

```
Iterations: 17
Func Evals: 32
Algorithm: interior-point
exit flag = 1
error = ( 1.985087e-10 )
a0y = ( 0.142019 )
a0z = ( 0.854302 )
a1x = ( -0.636907 )
a1y = ( 0.616890 )
a1z = ( 0.462390 )
a1sx = ( 0.883578 )
a1sy = ( -0.454920 )
a1sz = ( 0.111120 )
Elapsed time is 23.561028 seconds.
The solution is complete.
>>
```

FIGURE E.9.20 Output from the Appendix K.10 MATLAB file *a0_a1_a1s_interior_point.m* (with $b_{0x} = a_{0x} = -0.5$) for Example 9.6.

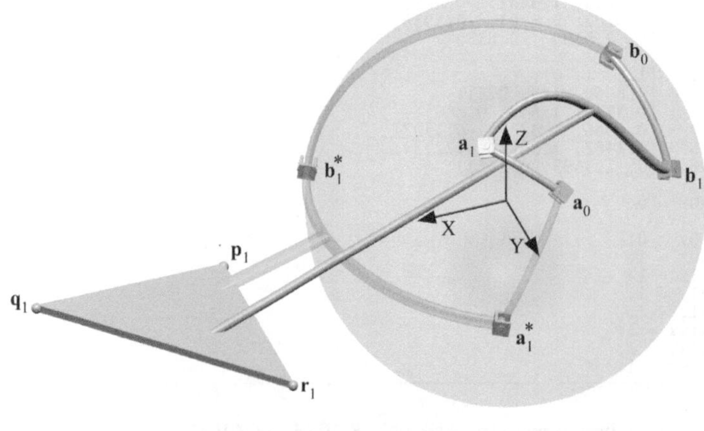

FIGURE E.9.21 Synthesized adjustable 4R Spherical mechanism.

Example 9.7

Problem Statement: Using the Appendix K.11 MATLAB files, synthesize an adjustable-length fixed-pivot adjustable 4R Spherical mechanism for Example 9.6.

Known Information: Table E 9.4 and Appendix K.11 MATLAB files

Solution Approach: In this example, adjustable planar four-bar dyad variables a_{0x} and b_{0x} were specified to $a_{0x} = 0.5$ and $b_{0x} = 0.6$, respectively. Figure E.9.22 includes the input in bold type in the Appendix K.11 MATLAB file *a0_a1_a1s_interior_point.m* when $a_{0x} = 0.5$. Figure E.9.23

```
M = 3;
N = 4;

P1 = [1.5, 0, 0];
P2 = [1.299, 0.75, 0];
P3 = [0.75, 1.299, 0];
P4 = [1.299, 0.6797, -0.317];
P5 = [0.75, 1.1773, -0.549];
P6 = [0, 1.3595, -0.6339];

Q1 = [2.5, 0, 0.0175];
Q2 = [2.1651, 1.25, 0];
Q3 = [1.25, 2.1651, 0];
Q4 = [2.0235, 1.3481, -0.4858];
Q5 = [1.0202, 2.1391, -0.5936];
Q6 = [0, 2.3595, -0.6339];

R1 = [1.5, 1, 0];
R2 = [0.8660, 1.5, 0.5];
R3 = [0.3170, 1.549, 0.866];
R4 = [0.6566, 1.4231, -0.1307];
R5 = [0.1462, 1.3827, 0.2213];
R6 = [0, 1.3595, 0.3661];

a0x = 0.5;

a0y = 0;
a0z = 1;
a1x = 1;
a1y = 1;
a1z = 1;
a0sx = 0;
a0sy = 0;
a0sz = 1;
```

FIGURE E.9.22 Input in bold type in the Appendix K.11 MATLAB file *a0_a1_a1s_interior_point.m* (with $a_{0x} = 0.5$) for Example 9.7.

```
Iterations: 16
Func Evals: 39
Algorithm: interior-point
exit flag = 2
error = ( 9.383887e-04 )
a0y = ( 0.197802 )
a0z = ( 0.843134 )
a1x = ( 0.925255 )
a1y = ( 0.370271 )
a1z = ( 0.082473 )
a0sx = ( -0.281027 )
a0sy = ( 0.512419 )
a0sz = ( 0.811450 )
Elapsed time is 25.346185 seconds.
Solution calculation is complete.
>>
```

FIGURE E.9.23 Output from the Appendix K.11 MATLAB file *a0_a1_a1s_interior_point.m* (with $a_{0x} = 0.5$) for Example 9.7.

includes the output from the same file. Figure E.9.24 includes the output from the Appendix K.11 MATLAB file *a0_a1_a1s_interior_point.m* when $b_{0x} = a_{0x} = 0.6$. Figure E.9.25 illustrates the resulting adjustable 4R Spherical mechanism. A kinematic analysis of this solution is presented in Example 9.14 in Section 9.7. The equation systems formulated from Equations (9.21) through (9.32) include no constraints to ensure that branch- and order defect-free solutions are calculated. With these equation systems, branch- and order defect-free solutions can be calculated through trial and error.

```
Iterations: 15
Func Evals: 24
Algorithm: interior-point
exit flag = 2
error = ( 2.241022e-03 )
a0y = ( 0.312932 )
a0z = ( 0.736256 )
a1x = ( 0.899331 )
a1y = ( 0.430077 )
a1z = ( 0.078974 )
a0sx = ( -0.265025 )
a0sy = ( 0.532580 )
a0sz = ( 0.803816 )
Elapsed time is 27.616105 seconds.
Solution calculation is complete.
>>
```

FIGURE E.9.24 Output from the Appendix K.11 MATLAB file *a0_a1_a1s_interior_point.m* (with $b_{0x} = a_{0x} = 0.6$) for Example 9.7.

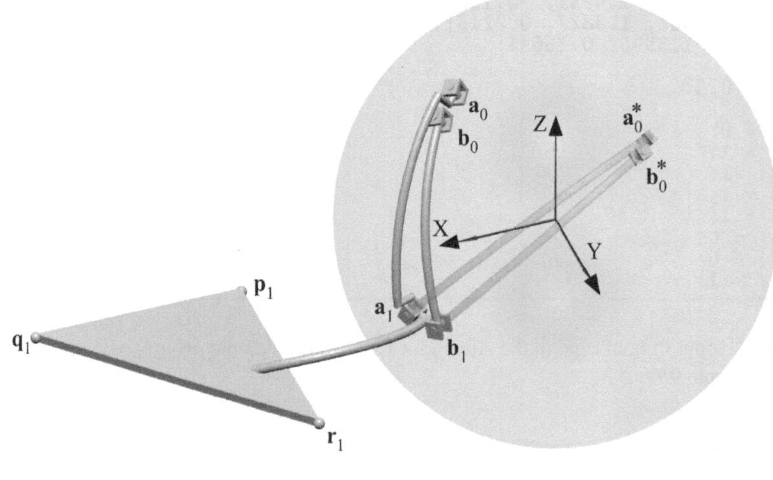

FIGURE E.9.25 Synthesized adjustable 4R Spherical mechanism.

9.6 ADJUSTABLE 4R SPHERICAL MOTION AND PATH GENERATION: CONSTRAINED OPTIMIZATION

A constrained nonlinear optimization model similar in form to the model presented in Section 8.4 can be formulated for adjustable 4R Spherical motion or path generation. To synthesize adjustable 4R Spherical motion generators to approximate an indefinite number of precision positions (two groups of positions where $j = 2,3,\ldots,M$ in one group and $k = M+1, M+2,\ldots,N$ in the other group), the objective function in Equation (9.33) is formed.

$$f(\mathbf{X}) = \sum_{j=2}^{M}\left\{\left\|\mathbf{p}_j-\mathbf{p}\right\|^2+\left\|\mathbf{q}_j-\mathbf{q}\right\|^2+\left\|\mathbf{r}_j-\mathbf{r}\right\|^2\right\}+\sum_{k=M+1}^{N}\left\{\left\|\mathbf{p}_k^*-\mathbf{p}^*\right\|^2+\left\|\mathbf{q}_k^*-\mathbf{q}^*\right\|^2+\left\|\mathbf{r}_k^*-\mathbf{r}^*\right\|^2\right\} \quad (9.33)$$

TABLE 9.2
Optimization Model: Constant-Length, Moving-Pivot Adjustable 4R Spherical Motion and Path Generation

Equation Type	Equation Number
Objective function	9.33[a]
Equality constraint	8.67–8.70, 9.8–9.13[b], 9.35, and 9.36
Inequality constraint	8.65, 8.66, 9.14, and 9.34

[a] For the synthesis of constant-length moving-pivot adjustable 4R Spherical path generators, only the squared norms $\|\mathbf{p}_j-\mathbf{p}\|^2$ and $\|\mathbf{p}_k^*-\mathbf{p}^*\|^2$ are kept in Equation (9.33).

[b] Equations (9.8) through (9.13) should appear in their spatial forms (including fixed- and moving-pivot x-, y-, and z-components).

where $\mathbf{X}=\left(\mathbf{a}_0,\mathbf{a}_1,\mathbf{a}_1^*,\mathbf{b}_0,\mathbf{b}_1,\mathbf{b}_1^*,\theta_2\ldots\theta_M,\theta_{M+1}\ldots\theta_N,\alpha_2\ldots\alpha_M,\alpha_{M+1}\ldots\alpha_N\right)$ for moving-pivot adjustments. In Equation (9.33), $\mathbf{p}_j$, $\mathbf{q}_j$, and $\mathbf{r}_j$ are the precision position variables in the first group, while $\mathbf{p}$, $\mathbf{q}$, and $\mathbf{r}$ are the coupler points achieved by the synthesized mechanism (using Equation (8.13)). In Equation (9.33), $\mathbf{p}_k^*$, $\mathbf{q}_k^*$, and $\mathbf{r}_k^*$ are the precision position variables in the second group, while $\mathbf{p}^*$, $\mathbf{q}^*$, and $\mathbf{r}^*$ are the coupler points achieved by the synthesized mechanism (also using Equation (8.13)). Angles θ_j and α_j are the crank and coupler link displacement angles for the mechanism configuration associated with the first group of precision positions, respectively, while angles θ_k and α_k are for the mechanism configuration associated with the second group of precision positions.

Along with Equation (9.33) as an objective function, the spatial forms of Equations (9.8) through (9.13) are included (for precision positions $2\ldots M$ and $M+1\ldots N$) as constraints for constant link length moving-pivot adjustments. Inequality (8.64), Inequality (8.65), and Equations (8.66) through (8.69) are also included for order defect elimination, branch defect elimination, and unit vector constraints (for precision positions $2\ldots M$). Lastly, Inequality (9.14), Inequality (9.34), Equation (9.35), and Equation (9.36) are also included for order defect elimination, branch defect elimination, and unit vector constraints (for precision positions $M+1\ldots N$).

$$\left[\left(\mathbf{b}^*-\mathbf{b}_0\right)\times\left(\mathbf{a}^*-\mathbf{b}_0\right)\right]\cdot\left[\left(\mathbf{b}_1^*-\mathbf{b}_0\right)\times\left(\mathbf{a}_1^*-\mathbf{b}_0\right)\right]>0 \tag{9.34}$$

$$\left(\mathbf{a}_1^*\right)^T\left(\mathbf{a}_1^*\right)-1=0 \tag{9.35}$$

$$\left(\mathbf{b}_1^*\right)^T\left(\mathbf{b}_1^*\right)-1=0 \tag{9.36}$$

Table 9.2 includes the constrained nonlinear optimization model for constant-length, moving-pivot adjustable 4R Spherical motion and path generation. Appendices K.12 and K.13 include MATLAB file user instructions for the adjustable 4R Spherical motion and path generation optimization models presented in this section, respectively.

Example 9.8

Problem Statement: Using the Appendix K.12 MATLAB files, synthesize an adjustable-length moving-pivot adjustable 4R Spherical mechanism for Example 9.6.

Known Information: Table E 9.4 and Appendix K.12 MATLAB files

Solution Approach: Figure E.9.26 includes the input in bold type in the Appendix K.12 MATLAB file *motion_interior_point*. Figure E.9.27 includes the output from the same file. Figure E.9.28 illustrates the resulting adjustable planar four-bar mechanism. A kinematic analysis of this solution is presented in Example 9.15 in Section 9.7.

```
M = 3; N = 4;

P1 = [1.5, 0, 0];
P2 = [1.299, 0.75, 0];
P3 = [0.75, 1.299, 0];
P4 = [1.299, 0.6797, -0.317];
P5 = [0.75, 1.1773, -0.549];
P6 = [0, 1.3595, -0.6339];

Q1 = [2.5, 0, 0.0175];
Q2 = [2.1651, 1.25, 0];
Q3 = [1.25, 2.1651, 0];
Q4 = [2.0235, 1.3481, -0.4858];
Q5 = [1.0202, 2.1391, -0.5936];
Q6 = [0, 2.3595, -0.6339];

R1 = [1.5, 1, 0];
R2 = [0.8660, 1.5, 0.5];
R3 = [0.3170, 1.549, 0.866];
R4 = [0.6566, 1.4231, -0.1307];
R5 = [0.1462, 1.3827, 0.2213];
R6 = [0, 1.3595, 0.3661];

a0x = 1; a0y = 1; a0z = 1;

a1x = 1; a1y = 1; a1z = 1;

a1sx = 1; a1sy = 1; a1sz = 1;

b0x = 1; b0y = 1; b0z = 1;

b1x = 1; b1y = 1; b1z = 1;

b1sx = 1; b1sy = 1; b1sz = 1;

alpha2 = 10 *pi/180;
alpha3 = 20 *pi/180;
alpha4 = 20 *pi/180;
alpha5 = 30 *pi/180;
alpha6 = 40 *pi/180;

theta2 = 10 *pi/180;
theta3 = 20 *pi/180;
theta4 = 20 *pi/180;
theta5 = 30 *pi/180;
theta6 = 40 *pi/180;
```

FIGURE E.9.26 Input in bold type in the Appendix K.12 MATLAB file *motion_interior_point.m* for Example 9.8 (counterclockwise crank rotation constraints applied).

Example 9.9

Problem Statement: Using the Appendix K.13 MATLAB files, synthesize a fixed-length moving-pivot adjustable 4R Spherical mechanism for the two groups of coupler point coordinates given in Table E 9.5.

Known Information: Table E.9.5 and Appendix F.13 MATLAB files

Solution Approach: Figure E.9.29 includes the input in bold type in the Appendix K.13 MATLAB file *path_interior_point*. Figure E.9.30 includes the output from the same file. Figure E.9.31 illustrates the resulting adjustable 4R Spherical mechanism.

```
Iterations: 48
Func Evals: 165
Algorithm: interior-point
exit flag = 2
error = ( 2.639398e-01 )
a0 = [0.317154, 0.491783, 0.810914]
a1 = [0.189719, -0.913223, -0.360612]
a1s = [0.451159,-0.623259,-0.638722]
b0 = [0.391604, -0.861513, 0.323032]
b1 = [-0.676530, -0.455148, -0.578933]
b1s = [0.403472,0.531945,0.744480]
alpha2 = ( 26.470941 )
alpha3 = ( 41.028958 )
alpha4 = ( 5.109626 )
alpha5 = ( 8.481845 )
alpha6 = ( 53.661968 )
theta2 = ( 55.185815 )
theta3 = ( 107.707098 )
theta4 = ( 41.170378 )
theta5 = ( 75.559060 )
theta6 = ( 143.463885 )
Elapsed time is 1993.823152 seconds.
Solution calculation is complete.
>>
```

FIGURE E.9.27 Output from the Appendix K.12 MATLAB file *motion_interior_point.m* for Example 9.8.

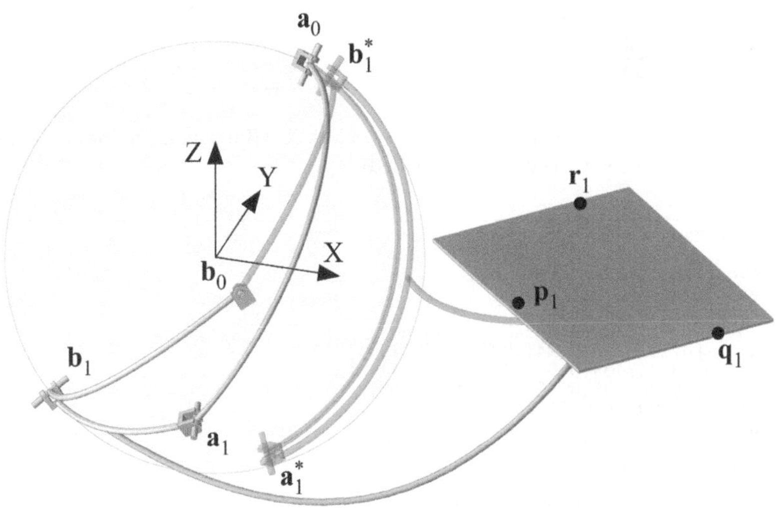

FIGURE E.9.28 Synthesized adjustable 4R Spherical mechanism.

TABLE E.9.5
Coupler Point Coordinates for Adjustable Four-Bar Mechanism

Adjustment 1		Adjustment 2	
Precision Point	$\mathbf{p}_j$	**Precision Point**	$\mathbf{p}_k$
1	1.5, 0, 0	1	1.5, 0, 0
2	1.299, 0.75, 0	4	1.299, 0.6797, −0.317
3	0.75, 1.299, 0	5	0.75, 1.1773, −0.549

```
M = 3;
N = 3;

P1 = [1.5, 0, 0];
P2 = [1.299, 0.75, 0];
P3 = [0.75, 1.299, 0];
P4 = [1.299, 0.6797, -0.317];
P5 = [0.75, 1.1773, -0.549];

a0x = 1;
a0y = 1;
a0z = 1;

a1x = 1;
a1y = 1;
a1z = 1;

a1sx = 1;
a1sy = 1;
a1sz = 1;

b0x = 1;
b0y = 1;
b0z = 1;

b1x = 1;
b1y = 1;
b1z = 1;

b1sx = 1;
b1sy = 1;
b1sz = 1;

alpha2 = 20 *pi/180;
alpha3 = 20 *pi/180;
alpha4 = 20 *pi/180;
alpha5 = 20 *pi/180;

theta2 = 20 *pi/180;
theta3 = 20 *pi/180;
theta4 = 20 *pi/180;
theta5 = 30 *pi/180;
```

FIGURE E.9.29 Input in bold type in the Appendix K.13 MATLAB file *path_interior_point.m* for Example 9.9 (counterclockwise crank rotation constraints applied).

```
Iterations: 41
Func Evals: 142
Algorithm: interior-point
exit flag = 1
error = ( 8.680924e-06 )
a0 = [0.021610,-0.003483,0.999760]
a1 = [0.132922,-0.941592,-0.309413]
a1s = [0.942403,0.085639,-0.323331]
b0 = [-0.723507,0.662031,-0.195580]
b1 = [0.009889,0.017730,0.999794]
b1s = [-0.528999,-0.845672,0.070710]
alpha2 = ( -0.289619 )
alpha3 = ( -0.817327 )
alpha4 = ( 57.547350 )
alpha5 = ( 87.966899 )
theta2 = ( 29.898714 )
theta3 = ( 59.803114 )
theta4 = ( 41.391016 )
theta5 = ( 72.384694 )
Elapsed time is 55.712033 seconds.
Solution calculation is complete.
>>
```

FIGURE E.9.30 Output from the Appendix K.13 MATLAB file *path_interior_point.m* for Example 9.9.

9.7 FROM ADJUSTABLE MECHANISM DIMENSIONAL SYNTHESIS TO KINEMATIC ANALYSIS

As noted in prior chapters throughout this book, Appendices B, C, H, and I include user instructions for the planar and spatial mechanisms in this book in MATLAB and Simscape Multibody. In this section, the Appendix H files are used to conduct kinematic displacement analyses on the planar four-bar and 4R Spherical mechanism solutions calculated in this chapter.

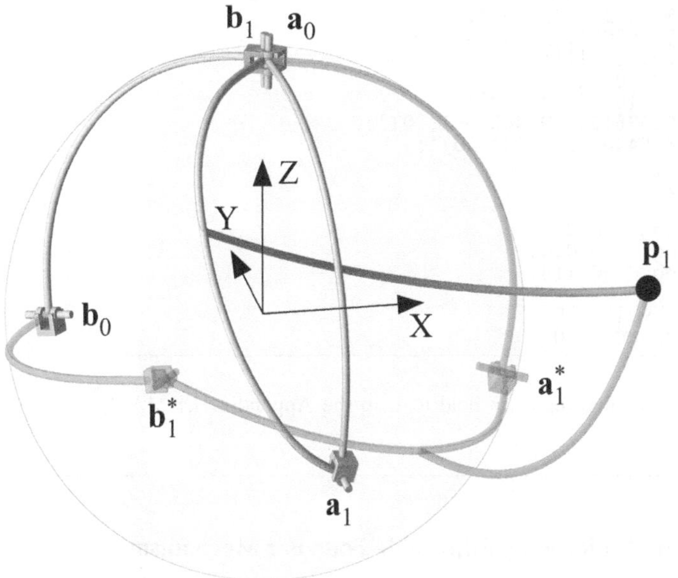

FIGURE E.9.31 Synthesized adjustable 4R Spherical mechanism.

Example 9.10

Problem Statement: Using the Appendix H.1 MATLAB files, calculate the coupler positions achieved by the planar four-bar mechanism solution in Example 9.1. Also confirm that this solution is order- and branch defect-free.

Known Information: Example 9.1 solution and Appendix H.1 MATLAB file.

Solution Approach: Figure E.9.32 includes the input specified (in bold text) in the Appendix H.1 MATLAB file for mechanism adjustment 1. Figure E.9.33 includes the input specified (in bold text) in the Appendix H.1 MATLAB file for mechanism adjustment 2. Table E.9.6 includes the coupler positions achieved by the adjustable mechanism.[2] In this example, the calculated results for $\mathbf{a}_0$ and $\mathbf{a}_1$ are required for the driver link in the first mechanism adjustment, and the calculated results for $\mathbf{b}_0$ and $\mathbf{b}_1^*$ are required for the driver link in the second mechanism adjustment. As indicated by the continually ascending order of the crank angle solutions in Table E.9.6, the order of the achieved coupler positions, and the accuracy of these positions, the adjustable planar four-bar motion generator solution is order- and branch defect-free.

```
a0 = [0, 16.280226, 0]';
a1 = [8.482601, 9.273511, 0]';
ua0 = [0, 0, 1]';
ua1 = [0, 0, 1]';
b0 = [5, 4.237433, 0]';
b1 = [6.314356, 8.595783, 0]';
p1 = [6.5450, 20.1547, 0]';
q1 = p1;
r1 = p1;

start_ang   = 0;
step_ang    = -0.01;
stop_ang    = -20;

angular_vel  = 0;
angular_acc  = 0;
```

FIGURE E.9.32 Specified input (in bold text) in the Appendix H.1 MATLAB file for Example 9.10 for Mechanism Adjustment 1.

```
a0 = [5, 4.237433, 0]';
a1 = [9.398057, 5.412097, 0]';
ua0 = [0, 0, 1]';
ua1 = [0, 0, 1]';
b0 = [0, 16.280226, 0]';
b1 = [8.996460, 9.9467991, 0]';
p1 = [6.5450, 20.1547, 0]';
q1 = p1;
r1 = p1;

start_ang   = 0;
step_ang    = 0.01;
stop_ang    = 114;

angular_vel  = 0;
angular_acc  = 0;
```

FIGURE E.9.33 Specified input (in bold text) in the Appendix H.1 MATLAB file for Example 9.10 for Mechanism Adjustment 2.

TABLE E.9.6
Coupler Positions Achieved by Adjustable Four-Bar Mechanism

Adjustment 1				Adjustment 2			
Precision Position	θ_j	$\mathbf{p}_j$	α_j	**Precision Position**	θ_k	$\mathbf{p}_k$	α_k
1		6.5450, 20.1547		1		6.5450, 20.1547	
2	−9.44°	12.4400, 17.7180	−38.2871°	4	74.66°	18.4720, 15.4830	−74.4792°
3	−19.48°	15.6260, 11.6260	−74.4675°	5	113.19°	17.1850, 8.5774	−98.0528°

Example 9.11

Problem Statement: Using the Appendix H.1 MATLAB files, calculate the coupler positions achieved by the planar four-bar mechanism solution in Example 9.3. Also confirm that this solution is order- and branch defect-free.

Known Information: Example 9.3 solution and Appendix H.1 MATLAB file.

Solution Approach: Figure E.9.34 includes the input specified (in bold text) in the Appendix H.1 MATLAB file for mechanism adjustment 1. Figure E.9.35 includes the input specified (in bold text) in the Appendix H.1 MATLAB file for mechanism adjustment 2. Table E.9.7 includes the coupler positions achieved by the adjustable mechanism. As indicated by the continually ascending order of the crank angle solutions in Table E.9.7, the order of the achieved coupler positions, and the accuracy of these positions, the adjustable planar four-bar motion generator solution is order- and branch defect-free.

```
a0 = [-5.162407, 4.682209, 0]';
a1 = [4.649614, 7.131542, 0]';
ua0 = [0, 0, 1]';
ua1 = [0, 0, 1]';
b0 = [11.712962, -1.874513, 0]';
b1 = [8.900739, 9.061422, 0]';
p1 = [6.5450, 20.1547, 0]';
q1 = [17.3726, 20.1547, 0]';
r1 = q1;

start_ang   = 0;
step_ang    = 0.01;
stop_ang    = 26.1;

angular_vel  = 0;
angular_acc  = 0;
```

FIGURE E.9.34 Specified input (in bold text) in the Appendix H.1 MATLAB file for Example 9.11 for Mechanism Adjustment 1.

```
a0 = [-5.162407, 4.682209, 0]';
a1 = [4.885667, 3.537139, 0]';
ua0 = [0, 0, 1]';
ua1 = [0, 0, 1]';
b0 = [11.712962, -1.874513, 0]';
b1 = [9.579078, 9.213761, 0]';
p1 = [6.5450, 20.1547, 0]';
q1 = [17.3726, 20.1547, 0]';
r1 = q1;

start_ang   = 0;
step_ang    = 0.01;
stop_ang    = 62.17;

angular_vel  = 0;
angular_acc  = 0;
```

FIGURE E.9.35 Specified input (in bold text) in the Appendix H.1 MATLAB file for Example 9.11 for Mechanism Adjustment 2.

TABLE E.9.7
Coupler Positions Achieved by Adjustable Four-Bar Mechanism

Adjustment 1				Adjustment 2			
Precision Position	θ_j	$\mathbf{p}_j$	$\mathbf{q}_j$	Precision Position	θ_k	$\mathbf{p}_k$	$\mathbf{q}_k$
1	0°	6.5450, 20.1550	17.3730, 20.1550	1	0°	6.5450, 20.1550	17.3730, 20.1550
2	18.8°	13.5440, 18.4700	21.5170, 11.1430	4	52.17°	18.3230, 14.9740	21.3530, 4.5785
3	26.1°	15.6790, 12.3780	18.1930, 1.8465	5	62.17°	16.6420, 8.6001	14.8190, –2.0729

Example 9.12

Problem Statement: Using the Appendix H.1 MATLAB files, calculate the coupler points achieved by the planar four-bar mechanism solution in Example 9.4. Also confirm that this solution is order- and branch defect-free.

Known Information: Example 9.4 solution and Appendix H.1 MATLAB file.

Solution Approach: Figure E.9.36 includes the input specified (in bold text) in the Appendix H.1 MATLAB file for mechanism adjustment 1. Figure E.9.37 includes the input specified (in bold text) in the Appendix H.1 MATLAB file for mechanism adjustment 2. Table E.9.8 includes the coupler points achieved by the adjustable mechanism. As indicated by the continually ascending

```
a0 = [-0.250142, -1.967835, 0]';
a1 = [-2.181462, -12.647853, 0]';
ua0 = [0, 0, 1]';
ua1 = [0, 0, 1]';
b0 = [3.253501, 2.761649, 0]';
b1 = [17.438341, -8.098895, 0]';
p1 = [0, 0, 0]';
q1 = p1;
r1 = p1;

start_ang   = 0;
step_ang    = 0.01;
stop_ang    = 48.34;

angular_vel  = 0;
angular_acc  = 0;
```

FIGURE E.9.36 Specified input (in bold text) in the Appendix H.1 MATLAB file for Example 9.12 for Mechanism Adjustment 1.

```
a0 = [-0.250142, -1.967835, 0]';
a1 = [-3.484017, -12.328292, 0]';
ua0 = [0, 0, 1]';
ua1 = [0, 0, 1]';
b0 = [3.253501, 2.761649, 0]';
b1 = [14.569997, -11.067933, 0]';
p1 = [0, 0, 0]';
q1 = p1;
r1 = p1;

start_ang   = 0;
step_ang    = 0.01;
stop_ang    = 54.88;

angular_vel  = 0;
angular_acc  = 0;
```

FIGURE E.9.37 Specified input (in bold text) in the Appendix H.1 MATLAB file for Example 9.12 for Mechanism Adjustment 2.

TABLE E.9.8
Coupler Points Achieved by Adjustable Four-Bar Mechanism

Adjustment 1			Adjustment 2		
Precision Point	θ_j	$\mathbf{p}_j$	**Precision Point**	θ_k	$\mathbf{p}_k$
1	0°	0, 0	1	0°	0, 0
2	16.77°	0.8533, 0.0831	5	21.03°	1.1579, −0.0072
3	33.88°	1.6097, 0.6796	6	42.05°	2.2639, 0.8232
4	48.34°	1.9656, 1.5177	7	54.88°	2.7009, 1.7031

order of the crank angle solutions in Table E.9.8, the order of the achieved coupler points, and the accuracy of these points, the adjustable planar four-bar path generator solution is order- and branch defect-free.

Example 9.13

Problem Statement: Using the Appendix H.2 MATLAB files, calculate the crank and follower displacement angles achieved by the planar four-bar mechanism solution given in Example 9.5. Also confirm that this solution is order- and branch defect-free.

Known Information: Example 9.5 solution and Appendix H.2 MATLAB file.

Solution Approach: Figure E.9.38 includes the input specified (in bold text) in the Appendix H.2 MATLAB file for mechanism adjustment 1. Figure E.9.39 includes the input specified (in bold

```
a0 = [0, 0, 0]';
a1 = [-0.262848, 0.051665, 0]';
ua0 = [0, 0, 1]';
b0 = [1, 0, 0]';
b1 = [0.686373, -0.154325, 0]';
ub0 = [0, 0, 1]';'

start_ang   = 0;
step_ang    = 1;
stop_ang    = 180;

angular_vel  = 0;
angular_acc  = 0;
```

FIGURE E.9.38 Specified input (in bold text) in the Appendix H.2 MATLAB file for Example 9.13 for Mechanism Adjustment 1.

```
a0 = [0, 0, 0]';
a1 = [-0.262848, 0.051665, 0]';
ua0 = [0, 0, 1]';
b0 = [1, 0, 0]';
b1 = [0.699519, -0.079874, 0]';
ub0 = [0, 0, 1]';

start_ang   = 0;
step_ang    = 1;
stop_ang    = 180;

angular_vel  = 0;
angular_acc  = 0;
```

FIGURE E.9.39 Specified input (in bold text) in the Appendix H.2 MATLAB file for Example 9.13 for Mechanism Adjustment 2.

text) in the Appendix H.2 MATLAB file for mechanism adjustment 2. Table E.9.9 includes the crank and follower displacement angles achieved by the adjustable mechanism. As indicated by the continually ascending order of the crank angle solutions in Table E.9.9, the order of the achieved displacement angles, and the accuracy of these points, the adjustable planar four-bar path generator solution is order- and branch defect-free.

TABLE E.9.9
Crank and Follower Displacement Angles Achieved by Adjustable Four-Bar Mechanism

Adjustment 1			Adjustment 2		
Precision Point	θ_j	ϕ_j	Precision Point	θ_k	ϕ_k
1	0°	0°	1	0°	0°
2	90°	50°	4	90°	60°
3	180°	100°	5	180°	120°

Example 9.14

Problem Statement: Using the Appendix H.1 MATLAB files, calculate the coupler positions achieved by the 4R Spherical mechanism solution in Example 9.7. Also confirm that this solution is order- and branch defect-free.

Known Information: Example 9.7 solution and Appendix H.1 MATLAB file.

Solution Approach: Figure E.9.40 includes the input specified (in bold text) in the Appendix H.1 MATLAB file for mechanism adjustment 1. Figure E.9.41 includes the input specified (in bold text) in the Appendix H.1 MATLAB file for mechanism adjustment 2. Table E.9.10 includes the coupler positions

```
a0 = [0.5, 0.197802, 0.843134]';
a1 = [0.925255, 0.370271, 0.082473]';
ua0 = a0;
ua1 = a1;
b0 = [0.6, 0.312932, 0.736256]';
b1 = [0.899331, 0.430077, 0.078974]';
p1 = [1.5, 0, 0]';
q1 = [2.5, 0, 0.0175]';
r1 = [1.5, 1, 0]';

start_ang   = 0;
step_ang    = 0.01;
stop_ang    = 58;

angular_vel  = 0;
angular_acc  = 0;
```

FIGURE E.9.40 Specified input (in bold text) in the Appendix H.1 MATLAB file for Example 9.14 for Mechanism Adjustment 1.

```
a0 = [-0.281027, 0.512419, 0.811450]';
a1 = [0.925255, 0.370271, 0.082473]';
ua0 = a0;
ua1 = a1;
b0 = [-0.265025, 0.532580, 0.803816]';
b1 = [0.899331, 0.430077, 0.078974]';
p1 = [1.5, 0, 0]';
q1 = [2.5, 0, 0.0175]';
r1 = [1.5, 1, 0]';

start_ang    = 0;
step_ang     = 0.01;
stop_ang     = 78;

angular_vel  = 0;
angular_acc  = 0;
```

FIGURE E.9.41 Specified input (in bold text) in the Appendix H.1 MATLAB file for Example 9.14 for Mechanism Adjustment 2.

TABLE E.9.10
Coupler Positions Achieved by the Adjustable 4R Spherical Mechanism

Precision Position	θ	p	q	r
1	0°	1.5, 0, 0	2.5, 0, 0.0175	1.5, 1, 0
2	35.33°	1.2907, 0.7618, –0.0613	2.1547, 1.265, –0.0856	0.8227, 1.5812, 0.2697
3	57.84°	0.7128, 1.3127, –0.1368	1.2022, 2.1809, –0.2206	0.3800, 1.5874, 0.7653
4	33.40°	1.2857, 0.6367, –0.4378	2.1460, 1.0660, –0.7131	0.8052, 1.4998, –0.5935
5	56.79°	0.8851, 0.9266, –0.7798	1.4802, 1.5520, –1.2847	0.1321, 1.5800, –0.8580
6	77.05°	0.4105, 1.0028, –1.0372	0.6908, 1.6817, –1.7161	–0.4747, 1.4590, –0.9465

achieved by the adjustable mechanism. As indicated by the continually ascending order of the crank angle solutions in Table E.9.10, the order of the achieved displacement angles, and the accuracy of these points, the adjustable 4R Spherical motion generator solution is order- and branch defect-free.

Example 9.15

Problem Statement: Using the Appendix H.1 MATLAB files, calculate the coupler positions achieved by the 4R Spherical mechanism solution in Example 9.8. Also confirm that this solution is order- and branch defect-free.

Known Information: Example 9.8 solution and Appendix H.1 MATLAB file.

Solution Approach: Figure E.9.42 includes the input specified (in bold text) in the Appendix H.1 MATLAB file for mechanism adjustment 1. Figure E.9.43 includes the input specified (in bold

```
a0 = [0.317154, 0.491783, 0.810914]';
a1 = [0.189719, -0.913223, -0.360612]';
ua0 = a0;
ua1 = a1;
b0 = [0.391604, -0.861513, 0.323032]';
b1 = [-0.676530, -0.455148, -0.578933]';
p1 = [1.5, 0, 0]';
q1 = [2.5, 0, 0.0175]';
r1 = [1.5, 1, 0]';

start_ang    = 0;
step_ang     = 0.01;
stop_ang     = 108;

angular_vel  = 0;
angular_acc  = 0;
```

FIGURE E.9.42 Specified input (in bold text) in the Appendix H.1 MATLAB file for Example 9.15 for Mechanism Adjustment 1.

```
a0 = [0.317154, 0.491783, 0.810914]';
a1 = [0.451159, -0.623259, -0.638722]';
ua0 = a0;
ua1 = a1;
b0 = [0.391604, -0.861513, 0.323032]';
b1 = [0.403472, 0.531945, 0.744480]';
p1 = [1.5, 0, 0]';
q1 = [2.5, 0, 0.0175]';
r1 = [1.5, 1, 0]';

start_ang   = 0;
step_ang    = 0.01;
stop_ang    = 144;

angular_vel  = 0;
angular_acc  = 0;
```

FIGURE E.9.43 Specified input (in bold text) in the Appendix H.1 MATLAB file for Example 9.15 for Mechanism Adjustment 2.

TABLE E.9.11
Coupler Positions Achieved by the Adjustable 4R Spherical Mechanism

Precision Position	θ	p	q	r
1	0°	1.5, 0, 0	2.5, 0, 0.0175	1.5, 1, 0
2	55.19°	1.2971, 0.7534, 0.0002	2.1660, 1.2484, 0.0157	0.8574, 1.5102, 0.4840
3	107.71°	0.7429, 1.3030, −0.0131	1.2516, 2.1642, −0.0134	0.3286, 1.5481, 0.8635
4	41.17°	1.2374, 0.7762, −0.3411	2.0687, 1.2906, −0.5525	0.8029, 1.6141, −0.0107
5	75.56°	0.6698, 1.2778, −0.4107	1.1282, 2.1275, −0.6717	0.0832, 1.7875, 0.2186
6	143.46°	−0.0038, 1.4268, −0.4627	0.0112, 2.3781, −0.7712	−0.0063, 1.7353, 0.4885

text) in the Appendix H.1 MATLAB file for mechanism adjustment 2. Table E.9.11 includes the coupler positions achieved by the adjustable mechanism. As indicated by the continually ascending order of the crank angle solutions in Table E.9.11, the order of the achieved displacement angles, and the accuracy of these points, the adjustable 4R Spherical motion generator solution is order- and branch defect-free.

9.8 SUMMARY

In comparison to classical dimensional synthesis (where the dimensions of a single mechanism configuration are calculated), dimensional synthesis of an adjustable mechanism involves calculating the dimensions of a mechanism that can be adjusted to multiple configurations—each configuration approximating a particular group of precision positions or precision points. Adjustable dimensional synthesis offers the advantages of *design flexibility* and subsequently *application flexibility* over classical dimensional synthesis. Because the product of adjustable dimensional synthesis is a single mechanism with fixed and/or moving pivots that can be adjusted to achieve multiple mechanism configurations, such a mechanism offers a greater degree of design flexibility over a non-adjustable mechanism. Also, since each mechanism configuration approximates a specific group of precision positions or precision points, the adjustable mechanism offers a greater degree of application flexibility over a non-adjustable mechanism.

There are up to four possible dyad adjustment options for a four-bar planar mechanical or 4R Spherical mechanism. These adjustment options are *constant-length* and *adjustable-length fixed-pivot adjustments* and *constant-length* and *adjustable-length moving-pivot adjustments*. This chapter includes nonlinear simultaneous equation systems, unconstrained and constrained nonlinear optimization models for adjustable planar four-bar mechanisms and adjustable 4R Spherical mechanisms for motion, and path and function generation given two groups of precision positions or precision points. The equation systems presented in this chapter are solved using the Appendix K MATLAB files.

PROBLEMS

1. Using the Appendix K.1 MATLAB files, synthesize an order-, branch- and circuit defect-free adjustable planar four-bar mechanism for the two groups of unloading bucket positions given in Figure P.9.1. Order, branch, and circuit defect elimination can be verified through a displacement analysis using the Appendix H.1 or I.1 files.

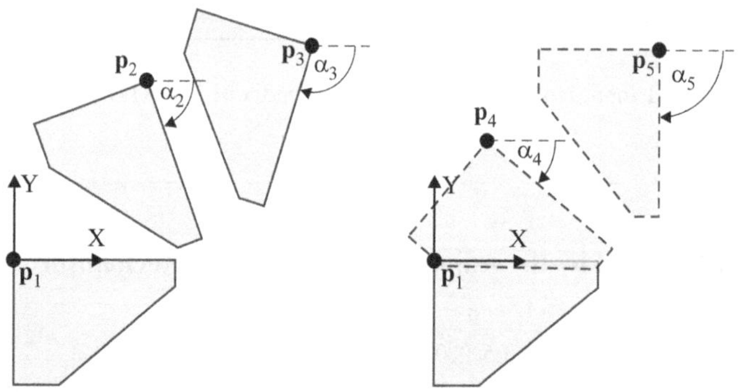

Group #1

p_1=(0, 0)

p_2=(2.4696, 3.1993) α_2= –70°

p_3=(5.5634, 3.85) α_3= –107°

Group #2

p_1=(0, 0)

p_4=(0.9931, 2.2087) α_4= –40°

p_5=(4.2134, 3.8716) α_5= –90°

FIGURE P.9.1 Two groups of unloading bucket positions.

2. Using the Appendix K.1 MATLAB files, synthesize an order-, branch-, and circuit defect-free adjustable planar four-bar mechanism for the two groups of digger bucket positions given in Figure P.9.2. Order, branch, and circuit defect elimination can be verified through a displacement analysis using the Appendix H.1 or I.1 files.

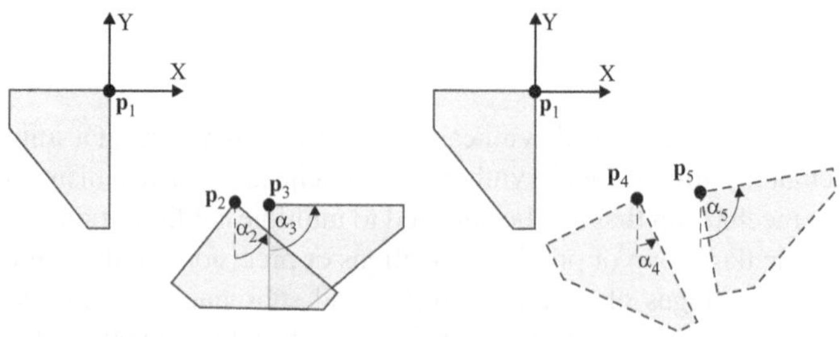

Group #1

p_1=(0, 0)

p_2=(2.1897, –1.8574) α_2= 50°

p_3=(2.7935, –1.9122) α_3= 90°

Group #2

p_1=(0, 0)

p_4=(1.8544, –1.9327) α_4= 27°

p_5=(3.0202, –1.8202) α_5= 100°

FIGURE P.9.2 Two groups of digger bucket positions.

3. Repeat Problem 1 using the Appendix K.2 MATLAB files. Order, branch, and circuit defect elimination can be verified through a displacement analysis using the Appendix H.1 or I.1 files.

4. Repeat Problem 1 using the Appendix K.3 MATLAB files. Order, branch, and circuit defect elimination can be verified through a displacement analysis using the Appendix H.1 or I.1 files.
5. Repeat Problem 2 using the Appendix K.3 MATLAB files. Order, branch, and circuit defect elimination can be verified through a displacement analysis using the Appendix H.1 or I.1 files.
6. Using the Appendix K.4 MATLAB files, synthesize an order-, branch-, and circuit defect-free adjustable planar four-bar mechanism for the two groups of precision positions given in Table P.9.1. Order, branch, and circuit defect elimination can be verified through a displacement analysis using the Appendix H.1 or I.1 files.

TABLE P.9.1
Two Groups of Five Precision Positions

Group #1			Group #2		
Precision Position	$\mathbf{p}_j$	α_j (deg)	Precision Position	$\mathbf{p}_j$	α_j (deg)
1	–0.443, 2.626		1	–0.443, 2.626	
2	–1.594, 1.930	20	6	–0.707, 2.548	5
3	–1.898, 1.412	30	7	–1.199, 2.298	15
4	–2.026, 0.895	40	8	–1.919, 1.509	35
5	–2.019, 0.423	50	9	–2.167, 0.101	65

7. Using the Appendix K.5 MATLAB files, synthesize an order-, branch-, and circuit defect-free adjustable planar four-bar mechanism for the two groups of unloading bucket positions given in Figure P.9.3. Order, branch, and circuit defect elimination can be verified through a displacement analysis using the Appendix H.1 or I.1 files.

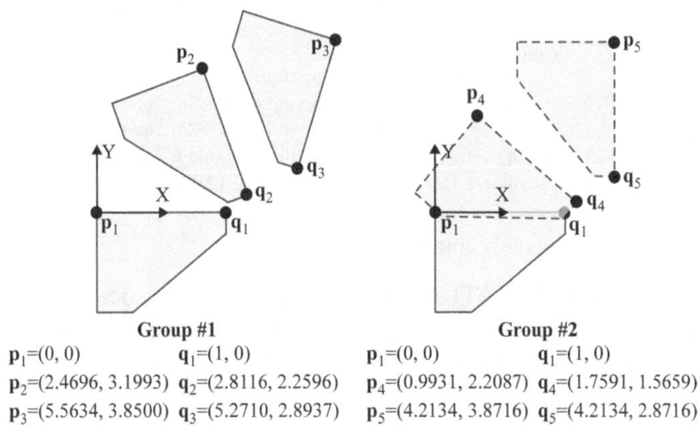

FIGURE P.9.3 Two groups of unloading bucket positions.

8. Using the Appendix K.5 MATLAB files, synthesize an order-, branch-, and circuit defect-free adjustable planar four-bar mechanism for the two groups of digger bucket positions given in Figure P.9.4. Order, branch, and circuit defect elimination can be verified through a displacement analysis using the Appendix H.1 or I.1 files.

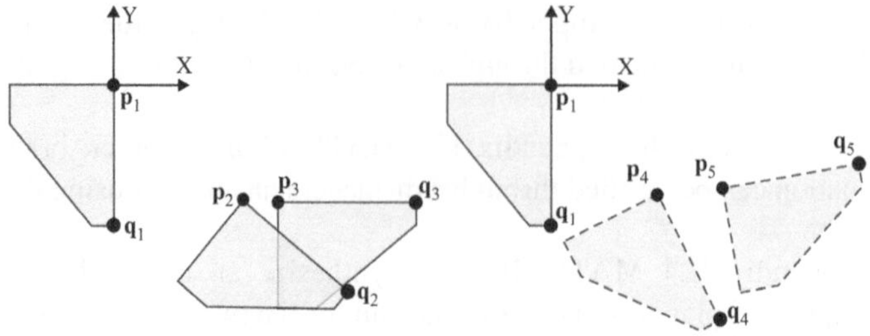

Group #1		Group #2	
$p_1=(0, 0)$	$q_1=(0, -1)$	$p_1=(0, 0)$	$q_1=(0, -1)$
$p_2=(2.1897, -1.8574)$	$q_2=(2.9557, -2.5002)$	$p_4=(1.8544, -1.9327)$	$q_4=(2.3084, -2.8237)$
$p_3=(2.7935, -1.9122)$	$q_3=(3.7935, -1.9122)$	$p_5=(3.0202, -1.8202$	$q_5=(4.0050, -1.6466)$

FIGURE P.9.4 Two groups of digger bucket positions.

9. Using the Appendix K.5 MATLAB files, synthesize an order-, branch-, and circuit defect-free adjustable planar four-bar mechanism for the two groups of assembly component positions given in Figure P.9.5. Order, branch, and circuit defect elimination can be verified through a displacement analysis using the Appendix H.1 or I.1 files.

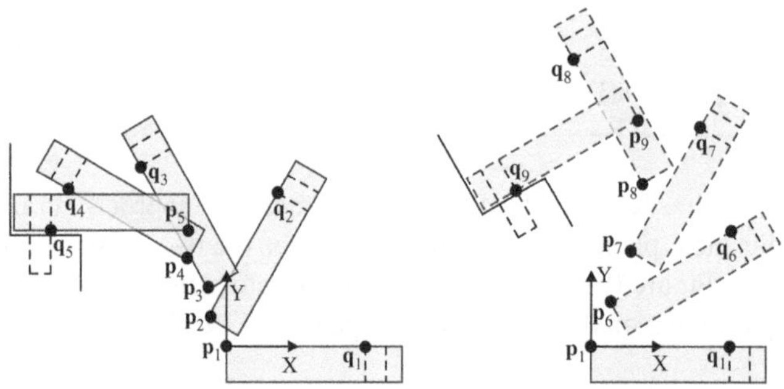

Group #1		Group #2	
$p_1=(0, 0)$	$q_1=(1, 0)$	$p_1=(0, 0)$	$q_1=(1, 0)$
$p_2=(-0.3627, 0.5346)$	$q_2=(0.1373, 1.4006)$	$p_6=(0.3761, 0.8216)$	$q_6=(1.2421, 1.3216)$
$p_3=(-0.3461, 1.0239)$	$q_3=(-0.8461, 1.8899)$	$p_7=(0.7423, 1.7723)$	$q_7=(1.2423, 2.6383)$
$p_4=(-0.7684, 1.5824)$	$q_4=(-1.6344, 2.0824)$	$p_8=(0.9758, 3.0450)$	$q_8=(0.4758, 3.9110)$
$p_5=(-0.7517, 2.1459)$	$q_5=(-1.7517, 2.1459)$	$p_9=(0.9136, 4.2590)$	$q_9=(0.0476, 3.7590)$

FIGURE P.9.5 Two groups of assembly component positions.

10. Using the Appendix K.5 MATLAB files, synthesize an order-, branch-, and circuit defect-free adjustable planar four-bar mechanism for the two groups of stamping tool positions given in Figure P.9.6. Order, branch, and circuit defect elimination can be verified through a displacement analysis using the Appendix H.1 or I.1 files.

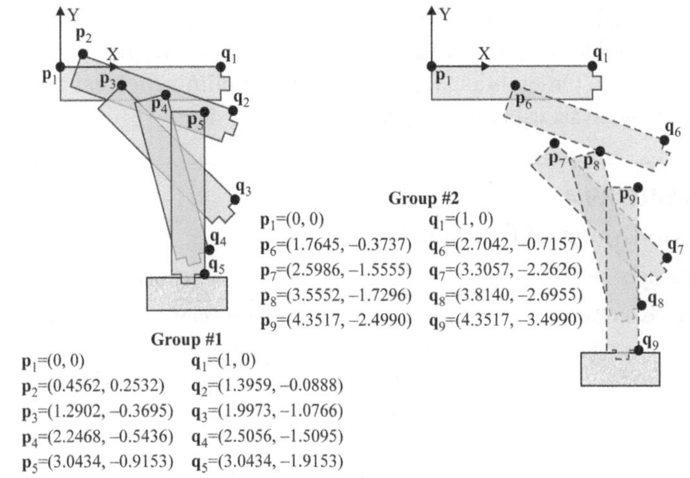

FIGURE P.9.6 Two groups of stamping tool positions.

11. Using the Appendix K.6 MATLAB files, synthesize an order-, branch-, and circuit defect-free adjustable planar four-bar mechanism for the two groups of precision points given in Table P.9.2. Order, branch, and circuit defect elimination can be verified through a displacement analysis using the Appendix H.1 or I.1 files.

TABLE P.9.2
Two Groups of Precision Points

Group 1		Group 2	
Precision Point	p_j	Precision Point	p_k
1	0, 0	1	0, 0
2	1.1790, 0.7252	5	2.3707, −0.5756
3	2.3580, 1.4504	6	3.5561, −0.8634
4	3.5370, 2.1756	7	4.7415, −1.1512

12. Using the Appendix K.6 MATLAB files, synthesize an order-, branch-, and circuit defect-free adjustable planar four-bar mechanism for the two groups of precision points given in Table P.9.3. Order, branch, and circuit defect elimination can be verified through a displacement analysis using the Appendix H.1 or I.1 files.

TABLE P.9.3
Two Groups of Precision Points

Group 1		Group 2	
Precision Point	p_j	Precision Point	p_k
1	0, 0	1	0, 0
2	1.2577, 0.3273	5	1.1599, 0.1651
3	1.9657, 1.1114	6	1.8302, 0.4505
4	2.0526, 1.5607	7	2.3327, 0.8432
		8	2.6035, 1.2423
		9	2.7076, 1.7128

13. Using the Appendix K.6 MATLAB files, synthesize an order-, branch-, and circuit defect-free adjustable planar four-bar mechanism for the two groups of precision points given in Table P.9.4. Order, branch, and circuit defect elimination can be verified through a displacement analysis using the Appendix H.1 or I.1 files.

TABLE P.9.4
Two Groups of Precision Points

Group 1		**Group 2**	
Precision Point	$\mathbf{p}_j$	**Precision Point**	$\mathbf{p}_k$
1	0, 0	1	0, 0
2	−0.6423, −0.1182	7	0.5551, 0.0364
3	−1.2017, −0.4585	8	1.1599, 0.1651
4	−1.5371, −0.8598	9	1.8302, 0.4505
5	−1.7366, −1.3160	10	2.3327, 0.8432
6	−1.8038, −1.8038	11	2.6035, 1.2423

14. Using the Appendix K.6 MATLAB files, synthesize an order-, branch-, and circuit defect-free adjustable planar four-bar mechanism for the two groups of precision points given in Table P.9.5. Order, branch, and circuit defect elimination can be verified through a displacement analysis using the Appendix H.1 or I.1 files.

TABLE P.9.5
Two Groups of Precision Points

Group 1		**Group 2**	
Precision Point	$\mathbf{p}_j$	**Precision Point**	$\mathbf{p}_k$
1	1, 1	1	1, 1
2	2, 2	6	2, 1.4142
3	3, 3	7	3, 1.7321
4	4, 4	8	4, 2
5	5, 5	9	5, 2.2361

15. Using the Appendix K.7 MATLAB files, synthesize an adjustable planar four-bar mechanism for 65° and 100° total wiper blade rotation ranges (Figure P.9.7) and a 50° total crank rotation range where the wiper blade is affixed to the follower link. Verify the solution using the Appendix H.2 or I.2 files.

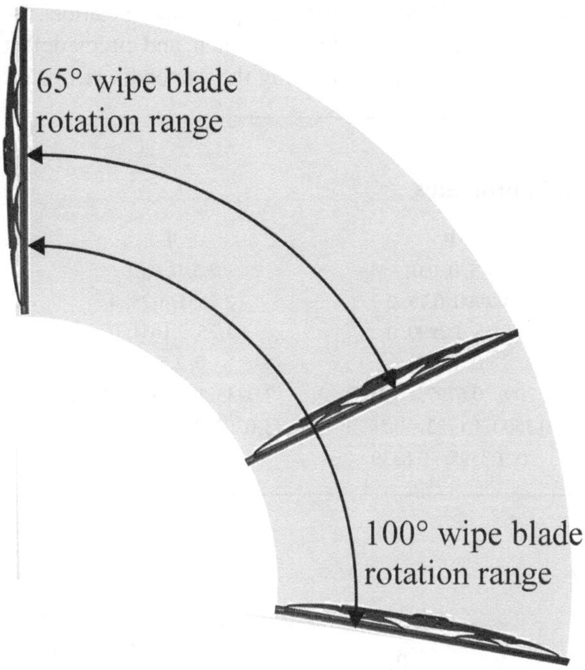

FIGURE P.9.7 Wiper blade rotation ranges.

16. Using the Appendix K.7 MATLAB files, synthesize an adjustable planar four-bar mechanism for the two groups of follower link precision points given in Table P.9.6. For both groups, $\theta_j = 7°, 14°, \ldots, 35°$ for the crank link. Verify the solution using the Appendix H.2 or I.2 files.

TABLE P.9.6
Two Groups of Follower Link Precision Points

Group 1		Group 2	
Precision Point	**ϕ_j (deg)**	**Precision Point**	**ϕ_j (deg)**
2	3.4260	7	28.4510
3	13.3688	8	58.7680
4	28.8550	9	82.4980
5	48.3688	10	101.9400
6	70	11	120

17. Using the Appendix K.7 MATLAB files, synthesize an adjustable planar four-bar mechanism for the two groups of follower link precision points given in Table P.9.6 in Problem 16. For Group 1, $\theta_j = 5°, 10°, \ldots, 25°$ for the crank link. For Group 2, $\theta_j = 12°, 24°, \ldots, 60°$ for the crank link. Verify the solution using the Appendix H.2 or I.2 files.
18. Using the Appendix K.7 MATLAB files, synthesize an adjustable planar four-bar mechanism for the two groups of follower link precision points given in Table P.9.6 in Problem 16. For Group 1, $\theta_j = 7°, 14°, \ldots, 35°$ for the crank link. For Group 2, $\theta_j = -7°, -14°, \ldots, -35°$ for the crank link. Verify the solution using the Appendix H.2 or I.2 files.
19. Using the Appendix K.8 MATLAB files, synthesize an order-, branch-, and circuit defect-free adjustable 4R Spherical mechanism for the two groups of solar panel positions

given in Table P.9.7 (Figure P.9.8). The first group includes positions 1–2–3–4, and the second group includes positions 1–5–6–7. Order, branch, and circuit defect elimination can be verified through a displacement analysis using the Appendix H.1 or I.1 files.

TABLE P.9.7
Solar Panel Position Coordinates

Precision Position	p	q	r
1	1.5, 0, 0.01	2.5, 0, 0.01	1.5, 1, 0.01
2	1.2990, 0.75, 0	2.1651, 1.25, 0	0.8660, 1.5, 0.5
3	0.75, 1.2990, 0	1.25, 2.1651, 0	0.3170, 1.5490, 0.8660
4	0, 1.5, 0	0, 2.5, 0	0, 1.5, 1
5	1.2990, 0.6797, –0.3170	2.0235, 1.3481, –0.4858	0.6566, 1.4231, –0.1307
6	0.7500, 1.1773, –0.5490	1.0202, 2.1391,–0.5936	0.1462, 1.3827, 0.2213
7	0, 1.3595, –0.6339	0, 2.3595, –0.6339	0, 1.3595, 0.3661

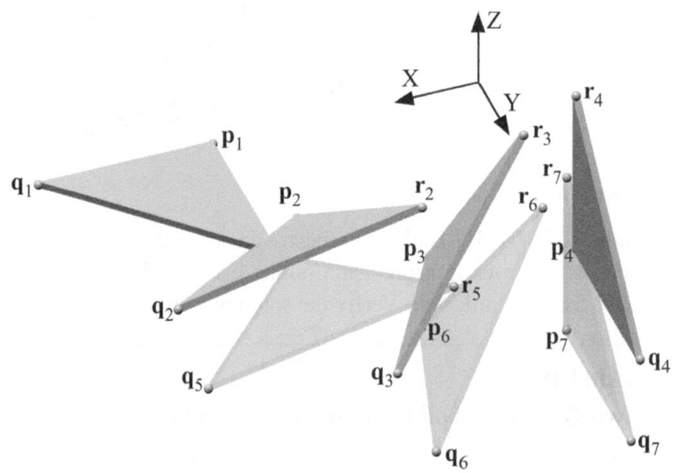

FIGURE P.9.8 Solar panel positions.

20. Using the Appendix K.9 MATLAB files, synthesize an order-, branch-, and circuit defect-free adjustable 4R Spherical mechanism for the two groups of solar panel positions given in Table P.9.8. The first group includes positions 1–2–3, and the second group includes positions 1–4–5–6. Order, branch, and circuit defect elimination can be verified through a displacement analysis using the Appendix H.1 or I.1 files.

TABLE P.9.8
Two Groups of Precision Position Coordinates

Precision Position	p	q	r
1	1.5, 0, 0.25	2.5, 0, 0	1.5, 1, 0
2	1.3491, 0.5103, –0.4817	2.0410, 0.8541, –1.1641	0.9324, 1.4522, –0.5213
3	0.9322, 0.9737, –0.7039	1.2470, 1.6059, –1.4548	0.2906, 1.7289, –0.4203
4	1.5150, 0.1308, –0.0089	2.4530, 0.2287, –0.4251	1.3861, 1.1327, –0.2141
5	1.4504, 0.3475, –0.2967	2.2622, 0.5971, –0.8808	1.1510, 1.3284, –0.4007
6	1.2304, 0.6895, –0.5686	1.8023, 1.1659, –1.2817	0.7281, 1.5832, –0.4619

21. Using the Appendix K.10 MATLAB files, synthesize an order-, branch-, and circuit defect-free adjustable 4R Spherical mechanism for the two groups of precision positions given in Table P.9.9. The first group includes positions 1–2–3, and the second group includes positions 1–4–5–6. Order, branch, and circuit defect elimination can be verified through a displacement analysis using the Appendix H.1 or I.1 files.

TABLE P.9.9
Two Groups of Precision Position Coordinates

Precision Position	p	q	r
1	1.5, 0, 0.25	2.5, 0, 0	1.5, 1, 0
2	1.3491, 0.5103, −0.4817	2.0410, 0.8541, −1.1641	0.9324, 1.4522, −0.5213
3	0.9322, 0.9737, −0.7039	1.2470, 1.6059, −1.4548	0.2906, 1.7289, −0.4203
4	1.4012, 0.3638, −0.4658	2.1379, 0.6100, −1.1434	1.0720, 1.3358, −0.5627
5	1.2546, 0.5710, −0.6422	1.8406, 0.9449, −1.4034	0.7932, 1.4926, −0.6269
6	0.7010, 1.0836, −0.8044	0.8276, 1.7561, −1.5753	0.0255, 1.7553, −0.4105

22. Using the Appendix K.11 MATLAB files, synthesize an order-, branch-, and circuit defect-free adjustable 4R Spherical mechanism for the two groups of precision positions given in Table P.9.10. The first group includes positions 1–2–3–4–5, and the second group includes positions 1–6–7–8–9. Order, branch, and circuit defect elimination can be verified through a displacement analysis using the Appendix H.1 or I.1 files.

TABLE P.9.10
Two Groups of Precision Position Coordinates

Precision Position	p	q	r
1	1.5, 0, 0.25	2.5, 0, 0	1.5, 1, 0
2	1.3916, 0.4422, −0.4248	2.1296, 0.7423, −1.0789	1.0189, 1.4002, −0.5012
3	1.2305, 0.6709, −0.5902	1.8056, 1.1163, −1.3205	0.7199, 1.5646, −0.5329
4	1.0443, 0.8720, −0.6794	1.4537, 1.4422, −1.4343	0.4386, 1.6817, −0.4791
5	0.8332, 1.0543, −0.7118	1.0665, 1.7353, −1.4496	0.1709, 1.7592, −0.3554
6	1.2822, 0.5917, −0.5642	1.8980, 1.0028, −1.2813	0.8310, 1.5168, −0.5085
7	1.0618, 0.8237, −0.7117	1.4699, 1.3830, −1.4753	0.5126, 1.6625, −0.4727
8	0.7866, 1.0433, −0.7780	0.9590, 1.7382, −1.5195	0.2025, 1.7617, −0.3249
9	0.5121, 1.2126, −0.7616	0.4639, 2.0072, −1.4163	−0.0238, 1.7995, −0.1052

23. Using the Appendix K.12 MATLAB files, synthesize an order-, branch-, and circuit defect-free adjustable 4R Spherical mechanism for the two groups of solar panel positions given in Table P.9.7 in Problem 19 (Figure P.9.8). Order, branch, and circuit defect elimination can be verified through a displacement analysis using the Appendix H.1 or I.1 files.
24. Using the Appendix K.12 MATLAB files, synthesize an order-, branch-, and circuit defect-free adjustable 4R Spherical mechanism for the two groups of landing gear positions given in Table P.9.11 (Figure P.9.9). The first group includes positions 1–2–3–4–5, and the second group includes positions 1–6–7–8–9. Order, branch, and circuit defect elimination can be verified through a displacement analysis using the Appendix H.1 or I.1 files.

TABLE P.9.11
Landing Gear Position Coordinates

Precision Position	p	q	r
1	1.5, 0, 0.01	2.5, 0, 0.01	1.5, 1, 0.01
2	1.4624, 0.3338, 0	2.4373, 0.5563, 0	1.2456, 1.2837, 0.2250
3	1.1727, 0.9352, 0	1.9546, 1.5587, 0	0.6882, 1.5428, 0.6293
4	0.6508, 1.3515, 0	1.0847, 2.2524, 0	0.2701, 1.5348, 0.9063
5	0, 1.5, 0	0, 2.5, 0	0, 1.5, 1
6	1.4624, 0.3338, 0	2.4373, 0.5563, 0	1.2496, 1.2661, −0.2924
7	1.1727, 0.9352, 0	1.9546, 1.5587, 0	0.5569, 1.7074, 0.1564
8	0.6508, 1.3515, 0	1.0847, 2.2524, 0	−0.0872, 1.7069, 0.5736
9	0, 1.5, 0	0, 2.5, 0	−0.5, 1.5, 0.8660

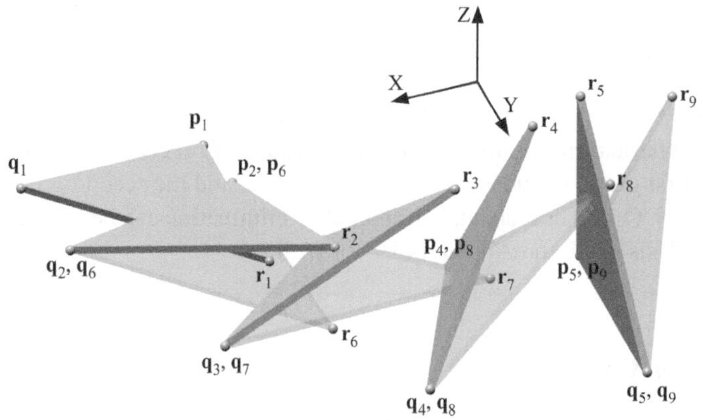

FIGURE P.9.9 Landing gear positions.

25. Using the Appendix K.12 MATLAB files, synthesize an order-, branch-, and circuit defect-free adjustable 4R Spherical mechanism for the two groups of folding wing positions given in Table P.9.12 (Figure P.9.10). The first group includes positions 1–2–3, and the second group includes positions 1–4–5. Order, branch, and circuit defect elimination can be verified through a displacement analysis using the Appendix H.1 or I.1 files.

TABLE P.9.12
Folding Wing Position Coordinates

Precision Position	p	q	r
1	0.9098, 1.1503, 0.3129	0.6249, 1.1154, 1.2705	1.5164, 1.9172, 0.5214
2	0.4861, 1.3632, 0.3918	0.0572, 1.2589, 1.2885	0.8103, 2.2720, 0.6527
3	0.1097, 1.3684, 0.6039	−0.4319, 1.7462, −0.1421	0.1855, 2.2788, 1.0074
4	0.2061, 1.4241, 0.4213	−0.0511, 1.1851, 1.3567	0.3437, 2.3735, 0.7016
5	−0.2046, 1.3550, 0.6123	−0.8007, 1.6109, −0.1446	−0.3386, 2.2563, 1.0210

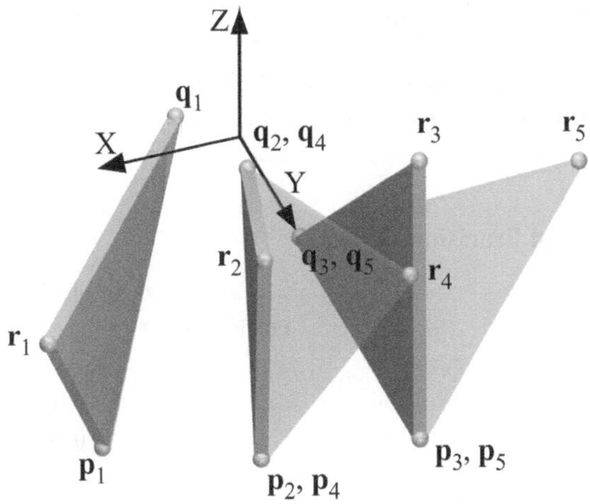

FIGURE P.9.10 Folding wing positions.

26. Using the Appendix K.12 MATLAB files, synthesize an order-, branch-, and circuit defect-free adjustable 4R Spherical mechanism for the two groups of assembly component positions given in Table P.9.13 (Figure P.9.11). The first group includes positions 1–2–3–4, and the second group includes positions 1–5–6. Order, branch, and circuit defect elimination can be verified through a displacement analysis using the Appendix H.1 or I.1 files.

TABLE P.9.13
Assembly Component Position Coordinates

Precision Position	p	q	r
1	1.5, 0, 0	2.5, 0, 0	1.5, 0, 1
2	1.4772, –0.2605, 0	2.4620, –0.4341, 0	1.4323, –0.5154, 0.9659
3	1.4095, –0.5130, 0	2.3492, –0.8551, 0	1.2385, –0.9829, 0.8660
4	1.2990, –0.7500, 0	2.1651, –1.2500, 0	0.9455, –1.3624, 0.7071
5	1.4095, –0.5130, 0	2.3534, –0.8326, –0.0842	1.4033, –0.7851, 0.9623
6	1.1491, –0.9642, 0	1.9593, –1.5307, –0.1504	0.9656, –1.4530, 0.8529

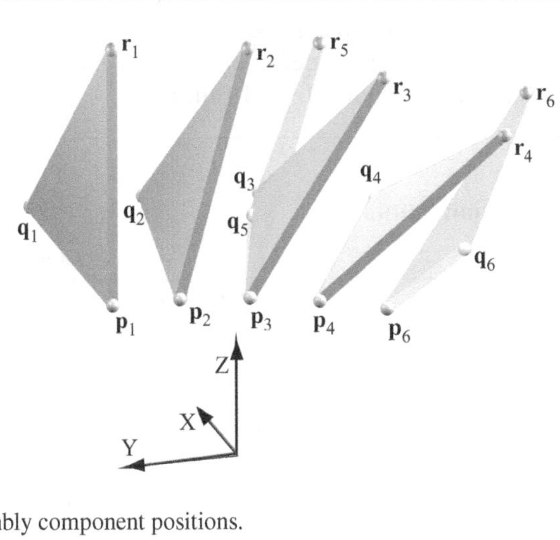

FIGURE P.9.11 Assembly component positions.

27. Using the Appendix K.13 MATLAB files, synthesize a branch- and order defect-free adjustable 4R Spherical mechanism to approximate the two groups of precision points in Table P.9.14. Order, branch, and circuit defect elimination can be verified through a displacement analysis using the Appendix H.1 or I.1 files.

TABLE P.9.14
Two Groups of Precision Points

Group 1		Group 2	
Precision Point	$\mathbf{p}_j$	**Precision Point**	$\mathbf{p}_k$
1	0, 0, 1	1	0, 0, 1
2	0.1240, 0, 0.9923	6	0.1074, 0.0620, 0.9923
3	0.2425, 0, 0.9701	7	0.2100, 0.1213, 0.9701
4	0.3511, 0, 0.9363	8	0.3041, 0.1756, 0.9363
5	0.4472, 0, 0.8944	9	0.3873, 0.2236, 0.8944

28. Using the Appendix K.13 MATLAB files, synthesize a branch- and order defect-free adjustable 4R Spherical mechanism to approximate the two groups of precision points in Table P.9.15. Order, branch, and circuit defect elimination can be verified through a displacement analysis using the Appendix H.1 or I.1 files.

TABLE P.9.15
Two Groups of Precision Points

Group 1		Group 2	
Precision Point	$\mathbf{p}_j$	**Precision Point**	$\mathbf{p}_k$
1	0, 0, 1	1	0, 0, 1
2	0.1240, 0, 0.9923	6	0.1272, −0.0273, 0.9915
3	0.2425, 0, 0.9701	7	0.3512, 0.1271, 0.9277
4	0.3511, 0, 0.9363	8	0.4044, 0.2821, 0.8700
5	0.4472, 0, 0.8944	9	0.3430, 0.3754, 0.8610

29. Using the Appendix K.13 MATLAB files, synthesize a branch- and order defect-free adjustable 4R Spherical mechanism to approximate the two groups of precision points in Table P.9.16. Order, branch, and circuit defect elimination can be verified through a displacement analysis using the Appendix H.1 or I.1 files.

TABLE P.9.16
Two Groups of Precision Points

Group 1		Group 2	
Precision Point	$\mathbf{p}_j$	**Precision Point**	$\mathbf{p}_k$
1	0, 0, 1	1	0, 0, 1
2	0.1054, 0.1070, 0.9887	6	0.2418, 0.0295, 0.9699
3	0.1527, 0.2693, 0.9509	7	0.1190, 0.2588, 0.9586
4	−0.1122, 0.1646, 0.9799	8	−0.0994, 0.3042, 0.9474
5	−0.2428, −0.0884, 0.9660	9	−0.1330, 0.0902, 0.9870

30. Using the Appendix K.13 MATLAB files, synthesize a branch- and order defect-free adjustable 4R Spherical mechanism to approximate the two groups of precision points given in Table P.9.17. Order, branch, and circuit defect elimination can be verified through a displacement analysis using the Appendix H.1 or I.1 files.

TABLE P.9.17
Two Groups of Precision Points

Group 1		Group 2	
Precision Point	$\mathbf{p}_j$	**Precision Point**	$\mathbf{p}_k$
1	0, 0, 1	1	0, 0, 1
2	0.1054, 0.1070, 0.9887	6	0.1501, 0.1918, 0.9699
3	0.1527, 0.2693, 0.9509	7	−0.0989, 0.2671, 0.9586
4	−0.1122, 0.1646, 0.9799	8	−0.2854, 0.1448, 0.9474
5	−0.2428, −0.0884, 0.9660	9	−0.1578, −0.0303, 0.9870

NOTES

1 The remaining two configurations for this adjustable mechanism are $\mathbf{a}_0 - \mathbf{a}_1 - \mathbf{b}_1^* - \mathbf{b}_0$ and $\mathbf{a}_0 - \mathbf{a}_1^* - \mathbf{b}_1 - \mathbf{b}_0$.

2 The alpha values in Table E.9.6 were calculated using $\alpha = \mathrm{acos}\left(\mathbf{V}_1 \bullet \mathbf{V}_j / \|\mathbf{V}_1\| \|\mathbf{V}_j\|\right)$, where $\mathbf{V}_1 = \mathbf{a}_1 - \mathbf{b}_1$ and $\mathbf{V} = \mathbf{a} - \mathbf{b}$. This is because the α values calculated from Equation (8.12) are the differences between angles $\angle \mathbf{b} - \mathbf{a} - \mathbf{a}_0$ and $\angle \mathbf{b}_1 - \mathbf{a}_1 - \mathbf{a}_0$.

REFERENCES

1. Zhou, H. and Cheung, E. H. M. 2004. Adjustable Four-Bar Linkages for Multi-phase Motion Generation. *Mechanism and Machine Theory*. 39: 261–279.
2. Zhou, H. 2009. Dimensional Synthesis of Adjustable Path Generation Linkages Using the Optimal Slider Adjustment. *Mechanism and Machine Theory*. 44: 983–990.
3. Zhou, H. 2009. Dimensional Synthesis of Adjustable Function Generation Linkages Using the Optimal Pivot Adjustment. *Mechanism and Machine Theory*. 44: 1866–1876.
4. Peng, C. and Sodhi, R. S. 2010. Optimal Synthesis of Adjustable Mechanisms Generating Multi-phase Approximate Paths. *Mechanism and Machine Theory*. 47: 989–996.
5. Russell, K. and Sodhi, R. S. 2005. On the Design of Slider-Crank Mechanisms Part I: Multi-phase Motion Generation. *Mechanism and Machine Theory*. 40: 285–299.
6. Russell, K. and Sodhi, R. S. 2005. On the Design of Slider-Crank Mechanisms Part II: Multi-Phase Path and Function Generation. *Mechanism and Machine Theory*. 40: 301–317.
7. Lee, W., Russell, K., Shen, Q. and Sodhi, R. S. 2009. On Adjustable Spherical Four-Bar Motion Generation for Expanded Prescribed Positions. *Mechanism and Machine Theory*. 44: 247–254.
8. Wang, S. J. and Sodhi, R. S. 1996. Kinematic Synthesis of Adjustable Moving Pivot Four-Bar Mechanisms for Multi-phase Motion Generation. *Mechanism and Machine Theory*. 31: 459–474.
9. Suh, C. H. and Radcliffe, C. W. 1978. *Kinematics and Mechanisms Design*. pages 51, 52, 129–132. New York: John Wiley & Sons.
10. Suh, C. H. and Radcliffe, C. W. 1978. *Kinematics and Mechanisms Design*. pages 167–169, 177–179. New York: John Wiley & Sons.

10 Introduction to Robotic Manipulators

CONCEPT OVERVIEW

In this chapter, the reader will gain a central understanding regarding the following:

1. Distinctions and disadvantages of *robotic manipulators* in comparison to classical linkages
2. Matrix-based formulation of displacement equation systems for *Cartesian*, *cylindrical*, *spherical*, *articulated,* and *SCARA* robots
3. *Forward* kinematics of *Cartesian*, *cylindrical*, *spherical*, *articulated,* and *SCARA* robots
4. *Inverse* kinematics and workspace of *Cartesian*, *cylindrical*, *spherical*, *articulated,* and *SCARA* robots

10.1 INTRODUCTION

As explained in Chapter 1 and demonstrated throughout this book, a *linkage* (also commonly called a *mechanism*) is an assembly of links and joints, where the motion of one link compels the motion of another link in a controlled manner. To enable controlled motion, the mechanism is either initially designed to have a single degree of freedom or ultimately configured (in the case of the geared five-bar mechanism) to have a single degree of freedom. Conventional planar and spatial linkages include the four-bar, geared five-bar, Stephenson, RRSS, RSSR, and 4R spherical linkages presented in Chapters 4 through 9.

Like the linkages presented in Chapters 4 and 9, a *robotic manipulator* (commonly called a *robot*) also includes an assembly of links and joints and is designed to produce a controlled output motion. In addition to links and joints, however, a robotic manipulator also includes *electronic circuitry* and *computer-controlled actuators* to compel link motion, and is guided by a *computer program.* Because robotic manipulators include both mechanical and electronic components, they are classified as *electro-mechanical systems.*

To achieve a controlled motion, each joint in a robotic manipulator can be controlled independently. As a result, robotic manipulators can be theoretically designed to have an indefinite number of degrees of freedom. In practice, however, robotic manipulators generally have up to six degrees of freedom.[1]

Another common distinction between linkages and robotic manipulators is in their overall design. The linkages presented in this book all have *closed-loop* designs. With this design, at least two joints in a linkage are connected to the ground (thus forming a closed loop of links). While robotic manipulators can have closed-loop designs, they often have *open loops*—where only one joint is connected to the ground.

Figure 10.1 includes a planar four-bar mechanism and a planar, 2 DOF, open-loop robotic manipulator. While coupler Point $\mathbf{p}_1$ on the four-bar mechanism (in the given location on the coupler link) can only trace the curve illustrated, Point $\mathbf{p}_1$ on the robotic manipulator can trace *any* path within the shaded annular area or *workspace*.[2] Because of the open-loop construction of the robotic manipulator and its mobility, this single manipulator can trace a greater variety of distinct paths than any number of planar four-bar mechanisms.

DOI: 10.1201/9781003591375-10

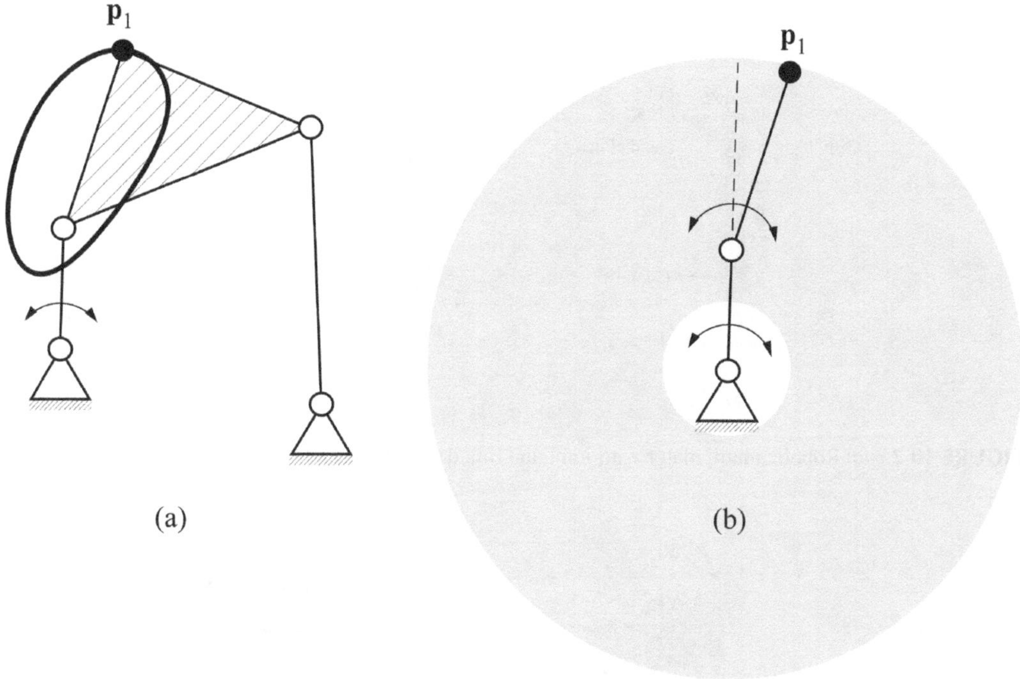

FIGURE 10.1 (a) Four-bar mechanism and coupler curve and (b) robotic manipulator and workspace.

Therefore, when compared to the linkages, robotic manipulators offer advantages such as greater variability—specifically for motion-specific and path-specific tasks. Being computer-controlled, robotic manipulators also offer advantages such as greater precision, accuracy, and repeatability. Lastly, robotic manipulators have the capacity for *remote operation* and *autonomous operation* since they can be guided by computer programs (as opposed to mechanisms which often require a degree of manual operation).

It is becoming increasingly difficult to find an industry where robotic manipulators are not employed either directly or indirectly. Common industries where robotic manipulators are widely employed (both in product manufacturing and operation) include *automotive*, *aerospace*, *defense*, *electronics*, and *medicine*. The applications for robotic systems are rapidly on the rise since new robotic manipulator capabilities and more practical robot manipulator designs are continually being developed and produced [1].

10.2 TERMINOLOGY AND NOMENCLATURE

To describe the spatial position and orientation of each link in a robotic manipulator, coordinate systems are assigned to each link. These coordinate systems are called *frames*. In Figure 10.2b, frames are assigned to Link 0 (called the *base frame*), Link 1, and Link 2 of the robotic manipulator shown in Figure 10.2a. As part of the kinematic equation formulation procedure, the positions and orientations between the frames are modeled using transformation matrices (Section 2.5). The general transformation matrix and its use in formulating kinematic equations for robotic manipulators will be further discussed in Sections 10.4 through 10.6.

A robotic manipulator often includes a component at its free end called an *end effector*. The end effector can serve to hold a tool or a component. It can also be the working end of a tool if the tool is rigidly affixed to the robotic manipulator and the tool itself is a rigid system. One common

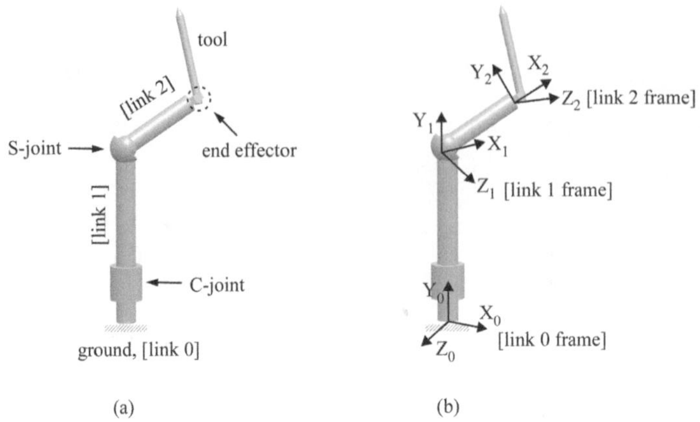

FIGURE 10.2 (a) Robotic manipulator with joint and link descriptions and (b) frames.

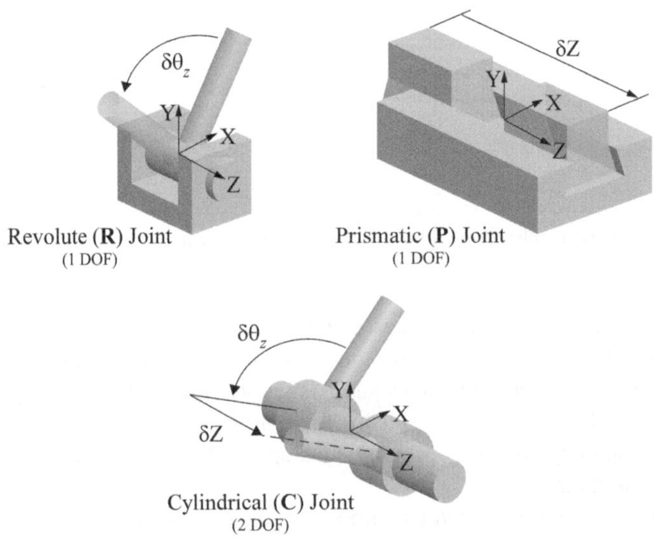

FIGURE 10.3 Example of robotic manipulator joint types.

end effector is a *gripper*. As shown in Figure 10.2, the end effector is a gripper that holds the tool. Because no frames are assigned to the gripper or the tool, the tool can be considered a rigid system that is rigidly affixed to the robotic manipulator through the gripper.[3]

10.3 ROBOTIC MANIPULATOR MOBILITY AND TYPES

In Chapter 3, it was noted that Gruebler's Equation is used to determine the mobility of a linkage. These equations can also be used to determine the mobility of robotic manipulators. For the robotic manipulators in this chapter, the spatial form of Gruebler's Equation becomes

$$\text{DOF}_{\text{SPATIAL}} = 6(L-1) - 5J_1 - 4J_2 \tag{10.1}$$

where only 1- and 2-DOF joints are used.[4]

Figure 10.3 includes the types of 1- and 2-DOF joints used in the robotic manipulators presented here. Although a robotic manipulator can be designed to include any joint type, the revolute, prismatic, and cylindrical joints are among those commonly used in practice.

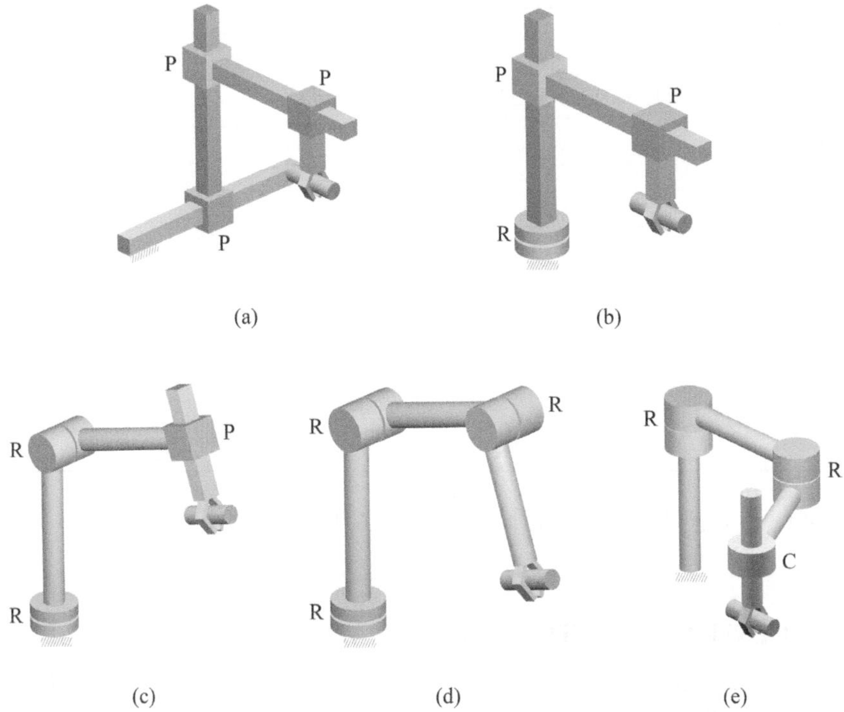

FIGURE 10.4 (a) Cartesian (P-P-P), (b) cylindrical (R-P-P), (c) spherical (R-R-P), (d) articulated (R-R-R), and (e) SCARA (R-R-C) robots.

Figure 10.4 illustrates the five spatial robotic manipulator types considered here. These particular robotic manipulator configurations are among those commonly utilized in industrial applications [2]. The designations P-P-P, R-P-P, R-R-P, R-R-R, and R-R-C given in Figure 10.4 denote the joint types and joint sequences used in the robotic manipulators.

The P-P-P robotic manipulator is commonly known as a *Cartesian robot* because its degrees of freedom are along the x, y and z-axes of the Cartesian frame. The R-P-P and R-R-P robotic manipulators are commonly known as *cylindrical* and *spherical robots,* respectively, because their motion is consistent with that of cylindrical and spherical joints, respectively. The R-R-R and R-R-C robotic manipulators are commonly known as *articulated* and *Selective Compliance Assembly/Articulated Robot Arm* (or *SCARA*) *robots*, respectively [2].

Example 10.1

Problem Statement: Calculate the mobility values for the R-R-R and R-R-C robotic manipulators

Known Information: Equation (10.1) and Figure 10.3

Solution Approach: The R-R-R robotic manipulator comprises four links interconnected with three revolute joints. Since there are four links and the revolute joint has a single degree of freedom, $L = 4$ and $J_1 = 3$ in Equation (10.1) for the R-R-R robotic manipulator.

The R-R-C comprises four links interconnected with two revolute joints and one cylindrical joint. Since there are four links and the cylindrical joint has 2 degrees of freedom, $L = 4$, $J_1 = 2$, and $J_2 = 1$ in Equation (10.1) for the R-R-C robotic manipulator.

Figure E.10.1 includes the calculation procedure in MATLAB's command window.

```
>> L = 4;
>> J1 = 3;
>> J2 = 0;
>> DOF_RRR = 6*(L - 1) -5*J1 - 4*J2

DOF_RRR =

     3

>> L = 4;
>> J1 = 2;
>> J2 = 1;
>> DOF_RRC = 6*(L - 1) - 5*J1 - 4*J2

DOF_RRC =

     4

>>
```

FIGURE E.10.1 Example 10.1 calculation procedure in MATLAB.

10.4 THE GENERAL TRANSFORMATION MATRIX

In Section 2.5, a general spatial transformation matrix was presented. This matrix is used to calculate point coordinates given in one link coordinate frame (Frame *j*) in reference to another link coordinate frame (Frame *i*) in a robotic manipulator. As previously shown, the general spatial transformation matrix can be expressed as

$$ {}_{j}^{i}[T]=\begin{bmatrix} R_{11} & R_{12} & R_{13} & \Delta_x \\ R_{21} & R_{22} & R_{23} & \Delta_y \\ R_{31} & R_{32} & R_{33} & \Delta_z \\ 0 & 0 & 0 & 1 \end{bmatrix} \tag{10.2} $$

where

$$ \begin{aligned} R_{11} &= \cos\delta_y \cos\delta_z \\ R_{12} &= \sin\delta_x \sin\delta_y \cos\delta_z - \cos\delta_x \sin\delta_z \\ R_{13} &= \cos\delta_x \sin\delta_y \cos\delta_z + \sin\delta_x \sin\delta_z \\ R_{21} &= \cos\delta_y \sin\delta_z \\ R_{22} &= \sin\delta_x \sin\delta_y \sin\delta_z + \cos\delta_x \cos\delta_z \\ R_{23} &= \cos\delta_x \sin\delta_y \sin\delta_z - \sin\delta_x \cos\delta_z \\ R_{31} &= -\sin\delta_y \\ R_{32} &= \sin\delta_x \cos\delta_y \\ R_{33} &= \cos\delta_x \cos\delta_y \end{aligned} $$

In this matrix, variables δ_x, δ_y, and δ_z are the rotations about a frame's *x*-, *y*-, and *z*-axes, respectively, and variables Δ_x, Δ_y, and Δ_z are the translations along a frame's *x*-, *y*-, and *z*-axes, respectively.[5]

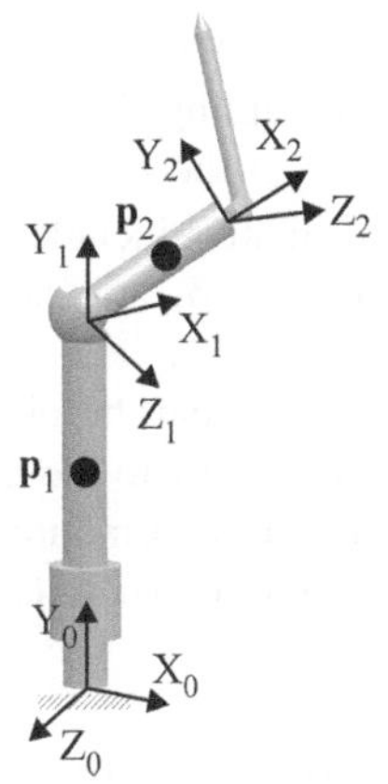

FIGURE 10.5 Robotic manipulator with frames and points.

Given ${}^{j}\{\mathbf{p}\}$, the spatial coordinates of a point $\mathbf{p}$ in Frame j, the coordinates of this point with respect to Frame i or ${}^{i}\{\mathbf{p}\}$ can be calculated as

$$ {}^{i}\{\mathbf{p}\} = {}^{i}_{j}[T]\,{}^{j}\{\mathbf{p}\} \tag{10.3} $$

where the spatial coordinates of $\mathbf{p}$ in frame j are ${}^{j}\{\mathbf{p}\} = \{\, p_x \quad p_y \quad p_z \quad 1 \,\}^T$.[6]

In Figure 10.5, the robotic manipulator in Figure 10.2b is again considered where Frames 0, 1, and 2 are attached to ground, Link 1, and Link 2, respectively. In this figure, Point $\mathbf{p}_1$ is attached to Link 1, and its coordinates are given with respect to Frame 1. Point $\mathbf{p}_2$ is attached to Link 2, and its coordinates are given with respect to Frame 2.

To calculate the coordinates of Point $\mathbf{p}_1$ with respect to the base frame, Equation (10.3) becomes

$$ {}^{0}\{\mathbf{p}_1\} = {}^{0}_{1}[T]\,{}^{1}\{\mathbf{p}_1\} \tag{10.4} $$

where ${}^{0}\{\mathbf{p}_1\}$ represents the coordinates of $\mathbf{p}_1$ with respect to the base frame (Frame X_0 - Y_0 - Z_0 in Figure 10.5).[7] The transformation matrix ${}^{0}_{1}[T]$ considers Frames 0 and 1. This matrix includes angular displacements δ_{1x}, δ_{1y}, and δ_{1z} and linear displacements Δ_{1x}, Δ_{1y}, and Δ_{1z}.

To calculate the value of Point $\mathbf{p}_2$ with respect to the base frame, Equation (10.4) becomes

$$ {}^{0}\{\mathbf{p}_2\} = {}^{0}_{1}[T]\,{}^{1}_{2}[T]\,{}^{2}\{\mathbf{p}_2\} = {}^{0}_{2}[T]\,{}^{2}\{\mathbf{p}_2\} \tag{10.5} $$

where ${}^{0}\{\mathbf{p}_2\}$ is the value of $\mathbf{p}_2$ with respect to the base frame. Because the transformation matrix ${}^{0}_{1}[T]$ already considers Frames 0 and 1, including the transformation matrix ${}^{1}_{2}[T]$ (which considers Frames 1 and 2) and taking the product of these two matrices produce a single transformation matrix that considers Frames 0 and 2. This matrix includes all of the displacement variables in ${}^{0}_{1}[T]$ as well as the additional angular displacements δ_{2x}, δ_{2y}, and δ_{2z} and linear displacements Δ_{2x}, Δ_{2y}, and Δ_{2z}.

Let's now assume an additional frame (X_3 - Y_3 - Z_3) is attached to the tool and the coordinates of a Point $\mathbf{p}_3$ are given with respect to this frame (Frame 3). To calculate the value of $\mathbf{p}_3$ with respect to the base frame, Equation (10.5) becomes

$$ {}^{0}\{\mathbf{p}_3\} = {}^{0}_{1}[T]\,{}^{1}_{2}[T]\,{}^{2}_{3}[T]\,{}^{3}\{\mathbf{p}_3\} = {}^{0}_{2}[T]\,{}^{2}_{3}[T]\,{}^{3}\{\mathbf{p}_3\} = {}^{0}_{3}[T]\,{}^{3}\{\mathbf{p}_3\} \tag{10.6} $$

where ${}^{0}\{\mathbf{p}_3\}$ is the value of $\mathbf{p}_3$ with respect to the base frame. Because the transformation matrix ${}^{0}_{2}[T]$ already considers Frames 0 and 2, including the transformation matrix ${}^{2}_{3}[T]$ (which considers Frames 2 and 3) and taking the product of these two matrices produce a single transformation matrix that considers Frames 0 and 3. This matrix includes all of the displacement variables in ${}^{0}_{2}[T]$ as well as the additional angular displacements δ_{3x}, δ_{3y}, and δ_{3z} and linear displacements Δ_{3x}, Δ_{3y}, and Δ_{3z}.

From Equations (10.4), (10.5), and (10.6), it can be observed that as an additional final frame is introduced, the transformation matrix to calculate the global coordinates of a point in the final frame becomes an increasing product of the transformation matrices for each frame. Knowing this, given a group of N frames, the transformation matrix to calculate the global coordinates of a point in the Nth frame can be expressed as

$$ {}^{0}_{N}[T] = {}^{0}_{1}[T]\,{}^{1}_{2}[T]\,{}^{2}_{3}[T]\cdots{}^{N-1}_{N}[T] \text{ [8]} \tag{10.7} $$

Therefore, given ${}^{N}\{\mathbf{p}_N\}$, the spatial coordinates of a Point $\mathbf{p}_N$ in Frame N, the global coordinates of this point with respect to the base frame or ${}^{0}\{\mathbf{p}_N\}$ can be calculated as

$$ {}^{0}\{\mathbf{p}_N\} = {}^{0}_{N}[T]\,{}^{N}\{\mathbf{p}_N\} \tag{10.8} $$

The coordinate frame x, y, and z rotations and translation variables presented in this section are analogous to *Denavit–Hartenberg* (or *DH*) parameters. Like the δ and Δ variables, DH parameters are also used to attach reference frames to links. In classical DH notation, only rotations about and translations along coordinate frame x- and z-axes are used. Therefore, with the y-axis rotations and translations eliminated, the general transformation matrix becomes

$$ {}^{i}_{j}[T] = \begin{bmatrix} \cos\delta_z & -\cos\delta_x \sin\delta_z & \sin\delta_x \sin\delta_z & \Delta_x \cos\delta_z \\ \sin\delta_z & \cos\delta_x \cos\delta_z & -\sin\delta_x \cos\delta_z & \Delta_x \sin\delta_z \\ 0 & \sin\delta_x & \cos\delta_x & \Delta_z \\ 0 & 0 & 0 & 1 \end{bmatrix} \text{ [9]} \tag{10.9} $$

Example 10.2

Problem Statement: Calculate the elements of transformation matrix ${}^{0}_{3}[T]$ for the coordinate frame displacement values given in Table E.10.1.

Known Information: Matrix (10.2) and Table E.10.1

Solution Approach: Because all variables in Matrix (10.2), except for those given in Table E.10.1, are 0, this matrix can be simplified. Figure E.10.2 includes the calculation procedure in MATLAB's command window.

TABLE E.10.1
Frame Displacement Variables (with Unitless Link Lengths)

Frames	δ_x	δ_y	δ_z	Δ_x	Δ_y	Δ_z
1 wrt 0	0	0	55°	0	0	0
2 wrt 1	0	0	0	0	0	1.75
3 wrt 2	0	0	0	3	0	0

```
>> d1z = 55*pi/180;
>> D2z = 1.75;
>> D3x = 3;
>> T01 = [cos(d1z), -sin(d1z), 0, 0; sin(d1z), cos(d1z), 0, 0;...
0, 0, 1, 0; 0, 0, 0, 1];
>> T12 = [1, 0, 0, 0; 0, 1, 0, 0; 0, 0, 1, D2z; 0, 0, 0, 1];
>> T23 = [1, 0, 0, D3x; 0, 1, 0, 0; 0, 0, 1, 0; 0, 0, 0, 1];
>> T03 = T01*T12*T23

T03 =

    0.5736   -0.8192         0    1.7207
    0.8192    0.5736         0    2.4575
         0         0    1.0000    1.7500
         0         0         0    1.0000

>>
```

FIGURE E.10.2 Example 10.2 calculation procedure in MATLAB.

10.5 FORWARD KINEMATICS

10.5.1 Definition and Application

In *forward kinematics*, the link dimensions and input joint motion of a robotic manipulator are known, and the corresponding output motion of the links (usually the end effector) is calculated [3]. By formulating kinematic displacement equations for a robotic manipulator to calculate the motion of specific link points and prescribing the link dimensions and input joint motions, the output link motion is calculated. The most common application for forward kinematics is for determining end-effector motion (e.g., tool paths and orientations).

10.5.2 P-P-P

Figure 10.6 includes the frames specified for the P-P-P robotic manipulator. The transformation matrix ${}^0_3[T]$ is required to formulate equations to calculate the global position of $\mathbf{p}_3$ on the end effector. To facilitate this procedure, Table 10.1 includes the displacement variables required to align Frame 0 to 1, Frame 1 to 2, and Frame 2 to 3. As this table indicates, a combination of three linear displacements is utilized in the P-P-P robotic manipulator.

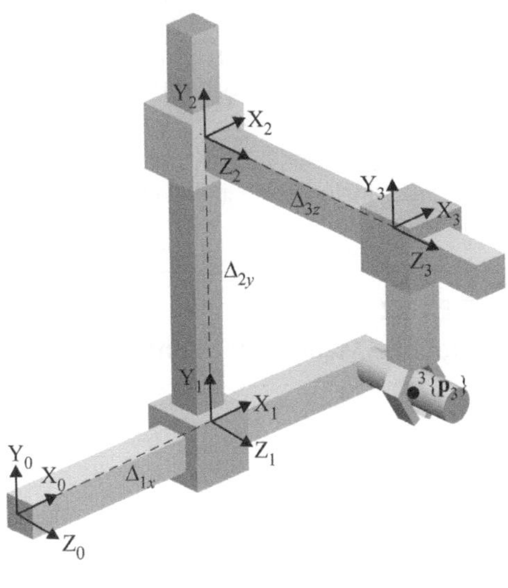

FIGURE 10.6 P-P-P robotic manipulator with frames and point.

TABLE 10.1
Frame Displacement Variables for the P-P-P Robotic Manipulator

Frames	δ_x	δ_y	δ_z	Δ_x	Δ_y	Δ_z
1 wrt 0	0	0	0	Δ_{1x}	0	0
2 wrt 1	0	0	0	0	Δ_{2y}	0
3 wrt 2	0	0	0	0	0	Δ_{3z}

From the information given in Table 10.1, the only nonzero variables are Δ_{1x}, Δ_{2y}, and Δ_{3z}. As a result, Matrix (10.7) becomes

$$ {}^0_3[T] = {}^0_1[T]\,{}^1_2[T]\,{}^2_3[T] = \begin{bmatrix} 1 & 0 & 0 & \Delta_{1x} \\ 0 & 1 & 0 & 0 \\ 0 & 0 & 1 & 0 \\ 0 & 0 & 0 & 1 \end{bmatrix} \begin{bmatrix} 1 & 0 & 0 & 0 \\ 0 & 1 & 0 & \Delta_{2y} \\ 0 & 0 & 1 & 0 \\ 0 & 0 & 0 & 1 \end{bmatrix} \begin{bmatrix} 1 & 0 & 0 & 0 \\ 0 & 1 & 0 & 0 \\ 0 & 0 & 1 & \Delta_{3z} \\ 0 & 0 & 0 & 1 \end{bmatrix} \tag{10.10} $$

Using this transformation matrix and Equation (10.8), Equation (10.11) is produced—the equation system required to calculate the global coordinates of $\mathbf{p}_3$.

$$ \begin{aligned} {}^0p_{3x} &= {}^3p_{3x} + \Delta_{1x} \\ {}^0p_{3y} &= {}^3p_{3y} + \Delta_{2y} \\ {}^0p_{3z} &= {}^3p_{3z} + \Delta_{3z} \end{aligned} \tag{10.11} $$

In the P-P-P robotic manipulator, the linear displacement variables Δ_{1x}, Δ_{2y}, and Δ_{3z} are not assigned constant values because they correspond to the translational displacements of the prismatic joints. Displacement ranges are assigned to these variables.

Example 10.3

Problem Statement: Using the joint displacements given in Table E.10.2, calculate the global path points achieved by the end effector of the P-P-P robotic manipulator. In this example, ${}^3\{\mathbf{p}_3\} = [0,-1,0]^T$.

Known Information: Equation (10.11) and Table E.10.2

Solution Approach: Table E.10.3 includes the global end-effector path point coordinates calculated using Equation (10.11).

TABLE E.10.2
P-P-P Robotic Manipulator Joint Displacements

Point	Δ_{1x}	Δ_{2y}	Δ_{3z}
1	0.5	1.1	–0.1
2	1	1.2	–0.15
3	1.5	1.3	–0.3
4	2	1.2	–0.45
5	2.5	1.1	–0.60

TABLE E.10.3
P-P-P Robotic Manipulator Path Point Coordinates

Point	${}^{0}p_{3x}$	${}^{0}p_{3y}$	${}^{0}p_{3z}$
1	0.5	0.1	−0.1
2	1	0.2	−0.15
3	1.5	0.3	−0.3
4	2	0.2	−0.45
5	2.5	0.1	−0.60

10.5.3 R-P-P

Figure 10.7 includes the frames specified for the R-P-P robotic manipulator. The transformation matrix ${}^{0}_{3}[T]$ is required to formulate equations to calculate the global position of $\mathbf{p}_3$ on the end effector. To facilitate this procedure, Table 10.2 includes the displacement variables required to align Frame 0 to 1, Frame 1 to 2, and Frame 2 to 3. As this table indicates, a combination of a single angular displacement and two linear displacements is utilized in the R-P-P robotic manipulator.

From the information given in Table 10.2, the only nonzero variables are δ_{1z}, Δ_{2z}, and Δ_{3x}. As a result, Matrix (10.7) becomes

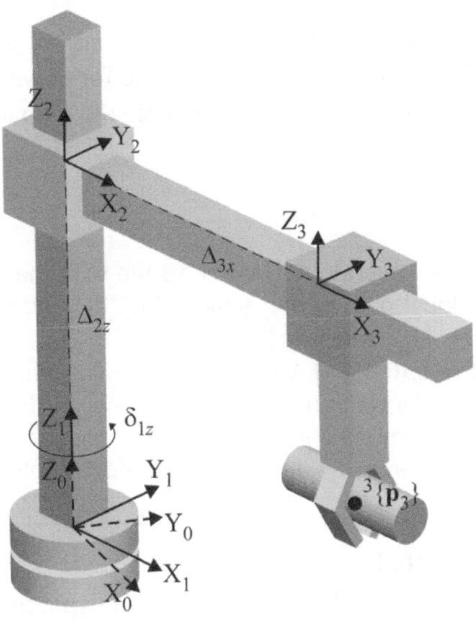

FIGURE 10.7 R-P-P robotic manipulator with frames and point.

TABLE 10.2
Frame Displacement Variables for the R-P-P Robotic Manipulator

Frames	δ_x	δ_y	δ_z	Δ_x	Δ_y	Δ_z
1 wrt 0	0	0	δ_{1z}	0	0	0
2 wrt 1	0	0	0	0	0	Δ_{2z}
3 wrt 2	0	0	0	Δ_{3x}	0	0

$$
{}^{0}_{3}[T] = {}^{0}_{1}[T]\,{}^{1}_{2}[T]\,{}^{2}_{3}[T] = \begin{bmatrix} \cos\delta_{1z} & -\sin\delta_{1z} & 0 & 0 \\ \sin\delta_{1z} & \cos\delta_{1z} & 0 & 0 \\ 0 & 0 & 1 & 0 \\ 0 & 0 & 0 & 1 \end{bmatrix} \begin{bmatrix} 1 & 0 & 0 & 0 \\ 0 & 1 & 0 & 0 \\ 0 & 0 & 1 & \Delta_{2z} \\ 0 & 0 & 0 & 1 \end{bmatrix} \begin{bmatrix} 1 & 0 & 0 & \Delta_{3x} \\ 0 & 1 & 0 & 0 \\ 0 & 0 & 1 & 0 \\ 0 & 0 & 0 & 1 \end{bmatrix} \tag{10.12}
$$

Using this transformation matrix and Equation (10.8), Equation (10.13) is produced—the equation system required to calculate the global coordinates of $\mathbf{p}_3$.

$$
\begin{aligned}
{}^{0}p_{3x} &= {}^{3}p_{3x}\cos\delta_{1z} - {}^{3}p_{3y}\sin\delta_{1z} + \Delta_{3x}\cos\delta_{1z} \\
{}^{0}p_{3y} &= {}^{3}p_{3x}\sin\delta_{1z} + {}^{3}p_{3y}\cos\delta_{1z} + \Delta_{3x}\sin\delta_{1z} \\
{}^{0}p_{3z} &= {}^{3}p_{3z} + \Delta_{2z}
\end{aligned} \tag{10.13}
$$

In the R-P-P robotic manipulator, the linear and angular displacement variables Δ_{2z}, Δ_{3x}, and Δ_{3z} are not assigned constant values because they correspond to the translational displacements, the prismatic joints, and rotational displacements of the revolute joint. Displacement ranges are assigned to these variables.

Appendix L.1 includes the MATLAB file user instructions for R-P-P robotic manipulator forward kinematics. In this MATLAB file (which is available for download at https://www.routledge.com/9781032956800). Equation (10.13) is used to calculate the global coordinates of $\mathbf{p}_3$.

Example 10.4

Problem Statement: Using the Appendix L.1 MATLAB file with the joint displacements given in Table E.10.4, calculate the global path points achieved by the end effector of the R-P-P robotic manipulator. In this example, ${}^{3}\{\mathbf{p}_3\} = [0,0,-1]^T$.

Known Information: Table E.10.4 and Appendix L.1 MATLAB file

Solution Approach: The data for columns δ_{1z} and Δ_{2z} and Δ_{3x} and δ_{1z} in Table E.10.4 are first specified in the file *RPP_Input.csv*. Figure E.10.3 includes the input specified (in bold text) in the Appendix L.1 MATLAB file, and Table E.10.5 includes the global end-effector path point coordinates calculated.

TABLE E.10.4
R-P-P Robotic Manipulator Joint Displacements

Point	δ_{1z} (deg)	Δ_{2z}	Δ_{3x}
1	12	1.1	–0.1
2	24	1.2	–0.15
3	36	1.3	–0.3
4	48	1.2	–0.45
5	60	1.1	–0.60

```
p3_3 = [0, 0, -1];
```

FIGURE E.10.3 Specified input (in bold text) in the Appendix L.1 MATLAB file for Example 10.4.

TABLE E.10.5
R-P-P Robotic Manipulator Path Point Coordinates

Point	${}^{0}p_{3x}$	${}^{0}p_{3y}$	${}^{0}p_{3z}$
1	–0.0978	–0.0208	0.1000
2	–0.1370	–0.0610	0.2000
3	–0.2427	–0.1763	0.3000
4	–0.3011	–0.3344	0.2000
5	–0.3000	–0.5196	0.1000

10.5.4 R-R-P

Figure 10.8 includes the frames specified for the R-R-P robotic manipulator. The transformation matrix ${}^{0}_{3}[T]$ is required to formulate equations to calculate the global position of $\mathbf{p}_3$ on the end effector. To facilitate this procedure, Table 10.3 includes the displacement variables required to align Frames 0 to 1, Frames 1 to 2, and Frames 2 to 3. As this table indicates, a combination of two angular displacements and three linear displacements is utilized in the R-R-P robotic manipulator.

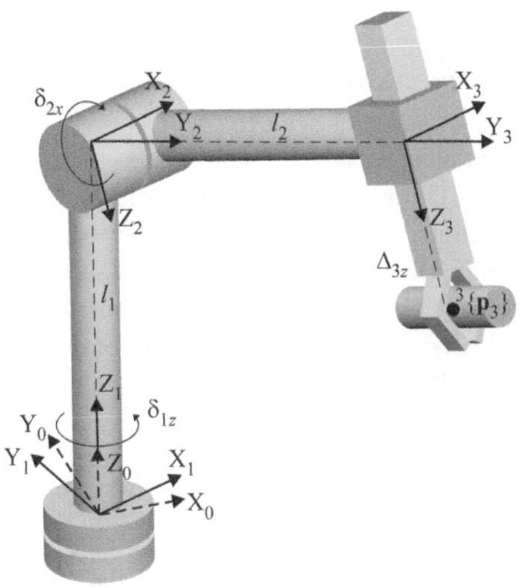

FIGURE 10.8 R-R-P robotic manipulator with frames and point.

TABLE 10.3
Frame Displacement Variables for the R-R-P Robotic Manipulator

Frames	δ_x	δ_y	δ_z	Δ_x	Δ_y	Δ_z
1 wrt 0	0	0	δ_{1z}	0	0	0
2 wrt 1	δ_{2x}	0	0	0	0	l_1
3 wrt 2	0	0	0	0	l_2	Δ_{3z}

From the information given in Table 10.3, the only nonzero variables are δ_{1z}, δ_{2x}, Δ_{3z}, and the terms representing Δ_{3z} and Δ_{3y}.[10] As a result, Matrix (10.7) becomes

$$
{}^0_3[T] = {}^0_1[T]\,{}^1_2[T]\,{}^2_3[T]
$$

$$
= \begin{bmatrix} \cos\delta_{1z} & -\sin\delta_{1z} & 0 & 0 \\ \sin\delta_{1z} & \cos\delta_{1z} & 0 & 0 \\ 0 & 0 & 1 & 0 \\ 0 & 0 & 0 & 1 \end{bmatrix} \begin{bmatrix} 1 & 0 & 0 & 0 \\ 0 & \cos\delta_{2x} & -\sin\delta_{2x} & 0 \\ 0 & \sin\delta_{2x} & \cos\delta_{2x} & l_1 \\ 0 & 0 & 0 & 1 \end{bmatrix} \begin{bmatrix} 1 & 0 & 0 & 0 \\ 0 & 1 & 0 & l_2 \\ 0 & 0 & 1 & \Delta_{3z} \\ 0 & 0 & 0 & 1 \end{bmatrix} \tag{10.14}
$$

Using the resulting transformation matrix and Equation (10.8), Equation (10.15) is produced—the equation system to calculate the global coordinates of $\mathbf{p}_3$.

$$
\begin{aligned}
{}^0p_{3x} &= {}^3p_{3x}\cos\delta_{1z} - {}^3p_{3y}\sin\delta_{1z}\cos\delta_{2x} + {}^3p_{3z}\sin\delta_{1z}\sin\delta_{2x} - l_2\sin\delta_{1z}\cos\delta_{2x} \\
&\quad + \Delta_{3z}\sin\delta_{1z}\sin\delta_{2x} \\
{}^0p_{3y} &= {}^3p_{3x}\sin\delta_{1z} + {}^3p_{3y}\cos\delta_{1z}\cos\delta_{2x} - {}^3p_{3z}\cos\delta_{1z}\sin\delta_{2x} + l_2\cos\delta_{1z}\cos\delta_{2x} \\
&\quad - \Delta_{3z}\cos\delta_{1z}\sin\delta_{2x} \\
{}^0p_{3z} &= {}^3p_{3y}\sin\delta_{2x} + {}^3p_{3z}\cos\delta_{2x} + l_2\sin\delta_{2x} + \Delta_{3z}\cos\delta_{2x} + l_1
\end{aligned} \tag{10.15}
$$

In the R-R-P robotic manipulator, variables l_1 and l_2 are assigned constant values because they represent constant link lengths. The angular and linear displacement variables δ_{1z}, δ_{2x}, and Δ_{3z}, however, correspond to the rotational displacements of the revolute joints and the translational displacements of the prismatic joint. Displacement ranges are assigned to these variables.

Appendix L.2 includes the MATLAB file user instructions for R-R-P robotic manipulator forward kinematics. In this MATLAB file (which is available for download at https://www.routledge.com/9781032956800), Equation (10.15) is used to calculate the global coordinates of $\mathbf{p}_3$.

Example 10.5

Problem Statement: Using the Appendix L.2 MATLAB file with the joint displacements given in Table E.10.6, calculate the global path points achieved by the end effector of the R-R-P robotic manipulator. In this example, $l_1 = l_2 = 0.5$ (unitless link lengths) and ${}^3\{\mathbf{p}_3\} = [0,0,0]^T$.

TABLE E.10.6
R-R-P Robotic Manipulator Joint Displacements

Point	δ_{1z} (deg)	δ_{2x} (deg)	Δ_{3z}
1	12	10	0.5
2	24	20	1
3	36	30	1.5
4	48	40	2
5	60	50	2.5

Known Information: Table E.10.6 and Appendix L.2 MATLAB file

Solution Approach: The data for columns δ_{1z}, δ_{2x}, and Δ_{3z} in Table E.10.6 are first specified in the file *RRP_Input.csv*. Figure E.10.4 includes the input specified (in bold text) in the Appendix L.2 MATLAB file, and Table E.10.7 includes the global end-effector path point coordinates calculated.

```
l1 = 0.5;
l2 = 0.5;
p3_3 = [0, 0, 0];
```

FIGURE E.10.4 Specified input (in bold text) in the Appendix L.2 MATLAB file for Example 10.5.

TABLE E.10.7
R-R-P Robotic Manipulator Path Point Coordinates

Point	${}^0p_{3x}$	${}^0p_{3y}$	${}^0p_{3z}$
1	−0.0843	0.3967	1.0792
2	−0.0520	0.1168	1.6107
3	0.1863	−0.2564	2.0490
4	0.6707	−0.6039	2.3535
5	1.3802	−0.7969	2.4900

10.5.5 R-R-R

Figure 10.9 includes the frames specified for the R-R-R robotic manipulator. The transformation matrix ${}^0_3[T]$ is required to formulate the equations required to calculate the global position of $\mathbf{p}_3$ on the end effector. To facilitate this procedure, Table 10.3 includes the displacement variables required to align Frames 0 to 1, Frames 1 to 2, and Frames 2 to 3. As this table indicates, a combination of three angular displacements and two linear displacements is utilized in the R-R-R robotic manipulator.

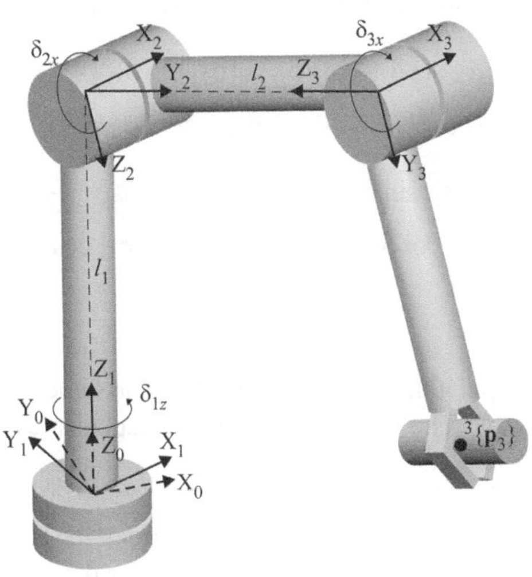

FIGURE 10.9 R-R-R robotic manipulator with frames and point.

TABLE 10.4
Frame Displacement Variables for the R-R-R Robotic Manipulator

Frames	δ_x	δ_y	δ_z	Δ_x	Δ_y	Δ_z
1 wrt 0	0	0	δ_{1z}	0	0	0
2 wrt 1	δ_{2x}	0	0	0	0	l_1
3 wrt 2	δ_{3x}	0	0	0	l_2	0

From the information given in Table 10.4, the only nonzero variables are δ_{1z}, δ_{2x}, δ_{3x}, and the terms representing Δ_{2z} and Δ_{3y}.[11] As a result, Matrix (10.7) becomes

$$
{}_3^0[T] = {}_1^0[T]\,{}_2^1[T]\,{}_3^2[T]
$$

$$
= \begin{bmatrix} \cos\delta_{1z} & -\sin\delta_{1z} & 0 & 0 \\ \sin\delta_{1z} & \cos\delta_{1z} & 0 & 0 \\ 0 & 0 & 1 & 0 \\ 0 & 0 & 0 & 1 \end{bmatrix} \begin{bmatrix} 1 & 0 & 0 & 0 \\ 0 & \cos\delta_{2x} & -\sin\delta_{2x} & 0 \\ 0 & \sin\delta_{2x} & \cos\delta_{2x} & l_1 \\ 0 & 0 & 0 & 1 \end{bmatrix} \begin{bmatrix} 1 & 0 & 0 & 0 \\ 0 & \cos\delta_{3x} & -\sin\delta_{3x} & l_2 \\ 0 & \sin\delta_{3x} & \cos\delta_{3x} & 0 \\ 0 & 0 & 0 & 1 \end{bmatrix} \quad (10.16)
$$

Using the resulting transformation matrix and Equation (10.8), Equation (10.17) is produced—the equation system required to calculate the global coordinates of $\mathbf{p}_3$.

$$
\begin{aligned}
{}^0p_{3x} &= {}^3p_{3x}\cos\delta_{1z} + {}^3p_{3y}\left(\sin\delta_{1z}\sin\delta_{2x}\sin\delta_{3x} - \sin\delta_{1z}\cos\delta_{2x}\cos\delta_{3x}\right) \\
&\quad + {}^3p_{3z}\left(\sin\delta_{1z}\cos\delta_{2x}\sin\delta_{3x} + \sin\delta_{1z}\sin\delta_{2x}\cos\delta_{3x}\right) - l_2\sin\delta_{1z}\cos\delta_{2x} \\
{}^0p_{3y} &= {}^3p_{3x}\sin\delta_{1z} + {}^3p_{3y}\left(\cos\delta_{1z}\cos\delta_{2x}\cos\delta_{3x} - \cos\delta_{1z}\sin\delta_{2x}\sin\delta_{3x}\right) \\
&\quad + {}^3p_{3z}\left(-\cos\delta_{1z}\cos\delta_{2x}\sin\delta_{3x} - \cos\delta_{1z}\sin\delta_{2x}\cos\delta_{3x}\right) + l_2\cos\delta_{1z}\cos\delta_{2x} \\
{}^0p_{3z} &= {}^3p_{3y}\left(\sin\delta_{2x}\cos\delta_{3x} + \cos\delta_{2x}\sin\delta_{3x}\right) + {}^3p_{3z}\left(\cos\delta_{2x}\cos\delta_{3x} - \sin\delta_{2x}\sin\delta_{3x}\right) \\
&\quad + l_2\sin\delta_{2x} + l_1
\end{aligned} \quad (10.17)
$$

In the R-R-R robotic manipulator, variables l_1 and l_2 are assigned constant values because they represent constant link lengths. The angular and linear displacement variables δ_{1z}, δ_{2x}, and δ_{3x}, however, correspond to the rotational displacements of the revolute joints. Displacement ranges are often assigned to these variables.

Appendix L.3 includes the MATLAB file user instructions for R-R-R robotic manipulator forward kinematics. In this MATLAB file (which is available for download at https://www.routledge.com/9781032956800), Equation (10.17) is used to calculate the global coordinates of $\mathbf{p}_3$.

Example 10.6

Problem Statement: Using the Appendix L.3 MATLAB file with the joint displacements given in Table E.10.8, calculate the global path points achieved by the end effector of the R-R-R robotic manipulator. In this example, $l_1 = l_2 = 0.5$ (unitless link lengths) and ${}^3\{\mathbf{p}_3\} = [0,1,0]^T$.

Known Information: Table E.10.8 and Appendix L.3 MATLAB file

Solution Approach: The data for columns δ_{1z}, δ_{2x}, and δ_{3x} in Table E.10.8 are first specified in the file *RRR_Input.csv*. Figure E.10.5 includes the input specified (in bold text) in the Appendix L.3 MATLAB file, and Table E.10.9 includes the global end-effector path point coordinates calculated using the Appendix L.3 MATLAB file.

TABLE E.10.8
R-R-R Robotic Manipulator Joint Displacements

Point	δ_{1z} (deg)	δ_{2x} (deg)	δ_{3x} (deg)
1	12	10	–5
2	24	20	–10
3	36	30	–15
4	48	40	–20
5	60	50	–25

```
l1 = 0.5;
l2 = 0.5;
p3_3 = [0, 1, 0];
```

FIGURE E.10.5 Specified input (in bold text) in the Appendix L.3 MATLAB file for Example 10.6.

TABLE E.10.9
R-R-R Robotic Manipulator Path Point Coordinates

Point	${}^0p_{3x}$	${}^0p_{3y}$	${}^0p_{3z}$
1	–0.3095	1.4561	0.6740
2	–0.5917	1.3289	0.8447
3	–0.8223	1.1318	1.0088
4	–0.9830	0.8851	1.1634
5	–1.0632	0.6139	1.3056

10.5.6 R-R-C

Figure 10.10 includes the frames specified for the R-R-C robotic manipulator. The transformation matrix ${}^0_3[T]$ is required to formulate equations to calculate the global position of $\mathbf{p}_3$ on the end effector. To facilitate this procedure, Table 10.5 includes the displacement variables required to align Frames 0 to 1, Frames 1 to 2, and Frames 2 to 3. As this table indicates, a combination of three angular displacements and four linear displacements is utilized in the R-R-C robotic manipulator.

Using the information given in Table 10.5, the nonzero variables are δ_{1z}, δ_{2z}, δ_{3z}, Δ_{3z}, and the terms representing Δ_{1z}, Δ_{2x}, and Δ_{3x}.[12] As a result, Matrix (10.7) becomes

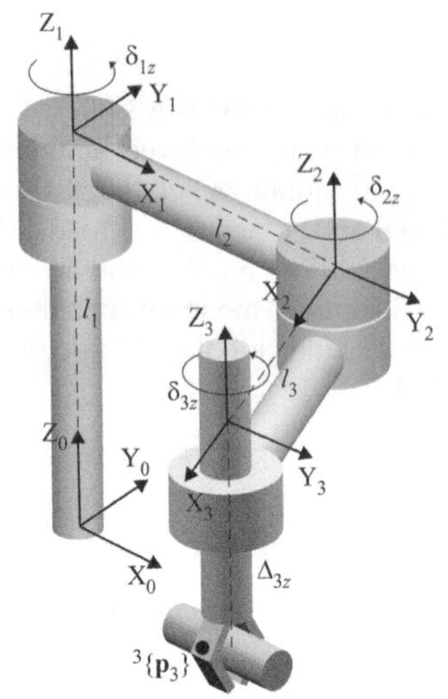

FIGURE 10.10 R-R-C robotic manipulator with frames and point.

TABLE 10.5
Frame Displacement Variables for the R-R-C Robotic Manipulator

Frames	δ_x	δ_y	δ_z	Δ_x	Δ_y	Δ_z
1 wrt 0	0	0	δ_{1z}	0	0	l_1
2 wrt 1	0	0	δ_{2z}	l_2	0	0
3 wrt 2	0	0	δ_{3z}	l_3	0	Δ_{3z}

$$
{}^{0}_{3}[T] = {}^{0}_{1}[T]\,{}^{1}_{2}[T]\,{}^{2}_{3}[T]
$$

$$
= \begin{bmatrix} \cos\delta_{1z} & -\sin\delta_{1z} & 0 & 0 \\ \sin\delta_{1z} & \cos\delta_{1z} & 0 & 0 \\ 0 & 0 & 1 & l_1 \\ 0 & 0 & 0 & 1 \end{bmatrix} \begin{bmatrix} \cos\delta_{2z} & -\sin\delta_{2z} & 0 & l_2 \\ \sin\delta_{2z} & \cos\delta_{2z} & 0 & 0 \\ 0 & 0 & 1 & 0 \\ 0 & 0 & 0 & 1 \end{bmatrix} \qquad (10.18)
$$

$$
\begin{bmatrix} \cos\delta_{3z} & -\sin\delta_{3z} & 0 & l_3 \\ \sin\delta_{3z} & \cos\delta_{3z} & 0 & 0 \\ 0 & 0 & 1 & \Delta_{3z} \\ 0 & 0 & 0 & 1 \end{bmatrix}
$$

Using the resulting transformation matrix and Equation (10.8), Equation (10.19) is produced—the equation system to calculate the global coordinates of $\mathbf{p}_3$.

$$
\begin{aligned}
{}^0p_{3x} &= {}^3p_{3x}\left[(\cos\delta_{1z}\cos\delta_{2z} - \sin\delta_{1z}\sin\delta_{2z})\cos\delta_{3z} + (-\cos\delta_{1z}\sin\delta_{2z} - \sin\delta_{1z}\cos\delta_{2z})\sin\delta_{3z}\right] \\
&\quad + {}^3p_{3y}\left[-(\cos\delta_{1z}\cos\delta_{2z} - \sin\delta_{1z}\sin\delta_{2z})\sin\delta_{3z} + (-\cos\delta_{1z}\sin\delta_{2z} - \sin\delta_{1z}\cos\delta_{2z})\cos\delta_{3z}\right] \\
&\quad + l_3(\cos\delta_{1z}\cos\delta_{2z} - \sin\delta_{1z}\sin\delta_{2z}) + l_2\cos\delta_{1z} \\
{}^0p_{3y} &= {}^3p_{3x}\left[(\sin\delta_{1z}\cos\delta_{2z} + \cos\delta_{1z}\sin\delta_{2z})\cos\delta_{3z} + (\cos\delta_{1z}\cos\delta_{2z} - \sin\delta_{1z}\sin\delta_{2z})\sin\delta_{3z}\right] \\
&\quad + {}^3p_{3y}\left[-(\sin\delta_{1z}\cos\delta_{2z} + \cos\delta_{1z}\sin\delta_{2z})\sin\delta_{3z} + (\cos\delta_{1z}\cos\delta_{2z} - \sin\delta_{1z}\sin\delta_{2z})\cos\delta_{3z}\right] \\
&\quad + l_3(\sin\delta_{1z}\cos\delta_{2z} + \cos\delta_{1z}\sin\delta_{2z}) + l_2\sin\delta_{1z} \\
{}^0p_{3z} &= {}^3p_{3z} + l_1 + \Delta_{3z}
\end{aligned}
\tag{10.19}
$$

In the R-R-C robotic manipulator, variables l_1, l_2, and l_3 are assigned constant values because they represent constant link lengths. The angular and linear displacement variables δ_{1z}, δ_{2z}, δ_{3z}, and Δ_{3y}, however, correspond to the rotational and translational displacements of the revolute and cylindrical joints, respectively. Displacement ranges are often assigned to these variables.

Appendix L.4 includes the MATLAB file user instructions for R-R-C robotic manipulator forward kinematics. In this MATLAB file (which is available for download at https://www.routledge.com/9781032956800), Equation (10.19) is used to calculate the global coordinates of $\mathbf{p}_3$.

Example 10.7

Problem Statement: Using the Appendix L.4 MATLAB file with the joint displacements given in Table E.10.10, calculate the global path points achieved by the end effector of the R-R-C robotic manipulator. In this example, $l_1 = 1$, $l_2 = l_3 = 0.5$ (unitless link lengths), and ${}^3\{\mathbf{p}_3\} = [1,0,0]^T$.

Known Information: Table E.10.10 and Appendix L.4 MATLAB file

Solution Approach: The data for columns δ_{1z}, δ_{2z}, δ_{3z}, and Δ_{3z} in Table E.10.10 are first specified in the file *RRC_Input.csv*. Figure E.10.6 includes the input specified (in bold text) in the Appendix L.4 MATLAB file, and Table E.10.11 includes the global end-effector path point coordinates calculated using the Appendix L.3 MATLAB file.

TABLE E.10.10
R-R-C Robotic Manipulator Joint Displacements

Point	δ_{1z} (deg)	δ_{2z} (deg)	δ_{3z} (deg)	Δ_{3z}
1	12	–5	15	–0.1000
2	24	–10	30	–0.2000
3	36	–15	45	–0.3000
4	48	–20	60	–0.2000
5	60	–25	75	–0.1000

```
l1 = 1;
l2 = 0.5;
l3 = 0.5;
p3_3 = [1, 0, 0];
```

FIGURE E.10.6 Specified input (in bold text) in the Appendix L.4 MATLAB file for Example 10.7.

TABLE E.10.11
R-R-C Robotic Manipulator Path Point Coordinates

Point	${}^{0}p_{3x}$	${}^{0}p_{3y}$	${}^{0}p_{3z}$
1	1.9125	0.5395	0.9000
2	1.6613	1.0190	0.8000
3	1.2780	1.3866	0.7000
4	0.8109	1.6057	0.8000
5	0.3176	1.6595	0.9000

10.6 INVERSE KINEMATICS

10.6.1 Definition and Application

While the robotic manipulator dimensions are known and the end-effector motion is calculated in forward kinematics, in *inverse kinematics*, the end-effector motion and link dimensions are known, and the joint motion required to achieve the end-effector motion is calculated [4–7]. Because of this, inverse kinematics is considered the reverse of forward kinematics. It is accomplished by calculating joint motion solutions (using the same forward kinematics-based equation system for a robotic manipulator) given the end-effector motion and link lengths. The most common application for inverse kinematics is for determining the joint motion required to achieve the required end-effector motion (e.g., the joint motion required to achieve the prescribed tool paths and orientations).[13]

10.6.2 P-P-P

Because Equation (10.11) includes three equations—each containing a single unknown variable (Δ_{1x}, Δ_{2y}, and Δ_{3z}, respectively), these unknowns can be calculated algebraically for the inverse kinematics of the P-P-P robotic manipulator.

Solving for Δ_{1x}, Δ_{2y}, and Δ_{3z} in Equation (10.11) produces

$$\begin{aligned} \Delta_{1x} &= {}^{0}p_{3x} - {}^{3}p_{3x} \\ \Delta_{2y} &= {}^{0}p_{3y} - {}^{3}p_{3y} \\ \Delta_{3z} &= {}^{0}p_{3z} - {}^{3}p_{3z} \end{aligned} \tag{10.20}$$

From Equation (10.20), the user can specify the global end-effector coordinates ${}^{0}\{\mathbf{p}_3\}$, the end effector coordinates in Frame 3 ${}^{0}\{\mathbf{p}_3\}$, and calculate the required displacements of each prismatic joint in the P-P-P robotic manipulator (Δ_{1x}, Δ_{2y}, and Δ_{3z}).

The global path point coordinates ${}^{0}\{\mathbf{p}_3\}$ should be prescribed from within the boundaries of the workspace of the P-P-P robotic manipulator. This ensures that the prescribed points are all within the reach of the robotic manipulator (and subsequently will be achieved). As shown in Figure 10.11, the P-P-P robotic manipulator has a prismatic (or rectangular cuboid) workspace with outer *x*, *y*, and *z* dimensions of $\Delta_{1x_{\max}}$, $\Delta_{2y_{\max}}$, and $\Delta_{3z_{\max}}$, respectively (the maximum prismatic joint translations).

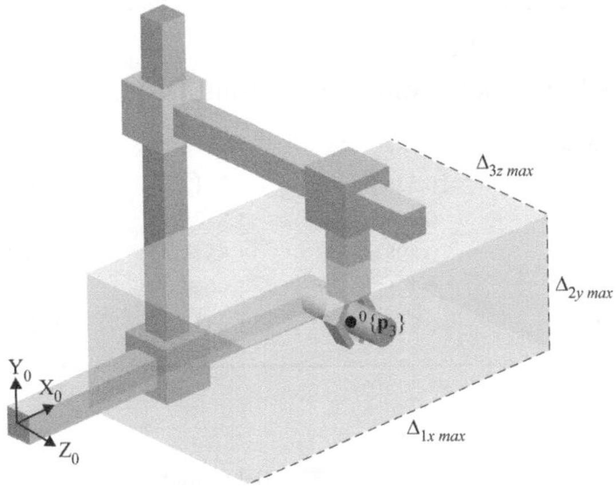

FIGURE 10.11 P-P-P robotic manipulator and workspace.

Example 10.8

Problem Statement: Calculate the P-P-P joint displacements required to achieve the global path points given in Table E.10.12. In this example, $^3\{\mathbf{p}_3\} = [0,-1,0]^T$.

TABLE E.10.12
P-P-P Robotic Manipulator End-Effector Path Point Coordinates

Point	$^0p_{3x}$	$^0p_{3y}$	$^0p_{3z}$
1	0.683	0.375	0.2165
2	0.6764	0.4521	0.2566
3	0.6569	0.5306	0.289
4	0.625	0.6083	0.3125
5	0.5817	0.6826	0.3266
6	0.5283	0.7514	0.3307
7	0.4665	0.8125	0.3248
8	0.3981	0.8641	0.309
9	0.3252	0.9047	0.2838
10	0.25	0.933	0.25

Known Information: Equation (10.20) and Table E.10.12

Solution Approach: Table E.10.13 includes the joint displacements calculated using Equation (10.20). Figure E.10.7 includes the initial and final positions of the P-P-P robotic manipulator over the range of prescribed global end-effector points.

TABLE E.10.13
P-P-P Robotic Manipulator Joint Displacements

Point	Δ_{1x}	Δ_{2y}	Δ_{3z}
1	0.683	1.375	0.2165
2	0.6764	1.4521	0.2566
3	0.6569	1.5306	0.289
4	0.625	1.6083	0.3125
5	0.5817	1.6826	0.3266

(Continued)

TABLE E.10.13 (*Continued*)
P-P-P Robotic Manipulator Joint Displacements

Point	Δ_{1x}	Δ_{2y}	Δ_{3z}
6	0.5283	1.7514	0.3307
7	0.4665	1.8125	0.3248
8	0.3981	1.8641	0.309
9	0.3252	1.9047	0.2838
10	0.25	1.933	0.25

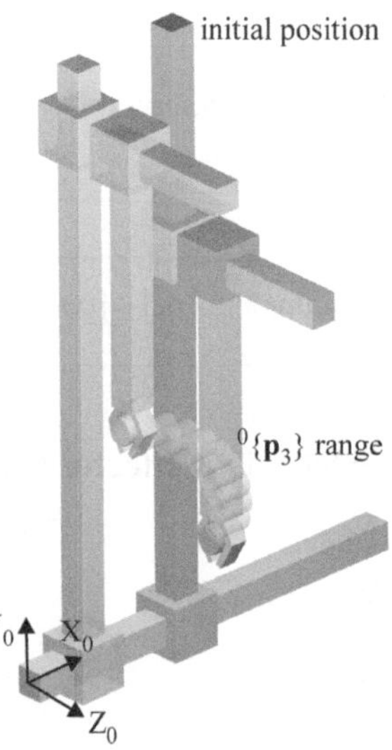

FIGURE E.10.7 P-P-P robotic manipulator in (semi-transparent) initial and final positions.

10.6.3 R-P-P

Unlike Equation (10.11), the joint displacement variables δ_{1z}, Δ_{2z}, and Δ_{3x} in Equation (10.13) cannot be calculated algebraically. With unknowns δ_{1z}, Δ_{2z}, and Δ_{3x}, Equation (10.13) becomes a set of three nonlinear simultaneous equations. Unconstrained nonlinear equation system optimization (see Section 2.7.1) is required for the inverse kinematics of the R-P-P robotic manipulator. Appendix L.5 includes the user instructions for the MATLAB file to calculate joint displacement solutions for Equation (10.13) given $^3\{\mathbf{p}_3\}$ and a prescribed range of values for $^0\{\mathbf{p}_3\}$.

As shown in Figure 10.12, the R-P-P robotic manipulator has a cylindrical workspace (having a center axis that is collinear with Δ_{2z}) with outer cylinder height and radius dimensions of $\Delta_{2z_{\max}}$ and $\Delta_{3x_{\max}}$, respectively (the maximum prismatic joint translations).

Example 10.9

Problem Statement: Using the Appendix L.5 MATLAB file, calculate the R-P-P joint displacements required to achieve the global path points given in Table E.10.12 in Example 10.8. In this example, $^3\{\mathbf{p}_3\} = [0,0,-1]^T$.

Known Information: Table E.10.12 and Appendix L.5 MATLAB file

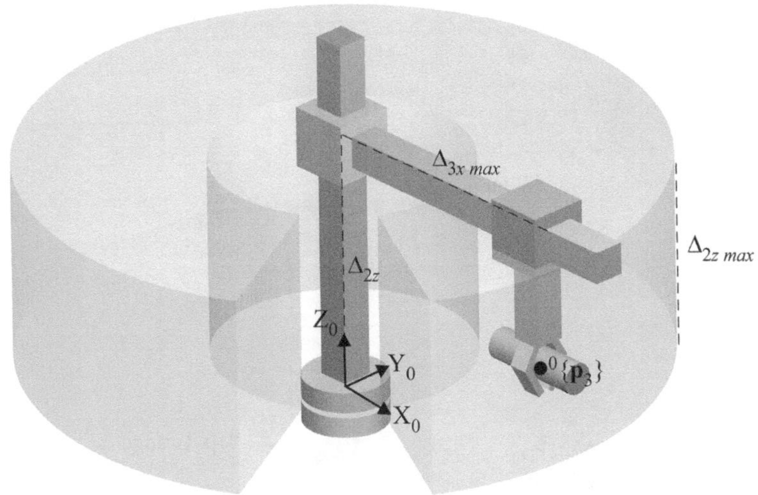

FIGURE 10.12 R-P-P robotic manipulator and workspace.

Solution Approach: The data for columns $^0p_{3x}$, $^0p_{3y}$, and $^0p_{3z}$ in Table E.10.12 are first specified in the file *RPP_Input.csv*. Figure E.10.8 includes the input specified (in bold text) in the Appendix L.5 MATLAB file, and Table E.10.14 includes the joint displacements calculated. Figure E.10.9 includes the initial and final positions of the R-P-P robotic manipulator over the range of prescribed global end-effector points.

```
p3_3 = [0, 0, -1];
```

FIGURE E.10.8 Specified input (in bold text) in the Appendix L.5 MATLAB file for Example 10.9.

TABLE E.10.14
R-P-P Robotic Manipulator Joint Displacements

Point	δ_{1z} (deg)	Δ_{2z}	Δ_{3x}
1	28.7689	1.2165	0.7792
2	33.7584	1.2566	0.8136
3	38.929	1.289	0.8444
4	44.2242	1.3125	0.8722
5	49.5629	1.3266	0.8968
6	54.8894	1.3307	0.9185
7	60.1375	1.3248	0.9369
8	65.264	1.309	0.9514
9	70.2287	1.2838	0.9614
10	74.9998	1.25	0.9659

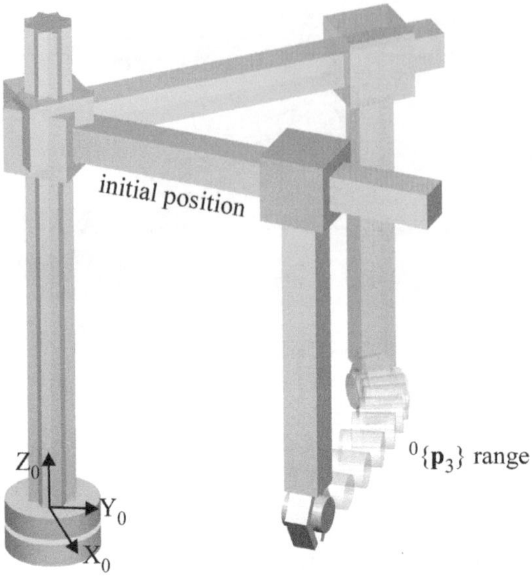

FIGURE E.10.9 R-P-P robotic manipulator in initial and final positions.

10.6.4 R-R-P

The joint displacement variables δ_{1z}, δ_{2x}, and Δ_{3z} in Equation (10.15) cannot be calculated algebraically also. With unknowns δ_{1z}, δ_{2x}, and Δ_{3z}, Equation (10.15) becomes a set of three nonlinear simultaneous equations. Like the R-P-P robotic manipulator, unconstrained nonlinear equation system optimization is also required for the inverse kinematics of the R-R-P robotic manipulator. Appendix L.6 includes the user instructions for the MATLAB file to calculate joint displacement solutions for Equation (10.15) given $^3\{\mathbf{p}_3\}$ and a prescribed range of values for $^0\{\mathbf{p}_3\}$.

As shown in Figure 10.13, the R-R-P robotic manipulator has a spherical workspace (having center coordinates center $= (0,0,l_1)$) with an outer radius dimension of $r_{\text{outer}} = \sqrt{(l_2)^2 + (\Delta_{3z_{\max}})^2}$.

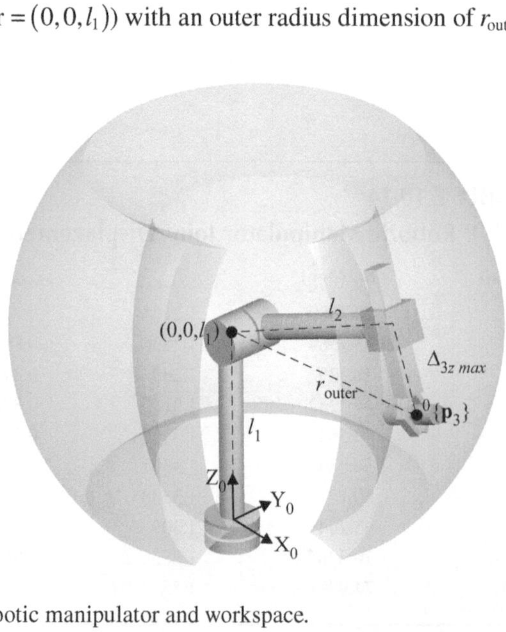

FIGURE 10.13 R-R-P robotic manipulator and workspace.

Example 10.10

Problem Statement: Using the Appendix L.6 MATLAB file, calculate the R-R-P joint displacements required to achieve the global path points given in Table E.10.12 in Example 10.8. In this example, $l_1 = l_2 = 0.5$ (unitless link lengths) and ${}^{3}\{\mathbf{p}_3\} = [0,0,0]^T$.

Known Information: Table E.10.12 and Appendix L.6 MATLAB file

Solution Approach: The data for columns ${}^{0}p_{3x}$, ${}^{0}p_{3y}$, and ${}^{0}p_{3z}$ in Table E.10.12 are first specified in the file *RRP_Input.csv*. Figure E.10.10 includes the input specified (in bold text) in the Appendix L.6 MATLAB file, and Table E.10.15 includes the joint displacements calculated. Figure E.10.11 includes the initial and final positions of the R-R-P robotic manipulator over the range of prescribed global end-effector points.

10.6.5 R-R-R

The joint displacement variables δ_{1z}, δ_{2x}, and δ_{3x} in Equation (10.17) cannot be calculated algebraically also. With unknowns δ_{1z}, δ_{2x}, and δ_{3x}, Equation (10.17) becomes a set of three nonlinear simultaneous equations. Like the R-P-P and R-R-P robotic manipulators, unconstrained nonlinear equation system optimization is also required for the inverse kinematics of the R-R-R robotic manipulator. Appendix L.7 includes the user instructions for the MATLAB file to calculate joint displacement solutions for Equation (10.17) given ${}^{3}\{\mathbf{p}_3\}$ and a prescribed range of values for ${}^{0}\{\mathbf{p}_3\}$.

As shown in Figure 10.14, the R-R-R robotic manipulator has a spherical workspace (having center coordinates center $= (0,0,l_1)$) with an outer radius dimension of $r_{\text{outer}} = l_2 + \left\| {}^{3}\{\mathbf{p}_3\} \right\|$, where ${}^{3}\{\mathbf{p}_3\} = \left[0, {}^{0}p_{3y}, {}^{0}p_{3z}\right]^T$.

```
l1 = 0.5;
l2 = 0.5;
p3_3 = [0, 0, 0];
```

FIGURE E.10.10 Specified input (in bold text) in the Appendix L.6 MATLAB file for Example 10.10.

TABLE E.10.15
R-R-P Robotic Manipulator Joint Displacements

Point	δ_{1z} (deg)	δ_{2x} (deg)	Δ_{3z}
1	−61.2311	32.9191	0.6614
2	−56.2416	37.2735	−0.6864
3	−51.0710	40.9089	−0.7124
4	−45.7758	43.7775	−0.7388
5	−40.4371	45.8699	−0.7644
6	−35.1105	47.1905	−0.7889
7	−29.8625	47.7679	−0.8115
8	−24.7360	47.6333	−0.8316
9	−19.7713	46.8339	−0.8491
10	−15.0002	45.4141	−0.8634

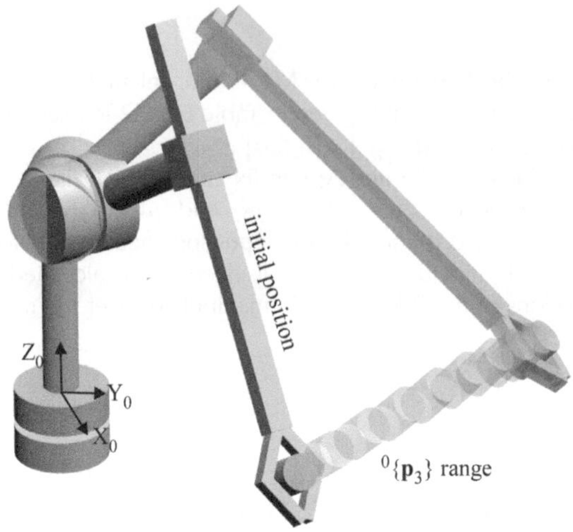

FIGURE E.10.11 R-R-P robotic manipulator in initial and final positions.

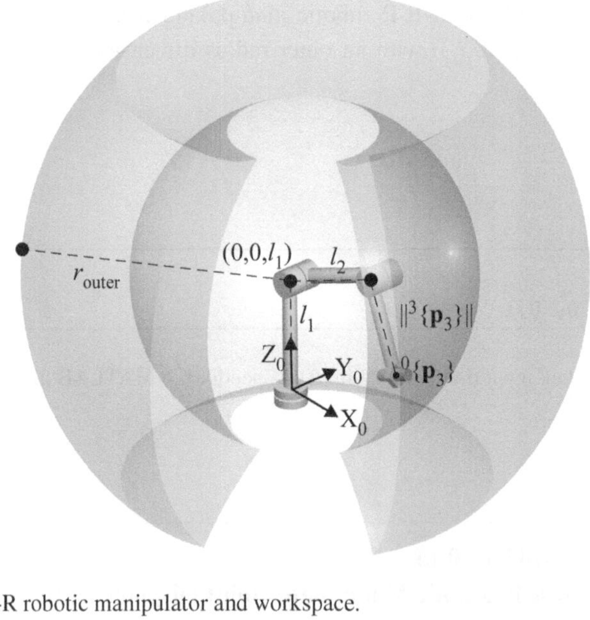

FIGURE 10.14 R-R-R robotic manipulator and workspace.

Example 10.11

Problem Statement: Using the Appendix L.7 MATLAB file, calculate the R-R-R joint displacements required to achieve the global path points given Table E.10.12 in Example 10.8. In this example, $l_1 = l_2 = 0.5$ (unitless link lengths) and $^3\{\mathbf{p}_3\} = [0,1,0]^T$.

Known Information: Table E.10.12 and Appendix L.7 MATLAB file

Solution Approach: The data for columns $^0p_{3x}$, $^0p_{3y}$, and $^0p_{3z}$ in Table E.10.12 are first specified in the file *RRR_Input.csv*. Figure E.10.12 includes the input specified (in bold text) in the Appendix L.7 MATLAB file, and Table E.10.16 includes the joint displacements calculated. Figure E.10.13 includes the initial and final positions of the R-R-R robotic manipulator.

10.6.6 R-R-C

The joint displacement variables δ_{1z}, δ_{2z}, δ_{3z}, and Δ_{3y} in Equation (10.19) cannot be calculated algebraically also. With unknowns δ_{1z}, δ_{2z}, δ_{3z}, and Δ_{3y}, Equation (10.19) becomes a set of three nonlinear simultaneous equations. Like the R-P-P, R-R-P, and R-R-R robotic manipulators, unconstrained nonlinear equation system optimization is also required for the inverse kinematics of the R-R-C robotic manipulator. Appendix L.8 includes the user instructions for the MATLAB file to calculate joint displacement solutions for Equation (10.19) given ${}^{3}\{\mathbf{p}_3\}$ and a prescribed range of values for ${}^{0}\{\mathbf{p}_3\}$.

```
l1 = 0.5;
l2 = 0.5;
p3_3 = [0, 1, 0];
```

FIGURE E.10.12 Specified input (in bold text) in the Appendix L.7 MATLAB file for Example 10.11.

TABLE E.10.16
R-R-R Robotic Manipulator Joint Displacements

Point	δ_{1z} (deg)	δ_{2x} (deg)	δ_{3x} (deg)
1	–61.2311	74.3302	–124.2298
2	–56.2416	75.2908	–121.9275
3	–51.0710	75.4719	–119.5001
4	–45.7758	74.9234	–117.0128
5	–40.4371	73.7565	–114.5580
6	–35.1106	72.0287	–112.1872
7	–29.8625	69.8373	–109.9699
8	–24.7360	67.2585	–107.9608
9	–19.7713	64.3666	–106.2017
10	–15.0002	61.2454	–104.7446

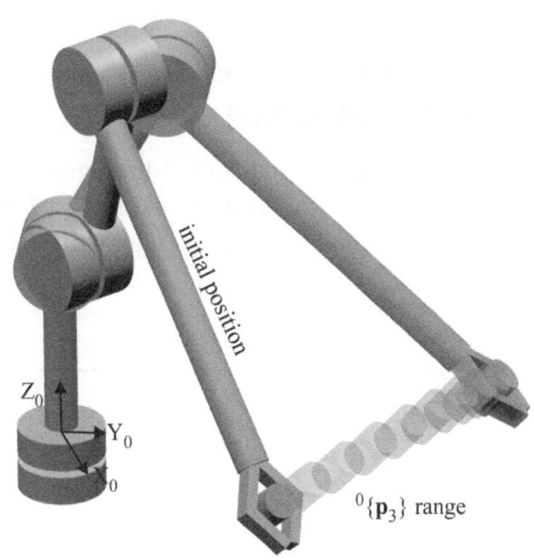

FIGURE E.10.13 R-R-R robotic manipulator in initial and final positions.

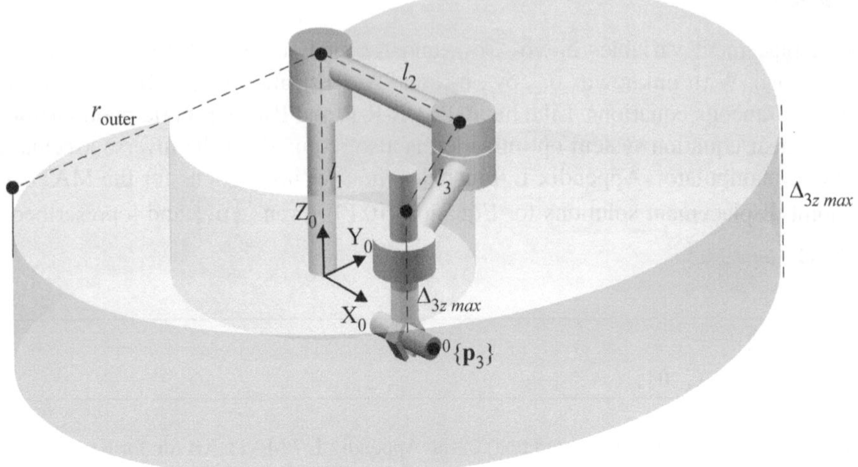

FIGURE 10.15 R-R-C robotic manipulator and workspace.

As shown in Figure 10.15, the R-R-C robotic manipulator has a cylindrical workspace (having a center axis that is collinear with l_1) with outer height and radius dimensions of $\Delta_{3z_{max}}$ and $r_{outer} = l_2 + l_3$, respectively.

Example 10.12

Problem Statement: Using the Appendix L.8 MATLAB file, calculate the R-R-C joint displacements required to achieve the global path points given in Table E.10.12 in Example 10.8. In this example, $l_1 = 1$, $l_2 = l_3 = 0.5$ (unitless link lengths), and ${}^3\{\mathbf{p}_3\} = [0,1,-1]^T$.

Known Information: Table E.10.12 and Appendix L.8 MATLAB file

Solution Approach: The data for columns ${}^0p_{3x}$, ${}^0p_{3y}$, and ${}^0p_{3z}$ in Table E.10.12 are first specified in the file *RRC_Input.csv*. Figure E.10.14 includes the input specified (in bold text) in the Appendix L.8 MATLAB file, and Table E.10.17 includes the joint displacements calculated. Figure E.10.15 includes the initial and final positions of the R-R-C robotic manipulator over the range of prescribed global end-effector points.

10.7 ROBOTIC MANIPULATOR KINEMATIC ANALYSIS AND MODELING IN SIMSCAPE MULTIBODY™

As has been noted throughout this chapter, Appendices L.1 through L.8 include MATLAB file user instructions for R-P-P, R-R-P, R-R-R, and R-R-C forward and inverse kinematics. In these files, the global end-effector coordinates are calculated for prescribed joint displacements (in Appendices L.1

```
l1 = 1;
l2 = 0.5;
l3 = 0.5;
p3_3 = [0, 1, -1];
```

FIGURE E.10.14 Specified input (in bold text) in the Appendix L.8 MATLAB file for Example 10.12.

TABLE E.10.17
R-R-C Robotic Manipulator Joint Displacements

Point	δ_{1z} (deg)	δ_{2z} (deg)	δ_{3z} (deg)	Δ_{3z}
1	−35.4860	−5.7728	46.9980	0.2165
2	−30.0170	−4.5237	44.2350	0.2566
3	−24.5330	−3.1631	41.6230	0.2890
4	−19.0690	−1.7162	39.1470	0.3125
5	−13.6890	−0.21255	36.8220	0.3266
6	−8.4370	1.3269	34.6550	0.3307
7	−3.3822	2.8823	32.6820	0.3248
8	1.4188	4.4435	30.9450	0.3090
9	5.9070	6.0036	29.4950	0.2838
10	10.0190	7.5616	28.3930	0.2500

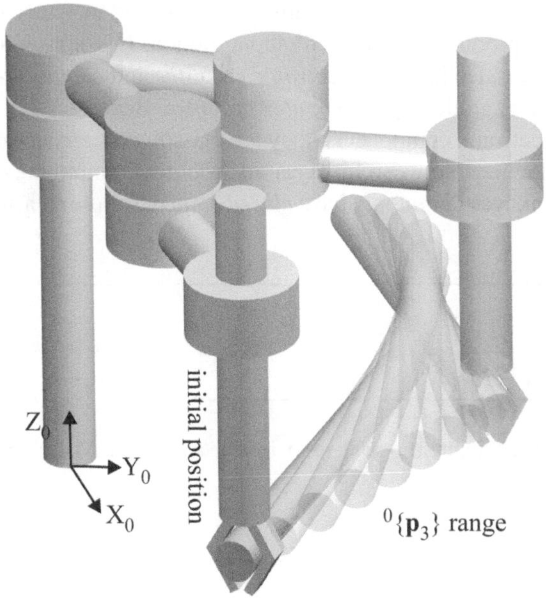

FIGURE E.10.15 R-R-C robotic manipulator in initial and final positions.

through L.4), and the joint displacements are calculated for prescribed global end-effector coordinates (in Appendices L.5 through L.8).

This book also utilizes Simscape Multibody as an alternate approach for simulation-based kinematic analysis. A library of Simscape Multibody files is also available for download (at https://www.routledge.com/9781032956800) for R-P-P, R-R-P, R-R-R, and R-R-C forward kinematics.[14] In addition to calculating the global end-effector coordinates for prescribed joint displacements, the motion of the robotic manipulator is also simulated over the joint displacements. The Simscape Multibody file user instructions for the forward kinematic analysis of the R-P-P, R-R-P, R-R-R, and R-R-C robotic manipulators are given in Appendices M.1 through M.4, respectively.

Example 10.13

Problem Statement: Using the Appendix M.4 Simscape Multibody files, calculate the R-R-C end-effector coordinates in Table E.10.12 in Example 10.12.

Known Information: Example 10.12 and Appendix M.4 Simscape Multibody files

Solution Approach: The data for columns δ_{1z}, δ_{2x}, δ_{3z}, and Δ_{3z} in Table E.10.17 are first specified in the file *RRC_Input.csv*. Figure E.10.16 includes the input specified (in bold text) in the Appendix M.4 Simscape Multibody file, and Table E.10.18 includes the global R-R-C end-effector coordinates calculated. These coordinates are identical to the prescribed end-effector coordinates given in Table E.10.12. Figure E.10.17 is an intermediate position snapshot of the robotic manipulator animation generated from the Appendix M.4 Simscape Multibody files.

Identical end-effector coordinates to those in Table E.10.18 that can be replicated for the R-P-P, R-R-P, and R-R-R robotic manipulators by using the joint displacements and dimension input data in Examples 10.9, 10.10, and 10.11, respectively (in the Appendix M.1, M.2, and M.3 Simscape Multibody files, respectively).

10.8 SUMMARY

Like a linkage, a *robotic manipulator* (commonly called a *robot*) includes an assembly of links and joints and is designed to produce a controlled output motion. In addition to links and joints, however, a robotic manipulator also includes *electronic circuitry* and *computer-controlled actuators* to compel link motion, and is guided by a *computer program*. Because robotic manipulators include both mechanical and electronic components, they are classified as *electro-mechanical systems*.

Other common distinctions between linkages and robotic manipulators pertain to their overall design and mobility. Linkages commonly have *closed-loop* designs and a single *degree of freedom*.

```
l1 = 1;
l2 = 0.5;
l3 = 0.5;
p3_3 = [0, 1, -1];
```

FIGURE E.10.16 Specified input (in bold text) in the Appendix M.4 Simscape Multibody file for Example 10.13.

TABLE E.10.18
Calculated R-R-C Robotic Manipulator End-Effector Coordinates

Point	${}^0p_{3x}$	${}^0p_{3y}$	${}^0p_{3z}$
1	0.6830	0.3749	0.2164
2	0.6764	0.4519	0.2565
3	0.6569	0.5305	0.2889
4	0.6251	0.6082	0.3125
5	0.5818	0.6825	0.3266
6	0.5284	0.7513	0.3307
7	0.4666	0.8124	0.3248
8	0.3983	0.8640	0.3090
9	0.3254	0.9046	0.2839
10	0.2502	0.9330	0.2501

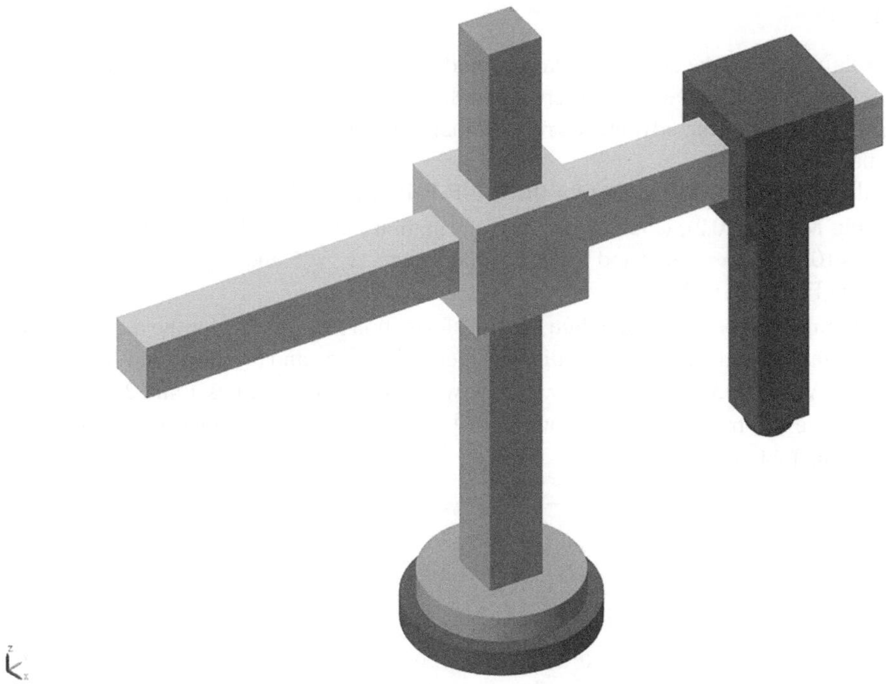

FIGURE E.10.17 Intermediate position of the R-R-C robotic manipulator in the Appendix M.4 Simscape Multibody animation for Example 10.13.

With a closed-loop design, at least two joints in a linkage are connected to the ground (thus forming a closed loop). With a single degree of freedom, full linkage motion control is achieved with a single driver. While robotic manipulators can have closed-loop designs, they often have *open loops*—where only one joint is connected to the ground. Full motion control is not achieved in a robotic manipulator by having a single degree of freedom. It is achieved by controlling each degree of freedom with its own driver.

The five spatial robotic manipulator types considered in this chapter are the *Cartesian, cylindrical, spherical, articulated,* and *SCARA* robots. They are commonly known as the P-P-P, R-P-P, R-R-P, R-R-R, and R-R-C robotic manipulators, respectively. By prescribing coordinate frames for each link in a robotic manipulator and establishing displacement variables between each frame, equation systems are formulated (using the general spatial transformation matrix) to calculate the motion of any link in the robotic manipulator.

This book includes a library of MATLAB files for the *forward kinematics* (Appendices L.1–L.4) and the *inverse kinematics* (Appendices L.5–L.8) of the R-P-P, R-R-P, R-R-R, and R-R-C robotic manipulators. In forward kinematics, the link dimensions and joint motion of a robotic manipulator are known, and the corresponding output motion of the links (usually the end effector) is calculated. In inverse kinematics, the end-effector motion and link dimensions are known, and the joint motion required to achieve the end-effector motion is calculated. Inverse kinematics is often described as *the reverse of forward kinematics.*

This book also includes a library of MATLAB and Simscape Multibody files for the *forward kinematics* (Appendices M.1 through M.4) of the R-P-P, R-R-P, R-R-R, and R-R-C robotic manipulators. In addition to calculating the global values for the end-effector Point $\mathbf{p}_3$ the motion of the robotic manipulator is also simulated in the Appendix M files.

PROBLEMS

1. Explain some of the design similarities and distinctions between robotic manipulators and the linkages presented in Chapters 4 and 8.
2. Explain some of the advantages robotic manipulators have over the linkages presented in Chapters 4 and 8.
3. Calculate the mobility values for the P-P-P, R-P-P, and R-R-P robotic manipulators.
4. Using Matrix (10.2), calculate the transformation matrix ${}^{0}_{1}[T]$ for angular displacements $\delta_{1x} = 10°$ and $\delta_{1z} = -40°$ and linear displacements $\Delta_{1x} = 5$ and $\Delta_{1y} = -3.75$ between coordinate Frames 0 and 1.
5. Including Matrix (10.2), calculate the transformation matrix ${}^{0}_{2}[T]$ for the displacement data given in Problem 4 and angular displacements $\delta_{2y} = 35°$ and $\delta_{2z} = 10°$ and linear displacements $\Delta_{2x} = 10$, $\Delta_{2y} = -5$, and $\Delta_{2z} = 5$ between coordinate Frames 1 and 2.
6. Calculate the transformation matrix ${}^{0}_{3}[T]$ for the coordinate frame displacement values given in Table P.10.1.

TABLE P.10.1
Frame Displacement Variables

Frames	δ_x	δ_y	δ_z	Δ_x	Δ_y	Δ_z
1 wrt 0	25°	0	0	0	2	0
2 wrt 1	0	0	12°	0	0	−1.75
3 wrt 2	0	−60°	0	0	5	0

7. Explain the distinction between a transformation matrix formulated using *Denavit–Hartenberg* parameters and the transformation matrix given in Matrix (10.2).
8. Calculate the ${}^{0}\{\mathbf{p}_3\}$ coordinates produced by the P-P-P robotic manipulator given the joint displacements

$$\Delta_{1x} = 1 + t, \quad \Delta_{2y} = t^{1.25}, \quad \Delta_{3z} = -2t^{0.75}, \quad 0 \le t \le 10$$

Consider at least 10 equally spaced curve points. Also, ${}^{3}\{\mathbf{p}_3\} = [0, -1, 0]^T$.

9. Using the Appendix L.1 or Appendix M.1 files, calculate the ${}^{0}\{\mathbf{p}_3\}$ coordinates produced by the R-P-P robotic manipulator given the joint displacements

$$\delta_{1z} = t, \quad \Delta_{2z} = t/45°, \quad \Delta_{3x} = \left(t/45°\right)^{1.5}, \quad 0 \le t \le 90°$$

Consider at least 10 equally spaced curve points. Also, ${}^{3}\{\mathbf{p}_3\} = [0, 0, -1]^T$.

10. Using the Appendix L.1 or Appendix M.1 files, calculate the ${}^{0}\{\mathbf{p}_3\}$ coordinates produced by the R-P-P robotic manipulator given the joint displacements

$$\delta_{1z} = (90°)\sin t, \quad \Delta_{2z} = t/45°, \quad \Delta_{3x} = \left(t/45°\right)^{1.5}, \quad 0 \le t \le 180°$$

Consider at least 10 equally spaced curve points. Also, ${}^{3}\{\mathbf{p}_3\} = [2, 0, -1.5]^T$.

11. Using the Appendix L.2 or Appendix M.2 files, calculate the ${}^{0}\{\mathbf{p}_3\}$ coordinates produced by the R-R-P robotic manipulator given the joint displacements

$$\delta_{1z} = 2t, \quad \delta_{2x} = -t, \quad \Delta_{3z} = t/45°, \quad 0 \le t \le 45°$$

Consider at least 10 equally spaced curve points. Also, $l_1 = l_2 = 2$ and ${}^3\{\mathbf{p}_3\} = [0,0,0]^T$.

12. Using the Appendix L.2 or Appendix M.2 files, calculate the ${}^0\{\mathbf{p}_3\}$ coordinates produced by the R-R-P robotic manipulator given the joint displacements

$$\delta_{1z} = (60°)\sin 2t, \quad \delta_{2x} = -t/2, \quad \Delta_{3z} = t/45°, \quad 0 \le t \le 90°$$

Consider at least ten equally spaced curve points. Also, $l_1 = 2, l_2 = 3.5$, and ${}^3\{\mathbf{p}_3\} = [0,0,0]^T$.

13. Using the Appendix L.3 or Appendix M.3 files, calculate the ${}^0\{\mathbf{p}_3\}$ coordinates produced by the R-R-R robotic manipulator given the joint displacements

$$\delta_{1z} = 3t, \quad \delta_{2x} = -2t, \quad \delta_{3x} = t, \quad 0 \le t \le 30°$$

Consider at least ten equally spaced curve points. Also, $l_1 = 2,\ l_2 = 1.5$, and ${}^3\{\mathbf{p}_3\} = [0,0.25,0]^T$.

14. Using the Appendix L.3 or Appendix M.3 files, calculate the ${}^0\{\mathbf{p}_3\}$ coordinates produced by the R-R-R robotic manipulator given the joint displacements

$$\delta_{1z} = (60°)\sin 2t, \quad \delta_{2x} = (30°)\sin 2t, \quad \delta_{3x} = (45°)\sin 2t, \quad 0 \le t \le 90°$$

Consider at least ten equally spaced curve points. Also, $l_1 = 2,\ l_2 = 4.5$, and ${}^3\{\mathbf{p}_3\} = [1,0.25,2]^T$.

15. Using the Appendix L.4 or Appendix M.4 files, calculate the ${}^0\{\mathbf{p}_3\}$ coordinates produced by the R-R-C robotic manipulator given the joint displacements

$$\delta_{1z} = 4t, \quad \delta_{2z} = -2t, \quad \delta_{3z} = 1, \quad \Delta_{3z} = -t/120°, \quad 0 \le t \le 30°$$

Consider at least ten equally spaced curve points. Also, $l_1 = l_2 = l_3 = 1$ and ${}^3\{\mathbf{p}_3\} = [0,1,-0.1]^T$.

16. Using the Appendix L.4 or Appendix M.4 files, calculate the ${}^0\{\mathbf{p}_3\}$ coordinates produced by the R-R-C robotic manipulator given the joint displacements

$$\delta_{1z} = (60°)\sin 4t, \quad \delta_{2z} = (90°)\sin 2t, \quad \delta_{3z} = t/2, \quad \Delta_{3z} = -t/120°, \quad 0 \le t \le 90°$$

Consider at least ten equally spaced curve points. Also, $l_1 = l_2 = l_3 = 1$ and ${}^3\{\mathbf{p}_3\} = [2,1,-2]^T$.

17. In robotic manipulator design, what are the objectives in and applications for *forward* and *inverse* kinematics?
18. Describe the shape and dimensions of the workspace of a P-P-P robotic manipulator with maximum translation distances (in meters) of $\Delta_{1x_{\max}} = 1.65$, $\Delta_{2y_{\max}} = 2.85$, and $\Delta_{3z_{\max}} = 3.2$.
19. Calculate the joint displacements of the P-P-P robotic manipulator required to achieve the curve

$${}^0p_{3x} = 1.5 + \cos(t), \quad {}^0p_{3y} = 2.75 + \sin(t), \quad {}^0p_{3z} = t/2\pi, \quad 0 \le t \le 2\pi$$

Consider at least ten equally spaced curve points. Also, ${}^3\{\mathbf{p}_3\} = [0,-1,0]^T$.

20. Describe the shape and outer dimensions of the workspace of an R-P-P robotic manipulator with maximum translation distances (in meters) of $\Delta_{2z_{\max}} = 8$ and $\Delta_{3x_{\max}} = 5.75$.

21. Describe the shape and outer dimensions of the workspace of an R-P-P robotic manipulator with maximum translation distances (in meters) of $\Delta_{2z_{\max}} = 7.5$ and $\Delta_{3x_{\max}} = 3.75$.
22. Using the Appendix L.5 files, calculate the joint displacements of the R-P-P robotic manipulator required to achieve the curve.

$$^{0}p_{3x} = 4\cos(t), \quad ^{0}p_{3y} = 6\sin(t), \quad ^{0}p_{3z} = t/90°, \quad 0 \le t \le 360°$$

Consider at least ten equally spaced curve points. Also, $^{3}\{\mathbf{p}_3\} = [2,-1,0]^T$.
23. Using the Appendix L.5 files, calculate the joint displacements of the R-P-P robotic manipulator required to achieve the curve

$$^{0}p_{3x} = t^{1.25}, \quad ^{0}p_{3y} = t^2, \quad ^{0}p_{3z} = 1 + t^{1.75}, \quad 0 \le t \le 18$$

Consider at least ten equally spaced curve points. Also, $^{3}\{\mathbf{p}_3\} = [-1,3,0]^T$.
24. Describe the shape and outer dimensions of the workspace of an R-R-P robotic manipulator with link lengths (in meters) $l_1 = l_2 = 10$ and a maximum translation distance of $\Delta_{3z_{\max}} = 8$.
25. Describe the shape and outer dimensions of the workspace of an R-R-P robotic manipulator with link lengths (in meters) $l_1 = 4.5$ and $l_2 = 3$ and a maximum translation distance of $\Delta_{3z_{\max}} = 1.75$.
26. Using the Appendix L.6 files, calculate the joint displacements of the R-R-P robotic manipulator required to achieve the curve

$$^{0}p_{3x} = 1.75\cos(t), \quad ^{0}p_{3y} = 5.5\sin(t), \quad ^{0}p_{3z} = \sin(t), \quad 0 \le t \le 360°$$

Consider at least ten equally spaced curve points. Also, $l_1 = l_2 = 2.5$ and $^{3}\{\mathbf{p}_3\} = [0,0,0]^T$.
27. Using the Appendix L.6 files, calculate the joint displacements of the R-R-P robotic manipulator required to achieve the curve

$$^{0}p_{3x} = 7.15 + t/360°, \quad ^{0}p_{3y} = 5.72\sin(t), \quad ^{0}p_{3z} = 0.52, \quad 0 \le t \le 360°$$

Consider at least ten equally spaced curve points. Also, $l_1 = 2.5, l_2 = 2$ and $^{3}\{\mathbf{p}_3\} = [0,0,0]^T$.
28. Describe the shape and outer dimensions of the workspace of an R-R-R robotic manipulator with link lengths (in meters) $l_1 = 12.5$ and $l_2 = 4$ and $^{3}\{\mathbf{p}_3\} = [0,2,0]^T$.
29. Describe the shape and outer dimensions of the workspace of an R-R-R robotic manipulator with link lengths (in meters) $l_1 = 8.25$ and $l_2 = 3.75$ and $^{3}\{\mathbf{p}_3\} = [0,2,1]^T$.
30. Using the Appendix L.7 files, calculate the joint displacements of the R-R-R robotic manipulator required to achieve the curve

$$^{0}p_{3x} = 3\sin(t), \quad ^{0}p_{3y} = 2.5\cos(t), \quad ^{0}p_{3z} = 1.5\cos(t), \quad 0 \le t \le 180°$$

Consider at least ten equally spaced curve points. Also, $l_1 = l_2 = 3.75$ and $^{3}\{\mathbf{p}_3\} = [3,1.5,0.75]^T$.
31. Using the Appendix L.7 files, calculate the joint displacements of the R-R-R robotic manipulator required to achieve the curve

$$^{0}p_{3x} = -2 + t/45°, \quad ^{0}p_{3y} = \sin(t), \quad ^{0}p_{3z} = 2t/45°, \quad 0 \le t \le 90°$$

Consider at least ten equally spaced curve points. Also, $l_1 = 2.5$, $l_2 = 3$, and $^{3}\{\mathbf{p}_3\} = [1.5,0,1.7]^T$.
32. Describe the shape and outer dimensions of the workspace of an R-R-C robotic manipulator with link lengths (in meters) $l_2 = 2$, $l_3 = 1.25$, and a maximum translation distance of $\Delta_{3z_{\max}} = 5$.

33. Describe the shape and outer dimensions of the workspace of an R-R-C robotic manipulator with link lengths (in meters) $l_1 = 1.25$, $l_2 = 2.95$ and a maximum translation distance of $\Delta_{3z_{max}} = 4.5$.
34. Using the Appendix L.8 files, calculate the joint displacements of the R-R-C robotic manipulator required to achieve the curve

$$^{0}p_{3x} = \frac{t}{45^\circ}\sin(t), \quad ^{0}p_{3y} = \frac{-t}{45^\circ}\cos(t), \quad ^{0}p_{3z} = t/30^\circ, \quad 0 \le t \le 180^\circ$$

Consider at least ten equally spaced curve points. Also, $l_1 = 3.25$, $l_2 = l_3 = 1.5$, and $^{3}\{\mathbf{p}_3\} = [0,1,-0.5]^T$.

35. Using the Appendix L.8 files, calculate the joint displacements of the R-R-C robotic manipulator required to achieve the curve

$$^{0}p_{3x} = \frac{-t}{120^\circ}\sin(t), \quad ^{0}p_{3y} = \frac{-t}{60^\circ}\cos(t), \quad ^{0}p_{3z} = t/45^\circ, \quad 0 \le t \le 180^\circ$$

Consider at least ten equally spaced curve points. Also, $l_1 = 2$, $l_2 = l_3 = 1.75$, and $^{3}\{\mathbf{p}_3\} = [0,0.5,-0.5]^T$.

NOTES

1 Robotic manipulators are generally limited to 6 DOFs because a spatial body has a maximum mobility of 6.
2 The workspace is the area or volume (for spatial robots) that includes all positions the robot can achieve.
3 Additional frames would be required for the gripper and tool if there was relative motion between the gripper and Link 2 (comparable to the human wrist and forearm) or if there was relative motion within the tool (like a movable multi-part tool).
4 As with mechanisms, for robotic manipulators, the ground link should also be counted among the total number of links (L) in Gruebler's Equation.
5 In a general *planar* transformation matrix, variables δ_x, δ_y, and Δ_z in the general spatial transformation matrix are all 0.
6 The *planar* coordinates of $\mathbf{p}$ in Frame j would be $^{j}\{\mathbf{p}\} = \{\ p_x \ \ p_y \ \ 0 \ \ 1\ \}^T$.
7 A value given with respect to the base frame is also called a *global value*.
8 When calculating the product of three or more matrices in Equation (10.7), it should be done in a right-to-left order.
9 In classical DH notation, the variables θ, α, d, and r are used instead of δ_z, δ_x, Δ_z, and Δ_z, respectively.
10 In Table 10.3, $\Delta_{2z} = l_1$ and $\Delta_{3y} = l_2$.
11 In Table 10.4, $\Delta_{2z} = l_1$ and $\Delta_{3y} = l_2$.
12 In Table 10.5, $\Delta_{1z} = l_1$, $\Delta_{2x} = l_2$, and $\Delta_{3x} = l_3$.
13 Inverse kinematics is like motion generation and path generation (see Chapters 5, 6, 8, and 9). The key distinction between dimensional synthesis and inverse kinematics is that with the latter, *joint displacements* are calculated (rather than link dimensions).
14 This library can also be downloaded from the authors' server at http://www.softalink.com/kruss/kin_dyn_mech_sys_3rd_ed/MATLAB_SIMMULT_FILES.zip.

REFERENCES

1. "Industrial Robots-Statistics," International Federation of Robotics, accessed May 18, 2015, https://www.ifr.org.
2. Lewis, F. L., Dawson, D. M. and Abdallah, C. T. 2004. *Robot Manipulator Control: Theory and Practice*, 2nd Edition. section 1.2. New York: Marcel Dekker.
3. Craig, J. J. 2005. *Introduction to Robotics: Mechanics and Control*, 3rd Edition. chapter 3. Upper Saddle River: Prentice-Hall.
4. Craig, J. J. 2005. *Introduction to Robotics: Mechanics and Control*, 3rd Edition. chapter 4. Upper Saddle River: Prentice-Hall.

5. Nikiv, S. B. 2001. *Introduction to Robotics: Analysis, Systems and Applications.* pages 76–82. Upper Saddle River: Prentice-Hall.
6. Jazar, R. N. 2007. *Theory of Applied Robotics: Kinematics, Dynamics and Control.* chapter 6. New York: Springer.
7. Kurfess, T. R. 2005. *Robotic and Automation Handbook.* chapter 3. New York: CRC Press.

ADDITIONAL READING

Sandor, G. N. and Erdman, A. G. 1984. *Advanced Mechanism Design: Analysis and Synthesis*, Volume 2. chapter 6. Englewood Cliffs: Prentice-Hall.

Wilson, C. E. and Peter Sadler, J. 2003. *Kinematics and Dynamics of Machinery*, 3rd Edition, chapter 12. Saddle River: Prentice Hall.

Appendix A
User Information and Instructions for MATLAB®

A.1 REQUIRED MATLAB TOOLKITS

To utilize the MATLAB and Simscape Multibody files that accompany this book (available for download at https://www.routledge.com/9781032956800), the following toolkits must be included when installing MATLAB:

- **Symbolic Math Toolbox:** Required for performing symbolic mathematics
- **Optimization Toolbox:** Required for solving equations (e.g., simultaneous equation sets)
- **Simscape Multibody, Simscape, and Simulink:** Required for running Simscape Multibody files

To view the installed toolkits, the user can type the command *ver* in the MATLAB command window. The MATLAB and Simscape Multibody™ files that accompany this book can also be downloaded from the author's server http://www.softalink.com/kruss/mech_des_vis_prog_app_2nd_ed/MATLAB_SIMMULT_FILES.zip.

A.2 DESCRIPTION OF MATLAB OPERATORS AND FUNCTIONS

For additional descriptions of the functions and operators given in Table A.1, please refer to www.mathworks.com/help/matlab/.

TABLE A.1
MATLAB functions and operators used throughout this book

Function or Operator	Description
+ − * / ^	Add, subtract multiply, divide, and exponent (respectively)
;	When this operator is used at the end of a command, the results from the command are not displayed.
i or *j*	Complex coefficient ($i = j = \sqrt{-1}$)
Pi	π (3.14159…)
sqrt(*x*)	Square root of *x*
atan2(*y*,*x*)	The angle between the positive *x*-axis and the point given by the coordinates (*x*,*y*)
abs(*x*)	Absolute value of a real value or magnitude of a complex value
norm(*x*)	Vector or matrix norm $\left(\lVert x \rVert_2\right)$
exp(*x*)	Exponent (e^x)
syms *var_names* real	Defines symbolic variables (e.g., syms *X Y Z* real makes variables *X*, *Y*, and *Z* symbolic variables).
syms *name*(*t*) *t* real	Defines time-based symbolic variables (e.g., syms *X*(*t*) *Y*(*t*) *Z*(*t*) *t* real makes variables *X*, *Y*, and *Z* time-based symbolic variables).
diff(*F*)	Calculates the first derivative of function *F* symbolically.

(*Continued*)

TABLE A.1 *(Continued)*
MATLAB functions and operators used throughout this book

Function or Operator	Description
diff(*V*, *N*)	Calculates the *N*th derivative of function *F* symbolically.
expand(expr)	Symbolic expansion (of polynomials and functions)
simplify(expr)	Algebraic simplification
sin(*x*)	Sine function
cos(*x*)	Cosine function
Name=[N1, N2, N3…]	Defines a row vector
Name(cell number)	Calls a vector cell (e.g., Name(1)=N1 in the previous vector)
Name [N11, N12, N13…; N21, N22, N23…; N31, N32, N33…]	Defines a matrix
Name (row#, col#)	Calls a matrix cell (e.g., Name(2,3)=N23 in previous matrix)
‘	When this operator is used at the end of a matrix or vector, the matrix or vector is transposed.
Inv(*A*)	The inverse of matrix A $\left([A]^{-1}\right)$
*A**b*	The solution to the matrix system $[\mathrm{A}]\mathbf{x}=\mathbf{b}\left(\mathbf{x}=[\mathrm{A}]^{-1}\mathbf{b}\right)$
…	When this operator is used at the end of a command, the command can continue on the next line.
f=@(var1, var2…) expression	When this operator is used before a function expression, an anonymous function is defined. (e.g., for $f=x^2+2y$, *f*=@(*x*, *y*) x^2+2**y*)
[*x*,fval]=fsolve(*fx*,*x*0)	Solves a system of linear or nonlinear equations *fx* (where *fx*=[eq1; eq2; eq3…]) for unknown variables *x* (where *x*=[var1, va2, var3…]). The variable *x*0 includes the initial unknown variable values.

A.3 PREPARING AND RUNNING FILES IN MATLAB AND OPERATIONS IN SIMSCAPE MULTIBODY

To execute commands in MATLAB, the user can simply type them line by line in the MATLAB command window (labeled in Figure A.1). This window appears by default in MATLAB. All the MATLAB commands illustrated throughout this book were implemented using this approach. After typing commands in one line, the user can use the “ENTER” key to advance to the text line.

As a set of commands becomes more extensive, the user may prefer to save them as a MATLAB file. To do this, the user can enter the commands in the MATLAB editor window (labeled in Figure A.1) and save them as a *script* (a *.*m* file).[1] The MATLAB editor window can be launched using the “CTRL+N” keys or either of the *New Script* buttons encircled in Figure A.1.

To run files in MATLAB, the user can either use the “F5” key or select the *Run* button. Figure A.2 illustrates the MATLAB Editor toolbar with the *Run* button encircled. This toolbar appears once a MATLAB file is opened.

There are two windows that appear automatically when running Simscape Multibody files: the *model window* and the *animation window*.[2] Figure A.3 illustrates the model window for Appendix C.1. In this window, Simscape Multibody models are constructed. Because the contents in the model window should never be edited in any way, the user should minimize this window whenever it appears.

Figure A.4 illustrates the animation window for Appendix C.1. In this window, the motion of the model (under the user-prescribed dimension and driving link parameters) is simulated. Figure A.5 illustrates the toolbar for the animation window. This toolbar includes options to adjust view settings, play speed, as well as animation recording settings (among others).

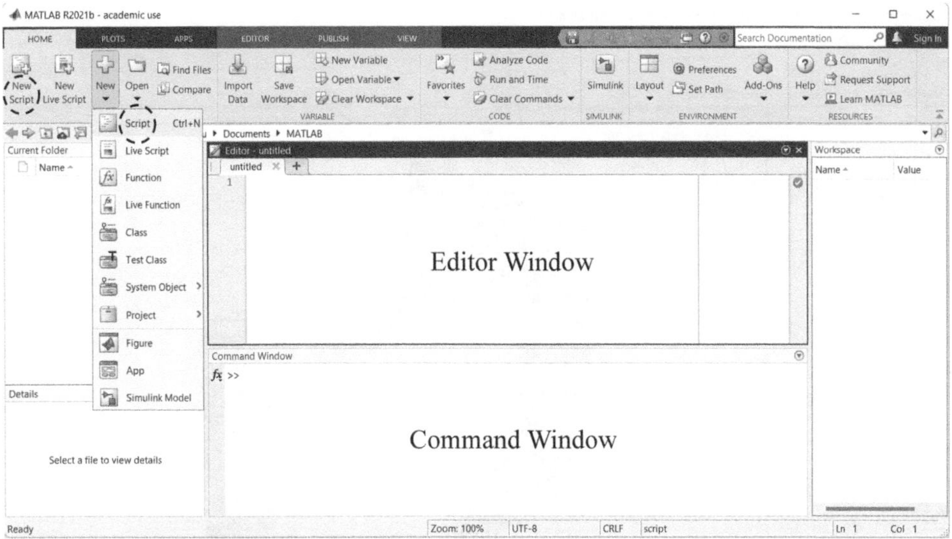

FIGURE A.1 MATLAB command and editor windows and *New Script* buttons (encircled).

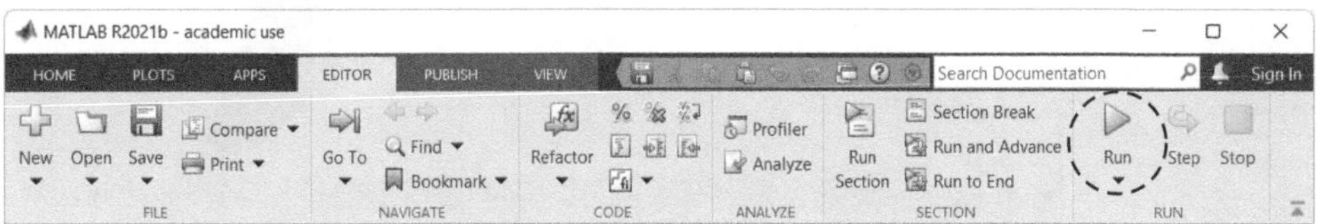

FIGURE A.2 MATLAB Editor toolbar with run button (encircled).

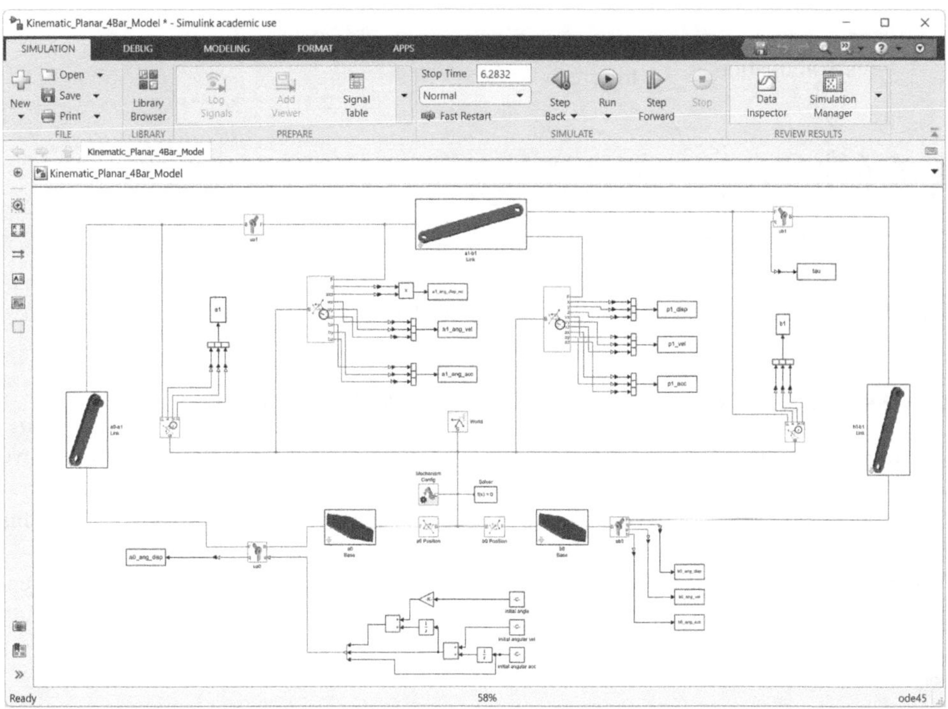

FIGURE A.3 Simscape Multibody model window (for Appendix C.1).

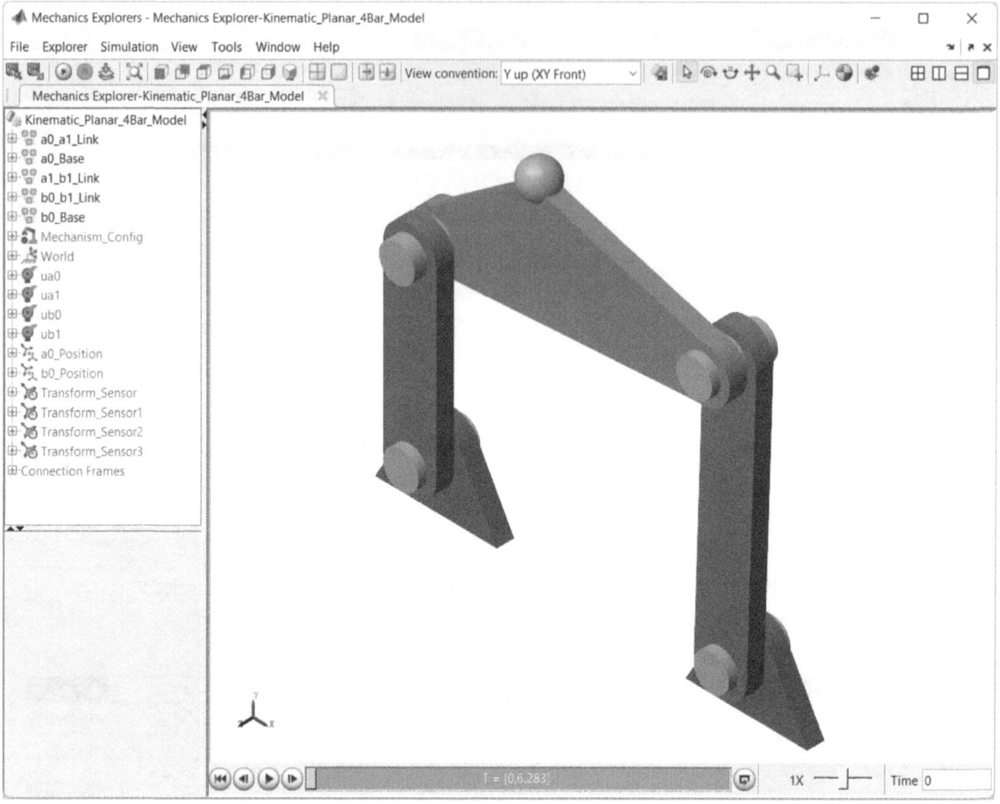

FIGURE A.4 Simscape Multibody animation window (for Appendix C.1).

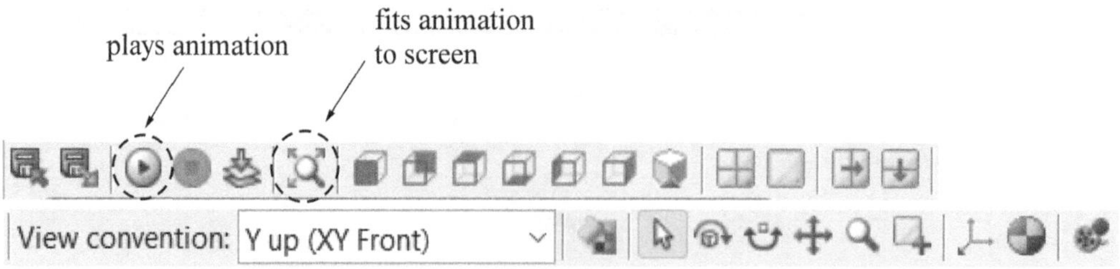

FIGURE A.5 Simscape Multibody animation window toolbar.

A.4 DESCRIPTION OF SIMSCAPE MULTIBODY FUNCTIONS

For additional descriptions of the functions given in Table A.2, please refer to www.mathworks.com/help/matlab/.

During the simulation process, a pop-up window will appear. This window includes a 3D animation of the mechanism rendered by the Simscape Multibody 3D engine.

At the end of the simulation, the simulation model is preconfigured to have the simulation engine invoke the *Post_Simulation_Task.m* file.

TABLE A.2
Simscape Multibody Interface Functions Used Throughout This Book

Function	Description
open_system(obj)	To open the Simscape Multibody environment and the simulation model name specified
set_param(Object, ParameterName, Value,… ParameterNameN, and ValueN)	To customize simulation parameters. In this book, the customized parameters are "StopTime" and "Refine"
sim(model)	To start the simulation
evalin(workspace,expression)	To retrieve simulation results data from the Simscape Multibody environment and make it visible in the MATLAB main workspace

A.5 RE-RUNNING MATLAB OR SIMSCAPE MULTIBODY FILES WITH EXISTING *.CSV FILES

To re-run a MATLAB or Simscape Multibody file when its *.csv file is open, the user must either (1) rename the *.csv file (to prevent it from being overwritten), (2) store the *.csv file in a different folder (also to prevent it from being overwritten), or (3) close the *.csv file in the same folder (so that it may be overwritten). Attempting to re-run a MATLAB or Simscape Multibody file with its originally named *.csv file open and stored in its original location will prevent the new results from being written (and produce an error message). This is because a new *.csv file cannot be written when a *.csv file having the same name is in use in the same folder.

A.6 MINIMUM PRECISION REQUIREMENT FOR APPENDIX FILE USER INPUT

The default precision tolerance for link vector loops in the MATLAB and Simscape Multibody files is 0.0002. Vector loop errors beyond this tolerance are not allowed. An error indicating this violation will appear in the command line in the Appendix B and C files. When specifying link vectors in any of the MATLAB and Simscape Multibody files (e.g., vector lengths, vector angles, or vector components), non-integer values should be specified to at least four decimal places.[3]

NOTES

1 Unlike commands in the MATLAB command line, commands in the MATLAB editor do not include *prompts* (the symbol >> that precedes every command in the MATLAB command window).
2 The Simscape Multibody files are associated with Appendices C, I, and M.
3 The only exceptions for this requirement are integer values or non-integer values that include a string of zeros before the fourth decimal place (e.g., ½ = 0.5000 = 0.5, ¼ = 0.2500 = 0.25, and 5/4 = 1.2500 = 1.25).

Appendix B
User Instructions for Chapter 4 MATLAB® Kinematic Analysis Files

B.1 PLANAR FOUR-BAR MECHANISM

The Appendix B.1 folder (which is available for download at https://www.routledge.com/9781032956800) includes the MATLAB file *Kinematic_Planar_4Bar.m* for the kinematic analysis of planar four-bar mechanisms.[1] To conduct a kinematic analysis, the user specifies the mechanism link dimensions and the crank motion parameters in this file. Values are specified for link variables $\mathbf{W}_1$, $\mathbf{V}_1$, $\mathbf{G}_1$, $\mathbf{U}_1$, and $\mathbf{L}_1$ (Figure B.1a). Values are also specified for the initial crank angle (*start_ang*), the crank rotation increment (*step_ang*), and the final crank angle (*stop_ang*). Lastly, values are specified for the angular velocity (*angular_vel*) and angular acceleration (*angular_acc*) of the crank.[2] Figure B.2 illustrates the user input sections of the file *Kinematic_Planar_4Bar.m* with sample values in bold type.[3] While most of the link vector values in this figure appear in the polar exponential form, they can all be specified in any of the rectangular and complex forms given in Equation (2.1).

After specifying the dimensions and driving link parameters in the file *Kinematic_Planar_4Bar.m*, the next step is to run this file. When running this file, one file (filename *Disp_Vel_Acc.csv*) is written in a folder named *Results* (in a format compatible with Microsoft® Excel) that includes the calculated mechanism output at each crank link rotation increment.[4] The calculated mechanism output is included in Figure B.1b.

B.2 GEARED 5-BAR MECHANISM (TWO GEARS)

The Appendix B.2 folder includes the MATLAB file *Kinematic_5Bar_2Gears.m* for the kinematic analysis of geared five-bar mechanisms having two gears. To conduct a kinematic analysis, the user specifies the mechanism link dimensions and the crank motion parameters in this file. Values are

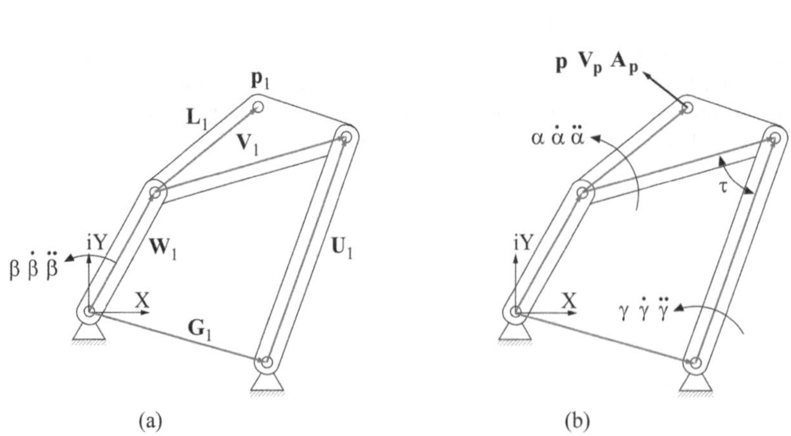

FIGURE B.1 (a) Planar four-bar mechanism and (b) output variables.

```
%-----------------------------------------------------------------
% Here, values for the planar 4-bar mechanism variables W1, V1,
% G1, U1 and L1 are assigned.

W1 = 0.5*exp(i*90*pi/180);
V1 = 0.75*exp(i*19.3737*pi/180);
G1 = 0.75 + i*0;
U1 = 0.75*exp(i*93.2461*pi/180);
L1 = 0.5*exp(-i*60.7834*pi/180);
%-----------------------------------------------------------------

%-----------------------------------------------------------------
% Here, values for the start, step and stop displacement angles
% for the crank link are assigned.

start_ang    = 0;
step_ang     = 1;
stop_ang     = 360;
%-----------------------------------------------------------------

%-----------------------------------------------------------------
% Here, values for crank link angular velocity and angular
% acceleration are assigned.

angular_vel   = 1.0*ones(N,1);
angular_acc   = 0.1*ones(N,1);
%-----------------------------------------------------------------
```

FIGURE B.2 Sections of *Kinematic_Planar_4Bar.m* with sample values in bold type.

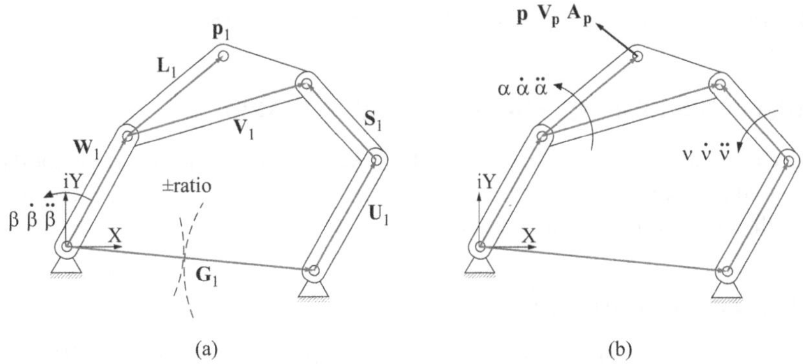

FIGURE B.3 (a) Geared five-bar mechanism and (b) output variables.

specified for link variables $\mathbf{W}_1$, $\mathbf{V}_1$, $\mathbf{G}_1$, $\mathbf{U}_1$, $\mathbf{L}_1$, $\mathbf{S}_1$, and the gear ratio (Figure B.3a). Values are also specified for the initial crank angle (*start_ang*), the crank rotation increment (*step_ang*), and the final crank angle (*stop_ang*). Lastly, values are specified for the angular velocity (*angular_vel*) and angular acceleration (*angular_acc*) for the crank.[5] Figure B.4 illustrates the user input sections of the file *Kinematic_Geared_5Bar.m* with sample values in bold type.[6] While most of the link vector values in this figure appear in the polar exponential form, they can all be specified in any of the rectangular and complex forms given in Equation (2.1).

After specifying the dimensions and driving link parameters in the file *Kinematic_Geared_5Bar.m*, the next step is to run this file. When running this file, one file (filename *Disp_Vel_Acc.csv*) is written in a folder named *Results* (in a format compatible with Microsoft™ Excel) that includes the calculated mechanism output at each crank link rotation increment.[7] The calculated mechanism output is included in Figure B.3b.

```
%-----------------------------------------------------------------
% Here, values for the geared 5-bar mechanism variables W1, V1,
% G1, U1, L1, S1 and gear ratio are assigned. Only negatiive
% gear ratio values should be specified.


W1 = 0.5*exp(i*90*pi/180);
V1 = 0.75*exp(i*32.7304*pi/180);
G1 = 0.75 + i*0;
U1 = 0.75*exp(i*45*pi/180);
L1 = 0.5*exp(-i*74.1400*pi/180);
S1 = 0.75*exp(i*149.9847*pi/180);
ratio = -2;
%-----------------------------------------------------------------

%-----------------------------------------------------------------
% Here, values for the start, step and stop displacement angles
% for the crank link are assigned.

start_ang   = 0;
step_ang    = 1;
stop_ang    = 360;
%-----------------------------------------------------------------

%-----------------------------------------------------------------
% Here, values for crank link angular velocity and angular
% acceleration are assigned.

angular_vel  = 1*ones(N,1);
angular_acc  = 0.1*ones(N,1);
%-----------------------------------------------------------------
```

FIGURE B.4 Sections of *Kinematic_Geared_5Bar.m* with sample values in bold type.

B.3 GEARED FIVE-BAR MECHANISM (THREE GEARS)

The Appendix B.3 folder includes the MATLAB file *Kinematic_5Bar_3Gears.m* for the kinematic analysis of geared five-bar mechanisms having three gears. The procedure to use this MATLAB file is identical to the procedure given in Section B.2.

B.4 STEPHENSON III MECHANISM

The Appendix B.4 folder includes the MATLAB file *Kinematic_Stephenson_III.m* for the kinematic analysis of Stephenson III mechanisms. To conduct a kinematic analysis, the user specifies the mechanism link dimensions and the crank motion parameters in this file. Values are specified for link variables $\mathbf{W}_1$, $\mathbf{V}_1$, $\mathbf{G}_1$, $\mathbf{U}_1$, $\mathbf{L}_1$, $\mathbf{V}_1^*$, $\mathbf{G}_1^*$, $\mathbf{U}_1^*$, and $\mathbf{L}_1^*$ (Figure B.5a). Values are also specified for the initial crank angle (*start_ang*), the crank rotation increment (*step_ang*), and the final crank angle (*stop_ang*). Lastly, values are specified for the angular velocity (*angular_vel*) and angular acceleration (*angular_acc*) for the crank.[8] Figure B.6 illustrates the user input sections of the file *Kinematic_Stephenson_III.m* with sample values in bold type.[9] While most of the link vector values in this figure appear in the polar exponential form, they can all be specified in any of the rectangular and complex forms given in Equation (2.1).

After specifying the dimensions and driving link parameters in the file *Kinematic_Stephenson_III.m*, the next step is to run this file. When running this file, one file (filename *Disp_Vel_Acc.csv*) is written in a folder named *Results* (in a format compatible with Microsoft™ Excel) that includes the calculated mechanism output at each crank link rotation increment.[10] The calculated mechanism output is included in Figure B.5b.

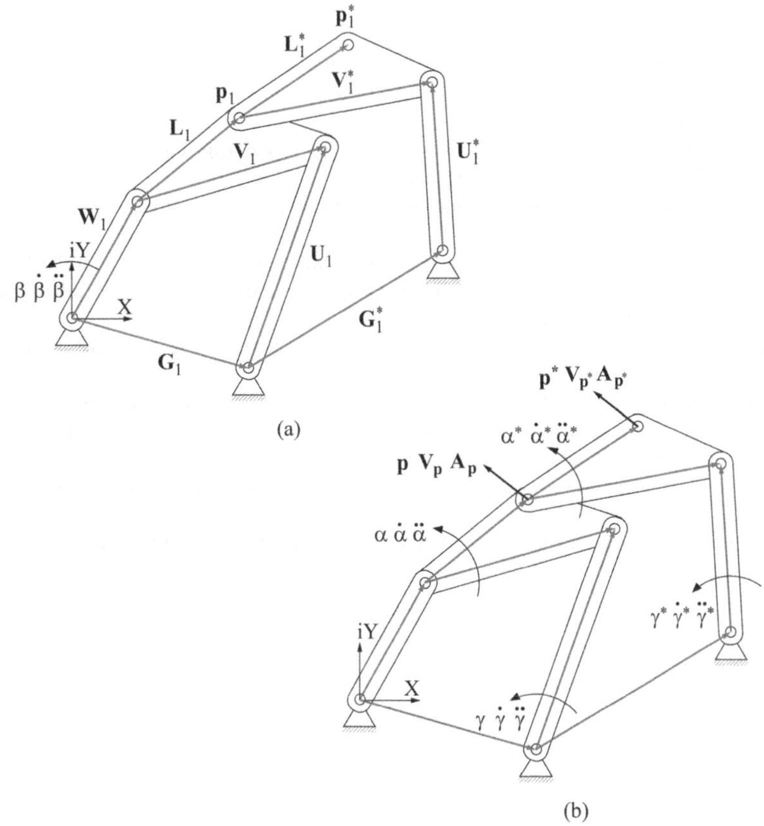

FIGURE B.5 (a) Stephenson III mechanism and (b) output variables.

```
%-------------------------------------------------------------
% Here, values for the Stephenson III mechanism variables W1,
% V1, G1, U1, L1, V1s, G1s, U1s and L1s are assigned.

W1 = 0.5*exp(i*90*pi/180);
V1 = 0.75*exp(i*19.3737*pi/180);
G1 = 0.75 + i*0;
U1 = 0.75*exp(i*93.2461*pi/180);
L1 = 0.5*exp(i*60.7834*pi/180);

V1s = 1*exp(i*17.1417*pi/180);
G1s = 0.2159 + i*0.2588;
U1s = 1*exp(i*76.4844*pi/180);
L1s = 0.5*exp(i*63.7091*pi/180);
%-------------------------------------------------------------

%-------------------------------------------------------------
% Here, values for the start, step and stop displacement angles
% for the crank link are assigned.

start_ang    = 0;
step_ang     = -1;
stop_ang     = -360;
%-------------------------------------------------------------

%-------------------------------------------------------------
% Here, values for crank link angular velocity and angular
% acceleration are assigned.

angular_vel  = -1*ones(N,1);
angular_acc  = -0.25*ones(N,1);
%-------------------------------------------------------------
```

FIGURE B.6 Sections of Kinematic_*Stephenson_III.m* with sample values in bold type.

NOTES

1 All of the Appendix B MATLAB files can be downloaded from the publisher's web page or from the authors' server http://www.softalink.com/kruss/mech_des_vis_prog_app_2nd_ed/MATLAB_SIMMULT_FILES.zip.
2 If no crank angular velocity or acceleration values are specified, time cannot be calculated, and the text *NaN* is written in the time column of the *.*csv* file.
3 The units for crank angular rotation, velocity, and acceleration are *degrees*, *radians/sec*, and *radians/sec*^{2}, respectively.
4 At crank rotation increments where circuit defects occur, the text *Inf* is written in the *.*csv* file.
5 If no crank angular velocity or acceleration values are specified, time cannot be calculated, and the text *NaN* is written in the time column of the *.*csv* file.
6 The units for crank angular rotation, velocity, and acceleration are *degrees*, *radians/sec,* and *radians/sec*^{2,} respectively.
7 At crank rotation increments where circuit defects occur, the text *Inf* is written in the *.*csv* file.
8 If no crank angular velocity or acceleration values are specified, time cannot be calculated, and the text *NaN* is written in the time column of the *.*csv* file.
9 The units for crank angular rotation, velocity, and acceleration are *degrees*, *radians/sec,* and *radians/sec*^{2,} respectively.
10 At crank rotation increments where circuit defects occur, the text *Inf* is written in the *.*csv* file.

Appendix C

User Instructions for Chapter 4 MATLAB® and Simscape Multibody™ Kinematic Analysis Files

C.1 PLANAR FOUR-BAR MECHANISM

The Appendix C.1 folder (which is available for download at https://www.routledge.com/9781032956800) includes three MATLAB and Simscape Multibody files for the kinematic analysis and simulation of planar four-bar mechanisms. These files are described in Table C.1.[1] To conduct a kinematic analysis, the user specifies the mechanism link dimensions and the crank motion parameters in the file *Kinematic_Planar_4Bar_Simulate.m*. Values are specified for link variables $\mathbf{W}_1$, $\mathbf{V}_1$, $\mathbf{G}_1$, $\mathbf{U}_1$, and $\mathbf{L}_1$ (Figure C.1a). Values are also specified for the initial crank angle (*start_ang*), the crank rotation increment (*step_ang*), and the final crank angle (*stop_ang*). Lastly, values are specified for the angular velocity (*angular_vel*) and angular acceleration (*angular_acc*) for the crank. Figure C.2 illustrates the user input sections of the file *Kinematic_Planar_4Bar_Simulate.m* with sample values in bold type.[2] While most of the link vector values in this figure appear in the polar exponential form, they can all be specified in any of the rectangular and complex forms given in Equation (2.1).

TABLE C.1
Appendix C.1 MATLAB and Simscape Multibody Files

Filename	Use of File
Kinematic_Planar_4Bar_Simulate.m	To specify mechanism link dimensions and crank link controls
Kinematic_Planar_4Bar_Model.slx	To calculate the mechanism output and simulate the mechanism motion
Post_Simulation_Task.m	To write the mechanism output (compatible with Microsoft Excel)

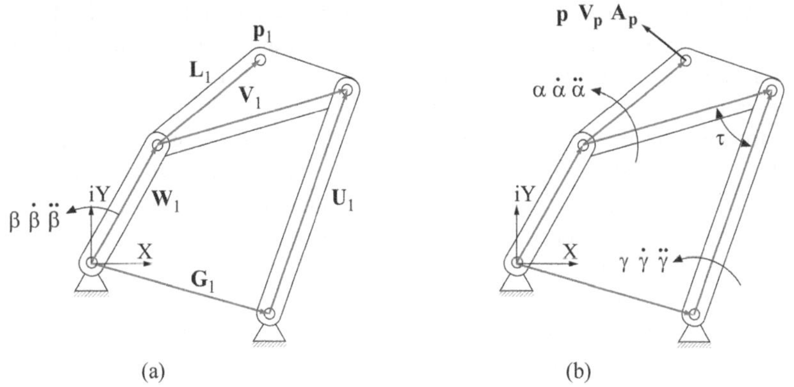

FIGURE C.1 (a) Planar four-bar mechanism and (b) output variables.

```
%-------------------------------------------------------------
% Here, values for the planar 4-bar mechanism variables W1, V1,
% G1, U1 and L1 are assigned.

W1 = 0.5*exp(i*90*pi/180);
V1 = 0.75*exp(i*19.3737*pi/180);
G1 = 0.75 + i*0;
U1 = 0.75*exp(i*93.2461*pi/180);
L1 = 0.5*exp(i*60.7834*pi/180);
%-------------------------------------------------------------

%-------------------------------------------------------------
% Here, values for the start, step and stop displacement angles
% for the crank link are assigned.

start_ang   = 0;
step_ang    = 1;
stop_ang    = 360;
%-------------------------------------------------------------

%-------------------------------------------------------------
% Here, values for crank link angular velocity and angular
% acceleration are assigned.

angular_vel  = 1.0;
angular_acc  = 0.1;
%-------------------------------------------------------------
```

FIGURE C.2 Sections of *Kinematic_Planar_4Bar_Simulate.m* with sample values in bold type.

After specifying the driving link parameters in the file *Kinematic_Planar_4Bar_Simulate.m*, the next step is to run this file. When running this file, one file (filename *Disp_Vel_Acc.csv*) is written in a folder named *Results* (in a format compatible with Microsoft® Excel) that includes the calculated mechanism output at each crank link rotation increment.[3] The calculated mechanism output is included in Figure C.1b.

When running *Kinematic_Planar_4Bar_Simulate.m*, a graphical user interface appears, where the motion of the planar four-bar mechanism is simulated over the defined crank rotation range (see Appendix A.3). Also, the folder *slprq* is written. This folder includes functions that preclude recompiling the model in future simulations. This optional folder can be deleted, however (e.g., if computer storage space is inadequate).

C.2 GEARED FIVE-BAR MECHANISM (TWO GEARS)

The Appendix C.2 folder includes three MATLAB and Simscape Multibody files for the kinematic analysis and simulation of geared five-bar mechanisms having two gears. These files are described in Table C.2. To conduct a dynamic force analysis, the user specifies the mechanism link dimensions and the crank motion parameters in the file *Kinematic_5Bar_2Gears_Simulate.m*. Values are specified for link variables $\mathbf{W}_1$, $\mathbf{V}_1$, $\mathbf{G}_1$, $\mathbf{U}_1$, $\mathbf{L}_1$, and $\mathbf{S}_1$ (Figure C.3a). Values are also specified for the initial crank angle (*start_ang*), the crank rotation increment (*step_ang*), and the final crank angle (*stop_ang*). Lastly, values are specified for the angular velocity (*angular_vel*) and angular acceleration (*angular_acc*) for the crank. Figure C.4 illustrates the user input sections of the file *Kinematic_5Bar_2Gears_Simulate.m* with sample values in bold type.[4] While most of the link vector values in this figure appear in the polar exponential form, they can all be specified in any of the rectangular and complex forms given in Equation (2.1).

After specifying the mechanism dimensions, loads, mass parameters, and driving link parameters in the file *Kinematic_5Bar_2Gear_Simulate.m*, the next step is to run this file. When running this file, one file (filename *Disp_Vel_Acc.csv*) is written in a folder named *Results* (in a format

TABLE C.2
Appendix C.2 MATLAB and Simscape Multibody Files

Filename	Use of File
Kinematic_5Bar_2Gears_Simulate.m	To specify mechanism link dimensions and crank link controls
Kinematic_5Bar_2Gears_Model.slx	To calculate the mechanism output and simulate the mechanism motion
Post_Simulation_Task.m	To write the mechanism output (compatible with Microsoft Excel)

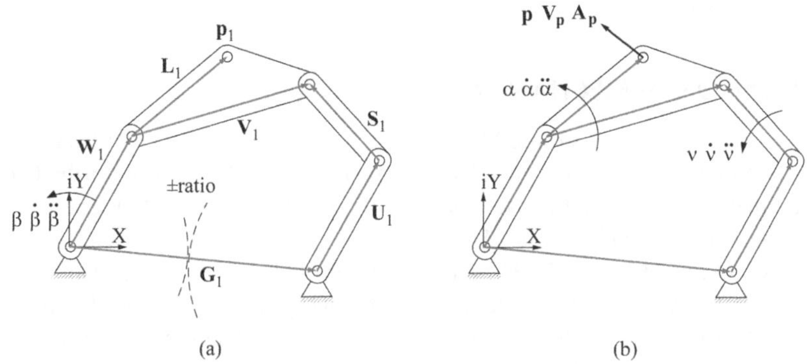

FIGURE C.3 (a) Geared five-bar mechanism and (b) output variables.

```
%-------------------------------------------------------------
% Here, values for the geared 5-bar mechanism variables W1, V1,
% G1, U1, L1 and S1 are assigned. Only negative gear ratio
% values should be specified.

W1 = 0.5*exp(i*90*pi/180);
V1 = 0.75*exp(i*32.7304*pi/180);
G1 = 0.75*exp(i*0*pi/180);
U1 = 0.75*exp(i*45*pi/180);
L1 = 0.5*exp(i*74.1400*pi/180);
S1 = 0.75*exp(i*149.9847*pi/180);
ratio = -2;
%-------------------------------------------------------------

%-------------------------------------------------------------
% Here, values for the start, step and stop displacement angles
% for the crank link are assigned.

start_ang   = 0;
step_ang    = -1;
stop_ang    = -360;
%-------------------------------------------------------------

%-------------------------------------------------------------
% Here, values for crank link angular velocity and angular
% acceleration are assigned.

angular_vel  = -1;
angular_acc  = 0;
%-------------------------------------------------------------
```

FIGURE C.4 Sections of *Kinematic_5Bar_2Gears_Simulate.m* with sample values in bold type.

compatible with Microsoft® Excel) that includes the calculated mechanism output at each crank link rotation increment.[5] The calculated mechanism output is included in Figure C.3b.

Also, when running *Kinematic_5Bar_2Gear_Simulate.m*, a graphical user interface appears, where the motion of the geared five-bar mechanism is simulated over the defined crank rotation range (see Appendix A.3). Also, the folder *slprq* is written. This folder includes functions that preclude recompiling the model in future simulations. This optional folder can be deleted, however (e.g., if computer storage space is inadequate).

C.3 GEARED FIVE-BAR MECHANISM (THREE GEARS)

The Appendix C.3 folder includes three MATLAB and Simscape Multibody files for the kinematic analysis and simulation of geared five-bar mechanisms having three gears. These files are described in Table C.3. The procedure to use these MATLAB and Simscape Multibody files is identical to the procedure given in Appendix C.2.

C.4 STEPHENSON III MECHANISM

The Appendix C.4 folder includes three MATLAB and Simscape Multibody files for the kinematic analysis and simulation of Stephenson III mechanisms. These files are described in Table C.4. To conduct a kinematic analysis, the user specifies the mechanism link dimensions and the crank motion parameters in the file *Kinematic_Stephenson_III_Simulate.m*. Values are specified for link variables $\mathbf{W}_1$, $\mathbf{V}_1$, $\mathbf{G}_1$, $\mathbf{U}_1$, $\mathbf{L}_1$, $\mathbf{V}_1^*$, $\mathbf{G}_1^*$, $\mathbf{U}_1^*$, and $\mathbf{L}_1^*$ (Figure C.5a). Values are also specified for the initial crank angle (*start_ang*), the crank rotation increment (*step_ang*), and the final crank angle (*stop_ang*). Lastly, values are specified for the angular velocity (*angular_vel*) and angular acceleration (*angular_acc*) for the crank. Figure C.6 illustrates the user input sections of the file *Kinematic_Stephenson_III_Simulate.m* with sample values in bold type.[6] While most of the link vector values in this figure appear in the polar exponential form, they can all be specified in any of the rectangular and complex forms given in Equation (2.1).

After specifying the mechanism dimensions, loads, mass parameters, and driving link parameters in the file *Kinematic_Stephenson_III_Simulate.m*, the next step is to run this file. When running this file, one file (filename *Disp_Vel_Acc.csv*) is written in a folder named *Results* (in a format compatible with Microsoft® Excel) that includes the calculated mechanism output at each crank link rotation increment.[7] The calculated mechanism output is included in Figure C.5b.

Also, when running *Kinematic_Stephenson_III_Simulate.m*, a graphical user interface appears, where the motion of the Stephenson III mechanism is simulated over the defined crank rotation

TABLE C.3
Appendix C.3 MATLAB and Simscape Multibody Files

Filename	Use of File
Kinematic_5Bar_3Gears_Simulate.m	To specify mechanism link dimensions and crank link controls
Kinematic_5Bar_3Gears_Model.slx	To calculate the mechanism output and simulate the mechanism motion
Post_Simulation_Task.m	To write the mechanism output (compatible with Microsoft Excel)

TABLE C.4
Appendix C.4 MATLAB and Simscape Multibody Files

Filename	Use of File
Kinematic_Stephenson_III_Simulate.m	To specify mechanism link dimensions and crank link controls
Kinematic_Stephenson_III_Model.slx	To calculate the mechanism output and simulate the mechanism motion
Post_Simulation_Task.m	To write the mechanism output (compatible with Microsoft Excel)

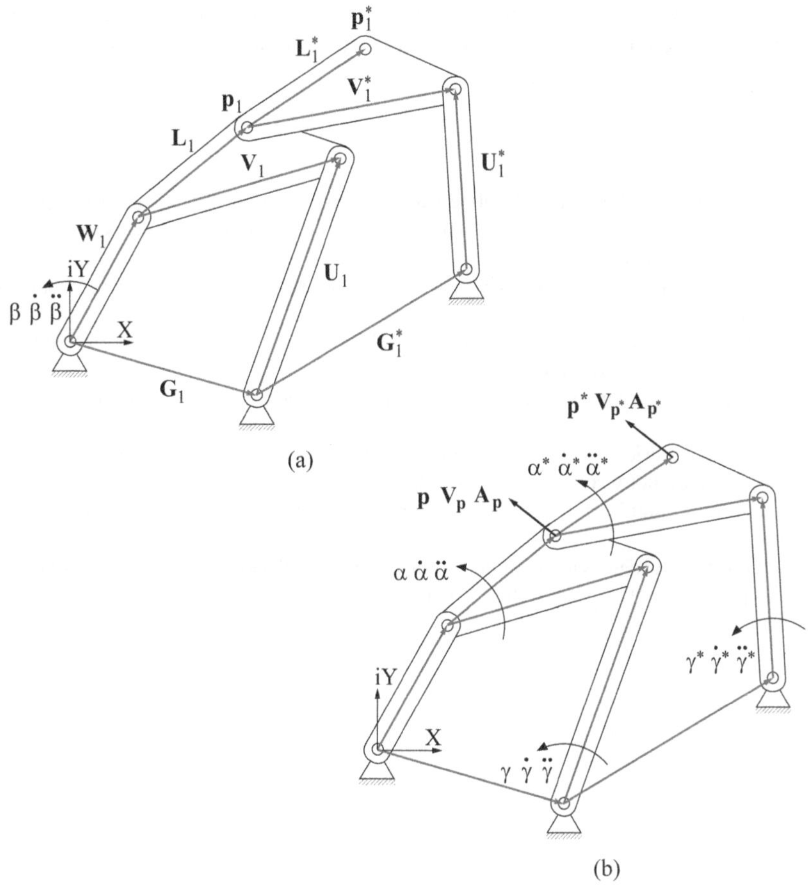

FIGURE C.5 (a) Stephenson III mechanism and (b) output variables.

```
%-----------------------------------------------------------------
% Here, values for the Stephenson III mechanism variables W1,
% V1, G1, U1, L1, V1s, G1s, U1s and L1s are assigned.

W1 = 0.5*exp(i*90*pi/180);
V1 = 0.75*exp(i*19.3737*pi/180);
G1 = 0.75 + i*0;
U1 = 0.75*exp(i*93.2461*pi/180);
L1 = 0.5*exp(i*60.7834*pi/180);

V1s = 1*exp(i*17.1417*pi/180);
G1s = 0.2159 + i*0.2588;
U1s = 1*exp(i*76.4844*pi/180);
L1s = 0.5*exp(i*63.7091*pi/180);
%-----------------------------------------------------------------

%-----------------------------------------------------------------
% Here, values for the start, step and stop displacement angles
% for the crank link are assigned.

start_ang   = 0;
step_ang    = -1;
stop_ang    = -360;
%-----------------------------------------------------------------

%-----------------------------------------------------------------
% Here, values for crank link angular velocity and angular
% acceleration are assigned.

angular_vel  = -1;
angular_acc  = -0.25;
%-----------------------------------------------------------------
```

FIGURE C.6 Sections of *Kinematic_Stephenson_III_Simulate.m* with sample values in bold type.

range (see Appendix A.3). Also, the folder *slprq* is written. This folder includes functions that preclude recompiling the model in future simulations. This optional folder can be deleted, however (e.g., if computer storage space is inadequate).

NOTES

1 All of the Appendix C Simscape Multibody files can be downloaded from the publisher's web page or from the authors' server http://www.softalink.com/kruss/mech_des_vis_prog_app_2nd_ed/MATLAB_SIMMULT_FILES.zip.
2 The units for crank angular rotation, velocity, and acceleration are *degrees, radians/sec*, and *radians/sec*^{2}, respectively.
3 If the mechanism experiences a circuit defect (or if the crank decelerates to a zero velocity state), results beyond this point are not written to the *.*csv* file.
4 The units for crank angular rotation, velocity, and acceleration are *degrees, radians/sec*, and *radians/sec*^{2}, respectively.
5 If the mechanism experiences a circuit defect (or if the crank decelerates to a zero velocity state), results beyond this point are not written to the *.*csv* file.
6 The units for crank angular rotation, velocity, and acceleration are *degrees, radians/sec*, and *radians/sec*^{2}, respectively.
7 If the mechanism experiences a circuit defect (or if the crank decelerates to a zero velocity state), results beyond this point are not written to the *.*csv* file.

Appendix D
Algorithm *for Analytical Four-Position Synthesis and MATLAB® File User Instructions*

D.1 ANALYTICAL PLANAR FOUR-BAR FOUR-POSITION SYNTHESIS ALGORITHM

For four-position synthesis, the standard-form equations for the **W** − **Z** dyad (Equation 5.3) form a set of three complex, linear equations. This equation set can be expressed in matrix form as

$$\begin{bmatrix} e^{i\beta_2}-1 & e^{i\alpha_2}-1 \\ e^{i\beta_3}-1 & e^{i\alpha_3}-1 \\ e^{i\beta_4}-1 & e^{i\alpha_4}-1 \end{bmatrix} \begin{bmatrix} W_1 e^{i\theta} \\ Z_1 e^{i\phi} \end{bmatrix} = \begin{bmatrix} P_{21}e^{i\delta_2} \\ P_{31}e^{i\delta_3} \\ P_{41}e^{i\delta_4} \end{bmatrix} \tag{D.1}$$

where α_j, P_{j1}, and δ_j ($j = 2, 3, 4$) are user-prescribed. As shown in Table 5.1, a single unknown variable can be prescribed in four-position synthesis. In order for Equation (D.1) to have simultaneous solutions for $\mathbf{W}_1$ and $\mathbf{Z}_1$, it must follow that

$$\det\left[\begin{array}{cc:c} e^{i\beta_2}-1 & e^{i\alpha_2}-1 & P_{21}e^{i\delta_2} \\ e^{i\beta_3}-1 & e^{i\alpha_3}-1 & P_{31}e^{i\delta_3} \\ e^{i\beta_4}-1 & e^{i\alpha_4}-1 & P_{41}e^{i\delta_4} \end{array}\right] = 0 \tag{D.2}$$

or the *rank* of the "augmented matrix" of the coefficients is 2.

Equation (D.2) can be expanded about its first column as

$$\Delta_2 e^{i\beta_2} + \Delta_3 e^{i\beta_3} + \Delta_4 e^{i\beta_4} + \Delta_1 = 0 \tag{D.3}$$

where

$$\Delta_2 = \begin{vmatrix} e^{i\alpha_3}-1 & P_{31}e^{i\delta_3} \\ e^{i\alpha_4}-1 & P_{41}e^{i\delta_4} \end{vmatrix}$$

$$\Delta_3 = \begin{vmatrix} e^{i\alpha_2}-1 & P_{21}e^{i\delta_2} \\ e^{i\alpha_4}-1 & P_{41}e^{i\delta_4} \end{vmatrix} \tag{D.4}$$

$$\Delta_4 = \begin{vmatrix} e^{i\alpha_2}-1 & P_{21}e^{i\delta_2} \\ e^{i\alpha_3}-1 & P_{31}e^{i\delta_3} \end{vmatrix}$$

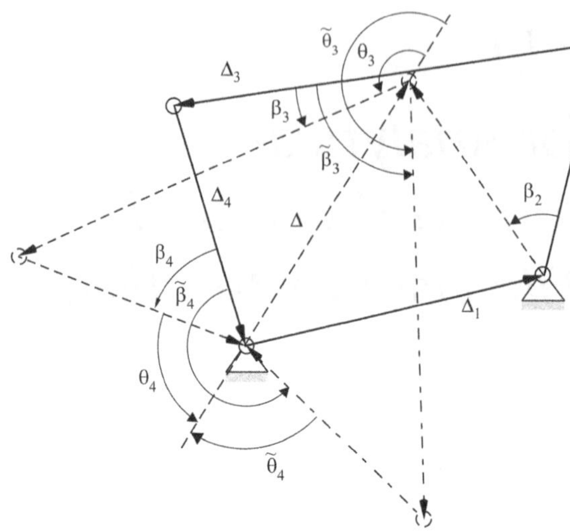

FIGURE D.1 Both branches of the compatibility linkage in starting and displaced positions.

and

$$\Delta_1 = -\Delta_2 - \Delta_3 - \Delta_4 \tag{D.5}$$

The Δs are known because they contain only user-prescribed data. Equation (D.3) is referred to as the *compatibility equation* because values of β_2, β_3, and β_4, which satisfy this equation, will render Equation (D.1) "compatible" (thus yielding simultaneous solutions for $\mathbf{W}_1$ and $\mathbf{Z}_1$).

The compatibility equation can be expressed geometrically as a *compatibility linkage* by regarding the Δs as link lengths where Δ_1, Δ_2, Δ_3, and Δ_4 are the ground, crank, coupler, and follower link lengths, respectively. Variables β_2, β_3, and β_4 are the rotations of Δ_2, Δ_3, and Δ_4, respectively. Equations (D.5) and (D.3) can be regarded as the "equation of closure" for the initial and displaced positions of the compatibility linkage. Figure D.1 illustrates both branches of the compatibility linkage in the starting and displaced positions.

Sandor and Erdman presented an analytical solution algorithm to calculate β_3, β_4, $\tilde{\beta}_3$, and $\tilde{\beta}_4$ for a given value of β_2.[1] This algorithm includes the computation schedule given in Equation (D.6). In Equation (D.6), the function ATAN2( ), a variation of the arctangent function, is widely available in many programming languages and commercial mathematical analysis codes. Angles β_3, β_4, θ_3, and θ_4 correspond to one branch of the compatibility linkage, and angles $\tilde{\beta}_3$, $\tilde{\beta}_4$, $\tilde{\theta}_3$, and $\tilde{\theta}_4$ correspond to the other branch (Figure D.1). After incorporating the β solutions in any two of the three standard-form equations in Equation (D.1), $\mathbf{W}_1$ and $\mathbf{Z}_1$ can be solved by applying Cramer's rule.[2] Circle and center point curves are produced by calculating solutions from Equation (D.6) over a range of prescribed β_2 values.

A procedure diagram for analytical four-position synthesis is illustrated in Figure D.2. Given four precision positions, Equations (D.4) and (D.5) are used to calculate Δ_1, Δ_2, Δ_3, and Δ_4. These calculated quantities, in addition to a prescribed range for angle β_2, are used in Equation (D.6) to ultimately calculate the corresponding ranges for angles β_3 and β_4 for one branch of a mechanism solution and angles $\tilde{\beta}_3$ and $\tilde{\beta}_4$ for the other branch of the mechanism solution. The β_2, β_3, and β_4 values and β_2, $\tilde{\beta}_3$, and $\tilde{\beta}_4$ values are used in Equation (D.1) to calculate $\mathbf{W}_1$ and $\mathbf{Z}_1$ for both mechanism branches (using any two of the three equations in Equation (D.1)).

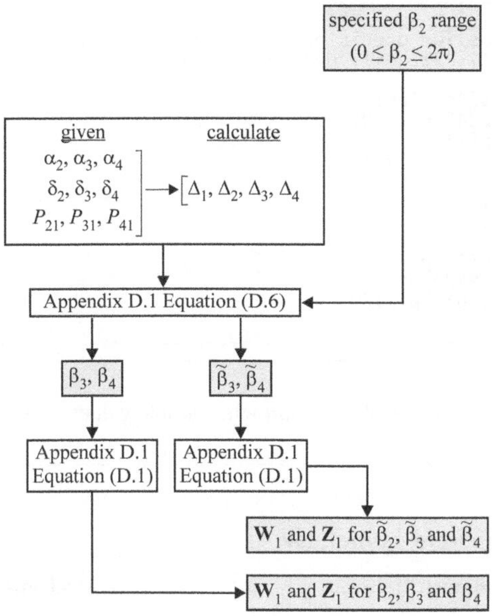

FIGURE D.2 Algorithm diagram for analytical four-position synthesis.

$$
\left\{
\begin{aligned}
&\Delta = \Delta_1 + \Delta_2 e^{i\beta_2} \\
&\cos\theta_3 = \frac{|\Delta_4|^2 - |\Delta_3|^2 - |\Delta|^2}{2|\Delta_3||\Delta|} \\
&\sin\theta_3 = \left|\sqrt{1-(\cos\theta_3)^2}\right| \geq 0 \\
&\theta_3 = \mathrm{ATAN2}(\sin\theta_3, \cos\theta_3) \quad \text{where} \quad 0 \leq \theta_3 \leq \pi \\
&\beta_3 = \arg\Delta + \theta_3 - \arg\Delta_3 \\
&\tilde{\theta}_3 = 2\pi - \theta_3 \\
&\tilde{\beta}_3 = \arg\Delta + \tilde{\theta}_3 - \arg\Delta_3 \\
&\cos\theta_4 = \frac{|\Delta_3|^2 - |\Delta_4|^2 - |\Delta|^2}{2|\Delta_4||\Delta|} \\
&\sin\theta_4 = \left|\sqrt{1-(\cos\theta_4)^2}\right| \geq 0 \\
&\theta_4 = \mathrm{ATAN2}(\sin\theta_4, \cos\theta_4) \quad \text{where} \quad 0 \leq \theta_4 \leq \pi \quad \text{and} \quad \tilde{\theta}_4 = -\theta_4 \\
&\beta_4 = \arg\Delta - \theta_4 - \arg\Delta_4 \\
&\tilde{\beta}_4 = \arg\Delta + \theta_4 - \arg\Delta_4 + \pi
\end{aligned}
\right. \tag{D.6}
$$

```
%-------------------------------------------------------------
% Here, values for precision position parameters alpha2, alpha3,
% alpha4, P21, P31, P41, delta2, delta3 and delta4 are assigned.

alpha2 = 35.55*pi/180;
alpha3 = 45.2345*pi/180;
alpha4 = 85*pi/180;

P21 = 0.5217;
P31 = 1.5421;
P41 = 2.3116;

delta2 = 29.0316*pi/180;
delta3 = -15.9829*pi/180;
delta4 = -13.6018*pi/180;

%-------------------------------------------------------------
```

FIGURE D.3 Section of file *Burm_Curve_4Pos.m* with sample values in bold type.

TABLE D.1
Column Layout of File *circle_center.csv*

Mechanism Branch 1				Mechanism Branch 2			
$(-Z_{1x})$	$(-Z_{1y})$	$(-Z_{1x}-W_{1x})$	$(-Z_{1y}-W_{1y})$	$(-\tilde{Z}_{1x})$	$(-\tilde{Z}_{1y})$	$(-\tilde{Z}_{1x}-\tilde{W}_{1x})$	$(-\tilde{Z}_{1y}-\tilde{W}_{1y})$

TABLE D.2
Column Layout of File *beta_vs_theta.csv*

	Mechanism Branch 1				Mechanism Branch 2			
β_2	θ_3	β_3	θ_4	β_4	$\tilde{\theta}_3$	$\tilde{\beta}_3$	$\tilde{\theta}_4$	$\tilde{\beta}_4$

D.2 MATLAB FILE USER INSTRUCTIONS FOR FOUR-POSITION SYNTHESIS

The Appendix D folder (which is available for download at https://www.routledge.com/9781032956800) includes a MATLAB file (filename *Burm_Curve_4Pos.m*) for circle and center point curve generation for planar four-bar four-position synthesis.[3] In this file, the user specifies the coupler displacement angles α_2, α_3, and α_4 (in degrees). The user also specifies the coupler point displacement magnitudes P_{21}, P_{31}, and P_{41} and the coupler point displacement line angles δ_2, δ_3, and δ_4 (in degrees). Figure D.3 illustrates the user input section of the file *Burm_Curve_4Pos.m* with sample values in bold type.

After running the file *Burm_Curve_4Pos.m*, two files are written in a folder named *Results* (in a format compatible with Microsoft® Excel). The file *circle_center.csv* includes the x and y-components of each circle point $(-\mathbf{Z}_1)$ and the x and y-components of each center point $(-\mathbf{Z}_1 - \mathbf{W}_1)$ for both mechanism branches over the β_2 range $\beta_2 = 1°, 2°, \ldots, 359°$. Table D.1 includes the column layout of file *circle_center.csv*. The file *beta_vs_theta.csv* includes each β and θ value in radians and each $\tilde{\beta}$ and $\tilde{\theta}$ value in radians for both mechanism branches over the β_2 range $\beta_2 = 1°, 2°, \ldots, 359°$. Table D.2 includes the column layout of file *beta_vs_theta.csv*.

NOTES

1. Sandor, G. N. and Erdman, A. G. 1984. *Advanced Mechanism Design: Analysis and Synthesis*, Volume 2. pages 180–183. Englewood Cliffs: Prentice-Hall.
2. Equation (D.1), when fully expanded, becomes Equation (5.8).
3. The Appendix D MATLAB file can be downloaded from the publisher's web page or from the authors' server http://www.softalink.com/kruss/mech_des_vis_prog_app_2nd_ed/MATLAB_SIMMULT_FILES.zip.

Appendix E
Algorithm for Analytical Five-Position Synthesis and MATLAB® File User Instructions

E.1 ANALYTICAL PLANAR FOUR-BAR FIVE-POSITION SYNTHESIS ALGORITHM

For five-position synthesis, the standard form equations for the **W** – **Z** dyad (Equation 5.3) form a set of four complex, linear equations. This equation set can be expressed in matrix form as

$$\begin{bmatrix} e^{i\beta_2}-1 & e^{i\alpha_2}-1 \\ e^{i\beta_3}-1 & e^{i\alpha_3}-1 \\ e^{i\beta_4}-1 & e^{i\alpha_4}-1 \\ e^{i\beta_5}-1 & e^{i\alpha_5}-1 \end{bmatrix} \begin{bmatrix} W_1 e^{i\theta} \\ Z_1 e^{i\phi} \end{bmatrix} = \begin{bmatrix} P_{21}e^{i\delta_2} \\ P_{31}e^{i\delta_3} \\ P_{41}e^{i\delta_4} \\ P_{51}e^{i\delta_5} \end{bmatrix} \tag{E.1}$$

where α_j, P_{j1}, and δ_j ($j=2, 3, 4, 5$) are user-prescribed. As shown in Table 5.1, no "free choice" is available in five-position synthesis. In order for Equation (E.1) to have simultaneous solutions for $\mathbf{W}_1$ and $\mathbf{Z}_1$, it must follow that

$$\det\left[\begin{array}{cc|c} e^{i\beta_2}-1 & e^{i\alpha_2}-1 & P_{21}e^{i\delta_2} \\ e^{i\beta_3}-1 & e^{i\alpha_3}-1 & P_{31}e^{i\delta_3} \\ e^{i\beta_4}-1 & e^{i\alpha_4}-1 & P_{41}e^{i\delta_4} \end{array}\right] = 0 \;\Rightarrow\; \Delta_2 e^{-i\beta_2} + \Delta_3 e^{-i\beta_3} + \Delta_4 e^{-i\beta_4} + \Delta_1 = 0 \tag{E.2}$$

and

$$\det\left[\begin{array}{cc|c} e^{i\beta_2}-1 & e^{i\alpha_2}-1 & P_{21}e^{i\delta_2} \\ e^{i\beta_3}-1 & e^{i\alpha_3}-1 & P_{31}e^{i\delta_3} \\ e^{i\beta_5}-1 & e^{i\alpha_5}-1 & P_{51}e^{i\delta_5} \end{array}\right] = 0 \;\Rightarrow\; \Delta_2' e^{-i\beta_2} + \Delta_3' e^{-i\beta_3} + \Delta_4' e^{-i\beta_4} + \Delta_1' = 0 \tag{E.3}$$

or the *rank* of the "augmented matrix" of the coefficients is 2. In comparison to four-position synthesis, there are two compatibility equations to be satisfied simultaneously in five-position synthesis.

In Equation (E.2), Δ_j ($j=1, 2, 3, 4$) are the same as in Equation (D.4) and in Equation (E.3):

$$
\begin{aligned}
\Delta_1' &= -\Delta_2' - \Delta_3' - \Delta_4 \\
\Delta_2' &= \begin{vmatrix} e^{i\alpha_3} - 1 & P_{31}e^{i\delta_3} \\ e^{i\alpha_5} - 1 & P_{51}e^{i\delta_5} \end{vmatrix} . \\
\Delta_3' &= \begin{vmatrix} e^{i\alpha_2} - 1 & P_{21}e^{i\delta_2} \\ e^{i\alpha_5} - 1 & P_{51}e^{i\delta_5} \end{vmatrix}
\end{aligned} \tag{E.4}
$$

The complex conjugates of Equations (E.2) and (E.3) also hold true:

$$\overline{\Delta}_2 e^{-i\beta_2} + \overline{\Delta}_3 e^{-i\beta_3} + \overline{\Delta}_4 e^{-i\beta_4} + \overline{\Delta}_1 = 0 \tag{E.5}$$

$$\overline{\Delta}_2' e^{-i\beta_2} + \overline{\Delta}_3' e^{-i\beta_3} + \overline{\Delta}_4' e^{-i\beta_4} + \overline{\Delta}_1' = 0. \tag{E.6}$$

Sandor and Erdman presented an analytical solution algorithm to calculate the coefficients of a fifth-order polynomial whose roots can be used to calculate values for β_2.[1] This algorithm includes the computation schedule in Equation (E.7).

$$
\begin{cases}
\mathbf{a}' = \overline{\Delta}_1\overline{\Delta}_1', \quad \mathbf{a}'' = \Delta_2'\Delta_3, \quad \mathbf{a}''' = \Delta_2\Delta_3', \quad \mathbf{a}'''' = \mathbf{a}'' - \mathbf{a}''' \\
\mathbf{a} = \mathbf{a}'\mathbf{a}'''' \\
\mathbf{b}' = \overline{\Delta}_1'\overline{\Delta}_2, \quad \mathbf{b}'' = \overline{\Delta}_1\overline{\Delta}_2', \quad \mathbf{b}''' = \overline{\Delta}_1\Delta_3, \quad \mathbf{b}'''' = \overline{\Delta}_1'\Delta_3' \\
n = -|\Delta_4|^2 + \sum_{k=1}^{3} |\Delta_k|^2, \quad n' = -|\Delta_4|^2 + \sum_{k=1}^{3} |\Delta_k'|^2 \\
\mathbf{b} = \mathbf{b}'\mathbf{a}'' - \mathbf{b}''\mathbf{a}''' + \mathbf{b}'''n' - \mathbf{b}''''n \\
\mathbf{c}' = \Delta_1'\Delta_3, \quad \mathbf{c}'' = \Delta_1\Delta_3', \quad \mathbf{c}''' = \overline{\Delta}_2\Delta_3, \quad \mathbf{c}'''' = \overline{\Delta}_2'\Delta_3' \\
\mathbf{c} = \mathbf{b}''\mathbf{c}' - \mathbf{b}'\mathbf{c}'' + \mathbf{c}'''n' - \mathbf{c}''''n \\
\mathbf{d}' = \overline{\Delta}_2\overline{\Delta}_2', \quad \mathbf{d}'' = \mathbf{c}' - \mathbf{c}'', \quad \mathbf{d} = \mathbf{d}'\mathbf{d}'' \\
\mathbf{u} = \overline{\Delta}_3\overline{\Delta}_3', \quad \mathbf{f}' = \overline{\Delta}_1\overline{\Delta}_2', \quad \mathbf{f} = \mathbf{f}'\mathbf{u} \\
\mathbf{h}' = \Delta_1'\overline{\Delta}_2, \quad \mathbf{h} = \mathbf{h}'\mathbf{u}, \quad \mathbf{k} = \mathbf{f} - \overline{\mathbf{h}}, \quad k = |\mathbf{k}| \\
\mathbf{g}' = \Delta_1'\overline{\Delta}_1, \quad \mathbf{g}'' = \Delta_2'\overline{\Delta}_2, \quad \mathbf{g}''' = \mathbf{g}' + \mathbf{g}'' \\
g_y = u_x g_y''' + u_y g_x''', \quad \mathbf{v} = i g_y (4\mathbf{k}) \\
m = -4g_y^2 - 2k^2, \quad \mathbf{p} = \mathbf{a}\overline{\mathbf{d}} \\
\mathbf{q} = \mathbf{a}\overline{\mathbf{c}} + \mathbf{b}\overline{\mathbf{d}} + \mathbf{k}\mathbf{k}, \quad \mathbf{s} = \mathbf{a}\overline{\mathbf{b}} + \mathbf{b}\overline{\mathbf{c}} + \mathbf{c}\overline{\mathbf{d}} + \mathbf{v} \\
t = \frac{1}{2}\left(|\mathbf{a}|^2 + |\mathbf{b}|^2 + |\mathbf{c}|^2 + |\mathbf{d}|^2 + m\right) \\
A_1 = -6p_y - 4q_y - 2s_y
\end{cases} \tag{E.7}
$$

$$\begin{cases} A_2 = -15p_x - 5q_x + s_x + 3t \\ A_3 = 20p_y - 4s_y, \quad A_4 = 15p_x - 5q_x - s_x + 3t \\ A_5 = -6p_y + 4q_y - 2s_y, \quad A_6 = -p_x + q_x - s_x + t \\ \text{Check}: A_0 = p_x + q_x + s_x + t = 0 \\ a_j = A_{j+1} / A_6 \quad j = 0,1,2,3,4 \end{cases}$$

The five a_j values are coefficients for the following fifth-order polynomial:

$$\tau^5 + a_4\tau^4 + a_3\tau^3 + a_2\tau^2 + a_1\tau + a_0 = 0 \tag{E.8}$$

Equation (E.8) has a trivial root τ_0 (which can be discarded). This trivial root is expressed as

$$\tau_0 = \frac{1 - \cos\alpha_2}{\sin\alpha_2} \tag{E.9}$$

Equation (E.8) will also have either zero, two, or four real roots. Subsequently, from the real roots, either zero, two, or four β_2 values are calculated. Using the real roots, the corresponding β_2 values are calculated from

$$\beta_2 = \arg\left(1 - \tau^2 + i2\tau\right) \tag{E.10}$$

The β_2 values can then be used in Equation (D.6) to calculate the corresponding values of **W** and **Z**.

A procedure diagram for analytical five-position synthesis is illustrated in Figure E.1. Given five precision positions, Equations (D.4) and (D.5) from Appendix D.1 are used to calculate Δ_1, Δ_2, Δ_3, and Δ_4, and Equation (E.4) is used to calculate Δ_1', Δ_2', and Δ_3'. These calculated quantities and their complex conjugates are used in Equations (E.7)–(E.10) to ultimately calculate values for β_2 (either two or four real values are possible).

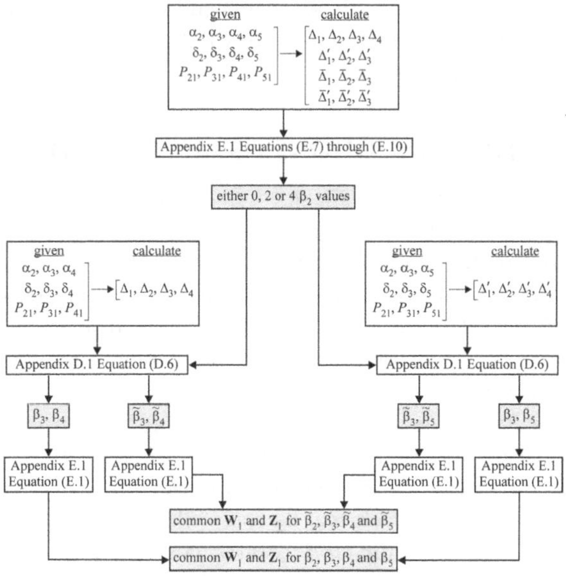

FIGURE E.1 Algorithm diagram for analytical five-position synthesis.

```
%-------------------------------------------------------------
% Here, values for precision position parameters alpha2, alpha3,
% alpha5, P21, P31, P41, P51, delta2, delta3 delta4 and delta5
% are assigned.

alpha2 = -20*pi/180;
alpha3 = -45*pi/180;
alpha4 = -75*pi/180;
alpha5 = -90*pi/180;

P21 = 0.5217;
P31 = 1.3421;
P41 = 2.3116;
P51 = 3.1780;

delta2 = 29.0316*pi/180;
delta3 = -15.9829*pi/180;
delta4 = -13.6018*pi/180;
delta5 = -16.7388*pi/180;

%-------------------------------------------------------------
```

FIGURE E.2 Section of file *Burm_Curve_5Pos.m* with sample values in bold type.

After calculating the β_2 values, the procedure is identical to the diagram illustrated in Figure D.2. Quantities Δ_1, Δ_2, Δ_3, and Δ_4 (along with the calculated values for β_2) are used in Equation (D.6) to calculate β_3 and β_4 for one branch of a mechanism solution and angles $\tilde{\beta}_3$ and $\tilde{\beta}_4$ for the other branch of the mechanism solution. The β_2, β_3, and β_4 values and β_2, $\tilde{\beta}_3$, and $\tilde{\beta}_4$ values are used in Equation (E.1) to calculate $\mathbf{W}_1$ and $\mathbf{Z}_1$ for both mechanism branches (using any two of the first three equations in Equation (E.1)). The calculated $\mathbf{W}_1 - \mathbf{Z}_1$ dyads correspond to precision positions 1, 2, 3, and 4.

To calculate dyad solutions for precision positions 1, 2, 3, and 5, quantities Δ_1', Δ_2', Δ_3', and Δ_4' (along with the calculated values for β_2) are used in Equation (D.6) to calculate β_3 and β_5 for one branch of a mechanism solution and angles $\tilde{\beta}_3$ and $\tilde{\beta}_5$ for the other branch of the mechanism solution. The β_2, β_3, β_4, and β_5 values and β_2, $\tilde{\beta}_3$, $\tilde{\beta}_4$, and $\tilde{\beta}_5$ values are used in Equation (E.1) to calculate $\mathbf{W}_1$ and $\mathbf{Z}_1$ for both mechanism branches (using any one of the first three equations in Equation (E.1) and the last equation).

The common $\mathbf{W}_1 - \mathbf{Z}_1$ solutions among those calculated for precision positions 1, 2, 3, and 4 and precision positions 1, 2, 3, and 5 will achieve positions 1, 2, 3, 4, and 5. Any two common $\mathbf{W}_1 - \mathbf{Z}_1$ solutions will produce a single mechanism.

E.2 MATLAB FILE USER INSTRUCTIONS FOR FIVE-POSITION SYNTHESIS

The Appendix E folder (which is available for download at https://www.routledge.com/9781032956800) includes a MATLAB file (filename *Burm_Curve_5Pos.m*) for circle and center-point pair calculation for five-position synthesis.[2] In this file, the user specifies the coupler displacement angles α_2, α_3, α_4, and α_5 (in degrees). The user also specifies the coupler point displacement magnitudes P_{21}, P_{31}, P_{41}, and P_{51} and the coupler point displacement line angles δ_2, δ_3, δ_4, and δ_5 (in degrees). Figure E.2 illustrates the user input section of the file *Burm_Curve_5Pos.m* with sample values in bold type.

After running the file *Burm_Curve_4Pos.m*, four files are written in a folder named *Results* (in a format compatible with Microsoft® Excel). The files *circle_center_pos1234.csv* and *circle_center_pos1235.csv* include the x and y-components of each circle point ($-\mathbf{Z}_1$) and the x and y-components of each center point ($-\mathbf{Z}_1 - \mathbf{W}_1$) for each real β_2 root. File *circle_center_pos1234.csv* considers precision positions 1, 2, 3, and 4, and file *circle_center_pos1235.csv* considers precision positions 1, 2, 3, and 5. Table E.1 includes the column layout of files *circle_center_pos1234.csv* and *circle_center_pos1235.csv*. The files *beta_vs_theta_pos1234.csv* and *beta_vs_theta_pos1235.csv* include

TABLE E.1
Column Layout of Files *circle_center_pos1234.csv* and *circle_center_pos1235.csv*

Mechanism Branch 1				Mechanism Branch 2			
Col. 1	Col. 2	Col. 3	Col. 4	Col. 5	Col. 6	Col. 7	Col. 8
$(-Z_{1x})$	$(-Z_{1y})$	$(-Z_{1x}-W_{1x})$	$(-Z_{1y}-W_{1y})$	$(-Z_{1x})$	$(-Z_{1y})$	$(-Z_{1x}-W_{1x})$	$(-Z_{1y}-W_{1y})$

TABLE E.2
Column Layout of File *beta_vs_theta_pos1234.csv*

	Mechanism Branch 1				Mechanism Branch 2			
Col. 1	Col. 2	Col. 3	Col. 4	Col. 5	Col. 6	Col. 7	Col. 8	Col. 9
β_2	θ_3	β_3	θ_4	β_4	$\tilde{\theta}_3$	$\tilde{\beta}_3$	$\tilde{\theta}_4$	$\tilde{\beta}_4$

TABLE E.3
Column Layout of File *beta_vs_theta_pos1235.csv*

	Mechanism Branch 1				Mechanism Branch 2			
Col. 1	Col. 2	Col. 3	Col. 4	Col. 5	Col. 6	Col. 7	Col. 8	Col. 9
β_2	θ_3	β_3	θ_5	β_5	$\tilde{\theta}_3$	$\tilde{\beta}_3$	$\tilde{\theta}_5$	$\tilde{\beta}_5$

each β and θ value in radians and each $\tilde{\beta}$ and $\tilde{\theta}$ value in radians for each real β_2 root. Tables E.2 and E.3 include the column layout of files *beta_vs_theta_pos1234.csv* and *beta_vs_theta_pos1235.csv*, respectively.

Because either zero, two, or four β_2 roots are possible as noted in Appendix E.1, the four files produced from *Burm_Curve_4Pos.m* will have either zero, two, or four rows of data. To select a mechanism solution, the user must select two circle and center point pairs that are common in the files *circle_center_pos1234.csv* and *circle_center_pos1235.csv* (and their corresponding angles in files *beta_vs_theta_pos1234.csv* and *beta_vs_theta_pos1235.csv*). Common circle and center point pairs in files *circle_center_pos1234.csv* and *circle_center_pos1235.csv* are dyad solutions that will achieve all five precision positions.

NOTES

1 Sandor, G. N. and Erdman, A. G. (1984). *Advanced Mechanism Design: Analysis and Synthesis*, Volume 2. pages 261–263. Prentice-Hall Inc., Englewood Cliffs, New Jersey.
2 The Appendix E MATLAB file can be downloaded from the publisher's web page or from the authors' server http://www.softalink.com/kruss/mech_des_vis_prog_app_2nd_ed/MATLAB_SIMMULT_FILES.zip.

Appendix F
User Instructions for Chapter 6 MATLAB® Dimensional Synthesis Files

F.1 PLANAR FOUR-BAR MOTION GENERATION, *N* PRECISION POSITIONS WITH ORDER CONSTRAINTS

The subfolder *Synthesis* within the folder *Appendix F.1* (which is available for download at https://www.routledge.com/9781032956800) includes MATLAB files (filenames *W_Z_prepare.m*, *W_Z_interior_point.m*, *U_S_prepare.m,* and *U_S_quasi_newton.m*) for calculating the unknowns $\mathbf{W}_1$, $\mathbf{Z}_1$, $\mathbf{U}_1$, $\mathbf{S}_1$, and β_j for motion generation given *N* precision positions.[1] The folder *Common* (which includes supplementary MATLAB files) must be included alongside the folder *Synthesis* to properly run the MATLAB files. Figure F.1 includes the five-step procedure for calculating $\mathbf{W}_1$, $\mathbf{Z}_1$, and β_j.

Synthesizing the $\mathbf{W}_1$ - $\mathbf{Z}_1$ dyad begins with the file *W_Z_prepare.m* (step 1 in Figure F.1). In this file, the number of precision positions is specified. After specifying the value of *N* (the number of precision positions) and running this file, the folder *WZfuncs* is produced. The folder *WZfuncs* includes functions required to run file *W_Z_interior_point.m.*

After exiting the file *W_Z_prepare.m* and opening the file *W_Z_interior_point.m*, step 2 in Figure F.1 is started. In this file, the user specifies the value of *N*—the number of precision positions (identical to the *N* value prescribed in the file *W_Z_prepare.m*). The user also includes the name of each coupler displacement angle (angle α) to be prescribed. Additionally, the user includes the specific precision position values (with angles specified in degrees). The number of *alphaN* and *PN* variables and values should match the value of *N*. Figure F.2 illustrates these user input sections in the file *W_Z_interior_point.m*, with sample data in bold type.

The user includes the initial values for the components of dyad $\mathbf{W}_1 - \mathbf{Z}_1$ (step 3 in Figure F.1). The user also includes the initial values for each dyad displacement angle β (in degrees). The number

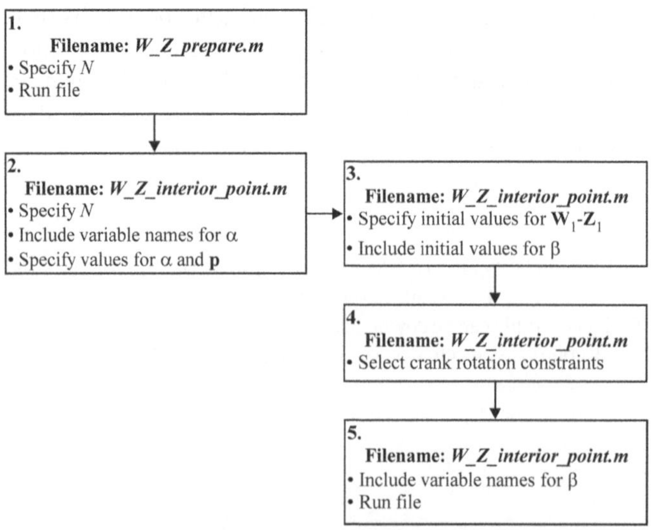

FIGURE F.1 Procedure for calculating $\mathbf{W}_1$, $\mathbf{Z}_1$, and β_j.

```
%-------------------------------------------------------------
% Here, the names for each coupler displacement angle are
%included(e.g., alpha2 alpha3 alpha4 alpha5 alpha6...).

global alpha2 alpha3 alpha4

%-------------------------------------------------------------

%-------------------------------------------------------------
% Here, values for precision position parameters alphaN and PN are
% assigned. Additional rows can be included to accommodate
% additionalalphaN and PN variablesand values.

alpha2 = 25 *pi/180;
alpha3 = 35 *pi/180;
alpha4 = 60 *pi/180;

P1 = [0,0];
P2 = [0.6397, -0.6005];
P3 = [1.4148, -1.0654];
P4 = [2.8010, -0.8675];

%-------------------------------------------------------------
```

FIGURE F.2 Sections (for step 2) of file *W_Z_interior_point.m*, with sample data in bold type.

```
%-------------------------------------------------------------
% Here, initial values for the components of dyad W1-Z1 are
% assigned.

W1x = 1;
W1y = 1;
Z1x = 1;
Z1y = 1;

%-------------------------------------------------------------

%-------------------------------------------------------------
% Here, values for dyad parameters betaN are assigned. Additional
%rows can be included to accommodate additional betaN variables
and % values.

beta2 = 10 *pi/180;
beta3 = 20 *pi/180;
beta4 = 30 *pi/180;

%-------------------------------------------------------------
```

FIGURE F.3 Sections (for step 3) of file *W_Z_interior_point.m*, with sample data in bold type.

of *beta N* variables and values should match the value of *N*. Figure F.3 illustrates these user input sections in the file *W_Z_interior_point.m*, with sample data in bold type.

In step 4 in Figure F.1, the user selects either clockwise or counterclockwise crank rotation constraints for order defect elimination. Counterclockwise constraints are selected in the file *W_Z_interior_point.m*. To select clockwise constraints, the characters *%{* must be included as instructed in the file (to deactivate the counterclockwise constraints). Also, the characters *%{* must be removed as instructed in the file (to activate the clockwise constraints).

In step 5 in Figure F.1, the user includes names for each dyad displacement angle (angle β) to be calculated. Figure F.4 illustrates this user input section in the file *W_Z_interior_point.m* with sample data in bold type.

Lastly, the file *W_Z_interior_point.m* is run (also in step 5). The calculated $\mathbf{W}_1 - \mathbf{Z}_1$ dyad components and parameters are displayed in the MATLAB command window after this file has been run. Figure F.5 illustrates sample output from the file *W_Z_interior_point.m*.

```
%-----------------------------------------------------------------
% Here, the names for each betaN displacement angle are included
% (e.g., beta2; beta3; beta4; beta5; beta6; .....]);

x0 = [W1x; W1y; Z1x; Z1y;
      beta2; beta3; beta4];

%-----------------------------------------------------------------
```

FIGURE F.4 Section (for step 5) of file *W_Z_interior_point.m* with sample data in bold type.

```
%%%%%%%%%%%%%%%%%%%%%%%%%%%%%%%%%%%%%%%%%%%%%%%%%%%%%%%%%%%%%%%%%%%
Solvingfor the components of W1-Z1
%%%%%%%%%%%%%%%%%%%%%%%%%%%%%%%%%%%%%%%%%%%%%%%%%%%%%%%%%%%%%%%%%%%
Iterations: 55
Func Evals: 85
Algorithm: interior-point
exit flag = 2
error = ( 1.454462e-030 )
W1x = ( 0.232320 )
W1y = ( 0.092013 )
Z1x = ( -1.952774 )
Z1y = ( -2.155586 )
beta2 = ( 131.035728 )
beta3 = ( 261.713783 )
beta4 = ( 297.939924 )
Elapsed time is 3.344318 seconds.
Solution calculation is complete.
```

FIGURE F.5 Sample output from the file *W_Z_interior_point.m.*

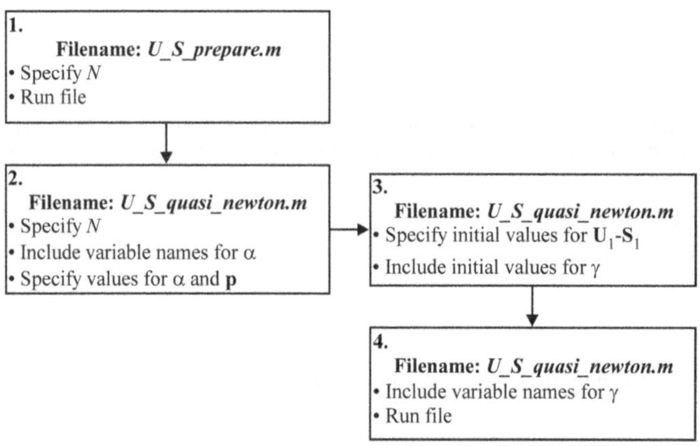

FIGURE F.6 Procedure for calculating $\mathbf{U}_1$ and $\mathbf{S}_1$.

Figure F.6 includes the four-step procedure for calculating the unknowns $\mathbf{U}_1$ and $\mathbf{S}_1$ for planar four-bar motion generation given *N* precision positions. Steps 1 through 3 in Figure F.6 are identical to those for calculating the $\mathbf{W}_1 - \mathbf{Z}_1$ dyad (where the components of $\mathbf{U}_1 - \mathbf{S}_1$ replace those of $\mathbf{W}_1 - \mathbf{Z}_1$ and the dyad displacement angle γ replaces β).

In step 4 in Figure F.6, the user includes names for each dyad displacement angle (angle γ) to be calculated. Figure F.7 illustrates this user input section in the file *U_S_quasi_newton.m* with sample data in bold type.

Lastly, the file *U_S_quasi_newton.m* is run (also in step 4). The calculated $\mathbf{U}_1 - \mathbf{S}_1$ components are parameters displayed in the MATLAB command window after this file has been run. Figure F.8 illustrates sample output from the file *U_S_quasi_newton.m.*

```
%-----------------------------------------------------------------
% Here, the names for each gammaN displacement angle are included
% (e.g., gamma2; gamma3; gamma4; gamma5; gamma6; .....]);

x0 = [U1x; U1y; S1x; S1y;
      gamma2; gamma3; gamma4];

%-----------------------------------------------------------------
```

FIGURE F.7 Section (for step 4) of file *U_S_quasi_newton.m* with sample data in bold type.

```
%%%%%%%%%%%%%%%%%%%%%%%%%%%%%%%%%%%%%%%%%%%%%%%%%%%%%%%%%%%%%%%%%%%%%%%%%%%%
Solving for the components and angles of U1-S1
%%%%%%%%%%%%%%%%%%%%%%%%%%%%%%%%%%%%%%%%%%%%%%%%%%%%%%%%%%%%%%%%%%%%%%%%%%%%
Iterations: 52
Func Evals: 53
Algorithm: large-scale: trust-region Newton
exit flag = 3
error = ( 1.362890e-020 )
U1x = ( -1.051176 )
U1y = ( -0.943605 )
S1x = ( -0.894853 )
S1y = ( -0.231639 )
gamma2 = ( 10.000000 )
gamma3 = ( 20.000000 )
gamma4 = ( 30.000000 )
Elapsed time is 1.471684 seconds.
Solution calculation is complete.
```

FIGURE F.8 Sample output from the file *U_S_quasi_newton.m*.

When calculating new solutions, repeating step 1 in Figures F.1 and F.6 is only necessary if the number of precision positions (N) for the new solution is different from the previous solution. If N is the same for the new solution, the user can begin at step 2 in Figures F.1 and F.6.

F.2 PLANAR FOUR-BAR PATH GENERATION, *N* PRECISION POINTS WITH ORDER CONSTRAINTS

The subfolder *Synthesis* within the folder *Appendix F.2* includes MATLAB files (filenames *W_Z_prepare.m*, *W_Z_interior_point.m*, *U_S_prepare.m* and *U_S_quasi_newton.m*) for calculating the unknowns $\mathbf{W}_1$, $\mathbf{Z}_1$, $\mathbf{U}_1$, $\mathbf{S}_1$, α_j, and β_j for path generation given N precision points. The folder *Common* (which includes supplementary MATLAB files) must be included alongside the folder *Synthesis* to properly run the MATLAB files. Figure F.9 includes the 5-step procedure for calculating $\mathbf{W}_1$, $\mathbf{Z}_1$, α_j, and β_j.

Synthesizing the $\mathbf{W}_1 - \mathbf{Z}_1$ dyad begins with the file *W_Z_prepare.m* (step 1 in Figure F.9). In this file, the number of precision points is specified. After specifying the value of N (the number of precision points) and running this file, the folder *WZfuncs* is produced. The folder *WZfuncs* includes functions required to run file *W_Z_interior_point.m*.

After exiting the file *W_Z_prepare.m* and opening the file *W_Z_interior_point.m*, step 2 in Figure F.9 is started. In this file, the user specifies the value of N—the number of precision points (identical to the N value prescribed in the file *W_Z_prepare.m*). The user also includes the specific precision point values. The number of *PN* variables and values should match the value of N. Figure F.10 illustrates this user input section in the file *W_Z_interior_point.m*, with sample data in bold type.

The user includes the initial values for the components of dyad $\mathbf{W}_1 - \mathbf{Z}_1$ (step 3 in Figure F.9). The user also includes the initial values for each dyad displacement angle α and β (in degrees). The number of *alphaN* and *betaN* variables and values should match the value of N. Figure F.11 illustrates this user input section in the file *W_Z_interior_point.m*, with sample data in bold type.

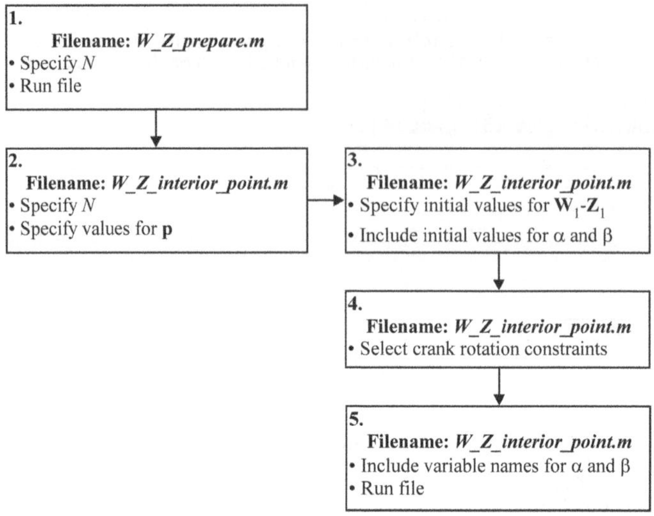

FIGURE F.9 Procedure for calculating $\mathbf{W}_1$, $\mathbf{Z}_1$, α_j and β_j.

```
%-----------------------------------------------------------------
% Here, values for precision points PN are assigned. Additional
%rows can be included to accommodate additional PN variables and
% values.

P1 = [0, 0];
P2 = [0.0574, 0.3339];
P3 = [0.2792, 0.6931];
P4 = [0.5965, 0.9150];
P5 = [1, 1];

%-----------------------------------------------------------------
```

FIGURE F.10 Section (for step 2) of file *W_Z_interior_point.m*, with sample data in bold type.

```
%-----------------------------------------------------------------
% Here, initial values for the components of dyad W1-Z1 are
% assigned.Also, values for dyad parameters alphaN and betaN are
% assigned.Additional rows can be included to accommodate
% additional alphaN and betaN variables and values.

W1x = -1;
W1y = -1;
Z1x = -1;
Z1y = -1;

alpha2 = 10 *pi/180;
alpha3 = 15 *pi/180;
alpha4 = 20 *pi/180;
alpha5 = 15 *pi/180;

beta2 = 10 *pi/180;
beta3 = 20 *pi/180;
beta4 = 30 *pi/180;
beta5 = 40 *pi/180;

%-----------------------------------------------------------------
```

FIGURE F.11 Section (for step 3) of file *W_Z_interior_point.m*, with sample data in bold type.

In step 4 in Figure F.9, the user selects either clockwise or counterclockwise crank rotation constraints for order defect elimination. Counterclockwise constraints are selected in the file *W_Z_interior_point.m*. To select clockwise constraints, the characters *%{* must be included as instructed

```
%-------------------------------------------------------------
% Here, the names for each alphaN and betaN displacement angle are
% included.
% (e.g., alpha2; alpha3; alpha4; alpha5; alpha6; alpha7; .....;)
% (e.g., beta2; beta3; beta4; beta5; beta6; beta7; .....];)

x0 = [W1x; W1y; Z1x; Z1y;
      alpha2; alpha3; alpha4; alpha5;
      beta2; beta3; beta4; beta5];

%-------------------------------------------------------------
```

FIGURE F.12 Section (for step 5) of file *W_Z_interior_point.m* with sample data in bold type.

```
%%%%%%%%%%%%%%%%%%%%%%%%%%%%%%%%%%%%%%%%%%%%%%%%%%%%%%%%%%%%%%%%%%%%%%%%%%%%%%
Solving for the components and angles of W1-Z1
%%%%%%%%%%%%%%%%%%%%%%%%%%%%%%%%%%%%%%%%%%%%%%%%%%%%%%%%%%%%%%%%%%%%%%%%%%%%%%
Iterations: 26
Func Evals: 51
Algorithm: interior-point
exit flag = 2
error = ( 5.731637e-020 )
W1x = ( 0.087400 )
W1y = ( -0.339335 )
Z1x = ( 0.080171 )
Z1y = ( -1.686248 )
alpha2 = ( -6.980074 )
alpha3 = ( 8.109669 )
alpha4 = ( 35.008123 )
alpha5 = ( 57.453049 )
beta2 = ( 74.217279 )
beta3 = ( 143.863122 )
beta4 = ( 215.697321 )
beta5 = ( 287.612341 )
Elapsed time is 1.719781 seconds.
Solution calculation is complete.
```

FIGURE F.13 Sample output from the file *W_Z_interior_point.m.*

in the file (to deactivate the counterclockwise constraints). Also, the characters *%{* must be removed as instructed in the file (to activate the clockwise constraints).

In step 5 in Figure F.9, the user includes the names for each dyad displacement angle (angles α and β) to be calculated. Figure F.12 illustrates this user input section in the file *W_Z_interior_point.m* with sample data in bold type.

Lastly, the file *W_Z_interior_point.m* is run (also in step 5). The calculated $\mathbf{W}_1 - \mathbf{Z}_1$ dyad components and parameters are displayed in the MATLAB command window after this file has been run. Figure F.13 illustrates sample output from the file *W_Z_interior_point.m.*

The four-step procedure in Figure F.6 is also used to calculate $\mathbf{U}_1 - \mathbf{S}_1$ dyad solutions for planar four-bar path generators. In step 2, the α values calculated from the Appendix F.2 file *W_Z_prepare.m* in are the α values to be specified in the Appendix F.2 file *U_S_quasi_newton.m.*

When calculating new solutions, repeating step 1 in Figures F.9 and F.6 is only necessary if the number of precision points (N) for the new solution is different from the previous solution. If N is the same for the new solution, the user can begin at step 2 in Figures F.9 and F.6.

F.3 GEARED FIVE-BAR MOTION GENERATION, *N* PRECISION POSITIONS WITH ORDER CONSTRAINTS

The subfolder *Synthesis* within the folder *Appendix F.3* includes MATLAB files (filenames *W_Z_prepare.m*, *W_Z_interior_point.m*, *U_S_T_prepare.m*, and *U_S_T_quasi_newton.m*) for calculating the unknowns $\mathbf{W}_1$, $\mathbf{Z}_1$, $\mathbf{U}_1$, $\mathbf{S}_1$, $\mathbf{T}_1$, β_j, and ν_j for motion generation given N precision positions.

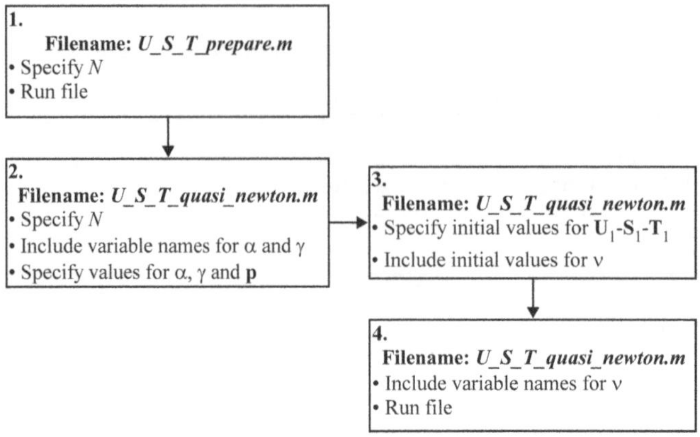

FIGURE F.14 Procedure for calculating $\mathbf{U}_1$, $\mathbf{S}_1$, $\mathbf{T}_1$, and ν_j.

```
%---------------------------------------------------------------
% Here, the names for each gammaN displacement angle are included
% (e.g., nu2;nu3; nu4;nu5; nu6; .....]);

x0 = [U1x; U1y; S1x; S1y; T1x; T1y;
      nu2; nu3; nu4];

%---------------------------------------------------------------
```

FIGURE F.15 Section (for step 4) of file *U_S_T_quasi_newton.m* with sample data in bold type.

The folder *Common* (which includes supplementary MATLAB files) must be included alongside the folder *Synthesis* to properly run the MATLAB files. The procedure for synthesizing the $\mathbf{W}_1 - \mathbf{Z}_1$ dyad is identical to the procedure given in Appendix F.1.

Figure F.14 includes the four-step procedure for calculating the unknowns $\mathbf{U}_1$, $\mathbf{S}_1$, $\mathbf{T}_1$, and ν_j for geared five-bar motion generation given N precision positions. Steps 1 through 3 in Figure F.14 are identical to those for calculating the $\mathbf{W}_1 - \mathbf{Z}_1$ dyad (where the components of $\mathbf{U}_1 - \mathbf{S}_1 - \mathbf{T}_1$ replace those of $\mathbf{W}_1 - \mathbf{Z}_1$, the triad displacement angle ν replaces β, and the dyad displacement angle γ is specified in addition to α).

In step 4 in Figure F.14, the user includes names for each triad displacement angle (angle ν) to be calculated. Figure F.15 illustrates this user input section in the file *U_S_T_quasi_newton.m* with sample data in bold type.

Lastly, the file *U_S_T_quasi_newton.m* is run (also in step 4). The calculated $\mathbf{U}_1 - \mathbf{S}_1 - \mathbf{T}_1$ components are parameters displayed in the MATLAB command window after this file has been run. Figure F.16 illustrates the sample output from the file *U_S_quasi_newton.m.*

When calculating new solutions, repeating step 1 in Figures F.1 and F.14 is only necessary if the number of precision positions (N) for the new solution is different from that of the previous solution. If N is the same for the new solution, the user can begin at step 2 in Figures F.1 and F.14.

F.4 GEARED FIVE-BAR PATH GENERATION, *N* PRECISION POINTS WITH ORDER CONSTRAINTS

The subfolder *Synthesis* within the folder *Appendix F.4* includes MATLAB files (filenames *W_Z_prepare.m, W_Z_interior_point.m, U_S_T_prepare.m,* and *U_S_T_quasi_newton.m*) for calculating the unknowns $\mathbf{W}_1$, $\mathbf{Z}_1$, $\mathbf{U}_1$, $\mathbf{S}_1$, $\mathbf{T}_1$, α_j, β_j, and ν_j for path generation given N precision points. The folder *Common* (which includes supplementary MATLAB files) must be included alongside the

```
%%%%%%%%%%%%%%%%%%%%%%%%%%%%%%%%%%%%%%%%%%%%%%%%%%%%%%%%%%%%%%%%%%%
Solving for the components and angles of U1-S1
%%%%%%%%%%%%%%%%%%%%%%%%%%%%%%%%%%%%%%%%%%%%%%%%%%%%%%%%%%%%%%%%%%%
Iterations: 63
Func Evals: 72
Algorithm: quasi-newton
exit flag = 5
error = ( 1.299668e-32 )
U1x = ( -0.929939 )
U1y = ( -0.491699 )
S1x = ( -0.240198 )
S1y = ( -0.090862 )
T1x = ( -0.929939 )
T1y = ( -0.491699 )
nu2 = ( 87.807294 )
nu3 = ( -1.912499 )
nu4 = ( 184.459929 )
Elapsed time is 0.426276 seconds.
Solution calculation is complete.
```

FIGURE F.16 Sample output from the file *U_S_T_quasi_newton.m*.

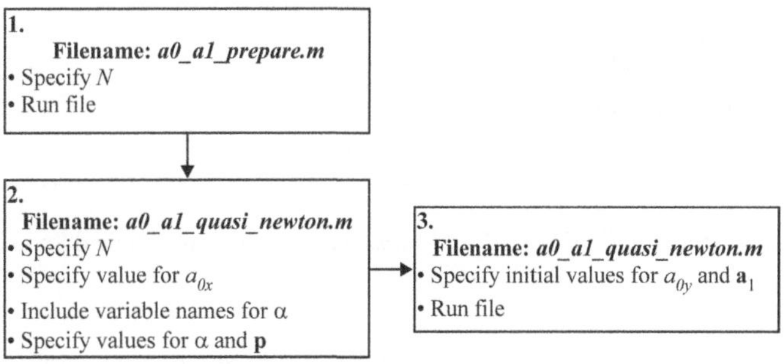

FIGURE F.17 Procedure for calculating a_{0y} and $\mathbf{a}_1$.

folder *Synthesis* to properly run the MATLAB files. The procedure for synthesizing the $\mathbf{W}_1 - \mathbf{Z}_1$ dyad is identical to the procedure given in Appendix F.2. The procedure for synthesizing the $\mathbf{U}_1 - \mathbf{S}_1 - \mathbf{T}_1$ triad is identical to the procedure given in Appendix F.3 where the α values calculated with the $\mathbf{W}_1 - \mathbf{Z}_1$ dyad are used as input to calculate the $\mathbf{U}_1 - \mathbf{S}_1 - \mathbf{T}_1$ triad.

F.5 PLANAR FOUR-BAR MOTION GENERATION, *N* PRECISION POSITIONS

The subfolder *Synthesis* within the folder *Appendix F.5* includes MATLAB files (filenames *a0_a1_prepare.m* and *a0_a1_quasi_newton.m*) for calculating the unknowns a_{0y} and $\mathbf{a}_1$ for motion generation given *N* precision positions. The folder *Common* (which includes supplementary MATLAB files) must be included alongside the folder *Synthesis* to properly run the MATLAB files. Figure F.17 includes the three-step procedure for calculating a_{0y} and $\mathbf{a}_1$.

Synthesizing the $\mathbf{a}_0 - \mathbf{a}_1$ dyad begins with the file *a0_a1_prepare.m* (step 1 in Figure F.17). In this file, the number of precision positions is specified. After specifying the value of *N* (the number of precision positions) and running this file, the folder *a0a1funcs* is produced. The folder *a0a1funcs* includes functions required to run file *a0_a1_quasi_newton.m*.

After exiting the file *a0_a1_prepare.m* and opening the file *a0_a1_quasi_newton.m*, step 2 in Figure F.17 is started. In this file, the user specifies the value of *N*—the number of precision positions (identical to the *N* value prescribed in the file *a0_a1_prepare.m*). The user specifies the value for the dyad parameter a_{0x}. The user also includes the name of each coupler displacement angle (angle α) to be prescribed. Additionally, the user includes the specific precision position values

```
%-------------------------------------------------------------
% Here, a value for the a0-a1 dyad component a0x is assigned.

global a0x
a0x =1;

%-------------------------------------------------------------

%-------------------------------------------------------------
% Here, the names for each coupler displacement angle are
% included(e.g., alpha2 alpha3 alpha4 alpha5 alpha6...).

global alpha2 alpha3 alpha4

%-------------------------------------------------------------

%-------------------------------------------------------------
% Here, values for precision position parameters alphaN and PN
% are assigned. Additional rows can be included to accommodate
% additional alphaN and PN variables and values.

alpha2 = -10 *pi/180;
alpha3 = -20 *pi/180;
alpha4 = -30 *pi/180;

P1 = [0, 0];
P2 = [0.4846, 0.3600];
P3 = [1.2073, 0.3876];
P4 = [1.9366, 0.4844];

%-------------------------------------------------------------
```

FIGURE F.18 Sections (for step 2) of file *a0_a1_quasi_newton.m* with sample data in bold type.

```
%-------------------------------------------------------------
% Here, initial values for the a0-a1 dyad components a0y, a1x and
% a1y are assigned.

a0y = 1;
a1x = 1;
a1y = 1;

%-------------------------------------------------------------
```

FIGURE F.19 Section (for step 3) of file *a0_a1_quasi_newton.m* with sample data in bold type.

(with angles specified in degrees). The number of *alphaN* and *PN* variables and values should match the value of *N*. Figure F.18 illustrates these user input sections in the file *a0_a1_quasi_newton.m*, with sample data in bold type.

The user includes the initial values for the dyad components a_{0y}, a_{1x}, and a_{1y} (step 3 in Figure F.17). Figure F.19 illustrates this user input section in the file *a0_a1_quasi_newton.m*, with sample data in bold type.

Lastly, the file *a0_a1_quasi_newton.m* is run (also in step 3). The calculated $\mathbf{a}_0 - \mathbf{a}_1$ dyad components are displayed in the MATLAB command window after this file has been run. Figure F.20 illustrates the sample output from the file *a0_a1_quasi_newton.m*. To calculate a solution for the $\mathbf{b}_0 - \mathbf{b}_1$ dyad, the user specifies different values for the components of $\mathbf{a}_0 - \mathbf{a}_1$ and reruns the file *a0_a1_quasi_newton.m*. The newly calculated solution will correspond to the $\mathbf{b}_0 - \mathbf{b}_1$ dyad.

When calculating new solutions, repeating step 1 in Figure F.17 is only necessary if the number of precision positions (*N*) for the new solution is different from that of the previous solution. If *N* is the same for the new solution, the user can begin at step 2 in Figure F.17.

```
%%%%%%%%%%%%%%%%%%%%%%%%%%%%%%%%%%%%%%%%%%%%%%%%%%%%%%%%%%%%%%%%
Solving for the components of a0-a1
%%%%%%%%%%%%%%%%%%%%%%%%%%%%%%%%%%%%%%%%%%%%%%%%%%%%%%%%%%%%%%%%
Iterations: 578
Func Evals: 579
Algorithm: large-scale: trust-region Newton
exit flag = 3
error = ( 3.952738e-020 )
a0y = ( -3.757746 )
a1x = ( 0.643977 )
a1y= ( -4.362003 )
Elapsed time is 3.421231 seconds.
Solution calculation is complete.
```

FIGURE F.20 Sample output from the file *a0_a1_quasi_newton.m.*

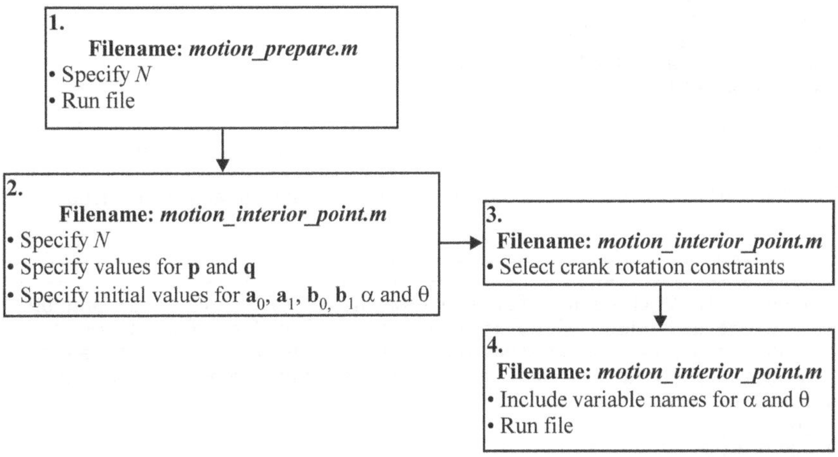

FIGURE F.21 Procedure for calculating $\mathbf{a}_0$, $\mathbf{a}_1$, $\mathbf{b}_0$, $\mathbf{b}_1$, α, and θ.

F.6 PLANAR FOUR-BAR MOTION GENERATION, *N* PRECISION POSITIONS WITH ORDER AND BRANCH CONSTRAINTS

The subfolder *Synthesis* within the folder *Appendix F.6* includes MATLAB files (filenames *motion_prepare.m* and *motion_interior_point.m*) for calculating the mechanism variables $\mathbf{a}_0$, $\mathbf{a}_1$, $\mathbf{b}_0$, $\mathbf{b}_1$, α, and θ for motion generation given *N* precision positions. The folder *Common* (which includes supplementary MATLAB files) must be included alongside the folder *Synthesis* to properly run the MATLAB files. Figure F.21 includes the four-step procedure for calculating $\mathbf{a}_0$, $\mathbf{a}_1$, $\mathbf{b}_0$, $\mathbf{b}_1$, α, and θ.

Synthesizing the mechanism begins with the file *motion_prepare.m* (step 1 in Figure F.21). In this file, the number of precision positions is specified. After specifying the value of *N* (the number of precision positions) and running this file, the folder *motionfuncs* is produced. The folder *motionfuncs* includes functions required to run file *motion_interior_point.m.*

After exiting the file *motion_prepare.m* and opening the file *motion_interior_point.m*, step 2 in Figure F.21 is started. In this file, the user specifies the value of *N*—the number of precision positions (identical to the *N* value prescribed in the file *motion_prepare.m*). The user also includes specific precision position values (defined by rigid-body points **p** and **q**, as described in Section 6.7). The number of *PN* and *QN* variables and values should match the value of *N*. Additionally, the user specifies initial values for $\mathbf{a}_0$, $\mathbf{a}_1$, $\mathbf{b}_0$, $\mathbf{b}_1$, α_j, and θ_j (in degrees). The number of *alphaN* and *thetaN* variables and values should match the value of *N*. Figures F.22 and F.23 illustrate the user input sections in the file *motion_interior_point.m*, with sample data in bold type.

```
%-------------------------------------------------------------
% Here, values for precision position parameter PN are assigned.
% Additional rows can be included to accommodate additional PN
% variables and values.

P1 = [0, 0];
P2 = [-5.2712, 0.2125];
P3 = [-8.9671, -4.2108];
P4 = [-7.7905, -9.6713];
P5 = [-1.6841, -12.539];

%-------------------------------------------------------------
%-------------------------------------------------------------
% Here, values for precision position parameter QN are assigned.
% Additional rows can be included to accommodate additional QN
% variables and values.

Q1 = [-0.3849, -0.3189];
Q2 = [-5.2308, -0.7107];
Q3 = [-8.5594, -4.5001];
Q4 = [-7.3259, -9.4870];
Q5 = [-1.6841, -12.0391];

%-------------------------------------------------------------
```

FIGURE F.22 Sections (for step 2) of file *motion_interior_point.m* with sample data in bold type.

```
%-------------------------------------------------------------
% Here, initial values for the components of dyads a0-a1 and b0-
% b1 are assigned. Also, initial values for dyad displacement
% angles alphaN and thetaN are assigned. Additional rows can be
% included to accommodate additional alphaN and thetaN variables
% and values.
a0x = 0;
a0y = 0;

a1x = 0;
a1y = -10;

b0x = -1;
b0y = 0;

b1x = -10;
b1y = -10;

alpha2 = 20 *pi/180;
alpha3 = 40 *pi/180;
alpha4 = 60 *pi/180;
alpha5 = 80 *pi/180;

theta2 = 60 *pi/180;
theta3 = 120 *pi/180;
theta4 = 180 *pi/180;
theta5 = 240 *pi/180;

%-------------------------------------------------------------
```

FIGURE F.23 Section (for step 2) of file *motion_interior_point.m* with sample data in bold type.

In step 3 in Figure F.21, the user selects either clockwise or counterclockwise crank rotation constraints for order defect elimination. Counterclockwise constraints are selected in the file *motion_interior_point.m*. To select clockwise constraints, the characters *%{* must be included as instructed in the file (to deactivate the counterclockwise constraints). Also, the characters *%{* must be removed as instructed in the file (to activate the clockwise constraints).

```
%-----------------------------------------------------------
% Here,the names for each alphaN and thetaN displacement angle
% are included.
% (e.g., alpha2; alpha3; alpha4; alpha5; alpha6; alpha7; .....;)
% (e.g., theta2; theta3; theta4; theta5; theta6; theta7; .....];)
x0 = [a0x; a0y; a1x; a1y; b0x; b0y; b1x; b1y;
      alpha2; alpha3; alpha4; alpha5;
      theta2; theta3; theta4; theta5];

%-----------------------------------------------------------
```

FIGURE F.24 Section (for step 4) of file *motion_interior_point.m* with sample data in bold type.

In step 4 in Figure F.21, the user includes the names for each dyad displacement angle (angles α and β) to be calculated. Figure F.24 illustrates this user input section in the file *motion_interior_point.m* with sample data in bold type.

Lastly, the file *motion_interior_point.m* is run (also in step 4). The calculated components of $\mathbf{a}_0$, $\mathbf{a}_1$, $\mathbf{b}_0$, and $\mathbf{b}_1$ and displacement angles α and θ are displayed in the MATLAB command window after this file has been run. Figure F.25 illustrates the sample output from the file *motion_interior_point.m*.

When calculating new solutions, repeating step 1 in Figure F.21 is only necessary if the number of precision positions (N) for the new solution is different from that of the previous solution. If N is the same for the new solution, the user can begin at step 2 in Figure F.21.

F.7 PLANAR FOUR-BAR PATH GENERATION, *N* PRECISION POINTS WITH ORDER AND BRANCH CONSTRAINTS

The subfolder *Synthesis* within the folder *Appendix F.7* includes MATLAB files (filenames *path_prepare.m* and *path_interior_point.m*) calculating the mechanism variables $\mathbf{a}_0$, $\mathbf{a}_1$, $\mathbf{b}_0$, $\mathbf{b}_1$, α, and θ for path generation given N precision points. The folder *Common* (which includes supplementary MATLAB files) must be included alongside the folder *Synthesis* to properly run the MATLAB files. Figure F.26 includes the four-step procedure for calculating $\mathbf{a}_0$, $\mathbf{a}_1$, $\mathbf{b}_0$, $\mathbf{b}_1$, α, and θ.

Synthesizing the mechanism begins with the file *path_prepare.m* (step 1 in Figure F.26). In this file, the number of precision points is specified. After specifying the value of N (the number of precision points) and running this file, the folder *pathfuncs* is produced. The folder *pathfuncs* includes functions required to run file *path_interior_point.m*.

After exiting the file *path_prepare.m* and opening the file *path_interior_point.m*, step 2 in Figure F.26 is started. In this file, the user specifies the value of N—the number of precision points (identical to the N value prescribed in the file *path_prepare.m*). The user also includes specific precision point values (defined by rigid-body point $\mathbf{p}$, as described in Section 6.7). The number of *PN* variables and values should match the value of N. Additionally, the user specifies initial values for $\mathbf{a}_0$, $\mathbf{a}_1$, $\mathbf{b}_0$, $\mathbf{b}_1$, α_j, and θ_j (in degrees). The number of *alphaN* and *thetaN* variables and values should match the value of N. Figure F.27 illustrates these user input sections in the file *path_interior_point.m*, with sample data in bold type.

In step 3 in Figure F.26, the user selects either clockwise or counterclockwise crank rotation constraints for order defect elimination. Counterclockwise constraints are selected in the file *path_interior_point.m*. To select clockwise constraints, the characters *%{* must be included as instructed in the file (to deactivate the counterclockwise constraints). Also, the characters *%{* must be removed as instructed in the file (to activate the clockwise constraints).

```
%%%%%%%%%%%%%%%%%%%%%%%%%%%%%%%%%%%%%%%%%%%%%%%%%%%%%%%%%%%%%%%%%%%%%%
Solvingfor the components of a0,a1,b0,b1and angles
%%%%%%%%%%%%%%%%%%%%%%%%%%%%%%%%%%%%%%%%%%%%%%%%%%%%%%%%%%%%%%%%%%%%%%
Iterations: 137
Func Evals: 326
Algorithm: interior-point
exit flag = 2
error = ( 3.444589e-001 )
a0x = ( -1.979732 )
a0y = ( -1.746697 )
a1x = ( 6.268255 )
a1y = ( -13.052440 )
b0x = ( -0.962500 )
b0y = ( -7.446623 )
b1x = ( -11.943966 )
b1y = ( -12.294918 )
alpha2 = ( -4.655522 )
alpha3 = ( -20.975649 )
alpha4 = ( -29.231648 )
alpha5 = ( -34.146667 )
theta2 = ( 121.177770 )
theta3 = ( 176.045532 )
theta4 = ( 215.631110 )
theta5 = ( 255.168283 )
Elapsed time is 10.991725 seconds.
Solution calculation is complete.
```

FIGURE F.25 Sample output from the file *motion_interior_point.m*.

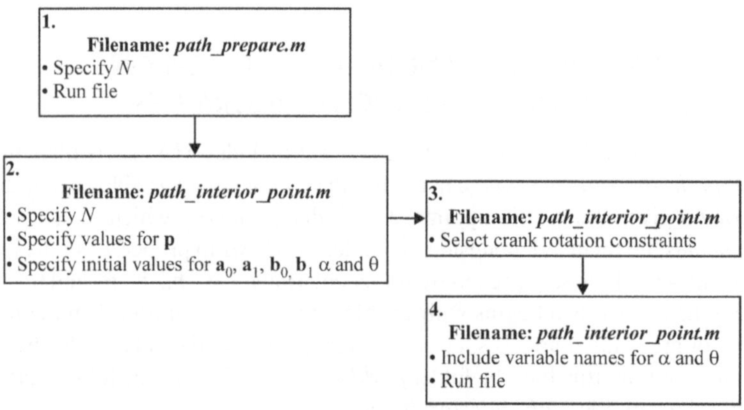

FIGURE F.26 Procedure for calculating $\mathbf{a}_0$, $\mathbf{a}_1$, $\mathbf{b}_0$, $\mathbf{b}_1$, α, and θ.

In step 4 in Figure F.26, the user includes the names for each dyad displacement angle (angles α and β) to be calculated. Figure F.28 illustrates this user input section in the file *path_interior_point.m* with sample data in bold type.

Lastly, the file *path_interior_point.m* is run (also in step 4). The calculated components of $\mathbf{a}_0 - \mathbf{a}_1$, $\mathbf{b}_0 - \mathbf{b}_1$, and dyad displacement angles α and θ are displayed in the MATLAB command window after this file has been run (see Figure F.25).

When calculating new solutions, repeating step 1 in Figure F.26 is only necessary if the number of precision points (N) for the new solution is different from that of the previous solution. If N is the same for the new solution, the user can begin at step 2 in Figure F.26.

```
%-----------------------------------------------------------------
% Here, values for precision point parameter PN are assigned.
% Additional rows can be included to accommodate additional PN
% variables and values.

P1 = [0, 0];
P2 = [1, -0.5];
P3 = [0, -0.8228];
P4 = [-0.7898, -0.8228];
P5 = [-0.9250, -0.3100];

%-----------------------------------------------------------------
%-----------------------------------------------------------------
% Here, initial values for the components of dyads a0-a1 and b0-b1
% areassigned. Also, initial values for dyad displacement angles
% alphaN andthetaN are assigned. Additional rows canbe included
% to accommodateadditional alphaN and thetaN variables and values.

a0x = 5;
a0y = 9;

a1x = 5;
a1y = 10;

b0x = 6;
b0y = 9;

b1x = 6;
b1y = 10;

alpha2 = 10 *pi/180;
alpha3 = 20 *pi/180;
alpha4 = 30 *pi/180;
alpha5 = 40 *pi/180;

theta2 = 70 *pi/180;
theta3 = 140 *pi/180;
theta4 = 210 *pi/180;
theta5 = 280 *pi/180;

%-----------------------------------------------------------------
```

FIGURE F.27 Sections (for step 2) of file *path_interior_point.m* with sample data in bold type.

```
%-----------------------------------------------------------------
% Here, the names for each alphaN and thetaN displacement angle
%areincluded
% (e.g., alpha2; alpha3; alpha4; alpha5; alpha6; alpha7; .....;)
% (e.g., theta2; theta3; theta4; theta5; theta6; theta7; .....];).

x0 = [a0x; a0y; a1x; a1y; b0x; b0y; b1x; b1y;
      alpha2; alpha3; alpha4; alpha5;
      theta2; theta3; theta4; theta5];

%-----------------------------------------------------------------
```

FIGURE F.28 Section (for step 4) of file *path_interior_point.m* with sample data in bold type.

NOTE

1 All of the Appendix F MATLAB files can be downloaded from the publisher's web page or from the authors' server http://www.softalink.com/kruss/mech_des_vis_prog_app_2nd_ed/MATLAB_SIMMULT_FILES.zip.

Appendix G

User Instructions for Chapter 7 MATLAB® Dimensional Synthesis Files

G.1 PLANAR FOUR-BAR FUNCTION GENERATION, *N* PRECISION POINTS

The subfolder *Synthesis* within the folder *Appendix G.1* (which is available for download at https://www.routledge.com/9781032956800) includes MATLAB files (filenames *UVW_prepare.m* and *UVW_quasi_newton.m*) for calculating the unknowns $\mathbf{W}_1$, $\mathbf{V}_1$, $\mathbf{U}_1$, and $\alpha_2 \ldots \alpha_N$ for planar four-bar function generation given *N* precision points.[1] The folder *Common* (which includes supplementary MATLAB files) must be included alongside the folder *Synthesis* to properly run the MATLAB files. Figure G.1 includes the four-step procedure for calculating $\mathbf{W}_1$, $\mathbf{V}_1$, $\mathbf{U}_1$, and $\alpha_2 \ldots \alpha_N$.

Synthesizing the function generator begins with the file *UVW_prepare.m* (step 1 in Figure G.1). In this file, the number of precision points is specified. After specifying the value of *N* (the number of precision points) and running this file, the folder *UVWfuncs* is produced. The folder *UVWfuncs* includes functions required to run file *UVW_quasi_newton.m.*

After exiting the file *UVW_prepare.m* and opening the file *UVW_quasi_newton.m*, step 2 in Figure G.1 is started. In this file, the user specifies the value of *N*—the number of precision points (which is identical to the *N* value prescribed in the file *UVW_prepare.m*). The user also includes the name of each crank and follower displacement angle variable to be prescribed. Additionally, the user includes the crank and follower displacement angle values in degrees. The number of *betaN* and *gammaN* variables and values should match the value of *N*. Figure G.2 illustrates these user input sections in the file *UVW_quasi_newton.m*, with sample data in bold type.

The user includes the initial values for $\mathbf{W}_1$, $\mathbf{V}_1$, $\mathbf{U}_1$, and $\alpha_2 \ldots \alpha_N$ (step 3 in Figure G.1). The user also includes the initial values for each coupler vector displacement angle α in degrees (also in step 3). The number of *alphaN* variables and values should match the value of *N*. Figure G.3 illustrates these user input sections in the file *W_Z_prepare.m*, with sample data in bold type.

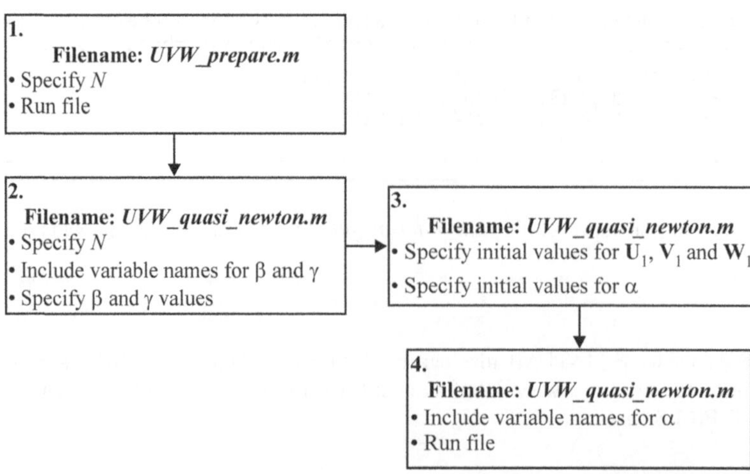

FIGURE G.1 Procedure for calculating $\mathbf{W}_1$, $\mathbf{V}_1$, $\mathbf{U}_1$, and $\alpha_2 \ldots \alpha_N$.

```
%------------------------------------------------------------------
% Here, a value for the number of precision points is assigned.
N = 6;
%------------------------------------------------------------------

%------------------------------------------------------------------
% Here, the names for each crank displacement angle are included
% (e.g., beta2 beta3 beta4 beta5 beta6 beta7...).
global beta2 beta3 beta4 beta5 beta6
%------------------------------------------------------------------

%----------------------------------------------------------------
% Here, values for precision point parameters betaN is assigned.
% Additional rows can be included to accommodate additional betaN
% variables and values.
beta2 = 12 *pi/180;
beta3 = 24 *pi/180;
beta4 = 36 *pi/180;
beta5 = 48 *pi/180;
beta6 = 60 *pi/180;
%------------------------------------------------------------------

%------------------------------------------------------------------
% Here, the names for each follower displacement angle are
%included
% (e.g., gamma2 gamma3 gamma4 gamma5 gamma6 gamma7...).
global gamma2 gamma3 gamma4 gamma5 gamma6
%------------------------------------------------------------------

%------------------------------------------------------------------
% Here, values for precision point parameters gammaN is assigned.
% Additional rows can be included to accommodate additional gammaN
% variables and values.
gamma2 = 7.636 *pi/180;
gamma3 = 17.455 *pi/180;
gamma4 = 29.455 *pi/180;
gamma5 = 43.636 *pi/180;
gamma6 = 60 *pi/180;
%------------------------------------------------------------------
```

FIGURE G.2 Sections (for step 2) of file *UVW_quasi_newton.m*, with the sample input in bold type.

In step 4 in Figure G.1, the user includes the names of each coupler vector displacement angle to be calculated. Figure G.4 illustrates this user input section in the file *UVW_quasi_newton.m* with sample data in bold type.

Lastly, the file *UVW_quasi_newton.m* is run (also in step 4). The calculated values of $\mathbf{W}_1$, $\mathbf{V}_1$, $\mathbf{U}_1$, and $\alpha_2 \ldots \alpha_N$ are displayed in the MATLAB command window after this file has been run. Figure G.5 illustrates the sample output from the file *UVW_quasi_newton.m*.

When calculating new solutions, repeating step 1 in Figure G.1 is only necessary if the number of precision points (N) will be different from the current value. If N is the same, the user can begin at step 2 in Figure G.1.

G.2 PLANAR FOUR-BAR FUNCTION GENERATION, *N* FSPS, AND *N* VELOCITY MSPS

The subfolder *Synthesis* within the folder *Appendix G.2* includes MATLAB files (filenames *UVW_prepare.m* and *UVW_interior_point.m*) for calculating unknowns $\mathbf{W}_1$, $\mathbf{V}_1$, $\mathbf{U}_1$, and $\alpha_2 \ldots \alpha_N$ for function generation given N precision points and N angular velocities. The folder *Common* (which includes

```
%------------------------------------------------------------
% Here, initial values for the components of U1-V1-W1 are
%assigned.

U1x = -1;
U1y = -1;
V1x = -1;
V1y = 1;
W1x = -1;
W1y = -1;

%------------------------------------------------------------

%------------------------------------------------------------
% Here, values for V1 parameter alphaN are assigned. Additional
%rows can be included to accommodate additional alphaNvariables
%and values.

alpha2 = 20 *pi/180;
alpha3 = 40 *pi/180;
alpha4 = 60 *pi/180;
alpha5 = 80 *pi/180;
alpha6 = 100 *pi/180;

%------------------------------------------------------------
```

FIGURE G.3 Sections (for step 3) of file *UVW_quasi_newton.m* with the sample input in bold type.

```
%------------------------------------------------------------
% Here, the names for each alphaN displacement angle are included
% (e.g., alpha2; alpha3; alpha4; alpha5; alpha6; alpha7; .....];).

x0 = [U1x; U1y; V1x; V1y; W1x; W1y;
      alpha2; alpha3; alpha4; alpha5; alpha6];

%------------------------------------------------------------
```

FIGURE G.4 Section (for step 4) of file *UVW_quasi_newton.m* with the sample input in bold type.

```
%%%%%%%%%%%%%%%%%%%%%%%%%%%%%%%%%%%%%%%%%%%%%%%%%%%%%%%%%%%%%%
Solve for the components of U1-V1-W1 and displacement angles of V1
%%%%%%%%%%%%%%%%%%%%%%%%%%%%%%%%%%%%%%%%%%%%%%%%%%%%%%%%%%%%%%
Iterations: 497
Func Evals: 621
Algorithm: quasi-newton
exit flag = 5
error = ( 2.233656e-09 )
U1x = ( -0.909576 )
U1y = ( -1.753528 )
V1x = ( -0.771127 )
V1y = ( 1.022948 )
W1x = ( 0.861360 )
W1y = ( -2.776222 )
alpha2 = ( 381.080477 )
alpha3 = ( 42.176577 )
alpha4 = ( 63.542708 )
alpha5 = ( -275.027338 )
alpha6 = ( -254.037446 )
Elapsed time is 38.925661 seconds.
Solution calculation is complete.
```

FIGURE G.5 Sample output from the file *UVW_quasi_newton.m*.

supplementary MATLAB files) must be included alongside the folder *Synthesis* to properly run the MATLAB files. Figure G.6 includes the four-step procedure for calculating $\mathbf{W}_1$, $\mathbf{V}_1$, $\mathbf{U}_1$, and $\alpha_2 \ldots \alpha_N$.

Synthesizing the function generator begins with the file *UVW_prepare.m* (step 1 in Figure G.6). In this file, the number of precision points is specified. After specifying the value of N (the number

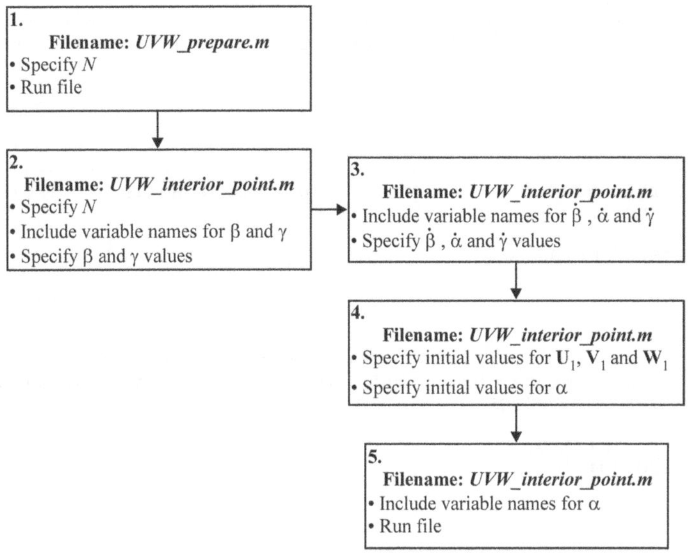

FIGURE G.6 Procedure for calculating $\mathbf{W}_1$, $\mathbf{V}_1$, $\mathbf{U}_1$, and $\alpha_2 \ldots \alpha_N$.

of precision points) and running this file, the folder *UVWfuncs* is produced. The folder *UVWfuncs* includes functions required to run file *UVW_interior_point.m*.

After exiting the file *UVW_prepare.m* and opening the file *UVW_interior_point.m*, step 2 in Figure G.6 is started. In this file, the user specifies the value of *N*—the number of precision points (identical to the *N* value prescribed in the file *UVW_prepare.m*). The user also includes the name of each crank and follower displacement angle to be prescribed. Additionally, the user includes the specific crank and follower displacement angle values in degrees. The number of *betaN* and *gammaN* variables and values should match the value of *N*. Figure G.7 illustrates these user input sections in the file *UVW_interior_point.m*, with sample data in bold type.

In step 3 in Figure G.6, the user includes the name of each crank, coupler, and follower angular velocity variable to be prescribed. The values for these variables are specified in rad/s. The number of *betapN*, *alphapN*, and *gammapN* variables and values should match the value of *N*. Figure G.8 illustrates these user input sections in the file *UVW_interior_point.m*, with sample data in bold type.

The user includes the initial values for $\mathbf{W}_1$, $\mathbf{V}_1$, and $\mathbf{U}_1$ (step 4 in Figure G.6). The user also includes the initial values for $\alpha_2 \ldots \alpha_N$ in degrees (also in step 4). The number of *alphaN* variables and values should match the value of *N*. Figure G.9 illustrates these user input sections in the file *UVW_interior_point.m*, with sample data in bold type.

In step 5 in Figure G.6, the user includes the names of each coupler vector displacement angle to be calculated. Figure G.10 illustrates this user input section in the file *UVW_interior_point.m* with sample data in bold type.

Lastly, the file *UVW_interior_point.m* is run (also in step 5). The calculated values of $\mathbf{W}_1$, $\mathbf{V}_1$, $\mathbf{U}_1$, and $\alpha_2 \ldots \alpha_N$ are displayed in the MATLAB command window after this file has been run. Figure G.11 illustrates sample output from the file *UVW_interior_point.m*.

When calculating new solutions, repeating step 1 in Figure G.6 is only necessary if the number of precision points (*N*) for the new solution is different from the current value. If *N* is the same, the user can begin at step 2 in Figure G.6.

```
%-------------------------------------------------------------------
% Here, a value for the number of precision points is assigned.

N = 3;

%-------------------------------------------------------------------

%-------------------------------------------------------------------
% Here, the names for each crank displacement angle are included
% (e.g., beta2 beta3 beta4 ...).

global beta2 beta3

%-------------------------------------------------------------------

%-------------------------------------------------------------------
% Here, values for precision point parameters betaN is assigned.
% Additional rows can be included to accommodate additional betaN
% variables and values.

beta2 = 10 *pi/180;
beta3 = 30 *pi/180;

%-------------------------------------------------------------------

%-------------------------------------------------------------------
% Here, the names for each follower displacement angle are
% included
% (e.g., gamma2 gamma3 gamma4 ...).

global gamma2 gamma3

%-------------------------------------------------------------------

%-------------------------------------------------------------------
% Here, values for precision point parameters gammaN is assigned.
% Additional rows can be included to accommodate additional gammaN
% variables and values.

gamma2 = 21.4323 *pi/180;
gamma3 = 51.742 *pi/180;

%-------------------------------------------------------------------
```

FIGURE G.7 Sections (for step 2) of file *UVW_interior_point.m*, with the sample input in bold type.

```
%-------------------------------------------------------------------
% Here, the names for each MSP are included
% (e.g., betap2 betap3 betap4 ..., gammap2 gammap3 gammap4 ...,
% alphap2alphap3 alphap4 ...).
% Also, values for MSP parameters betapN, gammapN and alphapN are
% assigned.Additional rows can be included to accommodate
% additional betapN,gammapN and alphapN variables and values.

global betap2 betap3

betap2 = 1.5;
betap3 = 1.5;

global gammap2 gammap3

gammap2 = 3.15;
gammap3 = 3.35;

global alphap2 alphap3

alphap2 = 0.75;
alphap3 = 0.75;

%-------------------------------------------------------------------
```

FIGURE G.8 Section (for step 3) of file *UVW_interior_point.m*, with the sample input in bold type.

```
%----------------------------------------------------------------
% Here, initial values for the components of U1-V1-W1 are
assigned.
U1x = 1;
U1y = 1;
V1x = 1;
V1y = 1;
W1x = 1;
W1y = 1;
%----------------------------------------------------------------

%----------------------------------------------------------------
% Here, values for V1 parameter alphaN are assigned. Additional
% rows can be included to accommodate additional alphaN variables
% and values.
alpha2 = 10 *pi/180;
alpha3 = 10 *pi/180;
%----------------------------------------------------------------
```

FIGURE G.9 Sections (for step 4) of file *UVW_interior_point.m*, with the sample input in bold type.

```
%----------------------------------------------------------------
% Here, the names for each alphaN displacement angle are included
% (e.g., alpha2; alpha3; alpha4; .....];).
x0 = [U1x; U1y; V1x; V1y; W1x; W1y;
    alpha2; alpha3];
%----------------------------------------------------------------
```

FIGURE G.10 Section (foe step 5) of file *UVW_interior_point.m* with the sample input in bold type.

```
%%%%%%%%%%%%%%%%%%%%%%%%%%%%%%%%%%%%%%%%%%%%%%%%%%%%%%%%%%%%%%%%%%%%
Solve for the components of U1-V1-W1 and displacement angles of V1
%%%%%%%%%%%%%%%%%%%%%%%%%%%%%%%%%%%%%%%%%%%%%%%%%%%%%%%%%%%%%%%%%%%%
Iterations: 56
Func Evals: 160
Algorithm: interior-point
exit flag = -2
error = ( 5.937578e-004 )
U1x = ( -0.506897 )
U1y = ( 0.059153 )
V1x = ( 3.077220 )
V1y = ( 0.014466 )
W1x = ( -2.600780 )
W1y = ( 0.044609 )
alpha2 = ( 364.740281 )
alpha3 = ( 16.916627 )
Elapsed time is 6.082023 seconds.
Solution calculation is complete.
```

FIGURE G.11 Sample output from the file *UVW_interior_point.m*.

NOTE

1 All of the Appendix G MATLAB files can be downloaded from the publisher's web page or from the authors' server http://www.softalink.com/kruss/mech_des_vis_prog_app_2nd_ed/MATLAB_SIMMULT_FILES.zip.

Appendix H

User Instructions for Chapter 8 MATLAB® Kinematic Analysis Files

H.1 RRSS MECHANISM

The Appendix H.1 folder (which is available for download at https://www.routledge.com/9781032956800) includes the MATLAB file *Kinematic_RRSS.m* for the kinematic analysis of RRSS mechanisms.[1] To conduct a kinematic analysis, the user specifies the mechanism point coordinates and the crank motion parameters in this file. Values are specified for link variables $\mathbf{a}_0$, $\mathbf{ua}_0$, $\mathbf{a}_1$, $\mathbf{ua}_1$, $\mathbf{b}_0$, $\mathbf{b}_1$, $\mathbf{p}_1$, $\mathbf{q}_1$, and $\mathbf{r}_1$ (Figure H.1a). Values are also specified for the initial crank angle (*start_ang*), the crank rotation increment (*step_ang*), and the final crank angle (*stop_ang*). Lastly, values are specified for the angular velocity (*angular_vel*) and angular acceleration (*angular_acc*) for the crank. Figure H.2 illustrates the user input sections of the file *Kinematic_RRSS.m,* with sample values in bold type.[2]

After specifying the dimensions and driving link parameters in the file *Kinematic_RRSS.m*, the next step is to run this file. When running this file, two files (filenames *RRSS_branch1.csv* and *RRSS_branch2.csv*) are written in a folder named *Results* (in a format compatible with Microsoft® Excel) that includes the calculated mechanism output at each crank link rotation increment for both RRSS branches.[3] The calculated mechanism output is included in Figure H.1b.

The initial RRSS mechanism configuration corresponds to the mechanism dimensions specified in the Appendix H.1 MATLAB file (see Figure H.2). The results for this configuration are included in the file *RRSS_branch1.csv*. The results for the second mechanism configuration are included in the file *RRSS_branch2.csv*.

The Appendix H.1 file can also be directly applied for the kinematic analysis of spherical four-bar mechanisms to calculate output coupler link motion.

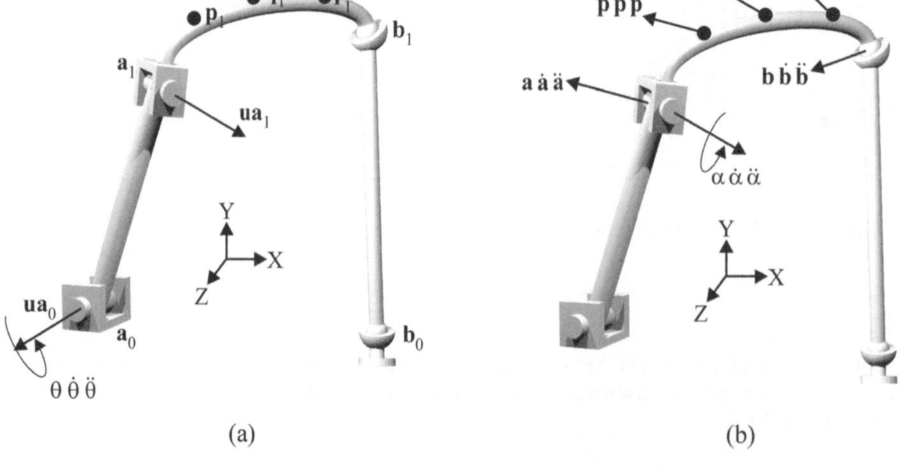

(a) (b)

FIGURE H.1 (a) RRSS mechanism and (b) output variables.

```
%------------------------------------------------------------
% Here, values for the RRSS mechanism variables a0, a1, ua0,
% ua1, b0, b1,p1, q1 and r1 are assigned.  They must all be
% defined as transposed row matrices (as shown).

a0 = [-0.0576, 0.2890, -1.4112]';
a1 = [0.1452, -2.5421, -1.1800]';
ua0 = [-0.0003, 0.0814, 0.9967]';
ua1 = [0.3040, 0.0992, 0.9475]';
b0 = [0.0851, 0.4570, 0.5096]';
b1 = [1.7725, 5.1566, 0.6499]';
p1 = [1.7321, 0, -1]';
q1 = [1.2321, 0, -1.8660]';
r1 = [1.9486, 0, -1.1250]';
%------------------------------------------------------------

%------------------------------------------------------------
% Here, values for the start, step and stop displacement angles
% for thecrank link are assigned.

start_ang = 0;
step_ang  = 10;
stop_ang  = 200;
%------------------------------------------------------------

%------------------------------------------------------------
% Here, values for crank link angular velocity and angular
% acceleration are assigned.

angular_vel = 1.0 * ones(N+1,1);
angular_acc = 0.1 * ones(N+1,1);
%------------------------------------------------------------
```

FIGURE H.2 Sections of *Kinematic_RRSS.m* with sample values in bold type.

H.2 RSSR MECHANISM

The Appendix H.2 folder includes the MATLAB file *Kinematic_RSSR.m* for the kinematic analysis of RSSR mechanisms. To conduct a kinematic analysis, the user specifies the mechanism point coordinates and the crank motion parameters in this file. Values are specified for link variables $\mathbf{a}_0$, $\mathbf{ua}_0$, $\mathbf{a}_1$, $\mathbf{b}_0$, $\mathbf{ub}_0$, and $\mathbf{b}_1$ (Figure H.3a). Values are also specified for the initial crank angle (*start_ang*), the crank rotation increment (*step_ang*), and the final crank angle (*stop_ang*). Lastly, values are specified for the angular velocity (*angular_vel*) and angular acceleration (*angular_acc*) for the crank. Figure H.4 illustrates the user input sections of the file *Kinematic_RSSR.m* with sample values in bold type.[4]

After specifying the dimensions and driving link parameters in the file *Kinematic_RSSR.m*, the next step is to run this file. When running this file, two files (filenames *RSSR_branch1.csv* and *RSSR_branch2.csv*) are written in a folder named *Results* (in a format compatible with Microsoft® Excel) that includes the calculated mechanism output at each crank link rotation increment for both RSSR branches.[5] The calculated mechanism output is included in Figure H.3b.

The initial RSSR mechanism configuration corresponds to the mechanism dimensions specified in the Appendix H.2 MATLAB file (see Figure H.4). The results for this configuration are included in the file *RSSR_branch1.csv*. The results for the second mechanism configuration are included in the file *RSSR_branch2.csv*.

The Appendix H.2 file can also be directly applied for the kinematic analysis of spherical four-bar mechanisms to calculate output follower link motion.

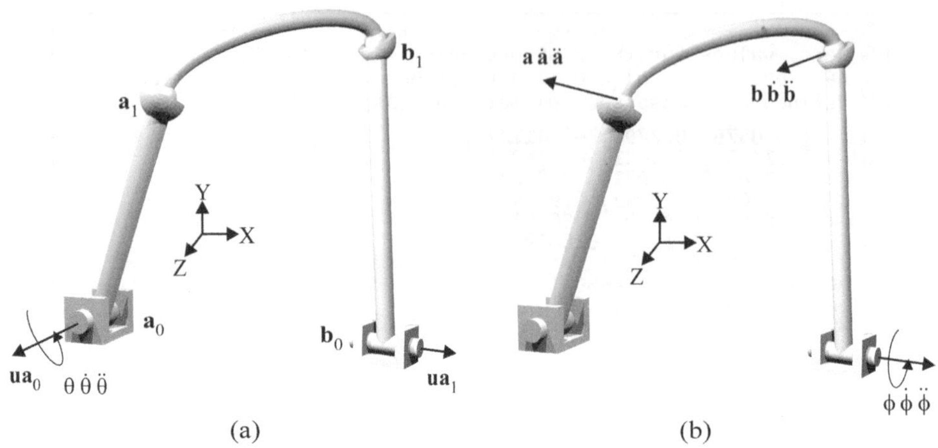

FIGURE H.3 (a) RSSR mechanism and (b) output variables.

```
%------------------------------------------------------------
% Here, values for the RSSR mechanism variables a0, a1, ua0, b0,
% b1 andub0 are assigned.  They must all be defined as
% transposed row matricesas shown).

a0 = [0, 0, -0.4023]';
a1 = [0.3356, -0.0708, -0.4023]';
ua0 = [0, 0, 1]';
b0 = [1, -2.3885, 0]';
b1 = [1, -2.3885, 0.7300]';
ub0 = [0, -1, 0]';
%------------------------------------------------------------

%------------------------------------------------------------
% Here, values for the start, step and stop displacement angles
% for thecrank link are assigned.

start_ang   = 0;
step_ang    = 10;
stop_ang    = 50;
%------------------------------------------------------------

%------------------------------------------------------------
% Here, values for crank link angular velocity and angular
% acceleration are assigned.

angular_vel  = 0 * ones(N+1,1);
angular_acc  = 0 * ones(N+1,1);
%------------------------------------------------------------
```

FIGURE H.4 Sections of *Kinematic_RSSR.m* with sample values in bold type.

NOTES

1 All of the Appendix H MATLAB files can be downloaded from the publisher's web page or from the authors' server http://www.softalink.com/kruss/mech_des_vis_prog_app_2nd_ed/MATLAB_SIMMULT_FILES.zip.
2 The units for crank angular rotation, velocity, and acceleration are *degrees*, *radians/sec*, and *radians/sec²*, respectively.
3 At crank rotation increments where circuit defects occur, the coupler link motion data written in the *.*csv* file appear as complex numbers.
4 The units for crank angular rotation, velocity, and acceleration are *degrees*, *radians/sec*, and *radians/sec²*, respectively.
5 At crank rotation increments where circuit defects occur, the follower link motion data written in the *.*csv* file appear as complex numbers.

Appendix I
User Instructions for Chapter 8 MATLAB® and Simscape Multibody™ Kinematic Analysis Files

I.1 RRSS MECHANISM

The Appendix I.1 folder (which is available for download at https://www.routledge.com/9781032956800) includes four MATLAB and Simscape Multibody files for the kinematic analysis and simulation of RRSS mechanisms.[1] These files are described in Table I.1. To conduct a kinematic analysis, the user specifies the mechanism point coordinates and the crank motion parameters in the file *Kinematic_RRSS_Simulate.m*. Values are specified for link variables $\mathbf{a}_0$, $\mathbf{ua}_0$, $\mathbf{a}_1$, $\mathbf{ua}_1$, $\mathbf{b}_0$, $\mathbf{b}_1$, $\mathbf{p}_1$, $\mathbf{q}_1$, and $\mathbf{r}_1$ (Figure I.1a). Values are also specified for the initial crank angle (*start_ang*), the crank rotation increment (*step_ang*), and the final crank angle (*stop_ang*). Lastly, values are specified for the angular velocity (*angular_vel*) and angular acceleration (*angular_acc*) for the crank. Figure I.2 illustrates the user input sections of the file *Kinematic_RRSS_Simulate.m,* with sample values in bold type.[2]

TABLE I.1
Appendix I.1 MATLAB and Simscape Multibody Files

Filename	Use of File
Kinematic_RRSS_Simulate.m	To specify mechanism link dimensions and crank link controls. File *M_xform.m* is used in the file *Kinematic_RRSS_Simulate.m.*
M_xform.m	
Kinematic_RRSS_Model.slx	To calculate the mechanism output and simulate mechanism motion
Post_Simulation_Task.m	To write the mechanism output (compatible with Microsoft Excel)

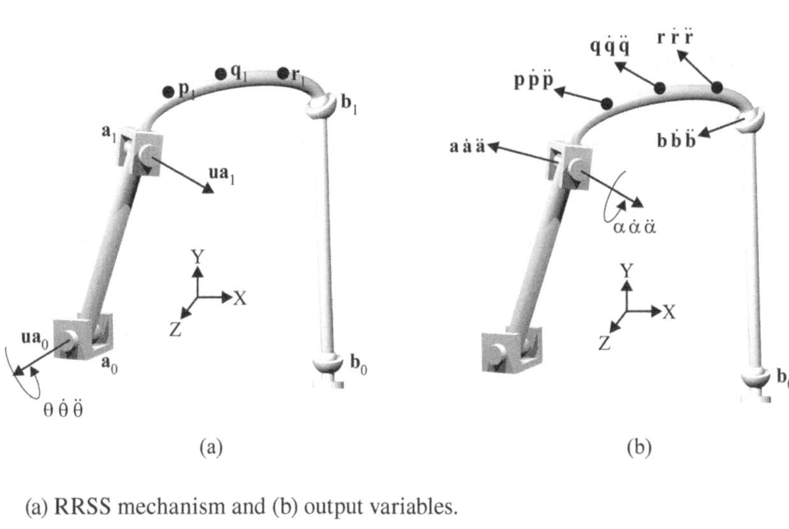

FIGURE I.1 (a) RRSS mechanism and (b) output variables.

```
%-----------------------------------------------------------
% Here, values for the RRSS mechanism variables a0, a1, ua0,
% ua1, b0, b1,p1, q1 and r1 are assigned.  They must all be
% defined as transposed row matrices (as shown).
a0 = [-0.0576, 0.2890, -1.4112]';
a1 = [0.1452, -2.5421, -1.1800]';
ua0 = [-0.0003, 0.0814, 0.9967]';
ua1 = [0.3040, 0.0992, 0.9475]';
b0 = [0.0851, 0.4570, 0.5096]';
b1 = [1.7725, 5.1566, 0.6499]';
p1 = [1.7321, 0, -1]';
q1 = [1.2321, 0, -1.8660]';
r1 = [1.9486, 0, -1.1250]';
%-----------------------------------------------------------

%-----------------------------------------------------------
% Here, values for the start, step and stop displacement angles
% for thecrank link are assigned.
start_ang = 0;
step_ang  = 10;
stop_ang  = 200;
%-----------------------------------------------------------

%-----------------------------------------------------------
% Here, values for crank link angular velocity and angular
% acceleration are assigned.
angular_vel = 1.0;
angular_acc = 0.1;
%-----------------------------------------------------------
```

FIGURE I.2 Sections of *Kinematic_RRSS_Simulate.m*, with sample values in bold type.

After specifying the dimensions and driving link parameters in the file *Kinematic_RRSS_Simulate.m*, the next step is to run this file. When running this file, one file (filename *Disp_Vel_Acc.csv*) is written in a folder named *Results* (in a format compatible with Microsoft® Excel) that includes the calculated mechanism output at each crank link rotation increment.[3] The calculated mechanism output is included in Figure I.1b.

When running *Kinematic_RRSS_Simulate.m*, a graphical user interface appears where the motion of the RRSS mechanism is simulated over the defined crank rotation range (see Appendix A.3). Also, the folder *slprq* is written. This folder includes functions that preclude recompiling the model in future simulations. This optional folder can be deleted however (e.g., if computer storage space is inadequate).

I.2 RSSR MECHANISM

The Appendix I.2 folder includes four MATLAB and Simscape Multibody files for the kinematic analysis and simulation of RSSR mechanisms. These files are described in Table I.2. To conduct a kinematic analysis, the user specifies the mechanism point coordinates and the crank motion parameters in the file *Kinematic_RSSR_Simulate.m*. Values are specified for link variables $\mathbf{a}_0$, $\mathbf{ua}_0$, $\mathbf{a}_1$, $\mathbf{b}_0$, $\mathbf{ub}_0$, and $\mathbf{b}_1$ (Figure I.3a). Values are also specified for the initial crank angle (*start_ang*), the crank rotation increment (*step_ang*), and the final crank angle (*stop_ang*). Lastly, values are specified for the angular velocity (*angular_vel*) and angular acceleration (*angular_acc*) for the crank. Figure I.4 illustrates the user input sections of the file *Kinematic_RSSR_Simulate.m,* with sample values in bold type.[4]

After specifying the dimensions and driving link parameters in the file *Kinematic_RSSR_Simulate.m*, the next step is to run this file. When running this file, one file (filenames *Disp_Vel_Acc.csv*) is written in a folder named *Results* (in a format compatible with Microsoft® Excel) that includes the calculated mechanism output at each crank link rotation increment.[5] The calculated mechanism output is included in Figure I.3b.

TABLE I.2
Appendix I.2 MATLAB and Simscape Multibody Files

Filename	Use of File
Kinematic_RSSR_Simulate.m	To specify mechanism link dimensions and crank link controls.
M_xform.m	File *M_xform.m* is used in the file *Kinematic_RSSR_Simulate.m.*
Kinematic_RSSR_Model.slx	To calculate the mechanism output and simulate mechanism motion
Post_Simulation_Task.m	To write the mechanism output (compatible with Microsoft Excel)

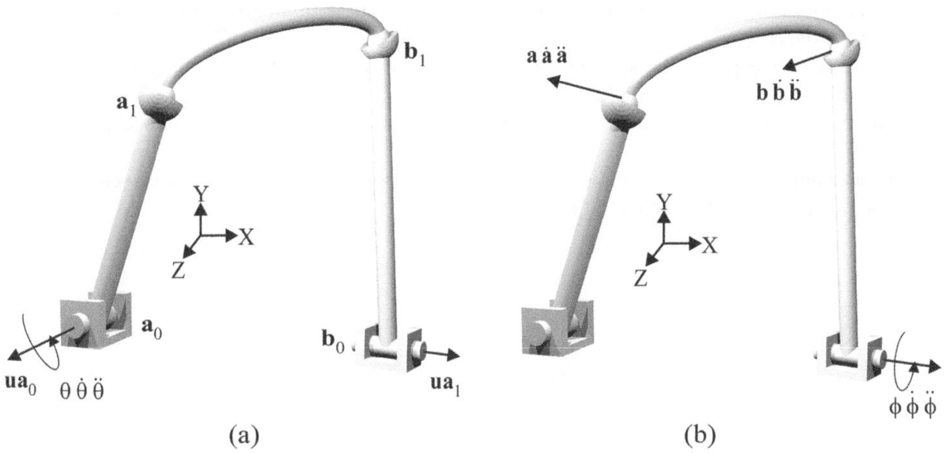

FIGURE I.3 (a) RSSR mechanism and (b) output variables.

```
%-----------------------------------------------------------------
% Here, values for the RSSR mechanism variables a0, a1, ua0, b0,
% b1andub0 are assigned. They must all be defined as
% transposed row matricesas shown).
a0 = [0, 0, -0.4023]';
a1 = [0.3356, -0.0708, -0.4023]';
ua0 = [0, 0, 1]';
b0 = [1, -2.3885, 0]';
b1 = [1, -2.3885, 0.7300]';
ub0 = [0, -1, 0]';
%-----------------------------------------------------------------

%-----------------------------------------------------------------
% Here, values for the start, step and stop displacement angles
% for the crank link are assigned.
start_ang   = 0;
step_ang    = 10;
stop_ang    = 50;
%-----------------------------------------------------------------

%-----------------------------------------------------------------
% Here, values for crank link angular velocity and angular
% acceleration are assigned.
angular_vel  = 0;
angular_acc  = 0;
%-----------------------------------------------------------------
```

FIGURE I.4 Sections of *Kinematic_RSSR_Simulate.m,* with sample values in bold type.

When running *Kinematic_RSSR_Simulate.m*, a graphical user interface appears where the motion of the RSSR mechanism is simulated over the defined crank rotation range (see Appendix A.3). Also, the folder *slprq* is written. This folder includes functions that preclude recompiling the model in future simulations. This optional folder can be deleted however (e.g., if computer storage space is inadequate).

NOTES

1 All of the Appendix I Simscape Multibody files can be downloaded from the publisher's web page or from the authors' server http://www.softalink.com/kruss/mech_des_vis_prog_app_2nd_ed/MATLAB_SIMMULT_FILES.zip.

2 The units for crank angular rotation, velocity, and acceleration are *degrees*, *radians/sec*, and *radians/sec²*, respectively.

3 If the mechanism experiences a circuit defect (or if the crank decelerates to a zero velocity state), results beyond this point are not written to the *.*csv* file.

4 The units for crank angular rotation, velocity, and acceleration are *degrees*, *radians/sec*, and *radians/sec²*, respectively.

5 If the mechanism experiences a circuit defect (or if the crank decelerates to a zero velocity state), results beyond this point are not written to the *.*csv* file.

Appendix J
User Instructions for Chapter 8 MATLAB® Dimensional Synthesis Files

J.1 MOTION GENERATION, R-R DYADS, *N* PRECISION POSITIONS

The subfolder *Synthesis* within the folder *Appendix J.1* (which is available for download at https://www.routledge.com/9781032956800) includes MATLAB files (filenames *RR_motion_prepare.m* and *RR_interior_point.m*) for calculating the R-R dyad variables $\mathbf{a}_0$, $\mathbf{a}_1$, $\mathbf{ua}_0$, and $\mathbf{ua}_1$ for motion generation given *N* precision positions.[1] The folder *Common* (which includes supplementary MATLAB files) must be included alongside the folder *Synthesis* to properly run the MATLAB files. Figure J.1 includes the three-step procedure for calculating $\mathbf{a}_0$, $\mathbf{a}_1$, $\mathbf{ua}_0$, and $\mathbf{ua}_1$.

Synthesizing the R-R dyad begins with the file *RR_motion_prepare.m* (step 1 in Figure J.1). In this file, the number of precision positions is specified. After specifying the value of *N* (the number of precision positions) and running this file, the folder *RRfuncs* is produced. The folder *RRfuncs* includes functions required to run file *RR_interior_point.m.*

After exiting the file *RR_motion_prepare.m* and opening the file *RR_interior_point.m*, step 2 in Figure J.1 is started. In this file, the user specifies the value of *N*—the number of precision positions (identical to the *N* value prescribed in the file *RR_motion_prepare.m*). The user also includes the name of each dyad displacement angle (angle θ) to be prescribed and the specific values of θ in degrees. Additionally, the user includes the specific precision position values (defined by rigid-body points $\mathbf{p}$, $\mathbf{q}$, $\mathbf{r}$, and $\mathbf{s}$, as given in Matrix (8.51)). The number of *thetaN*, *PN*, *QN*, *RN*, and *SN* variables and values should match the value of *N*. Figure J.2 illustrates these user input sections in the file *RR_interior_point.m*, with sample data in bold type.

The user includes the initial values for $\mathbf{a}_0$, $\mathbf{a}_1$, $\mathbf{ua}_0$, and $\mathbf{ua}_1$ (step 3 in Figure J.1). Figure J.3 illustrates this user input section in the file *RR_interior_point.m,* with sample data in bold type.

Lastly, the file *RR_interior_point.m* is run (also in step 3). The calculated R-R dyad components are displayed in the MATLAB command window after this file has been run. Figure J.4 illustrates the sample output from the file *RR_interior_point.m.*

When calculating new solutions, repeating step 1 in Figure J.1 is only necessary if the number of precision positions (*N*) for the new solution is different from that in the previous solution. If *N* is the same for the new solution, the user can begin at step 2 in Figure J.1.

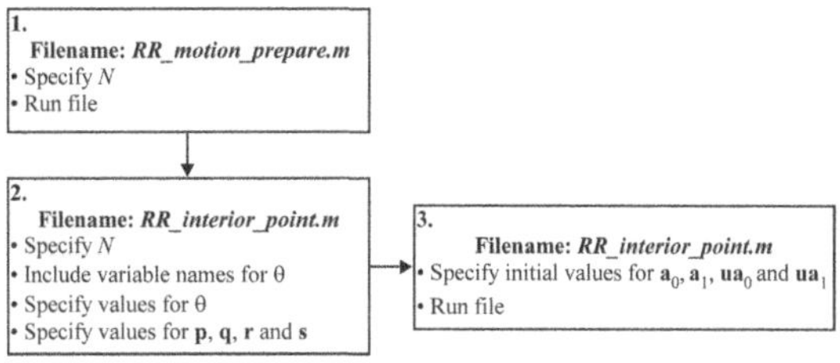

FIGURE J.1 Procedure for calculating $\mathbf{a}_0$, $\mathbf{a}_1$, $\mathbf{ua}_0$, and $\mathbf{ua}_1$.

```
%-----------------------------------------------------------------
% Here, the names for each thetaN displacement angle are included
% (e.g., theta2 theta3 theta4 ...)
% Also, initial values for dyad displacement angles thetaN are
% assigned.Additional rows can be included to accommodate
% additional thetaNvariables and values.

global theta2 theta3

theta2 = 50 *pi/180.0;
theta3 = 90 *pi/180.0;

%-----------------------------------------------------------------

%-----------------------------------------------------------------
% Here, values for precision position parameter PN are assigned.
% Additional rows can be included to accommodate additional PN
% variables and values.

P1 = [1.7321, 0, -1];
P2 = [1.2247, 1.2247, -1];
P3 = [0, 1.7321, -1];

%-----------------------------------------------------------------
                                 .
                                 .
                                 .
%-----------------------------------------------------------------
% Here,values for precision position parameter SN are assigned.
% Additional rows can be included to accommodate additional SN
% variables and values.

S1 = [1.7321, 2, -1];
S2 = [-0.1895, 2.639, -1];
S3 = [-1.9353, 1.2951, -0.7477];

%-----------------------------------------------------------------
```

FIGURE J.2 Sections (for step 2) of file *RR_interior_point.m,* with sample data in bold type.

```
%-----------------------------------------------------------------
% Here, initial values for the components of the R-R dyad are
assigned.

a0x = 0;
a0y = 0;
a0z = -1;

a1x = 0;
a1y = -2;
a1z = 1;

ua0x = 0;
ua0y = 0;
ua0z = 1;

ua1x = 0.7;
ua1y = 0;
ua1z = 0.7;

%-----------------------------------------------------------------
```

FIGURE J.3 Section (for step 3) of file *RR_interior_point.m,* with sample data in bold type.

```
%%%%%%%%%%%%%%%%%%%%%%%%%%%%%%%%%%%%%%%%%%%%%%%%%%%%%%%%%%%%%%%%%%%%%%
Solvingfor the components of the R-R dyad
%%%%%%%%%%%%%%%%%%%%%%%%%%%%%%%%%%%%%%%%%%%%%%%%%%%%%%%%%%%%%%%%%%%%%%
Iterations: 8
Func Evals: 14
Algorithm: interior-point
exit flag = 1
error = ( 1.059251e-012 )
a0x = ( 0.018226 )
a0y = ( 0.001700 )
a0z = ( 2.619934 )
a1x = ( -2.300520 )
a1y = ( 0.620569 )
a1z = ( 2.625484 )
ua0x = ( 0.002479 )
ua0y = ( 0.000322 )
ua0z = ( 0.999997 )
ua1x = ( 0.002479 )
ua1y = ( 0.000322 )
ua1z = ( 0.999997 )
Elapsed time is 14.181888 seconds.
Solution calculation is complete.
```

FIGURE J.4 Sample output from the file *RR_interior_point.m*.

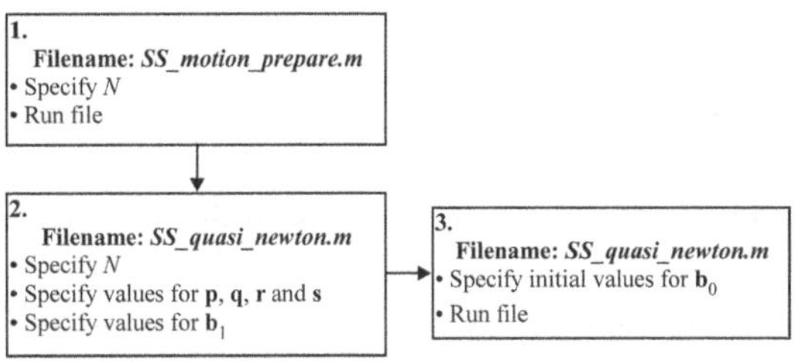

FIGURE J.5 Procedure for calculating $\mathbf{b}_0$.

J.2 MOTION GENERATION, S-S DYADS, *N* PRECISION POSITIONS

The subfolder *Synthesis* within the folder *Appendix J.2* includes MATLAB files (filenames *SS_motion_prepare.m* and *SS_quasi_newton.m*) for calculating the S-S dyad variable $\mathbf{b}_0$ for motion generation given *N* precision positions. The folder *Common* (which includes supplementary MATLAB files) must be included alongside the folder *Synthesis* to properly run the MATLAB files. Figure J.5 includes the three-step procedure for calculating $\mathbf{b}_0$.

Synthesizing the S-S dyad begins with the file *SS_motion_prepare.m* (step 1 in Figure J.5). In this file, the number of precision positions is specified. After specifying the value of *N* (the number of precision positions) and running this file, the folder *SSfuncs* is produced. The folder *SSfuncs* includes functions required to run file *SS_quasi_newton.m.*

After exiting the file *SS_motion_prepare.m* and opening the file *SS_quasi_newton.m*, step 2 in Figure J.5 is started. In this file, the user specifies the value of *N*—the number of precision positions (identical to the *N* value prescribed in the file *SS_motion_prepare.m*). The user also includes specific precision position values (defined by rigid-body points **p**, **q**, **r**, and **s**, as given in Matrix (8.51)). The number of *PN*, *QN*, *RN*, and *SN* variables and values should match the value of *N*. Additionally, the user includes specific values for $\mathbf{b}_1$. Figure J.6 illustrates these user input sections in the file *SS_quasi_newton.m*, with sample data in bold type.

The user includes the initial values for $\mathbf{b}_0$ (step 3 in Figure J.5). Figure J.7 illustrates this user input section in the file *SS_quasi_newton.m* with sample data in bold type.

```
%------------------------------------------------------------
% Here, values for precision position parameter PN are assigned.
% Additionalrows can be included to accommodate additional PN
% variables and values.

P1 = [-0.5, -0.5, -0.5];
P2 = [0, 0, 0];
P3 = [0.75, 0.75, 0.5];

%------------------------------------------------------------
                              .
                              .
                              .
%------------------------------------------------------------
% Here, values for precision position parameter SN are assigned.
% Additionalrows can be included to accommodate additional SN
% variables and values.

S1 = [-0.6677, -0.4551, 0.4848];
S2 = [0, 0, 1];
S3 = [0.6406, 1.1582, 1.4063];

%------------------------------------------------------------

%------------------------------------------------------------
% Here, values for dyad components b1x, b1y and b1z are assigned.

b1x = 0;
b1y = 0;
b1z = -5;

%------------------------------------------------------------
```

FIGURE J.6 Sections (for step 2) of file *SS_quasi_newton.m,* with sample data in bold type.

```
%------------------------------------------------------------
% Here, initial values for dyad components b0x, b0y and b0z are
%assigned.

b0x = -5;
b0y = 5;
b0z = -5;

%------------------------------------------------------------
```

FIGURE J.7 Section (for step 3) of file *SS_quasi_newton.m,* with sample data in bold type.

```
%%%%%%%%%%%%%%%%%%%%%%%%%%%%%%%%%%%%%%%%%%%%%%%%%%%%%%%%%%%%%%%%%%%%%%
Solvingfor the components of the S-S dyad
%%%%%%%%%%%%%%%%%%%%%%%%%%%%%%%%%%%%%%%%%%%%%%%%%%%%%%%%%%%%%%%%%%%%%%
Iterations: 17
Func Evals: 19
Algorithm: quasi-newton
exit flag = 5
error = ( 6.162976e-033 )
b0x = ( 0.071503 )
b0y = ( -0.032561 )
b0z = ( -4.091424 )
Elapsed time is 2.865181 seconds.
Solution calculation is complete.
```

FIGURE J.8 Sample output from the file *SS_quasi_newton.m.*

Lastly, the file *SS_quasi_newton.m* is run (also in step 3). The calculated S-S dyad components are displayed in the MATLAB command window after this file has been run. Figure J.8 illustrates the sample output from the file *SS_quasi_newton.m.*

When calculating new solutions, repeating step 1 in Figure J.5 is only necessary if the number of precision positions (N) for the new solution is different from that in the previous solution. If N is the same for the new solution, the user can begin at step 2 in Figure J.5.

J.3 MOTION GENERATION, SPHERICAL R-R DYADS, *N* PRECISION POSITIONS

The subfolder *Synthesis* within the folder *Appendix J.3* includes MATLAB files (filenames *RR_Sph_motion_prepare.m* and *RR_Sph_interior_point.m*) for calculating the spherical R-R dyad variables $\mathbf{a}_0$ and $\mathbf{a}_1$ for motion generation given N precision positions. The folder *Common* (which includes supplementary MATLAB files) must be included alongside the folder *Synthesis* to properly run the MATLAB files. Figure J.9 includes the three-step procedure for calculating $\mathbf{a}_0$ and $\mathbf{a}_1$.

Synthesizing the spherical R-R dyad begins with the file *RR_Sph_motion_prepare.m* (step 1 in Figure J.9). In this file, the number of precision positions is specified. After specifying the value of N (the number of precision positions) and running this file, the folder *RRSfuncs* is produced. The folder *RRSfuncs* includes functions required to run file *RR_Sph_interior_point.m.*

After exiting the file *RR_Sph_motion_prepare.m* and opening the file *RR_Sph_interior_point.m*, step 2 in Figure J.9 is started. In this file, the user specifies the value of N—the number of precision positions (identical to the N value prescribed in the file *RR_Sph_motion_prepare.m*). The user also includes specific precision position values (defined by rigid-body points $\mathbf{p}$, $\mathbf{q}$, and $\mathbf{r}$, as given in Matrix (8.57)). The number of *PN, QN,* and *RN* variables and values should match the value of N. Additionally, the user specifies the value for the dyad parameter a_{0x}. Figure J.10 illustrates these user input sections in the file *RR_Sph_interior_point.m*, with sample data in bold type.

The user includes the initial values for a_{0y}, a_{0z}, and $\mathbf{a}_1$ (step 3 in Figure J.9). Figure J.11 illustrates this user input section in the file *RR_Sph_interior_point.m* with sample data in bold type.

Lastly, the file *RR_Sph_interior_point.m* is run (also in step 3). The calculated spherical R-R dyad components are displayed in the MATLAB command window after this file has been run. Figure J.12 illustrates the sample output from the file *RR_Sph_interior_point.m.* To calculate an alternate spherical R-R dyad solution (e.g., the spherical R-R dyad $\mathbf{b}_0 - \mathbf{b}_1$ for the 4R Spherical mechanism), the user specifies different values for the components of $\mathbf{a}_0 - \mathbf{a}_1$ and reruns the file *RR_Sph_interior_point.m.* The newly calculated solution will correspond to the spherical R-R dyad $\mathbf{b}_0 - \mathbf{b}_1$.

When calculating new solutions, repeating step 1 in Figure J.9 is only necessary if the number of precision positions (N) for the new solution is different from that in the previous solution. If N is the same for the new solution, the user can begin at step 2 in Figure J.9.

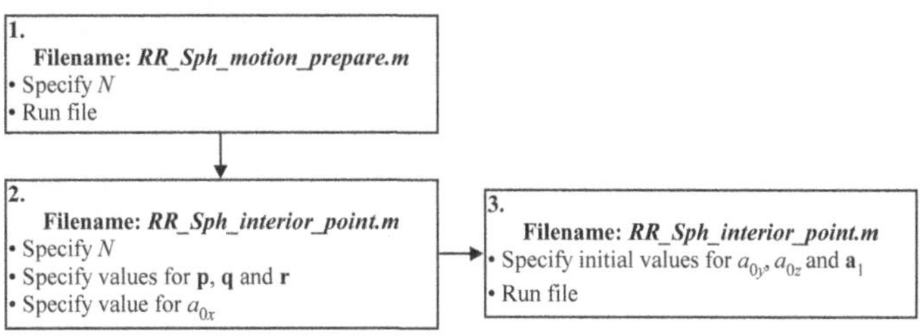

FIGURE J.9 Procedure for calculating $\mathbf{a}_0$, and $\mathbf{a}_1$.

```
%-----------------------------------------------------------------
% Here, values for precision position parameter PN are assigned.
% Additional rows can be included to accommodate additional PN
% variables and values.
P1 = [-0.2081, 0.1675, 0.8227];
P2 = [0.1442, 0.0859, 0.8486];
P3 = [0.5009, 0.0166, 0.705];
P4 = [0.7646, -0.0298, 0.3977];
%-----------------------------------------------------------------
                              .
                              .
                              .
%-----------------------------------------------------------------
% Here, values for precision position parameter RN are assigned.
% Additional rows can be included to accommodate additional RN
% variables and values.
R1 = [0, 0.3087, 0.808];
R2 = [0.2613, 0.2946, 0.7702];
R3 = [0.5104, 0.2616, 0.6475];
R4 = [0.7179, 0.2147, 0.4321];
%-----------------------------------------------------------------

%-----------------------------------------------------------------
% Here, value for dyad component a0x is assigned.
a0x = 0;
%-----------------------------------------------------------------
```

FIGURE J.10 Sections (for step 2) of file *RR_Sph_interior_point.m,* with sample data in bold type.

```
%-----------------------------------------------------------------
% Here, initial values for the components of the Spherical R-R dyad
% are assigned.
a0y = 1;
a0z = 1;

a1x = 1;
a1y = 1;
a1z = 1;
%-----------------------------------------------------------------
```

FIGURE J.11 Section (for step 3) of file *RR_Sph_interior_point.m,* with sample data in bold type.

```
%%%%%%%%%%%%%%%%%%%%%%%%%%%%%%%%%%%%%%%%%%%%%%%%%%%%%%%%%%%%%%%%%%%%%%%%
Solving for the components of the Spherical R-R dyad
%%%%%%%%%%%%%%%%%%%%%%%%%%%%%%%%%%%%%%%%%%%%%%%%%%%%%%%%%%%%%%%%%%%%%%%%
Iterations: 8
Func Evals: 19
Algorithm: interior-point
exit flag = 1
error = ( 1.164517e-006 )
a0y = ( 0.799764 )
a0z = ( 0.600315 )
a1x = ( 0.568437 )
a1y = ( 0.381875 )
a1z = ( 0.728733 )
Elapsed time is 6.428206 seconds.
Solution calculation is complete.
```

FIGURE J.12 Sample output from the file *RR_Sph_interior_point.m.*

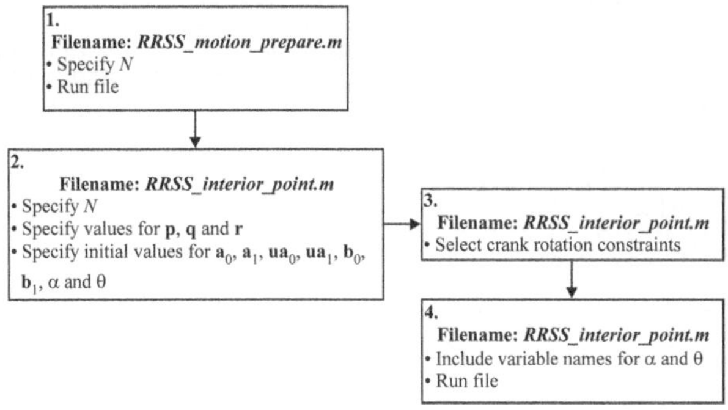

FIGURE J.13 Procedure for calculating $\mathbf{a}_0$, $\mathbf{a}_1$, $\mathbf{ua}_0$, $\mathbf{ua}_1$, $\mathbf{b}_0$, $\mathbf{b}_1$, $\alpha_2 \ldots \alpha_N$, and $\theta_2 \ldots \theta_N$.

J.4 RRSS MOTION GENERATION, *N* PRECISION POSITIONS WITH ORDER AND BRANCH CONSTRAINTS

The subfolder *Synthesis* within the folder *Appendix J.4* includes MATLAB files (filenames *RRSS_motion_prepare.m* and *RRSS_interior_point.m*) for calculating the RRSS mechanism variables $\mathbf{a}_0$, $\mathbf{a}_1$, $\mathbf{ua}_0$, $\mathbf{ua}_1$, $\mathbf{b}_0$, $\mathbf{b}_1$, $\alpha_2 \ldots \alpha_N$, and $\theta_2 \ldots \theta_N$ for motion generation given *N* precision positions. The folder *Common* (which includes supplementary MATLAB files) must be included alongside the folder *Synthesis* to properly run the MATLAB files. Figure J.13 includes the four-step procedure for calculating $\mathbf{a}_0$, $\mathbf{a}_1$, $\mathbf{ua}_0$, $\mathbf{ua}_1$, $\mathbf{b}_0$, $\mathbf{b}_1$, $\alpha_2 \ldots \alpha_N$, and $\theta_2 \ldots \theta_N$.

Synthesizing the mechanism begins with the file *RRSS_motion_prepare.m* (step 1 in Figure J.13). In this file, the number of precision positions is specified. After specifying the value of *N* (the number of precision positions) and running this file, the folder *RRSSfuncs* is produced. The folder *RRSSfuncs* includes functions required to run file *RRSS_interior_point.m.*

After exiting the file *RRSS_motion_prepare.m* and opening the file *RRSS_interior_point.m*, step 2 in Figure J.13 is started. In this file, the user specifies the value of *N*—the number of precision positions (identical to the *N* value prescribed in the file *RRSS_motion_prepare.m*). The user includes specific precision position values (defined by rigid-body points **p**, **q**, and **r**, as given in Equation (8.58)). The number of *PN*, *QN*, and *RN* variables and values should match the value of *N*. The user also specifies initial values for $\mathbf{a}_0$, $\mathbf{a}_1$, $\mathbf{ua}_0$, $\mathbf{ua}_1$, $\mathbf{b}_0$, $\mathbf{b}_1$, $\alpha_2 \ldots \alpha_N$, and $\theta_2 \ldots \theta_N$ (in degrees). The number of *alphaN* and *thetaN* variables and values should match the value of *N*. Figures J.14 and J.15 illustrate these user input sections in the file *RRSS_interior_point.m*, with sample data in bold type.

In step 3, the user specifies either clockwise or counterclockwise crank rotation constraints for order defect elimination. Counterclockwise constraints are specified in the file *RRSS_interior_point.m*. To select clockwise constraints, the characters *%{* must be included as instructed in the file (to deactivate the counterclockwise constraints). Also, the characters *%{* must be removed as instructed in the file (to activate the clockwise constraints).

In step 4 in Figure J.13, the user includes the names of each dyad displacement angle (angles α and θ) to be prescribed. Figure J.16 illustrates this user input section in the file *RRSS_interior_point.m*, with sample data in bold type.

Lastly, the file *RRSS_interior_point.m* is run (also in step 4). The calculated RRSS mechanism dyad components and displacement angles are displayed in the MATLAB command window after this file has been run. Figure J.17 illustrates the sample output from the file *RRSS_interior_point.m.*

When calculating new solutions, repeating step 1 in Figure J.13 is only necessary if the number of precision positions (*N*) for the new solution is different from that in the previous solution. If *N* is the same for the new solution, the user can begin at step 2 in Figure J.13.

```
%-------------------------------------------------------------
% Here, values for precision position parameter PN are assigned.
% Additional rows can be included to accommodate additional PN
% variables and values.

P1 = [-0.75, -0.75, -0.5];
P2 = [-0.375, -0.375, -0.25];
P3 = [0, 0, 0];
P4 = [0.375, 0.375, 0.25];
P5 = [0.75, 0.75, 0.5];

%-------------------------------------------------------------
                              .
                              .
                              .
%-------------------------------------------------------------
% Here, values for precision position parameter RN are assigned.
% Additional rows can be included to accommodate additional RN
% variables and values.

R1 = [-0.5154, 0.1254, -0.0774];
R2 = [-0.2476, 0.5929, -0.0336];
R3 = [0, 1, 0];
R4 = [0.2476, 1.3429, 0.0336];
R5 = [0.5154, 1.6254, 0.0774];

%-------------------------------------------------------------
```

FIGURE J.14 Sections (for step 2) of file *RRSS_interior_point.m,* with sample data in bold type.

```
%-------------------------------------------------------------
% Here, initial values for the RRSS variables are assigned.
% Additional rows can be included to accommodate additional
% alphaN and thetaN variables and values.

a0x = 0;
a0y = 0;
a0z = 0;

a1x = 0;
a1y = 1;
a1z = 0;

ua0x = 0;
ua0y = 0;
ua0z = 1;

ua1x = 0;
ua1y = 0;
ua1z = 1;

b0x = 1;
b0y = 0;
b0z = 0;

b1x = 1;
b1y = 1;
b1z = 0;

alpha2 = 10 *pi/180;
alpha3 = 10 *pi/180;
alpha4 = 10 *pi/180;
alpha5 = 10 *pi/180;

theta2 = 20 *pi/180;
theta3 = 35 *pi/180;
theta4 = 45 *pi/180;
theta5 = 55 *pi/180;

%-------------------------------------------------------------
```

FIGURE J.15 Section (for step 2) of file *RRSS_interior_point.m,* with sample data in bold type.

```
%-------------------------------------------------------------
% Here, the names for each alphaN and thetaN displacement angle
% are included
% (e.g., alpha2; alpha3; alpha4; alpha5; alpha6; .....;)
% (e.g., theta2; theta3; theta4; theta5; theta6; .....];).
x0 = [a0x; a0y; a0z; a1x; a1y; a1z;
      ua0x; ua0y; ua0z; ua1x; ua1y; ua1z;
      b0x; b0y; b0z; b1x; b1y; b1z;
      alpha2; alpha3; alpha4; alpha5;
      theta2; theta3; theta4; theta5];
%-------------------------------------------------------------
```

FIGURE J.16 Section (for step 4) of file *RRSS_interior_point.m,* with sample data in bold type.

```
%%%%%%%%%%%%%%%%%%%%%%%%%%%%%%%%%%%%%%%%%%%%%%%%%%%%%%%%%%%%%%%%%%%
Solving for the RRSS mechanism variables
%%%%%%%%%%%%%%%%%%%%%%%%%%%%%%%%%%%%%%%%%%%%%%%%%%%%%%%%%%%%%%%%%%%
Iterations: 17
Func Evals: 57
Algorithm: interior-point
exit flag = -2
error = ( 2.642219e-001 )
a0x = ( -2.152050 )
a0y = ( 1.976960 )
a0z = ( -0.367068 )
a1x = ( -1.715085 )
a1y = ( 5.264583 )
a1z = ( 0.708521 )
ua0x = ( -0.933127 )
ua0y = ( -0.010840 )
ua0z = ( 0.378485 )
ua1x = ( 0.959822 )
ua1y = ( -0.226166 )
ua1z = ( 0.282585 )
b0x = ( -2.012833 )
b0y = ( 2.113794 )
b0z = ( -1.162102 )
b1x = ( 1.215053 )
b1y = ( -0.008169 )
b1z = ( 0.308485 )
alpha2 = ( 4.303687 )
alpha3 = ( 8.231877 )
alpha4 = ( 12.427676 )
alpha5 = ( 16.654463 )
theta2 = ( 20.986132 )
theta3 = ( 37.755033 )
theta4 = ( 55.444431 )
theta5 = ( 75.194853 )
Elapsed time is 234.322570 seconds.
Solution calculation is complete.
```

FIGURE J.17 Sample output from the file *RRSS_interior_point.m.*

J.5 RRSS PATH GENERATION, *N* PRECISION POINTS WITH ORDER AND BRANCH CONSTRAINTS

The subfolder *Synthesis* within the folder *Appendix J.5* includes MATLAB files (filenames *RRSS_path_prepare.m* and *RRSS_interior_point.m*) for calculating the RRSS mechanism variables $\mathbf{a}_0$, $\mathbf{a}_1$, $\mathbf{ua}_0$, $\mathbf{ua}_1$, $\mathbf{b}_0$, $\mathbf{b}_1$, $\alpha_2 \ldots \alpha_N$, and $\theta_2 \ldots \theta_N$ for motion generation given *N* precision points. The folder *Common* (which includes supplementary MATLAB files) must be included alongside the folder *Synthesis* to properly run the MATLAB files. Figure J.18 includes the four-step procedure for calculating $\mathbf{a}_0$, $\mathbf{a}_1$, $\mathbf{ua}_0$, $\mathbf{ua}_1$, $\mathbf{b}_0$, $\mathbf{b}_1$, $\alpha_2 \ldots \alpha_N$, and $\theta_2 \ldots \theta_N$.

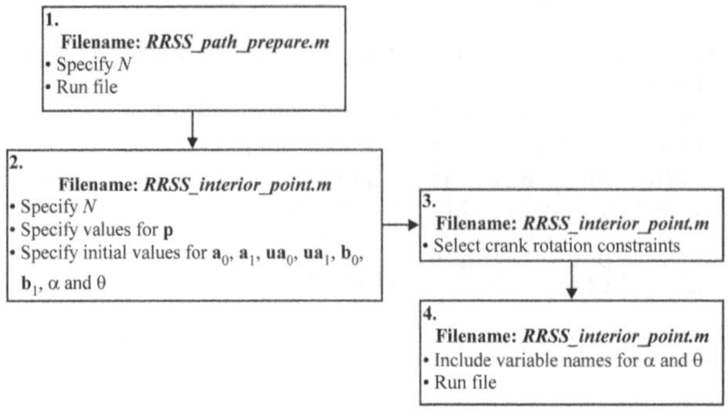

FIGURE J.18 Procedure for calculating $\mathbf{a}_0$, $\mathbf{a}_1$, $\mathbf{ua}_0$, $\mathbf{ua}_1$, $\mathbf{b}_0$, $\mathbf{b}_1$, $\alpha_2 \ldots \alpha_N$, and $\theta_2 \ldots \theta_N$.

```
%-------------------------------------------------------------
% Here, values for precision point parameter PN are assigned.
% Additional rows can be included to accommodate additional PN
% variables and values.
P1 = [0, 0, 0];
P2 = [0.5, 0.25, 0];
P3 = [1, 0.5, 0];
P4 = [1, 0.35, 1];
P5 = [0, -0.25, 0.5];

%-------------------------------------------------------------
```

FIGURE J.19 Section (for step 2) of file *RRSS_interior_point.m,* with sample data in bold type.

Synthesizing the dyads begins with the file *RRSS_path_prepare.m* (step 1 in Figure J.18). In this file, the number of precision points is specified. After specifying the value of *N* (the number of precision points) and running this file, the folder *RRSSfuncs* is produced. The folder *RRSSfuncs* includes functions required to run file *RRSS_interior_point.m.*

After exiting the file *RRSS_path_prepare.m* and opening the file *RRSS_interior_point.m*, step 2 in Figure J.18 is started. In this file, the user specifies the value of *N*—the number of precision points (identical to the *N* value prescribed in the file *RRSS_path_prepare.m*). The user includes specific precision point values (defined by rigid-body point **p**, as explained in Section 8.4). The number of *PN* variables and values should match the value of *N*. The user also specifies initial values for $\mathbf{a}_0$, $\mathbf{a}_1$, $\mathbf{ua}_0$, $\mathbf{ua}_1$, $\mathbf{b}_0$, $\mathbf{b}_1$, $\alpha_2 \ldots \alpha_N$, and $\theta_2 \ldots \theta_N$ (in degrees). The number of *alphaN* and *thetaN* variables and values should match the value of *N*. Figures J.19 and J.20 illustrate these user input sections in the file *RRSS_interior_point.m*, with sample data in bold type.

In step 3, the user specifies either clockwise or counterclockwise crank rotation constraints for order defect elimination. Counterclockwise constraints are specified in the file *RRSS_interior_point.m*. To select clockwise constraints, the characters *%{* must be included as instructed in the file (to deactivate the counterclockwise constraints). Also, the characters *%{* must be removed as instructed in the file (to activate the clockwise constraints).

In step 4 in Figure J.18, the user includes the names of each dyad displacement angle (angles α and θ) to be prescribed. Figure J.21 illustrates this user input section in the file *RRSS_interior_point.m,* with sample data in bold type.

Lastly, the file *RRSS_interior_point.m* is run (also in step 4). The calculated RRSS mechanism dyad components and displacement angles are displayed in the MATLAB command window after this file has been run. Figure J.22 illustrates sample output from the file *RRSS_interior_point.m.*

```
%-----------------------------------------------------------------
% Here, initial values for the RRSS variables are assigned.
% Additional rows can be included to accommodate additional alphaN
% and thetaN variables and values.
a0x = 0;
a0y = 0;
a0z = 0;

a1x = 0;
a1y = 1;
a1z = 0;

ua0x = 0;
ua0y = 0;
ua0z = 1;

ua1x = 0;
ua1y = 0;
ua1z = 1;

b0x = 1;
b0y = 0;
b0z = 0;

b1x = 1;
b1y = 1;
b1z = 0;

alpha2 = 20 *pi/180;
alpha3 = 40 *pi/180;
alpha4 = 60 *pi/180;
alpha5 = 80 *pi/180;

theta2 = 80 *pi/180;
theta3 = 160 *pi/180;
theta4 = 240 *pi/180;
theta5 = 320 *pi/180;

%-----------------------------------------------------------------
```

FIGURE J.20 Section (for step 2) of file *RRSS_interior_point.m,* with sample data in bold type.

```
%-----------------------------------------------------------------
% Here, the names for each alphaN and thetaN displacement angle are
% included
% (e.g., alpha2; alpha3; alpha4; alpha5; alpha6; .....;)
% (e.g., theta2; theta3; theta4; theta5; theta6; .....];).

x0 = [a0x; a0y; a0z; a1x; a1y; a1z;
      ua0x; ua0y; ua0z; ua1x; ua1y; ua1z;
      b0x; b0y; b0z; b1x; b1y; b1z;
      alpha2; alpha3; alpha4; alpha5;
      theta2; theta3; theta4; theta5;];

%-----------------------------------------------------------------
```

FIGURE J.21 Section (for step 4) of file *RRSS_interior_point.m,* with sample data in bold type.

When calculating new solutions, repeating step 1 in Figure J.18 is only necessary if the number of precision points (N) for the new solution is different from that in the previous solution. If N is the same for the new solution, the user can begin at step 2 in Figure J.18.

```
%%%%%%%%%%%%%%%%%%%%%%%%%%%%%%%%%%%%%%%%%%%%%%%%%%%%%%%%%%%%%%%%%%%%%
Solving for the RRSS mechanism variables
%%%%%%%%%%%%%%%%%%%%%%%%%%%%%%%%%%%%%%%%%%%%%%%%%%%%%%%%%%%%%%%%%%%%%
Iterations: 22
Func Evals: 89
Algorithm: interior-point
exit flag = 2
error = (3.049996e-003 )
a0x = ( -0.064540 )
a0y = ( 0.311683 )
a0z = ( 0.442293 )
a1x = ( 2.798472 )
a1y = ( 4.352184 )
a1z = ( 4.390796 )
ua0x = ( 0.834599 )
ua0y = ( -0.549161 )
ua0z = ( -0.043202 )
ua1x = ( -0.877998 )
ua1y = ( 0.194893 )
ua1z = ( 0.437191 )
b0x = ( 0.354333 )
b0y = ( 0.631174 )
b0z = ( 0.391535 )
b1x = ( 3.179748 )
b1y = ( 4.421619 )
b1z = ( -3.307704 )
alpha2 = ( 4.361910 )
alpha3 = ( 9.478536 )
alpha4 = ( 9.735462 )
alpha5 = ( 2.636086 )
theta2 = ( 115.353648 )
theta3 = ( 176.063348 )
theta4 = ( 251.969123 )
theta5 = ( 327.364874 )
Elapsed time is 110.134550 seconds.
Solution calculation is complete.
```

FIGURE J.22 Sample output from the file *RRSS_interior_point.m*.

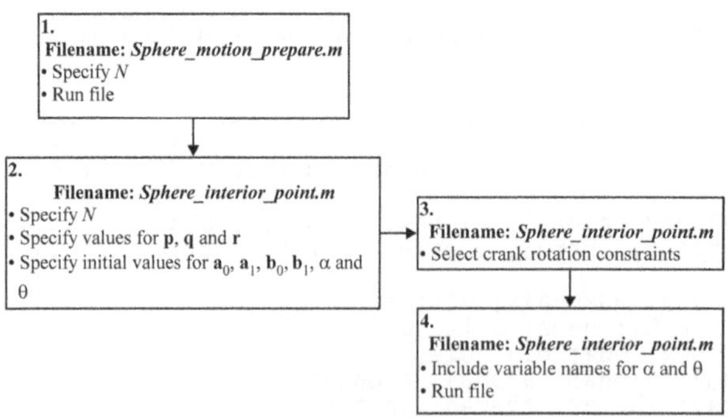

FIGURE J.23 Procedure for calculating $\mathbf{a}_0$, $\mathbf{a}_1$, $\mathbf{b}_0$, $\mathbf{b}_1$, $\alpha_2 \ldots \alpha_N$, and $\theta_2 \ldots \theta_N$.

J.6 4 R SPHERICAL MOTION GENERATION, *N* PRECISION POSITIONS WITH ORDER AND BRANCH CONSTRAINTS

The subfolder *Synthesis* within the folder *Appendix J.6* includes MATLAB files (filenames *Sphere_motion_prepare.m* and *Sphere_interior_point.m*) for calculating the 4R Spherical mechanism variables $\mathbf{a}_0$, $\mathbf{a}_1$, $\mathbf{b}_0$, $\mathbf{b}_1$, $\alpha_2 \ldots \alpha_N$, and $\theta_2 \ldots \theta_N$ for motion generation given *N* precision positions. The folder *Common* (which includes supplementary MATLAB files) must be included alongside the folder *Synthesis* to properly run the MATLAB files. Figure J.23 includes the four-step procedure for calculating $\mathbf{a}_0$, $\mathbf{a}_1$, $\mathbf{b}_0$, $\mathbf{b}_1$, $\alpha_2 \ldots \alpha_N$ and $\theta_2 \ldots \theta_N$.

Synthesizing the dyads begins with the file *Sphere_motion_prepare.m* (step 1 in Figure J.23). In this file, the number of precision positions is specified. After specifying the value of N (the number of precision positions) and running this file, the folder *Spherefuncs* is produced. The folder *Spherefuncs* includes functions required to run file *Sphere_interior_point.m*.

After exiting the file *Sphere_motion_prepare.m* and opening the file *Sphere_interior_point.m*, step 2 in Figure J.23 is started. In this file, the user specifies the value of N—the number of precision positions (identical to the N value prescribed in the file *Sphere_motion_prepare.m*). The user includes specific precision position values (defined by rigid-body points **p**, **q**, and **r**, as given in Equation (8.58)). The number of *PN, QN,* and *RN* variables and values should match the value of N. The user also specifies initial values for $\mathbf{a}_0$, $\mathbf{a}_1$, $\mathbf{b}_0$, $\mathbf{b}_1$, $\alpha_2 \ldots \alpha_N$, and $\theta_2 \ldots \theta_N$ (in degrees). The number of *alphaN* and *thetaN* variables and values should match the value of N. Figures J.24 and J.25 illustrate these user input sections in the file *Sphere_interior_point.m*, with sample data in bold type.

In step 3, the user specifies either clockwise or counterclockwise crank rotation constraints for order defect elimination. Counterclockwise constraints are specified in the file *Sphere_interior_point.m*. To select clockwise constraints, the characters *%{* must be included as instructed in the file (to deactivate the counterclockwise constraints). Also, the characters *%{* must be removed as instructed in the file (to activate the clockwise constraints).

In step 4 in Figure J.23, the user includes the names of each dyad displacement angle (angles α and θ) to be prescribed. Figure J.26 illustrates this user input section in the file *Sphere_interior_point.m*, with sample data in bold type.

Lastly, the file *Sphere_interior_point.m* is run (also in step 4). The calculated 4R Spherical mechanism dyad components and displacement angles are displayed in the MATLAB command window after this file has been run. Figure J.27 illustrates the sample output from the file *Sphere_interior_point.m*.

When calculating new solutions, repeating step 1 in Figure J.23 is only necessary if the number of precision positions (N) for the new solution is different from that in the previous solution. If N is the same for the new solution, the user can begin at step 2 in Figure J.23.

```
%------------------------------------------------------------------
% Here, values for precision position parameter PN are assigned.
% Additional rows can be included to accommodate additional PN
% variables and values.
P1 = [1.5, 0, 0];
P2 = [1.4624, 0.3338, 0];
P3 = [1.3515, 0.6508, 0];
P4 = [1.1727, 0.9352, 0];
P5 = [0.9352, 1.1727, 0];

%------------------------------------------------------------------
.
                              .
                              .
%------------------------------------------------------------------
% Here, values for precision position parameter RN are assigned.
% Additional rows can be included to accommodate additional RN
% variables and values.
R1 = [1.5, 1, 0];
R2 = [1.2456, 1.2837, 0.225];
R3 = [0.9615, 1.4606, 0.4384];
R4 = [0.6882, 1.5428, 0.6293];
R5 = [0.4539, 1.5566, 0.788];

%------------------------------------------------------------------
```

FIGURE J.24 Sections (for step 2) of file *Sphere_interior_point.m*, with sample data in bold type.

```
%------------------------------------------------------------------
% Here, initial values for the 4R Spherical variables are assigned.
% Additional rows can be included to accommodate additional alphaN
% and thetaN variables and values.

a0x = 1;
a0y = 1;
a0z = 1;

a1x = 1;
a1y = 1;
a1z = 1;

b0x = 0;
b0y = 1;
b0z = 0;

b1x = 0;
b1y = 0;
b1z = 1;

alpha2 = 10 *pi/180;
alpha3 = 10 *pi/180;
alpha4 = 10 *pi/180;
alpha5 = 10 *pi/180;

theta2 = 25 *pi/180;
theta3 = 35 *pi/180;
theta4 = 45 *pi/180;
theta5 = 55 *pi/180;

%------------------------------------------------------------------
```

FIGURE J.25 Section (for step 2) of file *Sphere_interior_point.m,* with sample data in bold type.

```
%-----------------------------------------------------------------
% Here, the names for each alphaN and thetaN displacement angle
% are included
% (e.g., alpha2; alpha3; alpha4; alpha5; alpha6; .....;)
% (e.g., theta2; theta3; theta4; theta5; theta6; .....];).

x0 = [a0x; a0y; a0z; a1x; a1y; a1z;
      b0x; b0y; b0z; b1x; b1y; b1z;
      alpha2; alpha3; alpha4; alpha5;
      theta2; theta3; theta4; theta5];

%-----------------------------------------------------------------
```

FIGURE J.26 Section (for step 4) of file *Sphere_interior_point.m,* with sample data in bold type.

J.7 4 R SPHERICAL PATH GENERATION, *N* PRECISION POINTS WITH ORDER AND BRANCH CONSTRAINTS

The subfolder *Synthesis* within the folder *Appendix J.7* includes MATLAB files (filenames *Sphere_path_prepare.m* and *Sphere_interior_point.m*) for calculating the 4R Spherical mechanism variables $\mathbf{a}_0$, $\mathbf{a}_1$, $\mathbf{b}_0$, $\mathbf{b}_1$, $\alpha_2 \ldots \alpha_N$, and $\theta_2 \ldots \theta_N$ for motion generation given *N* precision points. The folder *Common* (which includes supplementary MATLAB files) must be included alongside the folder *Synthesis* to properly run the MATLAB files. Figure J.28 includes the four-step procedure for calculating $\mathbf{a}_0$, $\mathbf{a}_1$, $\mathbf{b}_0$, $\mathbf{b}_1$, $\alpha_2 \ldots \alpha_N$, and $\theta_2 \ldots \theta_N$.

Synthesizing the dyads begins with the file *Sphere_path_prepare.m* (step 1 in Figure J.28). In this file, the number of precision points is specified. After specifying the value of *N* (the number of precision points) and running this file, the folder *Spherefuncs* is produced. The folder *Spherefuncs* includes functions required to run file *Sphere_interior_point.m.*

```
%%%%%%%%%%%%%%%%%%%%%%%%%%%%%%%%%%%%%%%%%%%%%%%%%%%%%%%%%%%%%%%%%
Solving for the 4R Spherical mechanism variables
%%%%%%%%%%%%%%%%%%%%%%%%%%%%%%%%%%%%%%%%%%%%%%%%%%%%%%%%%%%%%%%%%
Iterations: 50
Func Evals: 138
Algorithm: interior-point
exit flag = 2
error = ( 1.329472e-004 )
a0x = ( 0.515422 )
a0y = ( 0.229533 )
a0z = ( 0.825624 )
a1x = ( 0.392354 )
a1y = ( 0.379156 )
a1z = ( 0.838035 )
b0x = ( 0.551434 )
b0y = ( 0.296969 )
b0z = ( -0.779571 )
b1x = ( -0.950856 )
b1y = ( -0.147291 )
b1z = ( -0.272353 )
alpha2 = ( -24.449120 )
alpha3 = ( -45.327883 )
alpha4 = ( -67.527438 )
alpha5 = ( -92.072335 )
theta2 = ( 41.434574 )
theta3 = ( 80.050617 )
theta4 = ( 119.966262 )
theta5 = ( 160.948546 )
Elapsed time is 219.303009 seconds.
Solution calculation is complete.
```

FIGURE J.27 Sample output from the file *Sphere_interior_point.m*.

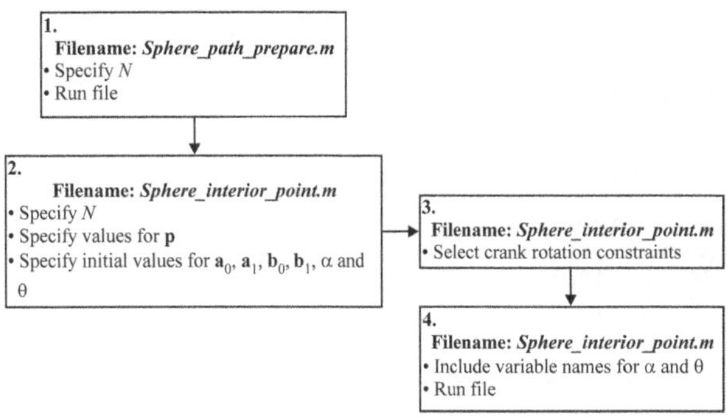

FIGURE J.28 Procedure for calculating $\mathbf{a}_0$, $\mathbf{a}_1$, $\mathbf{b}_0$, $\mathbf{b}_1$, $\alpha_2 \ldots \alpha_N$, and $\theta_2 \ldots \theta_N$.

After exiting the file *Sphere_path_prepare.m* and opening the file *Sphere_interior_point.m*, step 2 in Figure J.28 is started. In this file, the user specifies the value of *N*—the number of precision points (identical to the *N* value prescribed in the file *Sphere_path_prepare.m*). The user includes specific precision points values (defined by rigid-body points **p**, as explained in Section 8.4). The number of *PN* variables and values should match the value of *N*. The user also specifies initial values for $\mathbf{a}_0$, $\mathbf{a}_1$, $\mathbf{b}_0$, $\mathbf{b}_1$, $\alpha_2 \ldots \alpha_N$, and $\theta_2 \ldots \theta_N$ (in degrees). The number of *alphaN* and *thetaN* variables and values should match the value of *N*. Figures J.29 and J.30 illustrate these user input sections in the file *Sphere_interior_point.m*, with sample data in bold type.

In step 3, the user specifies either clockwise or counterclockwise crank rotation constraints for order defect elimination. Counterclockwise constraints are specified in the file *Sphere_interior_point.m*. To select clockwise constraints, the characters *%{* must be included as instructed in the file (to deactivate the counterclockwise constraints). Also, the characters *%{* must be removed as instructed in the file (to activate the clockwise constraints).

```
%-------------------------------------------------------------
% Here, values for precision position parameter PN are assigned.
% Additional rows can be included to accommodate additional PN
% variables and values.
P1 = [0, 1.25, 0];
P2 = [0.3382, 1.1548, 0.3382];
P3 = [0.8166, 0.4784, 0.8166];
P4 = [0.8839, 0, 0.8839];
P5 = [0, 0, 1.25];
%-------------------------------------------------------------
```

FIGURE J.29 Section (for step 2) of file *Sphere_interior_point.m,* with sample data in bold type.

```
%-------------------------------------------------------------
% Here, initial values for the 4R Spherical variables are
% assigned. Additional rows can be included to accommodate
% additional alphaN and thetaN variables and values.
a0x = 1;
a0y = 0;
a0z = 0;

a1x = 0;
a1y = 1;
a1z = 0;

b0x = 1;
b0y = 0;
b0z = 0;

b1x = -1;
b1y = 0;
b1z = 0;

alpha2 = 10 *pi/180;
alpha3 = 20 *pi/180;
alpha4 = 30 *pi/180;
alpha5 = 40 *pi/180;

theta2 = 50 *pi/180;
theta3 = 100 *pi/180;
theta4 = 150 *pi/180;
theta5 = 200 *pi/180;
%-------------------------------------------------------------
```

FIGURE J.30 Section (for step 2) of file *Sphere_interior_point.m,* with sample data in bold type.

```
%-------------------------------------------------------------
% Here, the names for each alphaN and thetaN displacement angle
% are included
% (e.g., alpha2; alpha3; alpha4; alpha5; alpha6; .....;)
% (e.g., theta2; theta3; theta4; theta5; theta6; .....];).
x0 = [a0x; a0y; a0z; a1x; a1y; a1z;
      b0x; b0y; b0z; b1x; b1y; b1z;
      alpha2; alpha3; alpha4; alpha5;
      theta2; theta3; theta4; theta5];
%-------------------------------------------------------------
```

FIGURE J.31 Section (for step 4) of file *Sphere_interior_point.m,* with sample data in bold type.

In step 4 in Figure J.28, the user includes the names each dyad displacement angle (angles α and θ) to be prescribed. Figure J.31 illustrates this user input section in the file *Sphere_interior_point.m,* with sample data in bold type.

```
%%%%%%%%%%%%%%%%%%%%%%%%%%%%%%%%%%%%%%%%%%%%%%%%%%%%%%%%%%%%%%%%%%%%%%%%
Solving for the 4R Spherical mechanism variables
%%%%%%%%%%%%%%%%%%%%%%%%%%%%%%%%%%%%%%%%%%%%%%%%%%%%%%%%%%%%%%%%%%%%%%%%
Iterations: 81
Func Evals: 243
Algorithm: interior-point
exit flag = 2
error = ( 6.409368e-005 )
a0x = ( 0.702586 )
a0y = ( 0.153157 )
a0z = ( -0.694921 )
a1x = ( 0.706315 )
a1y = ( -0.425487 )
a1z = ( -0.565756 )
b0x = ( 0.468321 )
b0y = ( -0.848630 )
b0z = ( 0.245974 )
b1x = ( -0.954529 )
b1y = ( -0.237369 )
b1z = ( 0.180364 )
alpha2 = ( -28.179977 )
alpha3 = ( -45.529297 )
alpha4 = ( -56.938552 )
alpha5 = ( 217.401101 )
theta2 = ( 48.426320 )
theta3 = ( 108.701173 )
theta4 = ( 140.698855 )
theta5 = ( 201.937242 )
Elapsed time is 186.359467 seconds.
Solution calculation is complete.
```

FIGURE J.32 Sample output from the file *Sphere_interior_point.m.*

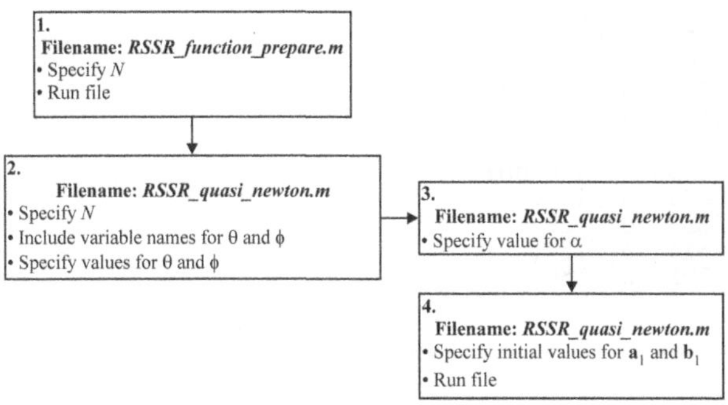

FIGURE J.33 Procedure for calculating $\mathbf{a}_1$ and $\mathbf{b}_1$.

Lastly, the file *Sphere_interior_point.m* is run (also in step 4). The calculated 4R Spherical mechanism dyad components and displacement angles are displayed in the MATLAB command window after this file has been run. Figure J.32 illustrates the sample output from the file *Sphere_interior_point.m.*

When calculating new solutions, repeating step 1 in Figure J.28 is only necessary if the number of precision points (N) for the new solution is different from that in the previous solution. If N is the same for the new solution, the user can begin at step 2 in Figure J.28.

J.8 RSSR FUNCTION GENERATION, *N* PRECISION POINTS

The subfolder *Synthesis* within the folder *Appendix J.8* includes MATLAB files (filenames *RSSR_function_prepare.m* and *RSSR_quasi_newton.m*) for calculating the RSSR mechanism moving pivots $\mathbf{a}_1$ and $\mathbf{b}_1$ for function generation given N precision points. The folder *Common* (which includes supplementary MATLAB files) must be included alongside the folder *Synthesis* to properly run the MATLAB files. Figure J.33 includes the four-step procedure for calculating $\mathbf{a}_1$ and $\mathbf{b}_1$.

Synthesizing the dyads begins with the file *RSSR_function_prepare.m* (step 1 in Figure J.33). In this file, the number of precision points is specified. After specifying the value of *N* (the number of precision points) and running this file, the folder *RSSRfuncs* is produced. The folder *RSSRfuncs* includes functions required to run file *RSSR_quasi_newton.m*.

After exiting the file *RSSR_function_prepare.m* and opening the file *RSSR_quasi_newton.m*, step 2 in Figure J.33 is started. In this file, the user specifies the value of *N*—the number of precision points (identical to the *N* value prescribed in the file *RSSR_function_prepare.m*). The user also includes the names for each crank and follower displacement angle (angles θ and ϕ, respectively) to be prescribed. The user also includes the specific precision points values in degrees (defined by rotation angles θ and ϕ, as given in Matrix (8.72)). The number of *thetaN* and *phiN* variables and values should match the value of *N*. Figure J.34 illustrates this user input section in the file *RSSR_quasi_newton.m*, with sample data in bold type.

In step 3, the user specifies values for the RSSR fixed pivot joint axis $\mathbf{ub}_0$. The user specifies the value for the angle α (which determines the orientation of $\mathbf{ub}_0$) in degrees. The user also specifies initial values for $\mathbf{a}_1$ and $\mathbf{b}_1$ (step 4 in Figure E.21). Figure J.35 illustrates these user input sections in the file *RSSR_quasi_newton.m*, with sample data in bold type.

```
%-------------------------------------------------------------
% Here, the names for each thetaN and phiN displacement angle are
% included
% (e.g., theta2 theta3 theta4 theta5 theta6 ...)
% (e.g., phi2 phi3 phi4 phi5 phi6 ...).
% Also, values for displacement angles thetaN and phiN are
% assigned. Additional rows can be included to accommodate
% additional thetaN and phiN variables and values.

global theta2 theta3 theta4 theta5
theta2 = 0.5763 *pi/180;
theta3 = 5.0559 *pi/180;
theta4 = 13.3329 *pi/180;
theta5 = 24.1473 *pi/180;

global phi2 phi3 phi4 phi5
phi2 = 0.0028 *pi/180;
phi3 = 0.2130 *pi/180;
phi4 = 1.4814 *pi/180;
phi5 = 4.8591 *pi/180;

%-------------------------------------------------------------
```

FIGURE J.34 Section (for step 2) of file *RSSR_quasi_newton.m,* with sample data in bold type.

```
%-------------------------------------------------------------
% Here, a value for joint axis angle ALPHA is assigned.

ALPHA = -30 *pi/180;

%-------------------------------------------------------------

%-------------------------------------------------------------
% Here, initial values for the RSSR moving pivot variables are
% assigned.

a1x = -1;
a1y = -1;
a1z = -1;

b1x = 1;
b1y = 1;
b1z = 1;

%-------------------------------------------------------------
```

FIGURE J.35 Sections (for steps 3 and 4) of file *RSSR_quasi_newton.m,* with sample data in bold type.

```
%%%%%%%%%%%%%%%%%%%%%%%%%%%%%%%%%%%%%%%%%%%%%%%%%%%%%%%%%%%%%%%%%%
Solving for the RSSR mechanism variables
%%%%%%%%%%%%%%%%%%%%%%%%%%%%%%%%%%%%%%%%%%%%%%%%%%%%%%%%%%%%%%%%%%
Iterations: 55
Func Evals: 64
Algorithm: quasi-newton
exit flag = 5
error = ( 8.537041e-015 )
a1x = ( -1.570977 )
a1y = ( -0.798988 )
a1z = ( 5.101747 )
b1x = ( 1.372694 )
b1y = ( 0.698019 )
b1z = ( 2.484452 )
Elapsed time is 97.603698 seconds.
Solution calculation is complete.
```

FIGURE J.36 Sample output from the file *RSSR_quasi_newton.m*.

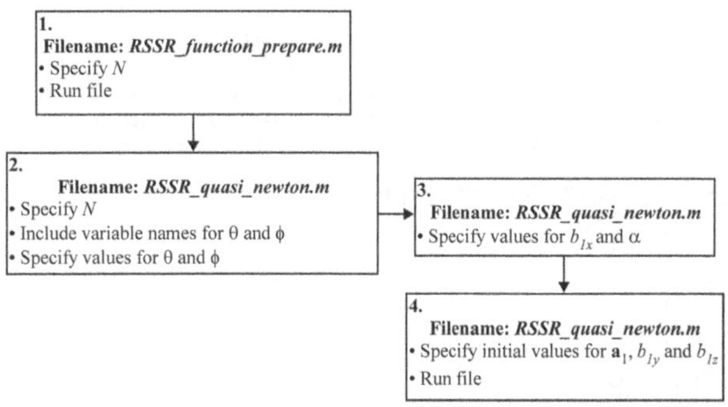

FIGURE J.37 Procedure for calculating $\mathbf{a}_1$, b_{1y}, and b_{1z}.

Lastly, the file *RSSR_quasi_newton.m* is run (also in step 4). The calculated RSSR mechanism moving pivots are displayed in the MATLAB command window after this file has been run. Figure J.36 illustrates the sample output from the file *RSSR_quasi_newton.m*.

When calculating new solutions, repeating step 1 in Figure J.33 is only necessary if the number of precision points (N) for the new solution is different from that in the previous solution. If N is the same for the new solution, the user can begin at step 2 in Figure J.33.

J.9 RSSR FUNCTION GENERATION (FOR RSSR STEERING LINKAGE), *N* PRECISION POINTS

The subfolder *Synthesis* within the folder *Appendix J.9* includes MATLAB files (filenames *RSSR_function_prepare.m* and *RSSR_quasi_newton.m*) for calculating the moving pivots $\mathbf{a}_1$, b_{1y}, and b_{1z} of an RSSR function generator steering linkage given N precision points. The folder *Common* (which includes supplementary MATLAB files) must be included alongside the folder *Synthesis* to properly run the MATLAB files. Figure J.37 includes the four-step procedure for calculating $\mathbf{a}_1$, b_{1y}, and b_{1z}.

Synthesizing the dyads begins with the file *RSSR_function_prepare.m* (step 1 in Figure J.37). In this file, the number of precision points is specified. After specifying the value of N (the number of precision points) and running this file, the folder *RSSRfuncs* is produced. The folder *RSSRfuncs* includes functions required to run file *RSSR_quasi_newton.m*.

After exiting the file *RSSR_function_prepare.m* and opening the file *RSSR_quasi_newton.m*, step 2 in Figure J.37 is started. In this file, the user specifies the value of N—the number of precision points (identical to the N value prescribed in the file *RSSR_function_prepare.m*). The user also includes the names for each crank and follower displacement angle (angles θ and ϕ, respectively) to be prescribed. The user also includes the specific precision points values in degrees (defined by rotation angles θ and ϕ, as given in Matrix (8.72)). The number of *thetaN* and *phiN* variables and values should match the value of *N*. Figure J.38 illustrates this user input section in the file *RSSR_quasi_newton.m*, with sample data in bold type.

In step 3, the user specifies values for the RSSR fixed pivot joint axis $\mathbf{ub}_0$ and the moving pivot component b_{1x}. The user also specifies value for b_{1x} and the angle α (which determines the orientation of $\mathbf{ub}_0$) in degrees. Additionally, the user specifies initial values for $\mathbf{a}_1$, b_{1y}, and b_{1z} (step 4 in Figure J.37). Figures J.39 illustrates these user input sections in the file *RSSR_quasi_newton.m*, with sample data in bold type.

Lastly, the file *RSSR_quasi_newton.m* is run (also in step 4). The calculated RSSR mechanism moving pivots are displayed in the MATLAB command window after this file has been run. Figure J.40 illustrates the sample output from the file *RSSR_quasi_newton.m*.

```
%------------------------------------------------------------
% Here, the names for each thetaN and phiN displacement angle are
% included
% (e.g., theta2 theta3 theta4 theta5 theta6 ...)
% (e.g., phi2 phi3 phi4 phi5 phi6 ...).
% Also, values for displacement angles thetaN and phiN are
% assigned. Additional rows can be included to accommodate
% additional thetaN and phiN variables and values.

global theta2 theta3 theta4 theta5
theta2 = 10 *pi/180;
theta3 = 20 *pi/180;
theta4 = 30 *pi/180;
theta5 = 40 *pi/180;

global phi2 phi3 phi4 phi5
phi2 = 15 *pi/180;
phi3 = 30 *pi/180;
phi4 = 45 *pi/180;
phi5 = 60 *pi/180;

%------------------------------------------------------------
```

FIGURE J.38 Section (for step 2) of file *RSSR_quasi_newton.m* with sample data in bold type.

```
%------------------------------------------------------------
% Here, a value for moving pivot component b1x is assigned.

b1x = 1;

%------------------------------------------------------------

%------------------------------------------------------------
% Here, a value for joint axis angle ALPHA is assigned.

ALPHA = 90 *pi/180;

%------------------------------------------------------------

%------------------------------------------------------------
% Here, initial values for the RSSR moving pivot variables are
assigned.

a1x = 1;
a1y = 1;
a1z = 1;

b1y = -1;
b1z = 1;

%------------------------------------------------------------
```

FIGURE J.39 Sections (for steps 3 and 4) of file *RSSR_quasi_newton.m* with sample data in bold type.

```
%%%%%%%%%%%%%%%%%%%%%%%%%%%%%%%%%%%%%%%%%%%%%%%%%%%%%%%%%%%%%%%%%%%%
Solving for the RSSR steering mechanism variables
%%%%%%%%%%%%%%%%%%%%%%%%%%%%%%%%%%%%%%%%%%%%%%%%%%%%%%%%%%%%%%%%%%%%
Iterations: 10001
Func Evals: 10002
Algorithm: quasi-newton
exit flag = 5
error = ( 1.183880e-007 )
a1x = ( 0.443664 )
a1y = ( -0.069476 )
a1z = ( -0.492154 )
b1y = ( -1.053214 )
b1z = ( 0.483358 )
Elapsed time is 77.336078 seconds.
Solution calculation is complete.
```

FIGURE J.40 Sample output from the file *RSSR_quasi_newton.m*.

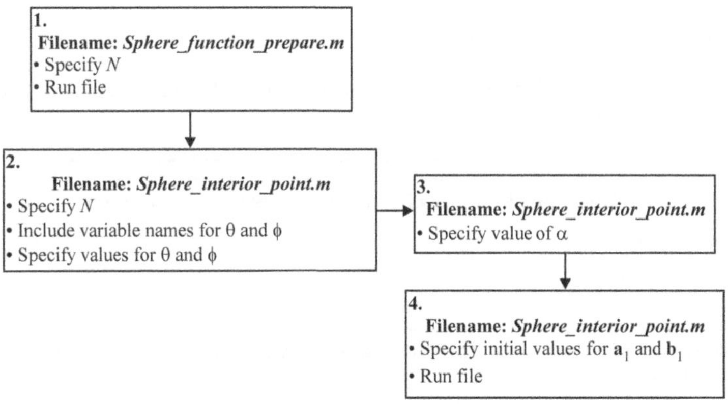

FIGURE J.41 Procedure for calculating $\mathbf{a}_1$ and $\mathbf{b}_1$.

When calculating new solutions, repeating step 1 in Figure J.37 is only necessary if the number of precision points (N) for the new solution is different from that in the previous solution. If N is the same for the new solution, the user can begin at step 2 in Figure J.37.

Once the RSSR steering linkage moving pivots $\mathbf{a}_1$ and $\mathbf{b}_1$ have been calculated, its fixed pivots $\mathbf{a}_0$ and $\mathbf{b}_0$ can be determined by $\mathbf{a}_0 = [0, 0, a_{1z}]$ and $\mathbf{b}_0 = [1, b_{1y}, 0]$.

J.10 4R SPHERICAL FUNCTION GENERATION, *N* PRECISION POINTS

The subfolder *Synthesis* within the folder *Appendix J.10* includes MATLAB files (filenames *Sphere_function_prepare.m* and *Sphere_interior_point.m*) for calculating the 4R Spherical mechanism moving pivots $\mathbf{a}_1$ and $\mathbf{b}_1$ for function generation given N precision points. The folder *Common* (which includes supplementary MATLAB files) must be included alongside the folder *Synthesis* to properly run the MATLAB files. Figure J.41 includes the four-step procedure for calculating $\mathbf{a}_1$ and $\mathbf{b}_1$.

Synthesizing the moving pivot solutions begins with the file *Sphere_function_prepare.m* (step 1 in Figure J.41). In this file, the number of precision points is specified. After specifying the value of N (the number of precision points) and running this file, the folder *Spherefuncs* is produced. The folder *Spherefuncs* includes functions required to run file *Sphere_interior_point.m*.

After exiting the file *Sphere_function_prepare.m* and opening the file *Sphere_interior_point.m*, step 2 in Figure J.41 is started. In this file, the user specifies the value of N—the number of precision points (identical to the N value prescribed in the file *Sphere_function_prepare.m*). The user

```
%---------------------------------------------------------------
% Here, the names for each thetaN and phiN displacement angle are
% included
% (e.g., theta2 theta3 theta4 theta5 theta6 ...)
% (e.g., phi2 phi3 phi4 phi5 phi6 ...).
% Also, values for displacement angles thetaN and phiN are
% assigned.Additional rows can be included to accommodate
% additional thetaN and phiN variables and values.

global theta2 theta3 theta4 theta5
theta2 = 45 *pi/180;
theta3 = 90 *pi/180;
theta4 = 135 *pi/180;
theta5 = 180 *pi/180;

global phi2 phi3 phi4 phi5
phi2 = 20 *pi/180;
phi3 = 40 *pi/180;
phi4 = 60 *pi/180;
phi5 = 80 *pi/180;

%---------------------------------------------------------------
```

FIGURE J.42 Sections (for steps 2) of *Sphere_interior_point.m* with sample data in bold type.

```
%---------------------------------------------------------------
% Here, a value for joint axis angle ALPHA is assigned.

ALPHA = 45;

%---------------------------------------------------------------

%---------------------------------------------------------------
% Here, initial values for the RSSR moving pivot components are
% assigned.

a1x = 0.7;
a1y = 0;
a1z = 0.7;

b1x = 0;
b1y = 0.7;
b1z = 1;

%---------------------------------------------------------------
```

FIGURE J.43 Sections (for steps 3 and 4) of *Sphere_interior_point.m* with sample data in bold type.

includes the names for each crank and follower displacement angle (angles θ and ϕ, respectively) to be prescribed. The user also includes the specific precision points values in degrees (defined by rotation angles θ and ϕ, as given in Matrix (8.76)). The number of *thetaN* and *phiN* variables and values should match the value of *N*. Figure J.42 illustrates this user input section in the file *Sphere_interior_point.m*, with sample data in bold type.

In step 3, the user specifies values for the 4R Spherical fixed pivot $\mathbf{b}_0$ as well as specify initial values for the 4R spherical moving pivots. The user specifies the value for the angle α (which corresponds to the location of $\mathbf{b}_0$) in degrees. The user also specifies initial values for $\mathbf{a}_1$ and $\mathbf{b}_1$ (step 4 in Figure J.41). Figure J.43 illustrates these user input sections in the file *Sphere_interior_point.m*, with sample data in bold type.

Lastly, the file *Sphere_interior_point.m* is run (also in step 4). The calculated 4R spherical mechanism moving pivots are displayed in the MATLAB command window after this file has been run. Figure J.44 illustrates the sample output from the file *Sphere_interior_point.m*.

```
%%%%%%%%%%%%%%%%%%%%%%%%%%%%%%%%%%%%%%%%%%%%%%%%%%%%%%%%%%%%%%%%%%%%
Solving for the 4R Spherical mechanism variables
%%%%%%%%%%%%%%%%%%%%%%%%%%%%%%%%%%%%%%%%%%%%%%%%%%%%%%%%%%%%%%%%%%%%
Iterations: 14
Func Evals: 51
Algorithm: interior-point
exit flag = 1
error = ( 1.240025e-004 )
a1x = ( 0.908306 )
a1y = ( -0.382413 )
a1z = ( 0.169528 )
b1x = ( -0.016554 )
b1y = ( -0.192797 )
b1z = ( 0.981096 )
Elapsed time is 7.379114 seconds.
Solution calculation is complete.
```

FIGURE J.44 Sample output from the file *Sphere_interior_point.m.*

When calculating new solutions, repeating step 1 in Figure J.41 is only necessary if the number of precision points (N) for the new solution is different from that in the previous solution. If N is the same for the new solution, the user can begin at step 2 in Figure J.41.

NOTE

1 All of the Appendix J MATLAB files can be downloaded from the publisher's web page or from the authors' server http://www.softalink.com/kruss/mech_des_vis_prog_app_2nd_ed/MATLAB_SIMMULT_FILES.zip.

Appendix K
User Instructions for Chapter 9 MATLAB® Dimensional Synthesis Files

K.1 FIXED-LENGTH, MOVING PIVOT-ADJUSTABLE PLANAR FOUR-BAR, MOTION GENERATION, *M* AND *N* PRECISION POSITIONS

The subfolder *Synthesis* within the folder *Appendix K.1* (which is available for download at https://www.routledge.com/9781032956800) includes MATLAB files (filenames *a0_a1_a1s_prepare.m* and *a0_a1_a1s_quasi_newton.m*) for calculating the fixed-length adjustable dyad variables a_{0y}, $\mathbf{a}_1$, and $\mathbf{a}_1^*$ for motion generation given *M* and *N* precision positions.[1] The folder *Common* (which includes supplementary MATLAB files) must be included alongside the folder *Synthesis* to properly run the MATLAB files. Figure K.1 includes the three-step procedure for calculating a_{0y}, $\mathbf{a}_1$, and $\mathbf{a}_1^*$.

Synthesizing the $\mathbf{a}_0 - \mathbf{a}_1$ and $\mathbf{a}_0 - \mathbf{a}_1^*$ dyads begins with the file *a0_a1_a1s_prepare.m* (step 1 in Figure K.1). In this file, the number of precision positions in two groups is specified. After specifying the values of *M* and *N* (the numbers of precision positions in groups 1 and 2, respectively) and running this file, the folder *a0a1a1sfuncs* is produced. The folder *a0a1a1sfuncs* includes functions required to run file *a0_a1_a1s_quasi_newton.m*.

After exiting the file *a0_a1_a1s_prepare.m* and opening the file *a0_a1_a1s_quasi_newton.m*, step 2 in Figure K.1 is started. In this file, the user specifies the values of *M* and *N*—the number of precision positions in groups 1 and 2, respectively (identical to the *M* and *N* values prescribed in the file *a0_a1_a1s_prepare.m*). The user specifies the value for the dyad parameter a_{0x}. The user also includes the name of each coupler displacement angle (angle α) to be prescribed. Additionally, the user includes the specific precision position values (defined by the coupler point **p** and the displacement angle α, as given in Matrix (6.16)). The number of *alphaM*, *alphaN*, *PM*, and *PN* variables and values should match the values of *M* and *N*. Figure K.2 illustrates these user input sections in the file *a0_a1_a1s_quasi_newton.m*, with sample data in bold type.

The user includes the initial values for the dyad components a_{0y}, a_{1x}, a_{1y}, a_{1x}^*, and a_{1y}^* (step 3 in Figure K.1). Figure K.3 illustrates this user input section in the file *a0_a1_a1s_quasi_newton.m*, with sample data in bold type.

Lastly, the file *a0_a1_a1s_quasi_newton.m* is run (also in step 3). The calculated $\mathbf{a}_0 - \mathbf{a}_1$ and $\mathbf{a}_0 - \mathbf{a}_1^*$ dyad components are displayed in the MATLAB command window after this file has

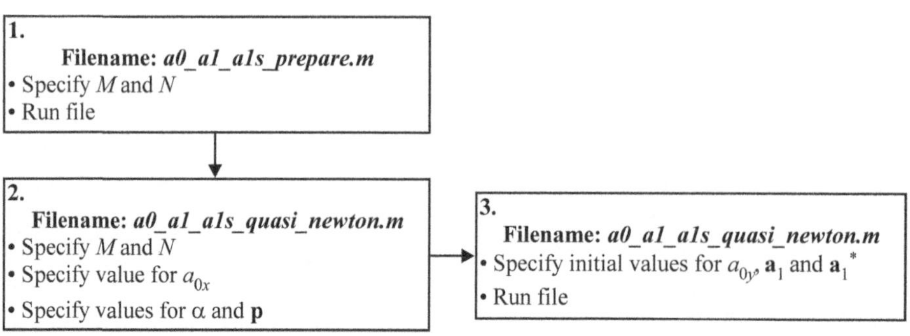

FIGURE K.1 Procedure for calculating a_{0y}, $\mathbf{a}_1$, and $\mathbf{a}_1^*$.

```
%-----------------------------------------------------------------
% Here, a value for the a0-a1 and a0-a1s dyad component a0x is
% assigned.

global a0x
a0x = 5;

%-----------------------------------------------------------------

%-----------------------------------------------------------------
% Here, values for precision position parameters alphaN and PN
% are assigned. Group 1 includes positions 1, 2...M and group 2
% includes positions 1, M+1...N. Additional rows can be included
% to accommodate additional alphaN and PN variables and values.

global alpha2 alpha3 alpha4 alpha5

alpha2 = -38.2965 *pi/180;
alpha3 = -74.4814 *pi/180;
alpha4 = -74.4814 *pi/180;
alpha5 = -98.0521 *pi/180;

P1 = [6.545,20.1547];
P2 = [12.4409, 17.7173];
P3 = [15.6261, 11.6238];
P4 = [18.4724, 15.483];
P5 = [17.1848, 8.5771];

%-----------------------------------------------------------------
```

FIGURE K.2 Sections (for step 2) of file *a0_a1_a1s_quasi_newton.m,* with sample data in bold type.

```
%-----------------------------------------------------------------
% Here, initial values for the a0-a1 and a0-a1s dyad components
% a0y, a1x, a1y, a1sx and a1sy are assigned.

a0y = -10;
a1x = 10;
a1y = 10;
a1sx = 10;
a1sy = 10;

%-----------------------------------------------------------------
```

FIGURE K.3 Section (for step 3) of file *a0_a1_a1s_quasi_newton.m,* with sample data in bold type.

been run. Figure K.4 illustrates the sample output from the file *a0_a1_a1s_quasi_newton.m.* To calculate solutions for the adjustable dyads $\mathbf{b}_0 - \mathbf{b}_1$ and $\mathbf{b}_0 - \mathbf{b}_1^*$, the user specifies a different value for a_{0x}, the components of $\mathbf{a}_0 - \mathbf{a}_1$ and $\mathbf{a}_0 - \mathbf{a}_1^*$, and reruns the file *a0_a1_a1s_quasi_newton.m.* The newly calculated solution will correspond to the $\mathbf{b}_0 - \mathbf{b}_1$ and $\mathbf{b}_0 - \mathbf{b}_1^*$ dyads.

When calculating new solutions, repeating step 1 in Figure K.1 is only necessary if the numbers of precision positions (*M* and *N*) for the new solution are different from those in the previous solution. If *M* and *N* are the same for the new solution, the user can begin at step 2 in Figure K.1.

K.2 FIXED-LENGTH, FIXED PIVOT-ADJUSTABLE PLANAR FOUR-BAR, MOTION GENERATION, *M* AND *N* PRECISION POSITIONS

The subfolder *Synthesis* within the folder *Appendix K.2* includes MATLAB files (filenames *a0_a1_a0s_prepare.m* and *a0_a1_a0s_quasi_newton.m*) for calculating the fixed-length adjustable dyad variables a_{0y}, $\mathbf{a}_1$, and $\mathbf{a}_0^*$ for motion generation given *M* and *N* precision positions.

```
%%%%%%%%%%%%%%%%%%%%%%%%%%%%%%%%%%%%%%%%%%%%%%%%%%%%%%%%%%%%%%%%%%%%%%%%%%
Solve for the components of a0-a1 and a0-a1s
%%%%%%%%%%%%%%%%%%%%%%%%%%%%%%%%%%%%%%%%%%%%%%%%%%%%%%%%%%%%%%%%%%%%%%%%%%
Iterations: 57
Func Evals: 58
Algorithm: quasi-newton
exit flag = 5
error = ( 1.668371e-019 )
a0y = ( 4.237433 )
a1x = ( 6.314356 )
a1y = ( 8.595783 )
a1sx = ( 9.398057 )
a1sy = ( 5.412097 )
Elapsed time is 2.248175 seconds.
Solution calculation is complete.
```

FIGURE K.4 Sample output from the file *a0_a1_a1s_quasi_newton.m.*

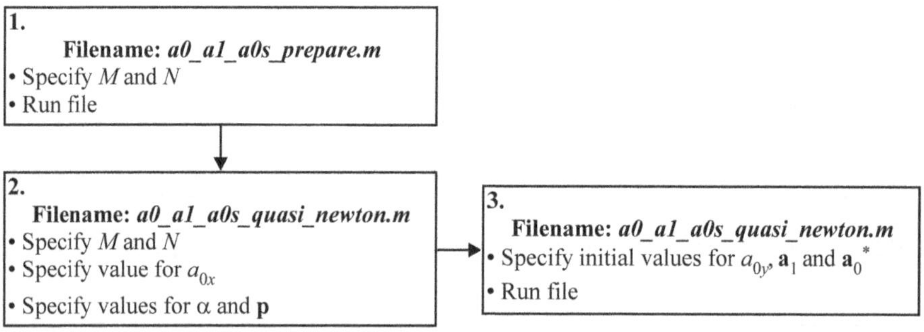

FIGURE K.5 Procedure for calculating a_{0y}, $\mathbf{a}_1$ and $\overset{*}{\mathbf{a}}_0$.

The folder *Common* (which includes supplementary MATLAB files) must be included alongside the folder *Synthesis* to properly run the MATLAB® files. Figure K.5 includes the three-step procedure for calculating a_{0y}, $\mathbf{a}_1$, and $\overset{*}{\mathbf{a}}_0$.

Synthesizing the $\mathbf{a}_0 - \mathbf{a}_1$ and $\overset{*}{\mathbf{a}}_0 - \mathbf{a}_1$ dyads begin with the file *a0_a1_a0s_prepare.m* (step 1 in Figure K.5). In this file, the numbers of precision positions in two groups are specified. After specifying the values of M and N (the numbers of precision positions in groups 1 and 2, respectively) and running this file, the folder *a0a1a0sfuncs* is produced. The folder *a0a1a0sfuncs* includes functions required to run file *a0_a1_a0s_quasi_newton.m.*

After exiting the file *a0_a1_a0s_prepare.m* and opening the file *a0_a1_a0s_quasi_newton.m*, step 2 in Figure K.5 is started. In this file, the user specifies the values of M and N—the number of precision positions in groups 1 and 2, respectively (identical to the M and N values prescribed in the file *a0_a1_a0s_prepare.m*). The user specifies the value for the dyad parameter a_{0x}. The user also includes the name of each coupler displacement angle (angle α) to be prescribed. Additionally, the user includes the specific precision position values (defined by the coupler point **p** and the displacement angle α, as given in Matrix (6.16)). The number of *alphaM, alphaN, PM,* and *PN* variables and values should match the values of M and N. Figure K.6 illustrates these user input sections in the file *a0_a1_a0s_quasi_newton.m*, with sample data in bold type.

The user includes the initial values for the dyad components a_{0y}, a_{1x}, a_{1y}, $\overset{*}{a}_{0x}$, and $\overset{*}{a}_{0y}$ (step 3 in Figure K.5). Figure K.7 illustrates this user input section in the file *a0_a1_a0s_quasi_newton.m*, with sample data in bold type.

Lastly, the file *a0_a1_a0s_quasi_newton.m* is run (also in step 3). The calculated $\mathbf{a}_0 - \mathbf{a}_1$ and $\overset{*}{\mathbf{a}}_0 - \mathbf{a}_1$ dyad components are displayed in the MATLAB command window after this file has been run. Figure K.8 illustrates the sample output from the file *a0_a1_a0s_quasi_newton.m.*

```
%------------------------------------------------------------
% Here, a value for the a0-a1 and a0-a1s dyad component a0x is
% assigned.

global a0x
a0x = 5;

%------------------------------------------------------------

%------------------------------------------------------------
% Here, values for precision position parameters alphaN and PN
% are assigned. Group 1 includes positions 1, 2...M and group 2
% includes positions 1, M+1...N. Additional rows can be included
% to accommodate additional alphaN and PN variables and values.

global alpha2 alpha3 alpha4 alpha5

alpha2 = -38.2965 *pi/180;
alpha3 = -74.4814 *pi/180;
alpha4 = -74.4814 *pi/180;
alpha5 = -98.0521 *pi/180;

P1 = [6.545,20.1547];
P2 = [12.4409, 17.7173];
P3 = [15.6261, 11.6238];
P4 = [18.4724, 15.483];
P5 = [17.1848, 8.5771];

%------------------------------------------------------------
```

FIGURE K.6 Sections (for step 2) of file *a0_a1_a0s_quasi_newton.m,* with sample data in bold type.

```
%------------------------------------------------------------
% Here, initial values for the a0-a1 and a0-a1s dyad components
% a0y, a1x, a1y, a1sx and a1sy are assigned.

a0y = -10;
a1x = -10;
a1y = -10;
a0sx = -10;
a0sy = -10;

%------------------------------------------------------------
```

FIGURE K.7 Section (for step 3) of file *a0_a1_a0s_quasi_newton.m,* with sample data in bold type.

```
%%%%%%%%%%%%%%%%%%%%%%%%%%%%%%%%%%%%%%%%%%%%%%%%%%%%%%%%%%%%%%%%%%%%%%
Solve for the components of a0-a1 and a0s-a1
%%%%%%%%%%%%%%%%%%%%%%%%%%%%%%%%%%%%%%%%%%%%%%%%%%%%%%%%%%%%%%%%%%%%%%
Iterations: 30
Func Evals: 32
Algorithm: quasi-newton
exit flag = 5
error = ( 1.969209e-018 )
a0y = ( 9.768974 )
a1x = ( 1.336122 )
a1y = ( 4.119853 )
a0sx = ( -1.476759 )
a0sy = ( 10.237391 )
Elapsed time is 13.751323 seconds.
Solution calculation is complete.
```

FIGURE K.8 Sample output from the file *a0_a1_a0s_quasi_newton.m.*

To calculate solutions for the adjustable dyads $\mathbf{b}_0 - \mathbf{b}_1$ and $\mathbf{b}_0^* - \mathbf{b}_1$, the user specifies a different value for a_{0x}, the components of $\mathbf{a}_0 - \mathbf{a}_1$ and $\mathbf{a}_0^* - \mathbf{a}_1$, and reruns the file *a0_a1_a0s_quasi_newton.m*. The newly calculated solution will correspond to the $\mathbf{b}_0 - \mathbf{b}_1$ and $\mathbf{b}_0^* - \mathbf{b}_1$ dyads.

When calculating new solutions, repeating step 1 in Figure K.5 is only necessary if the numbers of precision positions (*M* and *N*) for the new solution are different from those in the previous solution. If *M* and *N* are the same for the new solution, the user can begin at step 2 in Figure K.5.

K.3 ADJUSTABLE-LENGTH, MOVING PIVOT-ADJUSTABLE PLANAR FOUR-BAR, MOTION GENERATION, *M* AND *N* PRECISION POSITIONS

The subfolder *Synthesis* within the folder *Appendix K.3* includes MATLAB files (filenames *a0_a1_a1s_prepare.m* and *a0_a1_a1s_quasi_newton.m*) for calculating the adjustable-length adjustable dyad variables a_{0y}, $\mathbf{a}_1$, and $\mathbf{a}_1^*$ for motion generation given *M* and *N* precision positions. The folder *Common* (which includes supplementary MATLAB files) must be included alongside the folder *Synthesis* to properly run the MATLAB files. The procedure described in Appendix K.1 is identical to the procedure for the Appendix K.3 MATLAB files.

K.4 ADJUSTABLE-LENGTH, FIXED PIVOT-ADJUSTABLE PLANAR FOUR-BAR, MOTION GENERATION, *M* AND *N* PRECISION POSITIONS

The subfolder *Synthesis* within the folder *Appendix K.4* includes MATLAB files (filenames *a0_a1_a0s_prepare.m* and *a0_a1_a0s_quasi_newton.m*) for calculating the fixed-length adjustable dyad variables a_{0y}, $\mathbf{a}_1$, and $\mathbf{a}_0^*$ for motion generation given *M* and *N* precision positions. The folder *Common* (which includes supplementary MATLAB files) must be included alongside the folder *Synthesis* to properly run the MATLAB files. The procedure described in Appendix K.2 is identical to the procedure for the Appendix K.4 MATLAB files.

K.5 FIXED-LENGTH, MOVING PIVOT-ADJUSTABLE PLANAR FOUR-BAR, MOTION GENERATION, *M* AND *N* PRECISION POSITIONS WITH ORDER AND BRANCH CONSTRAINTS

The subfolder *Synthesis* within the folder *Appendix K.5* includes MATLAB files (filenames *motion_prepare.m* and *motion_interior_point.m*) for calculating the constant-length adjustable mechanism variables $\mathbf{a}_0$, $\mathbf{a}_1$, $\mathbf{a}_1^*$, $\mathbf{b}_0$, $\mathbf{b}_1$, $\mathbf{b}_1^*$, α, and θ for motion generation given *M* and *N* precision positions. The folder *Common* (which includes supplementary MATLAB files) must be included alongside the folder *Synthesis* to properly run the MATLAB files. Figure K.9 includes the four-step procedure for calculating $\mathbf{a}_0$, $\mathbf{a}_1$, $\mathbf{a}_1^*$, $\mathbf{b}_0$, $\mathbf{b}_1$, $\mathbf{b}_1^*$, α, and θ.

Synthesizing the adjustable mechanism begins with the file *motion_prepare.m* (step 1 in Figure K.9). In this file, the numbers of precision positions in two groups are specified. After specifying the values of *M* and *N* (the numbers of precision positions in groups 1 and 2, respectively) and running this file, the folder *motionfuncs* is produced. The folder *motionfuncs* includes functions required to run file *motion_interior_point.m.*

After exiting the file *motion_prepare.m* and opening the file *motion_interior_point.m*, step 2 in Figure K.9 is started. In this file, the user specifies the values of *M* and *N*—the number of precision positions in groups 1 and 2, respectively (identical to the *M* and *N* values prescribed in the file *motion_prepare.m*). The user includes specific precision position values (defined by coupler points **p** and **q**, as given in Equation (9.7)). The number of *PM*, *QM*, *PN*, and *QN* variables and values should match the values of *M* and *N*. The user also specifies initial values for $\mathbf{a}_0$, $\mathbf{a}_1$, $\mathbf{a}_1^*$, $\mathbf{b}_0$, $\mathbf{b}_1$, $\mathbf{b}_1^*$, α_j, α_k, θ_j, and θ_k (in degrees). The number of *alphaM*, *alphaN*, *thetaM,* and *thetaN* variables and

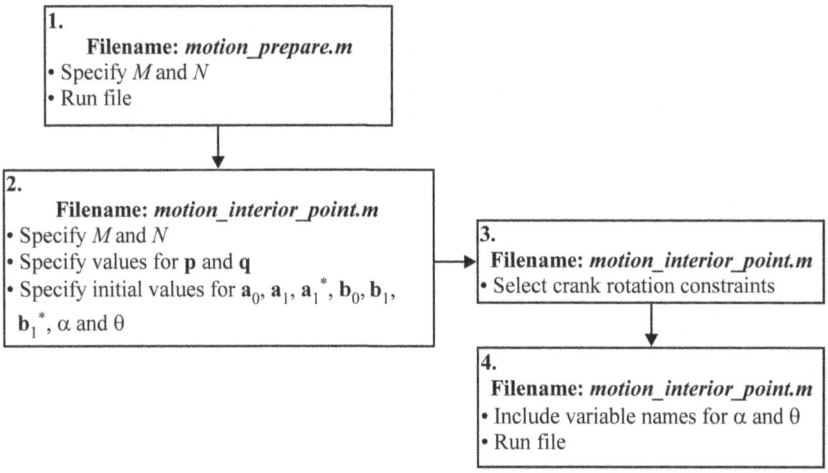

FIGURE K.9 Procedure for calculating $\mathbf{a}_0$, $\mathbf{a}_1$, $\mathbf{a}_1^*$, $\mathbf{b}_0$, $\mathbf{b}_1$, $\mathbf{b}_1^*$, α, and θ.

```
%-----------------------------------------------------------------
% Here, values for precision position parameters PN are assigned.
% Group 1 includes positions 1, 2...M and group 2 includes
% positions 1, M+1...N. Additional rows can be included to
% accommodate additional PN variables and values.

P1 = [0, 0];
P2 = [-5.2712, 0.2125];
P3 = [-8.9671, -4.2108];
P4 = [-4, 0.3];
P5 = [-6, -5];

%-----------------------------------------------------------------

%-----------------------------------------------------------------
% Here, values for precision position parameters QN are assigned.
% Group 1 includes positions 1, 2...M and group 2 includes
% positions 1, M+1...N. Additional rows can be included to
% accommodate additional PN variables and values.

Q1 = [-0.3849, -0.3189];
Q2 = [-5.2308, -0.7107];
Q3 = [-8.5594, -4.5001];
Q4 = [-3.8, -1];
Q5 = [-7.6841, -8.0391];

%-----------------------------------------------------------------
```

FIGURE K.10 Sections (for step 2) of file *motion_interior_point.m,* with sample data in bold type.

values should match the values of M and N. Figures K.10 and K.11 illustrate these user input sections in the file *motion_interior_point.m*, with sample data in bold type.

In step 3, the user specifies either clockwise or counterclockwise crank rotation constraints for the first and second group of precision positions for order defect elimination. Counterclockwise constraints are specified for both groups in the file *motion_interior_point.m*. To select clockwise constraints, the characters *%{* must be included as instructed in the file (to deactivate the counterclockwise constraints). Also, the characters *%{* must be removed as instructed in the file (to activate the clockwise constraints). The user has three options for selecting order constraints. These options are **clockwise (groups M and N)**, **counterclockwise (groups M and N)**, and **clockwise (group M)** and **counterclockwise (group N)**.

```
%----------------------------------------------------------------
% Here, initial values for the components of a0, a1, a1*, b0, b1
% and b1* are assigned. Also, initial values for displacement
% angles alphaM, alphaN, thetaM and thetaN are assigned.
% Additional rows can be included to accommodate additional
% alphaM, alphaN, thetaM and thetaN variables and values.

a0x = 0;
a0y = 0;

a1x = 0;
a1y = -10;

a1sx = 0;
a1sy = -9;

b0x = -1;
b0y = 0;

b1x = -10;
b1y = -10;

b1sx = -9;
b1sy = -9;

alpha2 = 20 *pi/180;
alpha3 = 40 *pi/180;
alpha4 = 60 *pi/180;
alpha5 = 80 *pi/180;

theta2 = 60 *pi/180;
theta3 = 120 *pi/180;
theta4 = 180 *pi/180;
theta5 = 240 *pi/180;

%----------------------------------------------------------------
```

FIGURE K.11 Section (for step 2) of file *motion_interior_point.m,* with sample data in bold type.

```
%----------------------------------------------------------------
% Here, the names for each alphaM, alphaN, thetaM and thetaN
% displacement angle are included
% (e.g., alpha2; alpha3; alpha4; alpha5; alpha6; .....;)
% (e.g., theta2; theta3; theta4; theta5; theta6; .....];).

x0 = [a0x; a0y; a1x; a1y; a1sx; a1sy;
      b0x; b0y; b1x; b1y; b1sx; b1sy;
      alpha2; alpha3; alpha4; alpha5;
      theta2; theta3; theta4; theta5];

%----------------------------------------------------------------
```

FIGURE K.12 Section (for step 4) of file *motion_interior_point.m,* with sample data in bold type.

In step 4 in Figure K.9, the user includes the names for each dyad displacement angle (angles α and β) to be calculated. Figure K.12 illustrates this user input section in the file *motion_interior_point.m,* with sample data in bold type.

Lastly, the file *motion_interior_point.m* is run (also in step 4). The calculated components of $\mathbf{a}_0$, $\mathbf{a}_1$, $\mathbf{a}_1^*$, $\mathbf{b}_0$, $\mathbf{b}_1$, and $\mathbf{b}_1^*$and dyad displacement angles α_j, α_k, θ_j, and θ_k are displayed in the MATLAB command window after this file has been run. Figure K.13 illustrates the sample output from the file *motion_interior_point.m.*

When calculating new solutions, repeating step 1 in Figure K.9 is only necessary if the numbers of precision positions (M and N) for the new solution are different from those in the previous solution. If M and N are the same for the new solution, the user can begin at step 2 in Figure K.9.

```
%%%%%%%%%%%%%%%%%%%%%%%%%%%%%%%%%%%%%%%%%%%%%%%%%%%%%%%%%%%%%%%%%%%%%%%%
Solve for a0, a1, a1*, b0, b1, b1* and displacement angles
%%%%%%%%%%%%%%%%%%%%%%%%%%%%%%%%%%%%%%%%%%%%%%%%%%%%%%%%%%%%%%%%%%%%%%%%
Iterations: 350
Func Evals: 417
Algorithm: interior-point
exit flag = 2
error = ( 4.690362e+000 )
a0x = ( -5.563627 )
a0y = ( -3.343083 )
a1x = ( -1.506736 )
a1y = ( -13.299624 )
a1sx = ( 3.643816 )
a1sy = ( -8.894124 )
b0x = ( 0.644369 )
b0y = ( -0.562492 )
b1x = ( -8.315927 )
b1y = ( -12.207285 )
b1sx = ( -8.664166 )
b1sy = ( -11.930837 )
alpha2 = ( 17.128270 )
alpha3 = ( 38.915816 )
alpha4 = ( 53.465270 )
alpha5 = ( 17.033596 )
theta2 = ( 22.183526 )
theta3 = ( 36.446008 )
theta4 = ( 152.777717 )
theta5 = ( 219.433792 )
Elapsed time is 21.094465 seconds.
Solution calculation is complete.
```

FIGURE K.13 Sample output from the file *motion_interior_point.m.*

K.6 FIXED-LENGTH, MOVING PIVOT-ADJUSTABLE PLANAR FOUR-BAR, PATH GENERATION, *M* AND *N* PRECISION POINTS WITH ORDER AND BRANCH CONSTRAINTS

The subfolder *Synthesis* within the folder *Appendix K.6* includes MATLAB files (filenames *path_prepare.m* and *path_interior_point.m*) for calculating the constant-length adjustable mechanism variables $\mathbf{a}_0$, $\mathbf{a}_1$, $\mathbf{a}_1^*$, $\mathbf{b}_0$, $\mathbf{b}_1$, $\mathbf{b}_1^*$, α, and θ for path generation given *M* and *N* precision points. The folder *Common* (which includes supplementary MATLAB files) must be included alongside the folder *Synthesis* to properly run the MATLAB files. Figure K.14 includes the four-step procedure for calculating $\mathbf{a}_0$, $\mathbf{a}_1$, $\mathbf{a}_1^*$, $\mathbf{b}_0$, $\mathbf{b}_1$, $\mathbf{b}_1^*$, α, and θ.

Synthesizing the adjustable mechanism begins with the file *path_prepare.m* (step 1 in Figure K.14). In this file, the numbers of precision points in two groups are specified. After specifying the values of *M* and *N* (the numbers of precision points in groups 1 and 2, respectively) and running this file, the folder *pathfuncs* is produced. The folder *pathfuncs* includes functions required to run file *path_interior_point.m.*

After exiting the file *path_prepare.m* and opening the file *path_interior_point.m*, step 2 in Figure K.14 is started. In this file, the user specifies the values of *M* and *N*—the number of precision points in groups 1 and 2, respectively (identical to the *M* and *N* values prescribed in the file *path_prepare.m*). The user includes specific precision point values (defined by coupler point **p**, as described in Section 9.3). The number of *PM* and *PN* variables and values should match the values of *M* and *N*. The user also specifies initial values for $\mathbf{a}_0$, $\mathbf{a}_1$, $\mathbf{a}_1^*$, $\mathbf{b}_0$, $\mathbf{b}_1$, $\mathbf{b}_1^*$, α_j, α_k, θ_j, and θ_k (in degrees). The number of *alphaM, alphaN, thetaM,* and *thetaN* variables and values should match the values of *M* and *N*. Figures K.15 and K.16 illustrate the user input sections in the file *path_interior_point.m*, with sample data in bold type.

In step 3, the user specifies either clockwise or counterclockwise crank rotation constraints for the first and second group of precision positions for order defect elimination. Counterclockwise constraints are specified for both groups in the file *path_interior_point.m.* To select clockwise

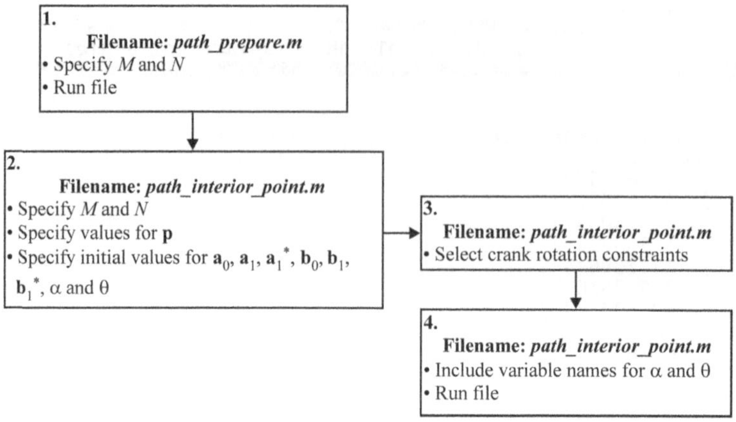

FIGURE K.14 Procedure for calculating $\mathbf{a}_0$, $\mathbf{a}_1$, $\mathbf{a}_1^*$, $\mathbf{b}_0$, $\mathbf{b}_1$, $\mathbf{b}_1^*$, α, and θ.

```
%-------------------------------------------------------------
% Here, values for precision point parameters PN are assigned.
% Group 1 includes points 1, 2...M and group 2 includes
% points 1, M+1...N. Additional rows can be included to
% accommodate additional PN variables and values.

P1 = [0, 0];
P2 = [-5.2712, 0.2125];
P3 = [-8.9671, -4.2108];
P4 = [-4, 0.3];
P5 = [-6, -5];

%-------------------------------------------------------------
```

FIGURE K.15 Section (for step 2) of file *path_interior_point.m,* with sample data in bold type.

```
%-------------------------------------------------------------
% Here, initial values for the components of a0, a1, a1*, b0, b1
% and b1* are assigned. Also, initial values for displacement
% angles alphaM, alphaN, thetaM and thetaN are assigned.
% Additional rows can be included to accommodate additional
% alphaM, alphaN, thetaM and thetaN variables and values.
a0x = 0;
a0y = 0;

a1x = 0;
a1y = -10;

a1sx = 0;
a1sy = -9;

b0x = -1;
b0y = 0;

b1x = -10;
b1y = -10;

b1sx = -9;
b1sy = -9;

alpha2 = 20 *pi/180;
alpha3 = 40 *pi/180;
alpha4 = 60 *pi/180;
alpha5 = 80 *pi/180;

theta2 = 60 *pi/180;
theta3 = 120 *pi/180;
theta4 = 180 *pi/180;
theta5 = 240 *pi/180;

%-------------------------------------------------------------
```

FIGURE K.16 Section (for step 2) of file *path_interior_point.m,* with sample data in bold type.

constraints, the characters *%{* must be included as instructed in the file (to deactivate the counter-clockwise constraints). Also, the characters *%{* must be removed as instructed in the file (to activate the clockwise constraints). The user has three options for selecting order constraints. These options are **clockwise (groups M and N)**, **counterclockwise (groups M and N)**, and **clockwise (group M)** and **counterclockwise (group N)**.

In step 4 in Figure K.14, the user includes the names for each dyad displacement angle (angles α and β) to be calculated. Figure K.17 illustrates this user input section in the file *path_interior_point.m,* with sample data in bold type.

Lastly, the file *path_interior_point.m* is run (also in step 4). The calculated components of $\mathbf{a}_0$, $\mathbf{a}_1$, $\mathbf{a}_1^*$, $\mathbf{b}_0$, $\mathbf{b}_1$, and $\mathbf{b}_1^*$ and dyad displacement angles α_j, α_k, θ_j, and θ_k are displayed in the MATLAB command window after this file has been run. Figure K.18 illustrates sample output from the file *path_interior_point.m.*

When calculating new solutions, repeating step 1 in Figure K.14 is only necessary if the numbers of precision points (*M* and *N*) for the new solution are different from those of the previous solution. If *M* and *N* are the same for the new solution, the user can begin at step 2 in Figure K.14.

```
%-------------------------------------------------------------
% Here, the names for each alphaM, alphaN, thetaM and thetaN
% displacement angle are included
% (e.g., alpha2; alpha3; alpha4; alpha5; alpha6; .....;)
% (e.g., theta2; theta3; theta4; theta5; theta6; .....];).
x0 = [a0x; a0y; a1x; a1y; a1sx; a1sy;
      b0x; b0y; b1x; b1y; b1sx; b1sy;
      alpha2; alpha3; alpha4; alpha5;
      theta2; theta3; theta4; theta5];

%-------------------------------------------------------------
```

FIGURE K.17 Section (for step 4) of file *path_interior_point.m,* with sample data in bold type.

```
%%%%%%%%%%%%%%%%%%%%%%%%%%%%%%%%%%%%%%%%%%%%%%%%%%%%%%%%%%%%%%%%%
Solve for a0, a1, a1*, b0, b1, b1* and displacement angles
%%%%%%%%%%%%%%%%%%%%%%%%%%%%%%%%%%%%%%%%%%%%%%%%%%%%%%%%%%%%%%%%%
Iterations: 40
Func Evals: 131
Algorithm: interior-point
exit flag = 2
error = ( 1.045316e-003 )
a0x = ( -4.597050 )
a0y = ( -6.284218 )
a1x = ( 1.402572 )
a1y = ( -15.460825 )
a1sx = ( 6.255864 )
a1sy = ( -7.839801 )
b0x = ( 0.099775 )
b0y = ( 1.115267 )
b1x = ( -5.863961 )
b1y = ( -11.382291 )
b1sx = ( -8.757532 )
b1sy = ( -9.529131 )
alpha2 = ( 7.157521 )
alpha3 = ( 20.946111 )
alpha4 = ( 79.759896 )
alpha5 = ( 52.328923 )
theta2 = ( 30.095691 )
theta3 = ( 54.654258 )
theta4 = ( 157.212257 )
theta5 = ( 202.265124 )
Elapsed time is 5.809867 seconds.
Solution calculation is complete.
```

FIGURE K.18 Sample output from the file *path_interior_point.m.*

K.7 MOVING PIVOT-ADJUSTABLE PLANAR FOUR-BAR, FUNCTION GENERATION, *M* AND *N* PRECISION POINTS

The subfolder *Synthesis* within the folder *Appendix K.7* includes MATLAB files (filenames *func_prepare.m* and *func_quasi_newton.m*) for calculating the adjustable mechanism variables $\mathbf{a}_1$, $\mathbf{b}_1$, and $\mathbf{b}_1^*$ for function generation given *M* and *N* precision points. The folder *Common* (which includes supplementary MATLAB files) must be included alongside the folder *Synthesis* to properly run the MATLAB files. Figure K.19 includes the three-step procedure for calculating $\mathbf{a}_1$, $\mathbf{b}_1$, and $\mathbf{b}_1^*$.

Synthesizing the adjustable mechanism solution begins with the file *func_prepare.m* (step 1 in Figure K.19). In this file, the numbers of precision points in two groups are specified. After specifying the values of *M* and *N* (the numbers of precision positions in groups 1 and 2, respectively) and running this file, the folder *funcfuncs* is produced. The folder *funcfuncs* includes functions required to run file *func_quasi_newton.m*.

After exiting the file *func_prepare.m* and opening the file *func_quasi_newton.m*, step 2 in Figure K.19 is started. In this file, the user specifies the values of *M* and *N*—the number of precision points in groups 1 and 2, respectively (identical to the *M* and *N* values prescribed in the file *func_prepare.m*). The user includes the name of each crank and follower displacement angle (angles θ and ϕ, respectively) to be prescribed. In addition, the user includes specific precision point values in degrees (defined by displacement angles θ and ϕ as given in Matrix (9.17)). The number of *thetaM, thetaN, phiM,* and *phiN* variables and values should match the values of *M* and *N*. Figure K.20 illustrates these user input sections in the file *func_quasi_newton.m*, with sample data in bold type.

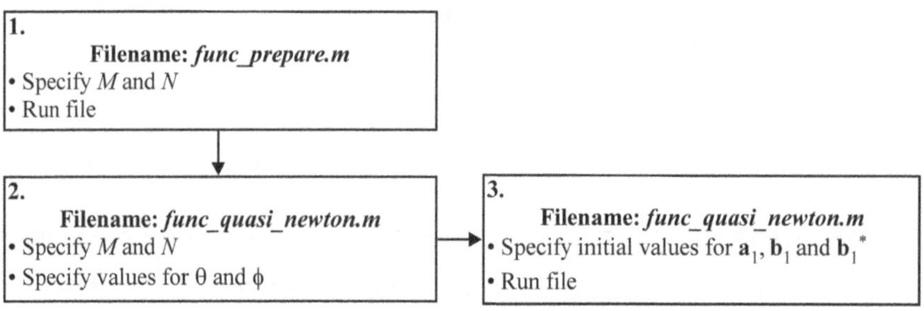

FIGURE K.19 Procedure for calculating $\mathbf{a}_1$, $\mathbf{b}_1$, and $\mathbf{b}_1^*$.

```
%------------------------------------------------------------
% Here, values for precision point parameters thetaM, thetaN,
% phiM and phiN are assigned. Group 1 includes positions 1, 2...M
% and group 2 includes positions 1, M+1...N. Additional rows can
% be included to accommodate additional thetaM, thetaN, phiM and
% phiN variables and values.
global theta2 theta3 theta4 theta5

theta2 = 90 *pi/180;
theta3 = 180 *pi/180;
theta4 = 90 *pi/180;
theta5 = 180 *pi/180;

global phi2 phi3 phi4 phi5

phi2 = 50 *pi/180;
phi3 = 100 *pi/180;
phi4 = 60 *pi/180;
phi5 = 120 *pi/180;

%------------------------------------------------------------
```

FIGURE K.20 Section (for step 2) of file *func_quasi_newton.m*, with sample data in bold type.

```
%---------------------------------------------------------------
% Here, initial values for the components of a1, b1 and b1* are
% assigned.

a1x = 1;
a1y = 1;

b1x = 1;
b1y = 1;

b1sx = 1;
b1sy = 1;

%---------------------------------------------------------------
```

FIGURE K.21 Section (for step 3) of file *func_quasi_newton.m*, with sample data in bold type.

```
%%%%%%%%%%%%%%%%%%%%%%%%%%%%%%%%%%%%%%%%%%%%%%%%%%%%%%%%%%%%%%%%%%%%%%%%%%%%%%%
Solve for a1, b1 and b1*
%%%%%%%%%%%%%%%%%%%%%%%%%%%%%%%%%%%%%%%%%%%%%%%%%%%%%%%%%%%%%%%%%%%%%%%%%%%%%%%
Iterations: 42
Func Evals: 44
Algorithm: quasi-newton
exit flag = 5
error = ( 1.468250e-020 )
a1 = [0.788503,0.667438]
b1 = [0.998167,1.303329]
b1s = [1.141919,1.236133]
Elapsed time is 2.108731 seconds.
Solution calculation is complete.
```

FIGURE K.22 Sample output from the file *func_quasi_newton.m*.

The user includes the initial values for the components of $\mathbf{a}_1$, $\mathbf{b}_1$, and $\mathbf{b}_1^*$ (step 3 in Figure K.19). Figure K.21 illustrates this user input section in the file *func_quasi_newton.m*, with sample data in bold type.

Lastly, the file *func_quasi_newton.m* is run (also in step 3). The calculated components of $\mathbf{a}_1$, $\mathbf{b}_1$, and $\mathbf{b}_1^*$ are displayed in the MATLAB command window after this file has been run. Figure K.22 illustrates the sample output from the file *func_quasi_newton.m*.

When calculating new solutions, repeating step 1 in Figure K.19 is only necessary if the numbers of precision points (M and N) for the new solution are different from those of the previous solution. If M and N are the same for the new solution, the user can begin at step 2 in Figure K.19.

K.8 FIXED-LENGTH, MOVING PIVOT-ADJUSTABLE 4R SPHERICAL, MOTION GENERATION, *M* AND *N* PRECISION POSITIONS

The subfolder *Synthesis* within the folder *Appendix K.8* includes MATLAB files (filenames *a0_a1_a1s_prepare.m* and *a0_a1_a1s_interior_point.m*) for calculating the constant-length adjustable spherical R-R dyad variables a_{0y}, a_{0z}, $\mathbf{a}_1$, and $\mathbf{a}_1^*$ for motion generation given M and N precision positions. The folder *Common* (which includes supplementary MATLAB files) must be included alongside the folder *Synthesis* to properly run the MATLAB files. Figure K.23 includes the three-step procedure for calculating a_{0y}, a_{0z}, $\mathbf{a}_1$, and $\mathbf{a}_1^*$.

Synthesizing the spherical $\mathbf{a}_0 - \mathbf{a}_1$ and $\mathbf{a}_0 - \mathbf{a}_1^*$ dyads begin with the file *a0_a1_a1s_prepare.m* (step 1 in Figure K.23). In this file, the numbers of precision positions in two groups are specified. After specifying the values of M and N (the numbers of precision positions in groups 1 and 2, respectively) and running this file, the folder *a0a1a1sfuncs* is produced. The folder *a0a1a1sfuncs* includes functions required to run file *a0_a1_a1s_interior_point.m*.

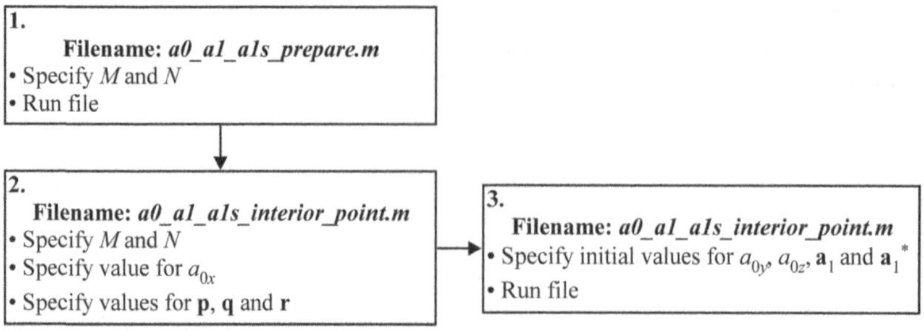

FIGURE K.23 Procedure for calculating a_{0y}, a_{0z}, $\mathbf{a}_1$, and $\mathbf{a}_0^*$.

```
%-----------------------------------------------------------------
% Here, values for precision position parameters PN are assigned.
% Group 1 includes positions 1, 2...M and group 2 includes
% positions 1, M+1...N. Additional rows can be included to
% accommodate additional PN variables % and values.
P1 = [1.4489, 0, 0.3882];
P2 = [1.299, 0.7386, 0.1302];
P3 = [0.75, 1.2941, 0.1132];
P4 = [1.3, 0.5, 0.2];
P5 = [1, 1, 0.15];
%-----------------------------------------------------------------
                              .
                              .
%-----------------------------------------------------------------
% Here, values for precision position parameters RN are assigned.
% Group 1 includes positions 1, 2...M and group 2 includes
% positions 1, M+1...N. Additional rows can be included to
% accommodate additional RN variables and values.
R1 = [1.4489, 1, 0.3882];
R2 = [0.866, 1.3904, 0.7529];
R3 = [0.317, 1.4677, 0.9977];
R4 = [1.5, 1, 0.5];
R5 = [1.6, 1.2, 0.7];
%-----------------------------------------------------------------
%-----------------------------------------------------------------
% Here, a value for the a0-a1 and a0-a1s dyad component a0x is
% assigned.
global a0x
a0x = 0.5;
%-----------------------------------------------------------------
```

FIGURE K.24 Sections (for step 2) of file *a0_a1_a1s_interior_point.m,* with sample data in bold type.

After exiting the file *a0_a1_a1s_prepare.m* and opening the file *a0_a1_a1s_interior_point.m*, step 2 in Figure K.23 is started. In this file, the user specifies the values of *M* and *N*—the number of precision positions in groups 1 and 2, respectively (identical to the *M* and *N* values prescribed in the file *a0_a1_a1s_prepare.m*). The user also includes specific precision position values (defined by coupler points **p**, **q**, and **r**, as given in Matrix (8.57)). The number of *PM, QM, RM, PN, QN,* and *RN* variables and values should match the values of *M* and *N*. Additionally, the user specifies the value for the dyad parameter a_{0x}. Figure F.24 illustrates these user input sections in the file *a0_a1_a1s_interior_point.m*, with sample data in bold type.

The user includes the initial values for the dyad components a_{0y}, a_{0z}, and the components of $\mathbf{a}_1$ and $\mathbf{a}_1^*$ (step 3 in Figure K.23). Figure K.25 illustrates this user input section in the file *a0_a1_a1s_interior_point.m*, with sample data in bold type.

```
%-------------------------------------------------------------------
% Here, initial values for the a0-a1 and a0-a1s dyad components
% a0y, a0z and the components of a1 and a1s are assigned.

a0y = 1;
a0z = 1;
a1x = 1;
a1y = 1;
a1z = 1;
a1sx = 1;
a1sy = 1;
a1sz = 1;

%-------------------------------------------------------------------
```

FIGURE K.25 Section (for step 3) of file *a0_a1_a1s_interior_point.m,* with sample data in bold type.

```
%%%%%%%%%%%%%%%%%%%%%%%%%%%%%%%%%%%%%%%%%%%%%%%%%%%%%%%%%%%%%%%%%%%%%%%
Solve for the components of a0-a1 and a0-a1s
%%%%%%%%%%%%%%%%%%%%%%%%%%%%%%%%%%%%%%%%%%%%%%%%%%%%%%%%%%%%%%%%%%%%%%%
Iterations: 71
Func Evals: 289
Algorithm: interior-point
exit flag = -2
error = ( 4.777356e-004 )
a0y = ( 0.636414 )
a0z = ( 0.587348 )
a1x = ( 0.713833 )
a1y = ( 0.664604 )
a1z = ( 0.220780 )
a1sx = ( 0.741892 )
a1sy = ( 0.640376 )
a1sz = ( 0.198782 )
Elapsed time is 9.860398 seconds.
Solution calculation is complete.
```

FIGURE K.26 Sample output from the file *a0_a1_a1s_interior_point.m.*

Lastly, the file *a0_a1_a1s_interior_point.m* is run (also in step 3). The calculated $\mathbf{a}_0 - \mathbf{a}_1$ and $\mathbf{a}_0 - \mathbf{a}_1^*$ dyad components are displayed in the MATLAB command window after this file has been run. Figure K.26 illustrates the sample output from the file *a0_a1_a1s_interior_point.m.* To calculate solutions for the adjustable spherical dyads $\mathbf{b}_0 - \mathbf{b}_1$ and $\mathbf{b}_0 - \mathbf{b}_1^*$, the user specifies a different value for a_{0x}, the components a_{0y}, a_{0z} and the components of $\mathbf{a}_1$, $\mathbf{a}_1^*$, and reruns the file *a0_a1_a1s_interior_point.m.* The newly calculated solution will correspond to the spherical $\mathbf{b}_0 - \mathbf{b}_1$ and $\mathbf{b}_0 - \mathbf{b}_1^*$ dyads.

When calculating new solutions, repeating step 1 in Figure K.23 is only necessary if the numbers of precision positions (M and N) for the new solution are different from those of the previous solution. If M and N are the same for the new solution, the user can begin at step 2 in Figure K.23.

K.9 FIXED-LENGTH, FIXED PIVOT-ADJUSTABLE 4R SPHERICAL, MOTION GENERATION, *M* AND *N* PRECISION POSITIONS

The subfolder *Synthesis* within the folder *Appendix K.9* includes MATLAB files (filenames *a0_a1_a0s_prepare.m* and *a0_a1_a0s_interior_point.m*) for calculating the constant-length adjustable spherical R-R dyad variables a_{0y}, a_{0z}, $\mathbf{a}_1$, and $\mathbf{a}_0^*$ for motion generation given M and N precision positions. The folder *Common* (which includes supplementary MATLAB files) must be included alongside the folder *Synthesis* to properly run the MATLAB files. Figure K.27 includes the three-step procedure for calculating a_{0y}, a_{0z}, $\mathbf{a}_1$, and $\mathbf{a}_0^*$.

Synthesizing the spherical $\mathbf{a}_0 - \mathbf{a}_1$ and $\mathbf{a}_0^* - \mathbf{a}_1$ dyads begin with the file *a0_a1_a0s_prepare.m* (step 1 in Figure K.27). In this file, the numbers of precision positions in two groups are specified.

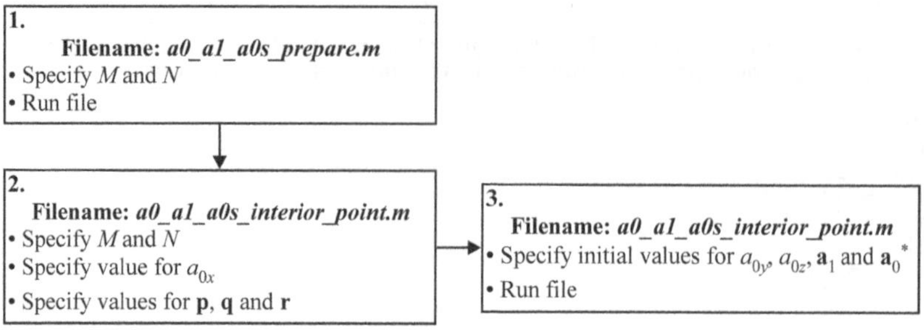

FIGURE K.27 Procedure for calculating a_{0y}, a_{0z}, $\mathbf{a}_1$, and $\mathbf{a}_0^*$.

```
%---------------------------------------------------------------
% Here, values for precision position parameters PN are assigned.
% Group 1 includes positions 1, 2...M and group 2 includes
% positions 1, M+1...N. Additional rows can be included to
% accommodate additional PN variables % and values.

P1 = [1.4489, 0, 0.3882];
P2 = [1.299, 0.7386, 0.1302];
P3 = [0.75, 1.2941, 0.1132];
P4 = [1.3, 0.5, 0.2];
P5 = [1.0, 1.0, 0.15];

%---------------------------------------------------------------
                              .
                              .
%---------------------------------------------------------------
% Here, values for precision position parameters RN are assigned.
% Group 1 includes positions 1, 2...M and group 2 includes
% positions 1, M+1...N. Additional rows can be included to
% accommodate additional RN variables and values.

R1 = [1.4489, 1, 0.3882];
R2 = [0.866, 1.3904, 0.7529];
R3 = [0.317, 1.4677, 0.9977];
R4 = [1.5, 1, 0.5];
R5 = [1.6, 1.2, 0.7];

%---------------------------------------------------------------

%---------------------------------------------------------------
% Here, a value for the a0-a1 and a0-a1s dyad component a0x is
% assigned.

global a0x
a0x = 0.5;

%---------------------------------------------------------------
```

FIGURE K.28 Sections (for step 2) of file *a0_a1_a0s_interior_point.m,* with sample data in bold type.

After specifying the values of M and N (the numbers of precision positions in groups 1 and 2, respectively) and running this file, the folder *a0a1a0sfuncs* is produced. The folder *a0a1a0sfuncs* includes functions required to run file *a0_a1_a0s_interior_point.m.*

After exiting the file *a0_a1_a0s_prepare.m* and opening the file *a0_a1_a0s_interior_point.m,* step 2 in Figure K.27 is started. In this file, the user specifies the values of M and N—the number of precision positions in groups 1 and 2, respectively (identical to the M and N values prescribed in the file *a0_a1_a0s_prepare.m*). The user also includes specific precision position values (defined by coupler points **p**, **q**, and **r**, as given in Matrix (8.57)). The number of PM, QM, RM, PN, QN, and RN variables and values should match the values of M and N. Additionally, the user specifies the value for the dyad parameter a_{0x}. Figure K.28 illustrates these user input sections in the file *a0_a1_a0s_interior_point.m*, with sample data in bold type.

```
%--------------------------------------------------------------------
% Here, initial values for the a0-a1 and a0s-a1 dyad components
% a0y, a0z and the components of a1 and a0s are assigned.

a0y = 1;
a0z = 1;
a1x = 1;
a1y = 1;
a1z = 1;
a0sx = 1;
a0sy = 1;
a0sz = 1;

%--------------------------------------------------------------------
```

FIGURE K.29 Section (for step 3) of file *a0_a1_a0s_interior_point.m,* with sample data in bold type.

```
%%%%%%%%%%%%%%%%%%%%%%%%%%%%%%%%%%%%%%%%%%%%%%%%%%%%%%%%%%%%%%%%%%%%%%%
Solve for the components of a0-a1 and a0s-a1
%%%%%%%%%%%%%%%%%%%%%%%%%%%%%%%%%%%%%%%%%%%%%%%%%%%%%%%%%%%%%%%%%%%%%%%
Iterations: 114
Func Evals: 302
Algorithm: interior-point
exit flag = -2
error = ( 9.606309e-003 )
a0y = ( 0.852872 )
a0z = ( -0.150366 )
a1x = ( 0.849025 )
a1y = ( 0.476872 )
a1z = ( 0.227485 )
a0sx = ( 0.430188 )
a0sy = ( 0.578941 )
a0sz = ( 0.692651 )
Elapsed time is 13.091067 seconds.
Solution calculation is complete.
```

FIGURE K.30 Sample output from the file *a0_a1_a0s_interior_point.m.*

The user includes the initial values for the dyad components a_{0y}, a_{0z}, and the components of $\mathbf{a}_1$ and $\mathbf{a}_0^*$ (step 3 in Figure K.27). Figure K.29 illustrates this user input section in the file *a0_a1_a0s_interior_point.m*, with sample data in bold type.

Lastly, the file *a0_a1_a0s_interior_point.m* is run (also in step 3). The calculated $\mathbf{a}_0 - \mathbf{a}_1$ and $\mathbf{a}_0^* - \mathbf{a}_1$ dyad components are displayed in the MATLAB command window after this file has been run. Figure K.30 illustrates the sample output from the file *a0_a1_a0s_interior_point.m.* To calculate solutions for the adjustable spherical dyads $\mathbf{b}_0 - \mathbf{b}_1$ and $\mathbf{b}_0^* - \mathbf{b}_1$, the user specifies different values for a_{0x}, the components a_{0y} and a_{0z}, and the components of $\mathbf{a}_1$, $\mathbf{a}_0^*$ and reruns the file *a0_a1_a0s_interior_point.m.* The newly calculated solution will correspond to the spherical $\mathbf{b}_0 - \mathbf{b}_1$ and $\mathbf{b}_0^* - \mathbf{b}_1$ dyads.

When calculating new solutions, repeating step 1 in Figure K.27 is only necessary if the numbers of precision positions (*M* and *N*) for the new solution are different from those of the previous solution. If *M* and *N* are the same for the new solution, the user can begin at step 2 in Figure K.27.

K.10 ADJUSTABLE-LENGTH, MOVING PIVOT-ADJUSTABLE 4R SPHERICAL, MOTION GENERATION, *M* AND *N* PRECISION POSITIONS

The subfolder *Synthesis* within the folder *Appendix K.10* includes MATLAB files (filenames *a0_a1_a1s_prepare.m* and *a0_a1_a1s_ interior_point.m*) for calculating the varying-length adjustable spherical R-R dyad variables a_{0y}, a_{0z}, $\mathbf{a}_1$, and $\mathbf{a}_1^*$ for motion generation given *M* and *N* precision positions. The folder *Common* (which includes supplementary MATLAB files) must be included alongside the folder *Synthesis* to properly run the MATLAB files. The procedure described in Appendix K.8 is identical to the procedure for the Appendix K.10 MATLAB files.

K.11 ADJUSTABLE-LENGTH, FIXED PIVOT-ADJUSTABLE 4R SPHERICAL, MOTION GENERATION, *M* AND *N* PRECISION POSITIONS

The subfolder *Synthesis* within the folder *Appendix K.11* includes MATLAB files (filenames *a0_a1_a0s_prepare.m* and *a0_a1_a0s_ interior_point.m*) for calculating the varying-length adjustable spherical R-R dyad variables a_{0y}, a_{0z}, $\mathbf{a}_1$, and $\mathbf{a}_0^*$ for motion generation given *M* and *N* precision positions. The folder *Common* (which includes supplementary MATLAB files) must be included alongside the folder *Synthesis* to properly run the MATLAB files. The procedure described in Appendix K.9 is identical to the procedure for the Appendix K.11 MATLAB files.

K.12 FIXED-LENGTH, MOVING PIVOT-ADJUSTABLE 4R SPHERICAL, MOTION GENERATION, *M* AND *N* PRECISION POSITIONS WITH ORDER AND BRANCH CONSTRAINTS

The subfolder *Synthesis* within the folder *Appendix K.12* includes MATLAB files (filenames *motion_prepare.m* and *motion_interior_point.m*) for calculating the constant-length adjustable mechanism variables $\mathbf{a}_0$, $\mathbf{a}_1$, $\mathbf{a}_1^*$, $\mathbf{b}_0$, $\mathbf{b}_1$, $\mathbf{b}_1^*$, α, and θ for motion generation given *M* and *N* precision positions. The folder *Common* (which includes supplementary MATLAB files) must be included alongside the folder *Synthesis* to properly run the MATLAB files. Figure K.31 includes the four-step procedure for calculating variables $\mathbf{a}_0$, $\mathbf{a}_1$, $\mathbf{a}_1^*$, $\mathbf{b}_0$, $\mathbf{b}_1$, $\mathbf{b}_1^*$, α, and θ.

Synthesizing the adjustable mechanism begins with the file *motion_prepare.m* (step 1 in Figure K.31). In this file, the numbers of precision positions in two groups are specified. After specifying the values of *M* and *N* (the numbers of precision positions in groups 1 and 2, respectively) and running this file, the folder *motionfuncs* is produced. The folder *motionfuncs* includes functions required to run file *motion_interior_point.m.*

After exiting the file *motion_prepare.m* and opening the file *motion_interior_point.m*, step 2 in Figure K.31 is started. In this file, the user specifies the values of *M* and *N*—the number of precision

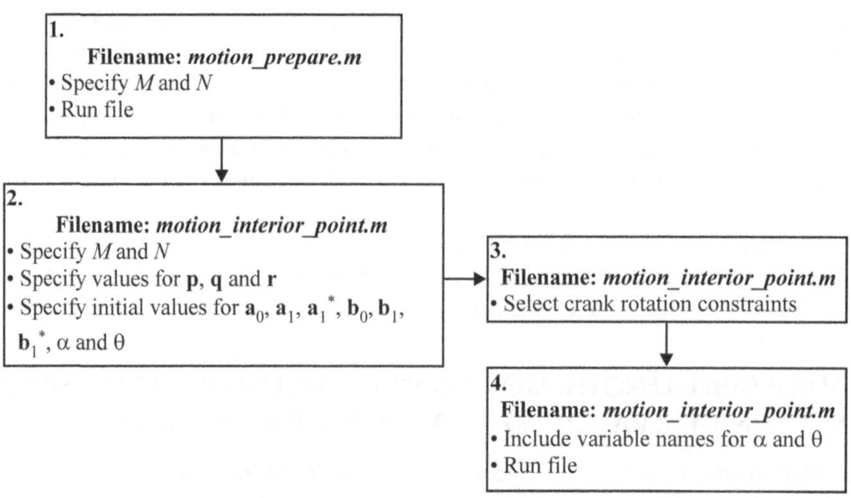

FIGURE K.31 Procedure for calculating variables $\mathbf{a}_0$, $\mathbf{a}_1$, $\mathbf{a}_1^*$, $\mathbf{b}_0$, $\mathbf{b}_1$, $\mathbf{b}_1^*$, α, and θ.

```
%------------------------------------------------------------
% Here, values for precision position parameters PN are assigned.
% Group 1 includes positions 1, 2...M and group 2 includes
% positions 1, M+1...N. Additional rows can be included to
% accommodate additional PM and PN variables and values.
P1 = [1.5, 0, 0];
P2 = [1.299, 0.75, 0];
P3 = [0.75, 1.299, 0];
P4 = [1.299, 0.6797, -0.317];
P5 = [0.75, 1.1773, -0.549];
P6 = [0, 1.3595, -0.6339];

%------------------------------------------------------------
                              .
                              .
                              .
%------------------------------------------------------------
% Here, values for precision position parameters RN are assigned.
% Group 1 includes positions 1, 2...M and group 2 includes
% positions 1, M+1...N. Additional rows can be included to
% accommodate additional RM and RN variables and values.

R1 = [1.5, 1, 0];
R2 = [0.8660, 1.5, 0.5];
R3 = [0.3170, 1.549, 0.866];
R4 = [0.6566, 1.4231, -0.1307];
R5 = [0.1462, 1.3827, 0.2213];
R6 = [0, 1.3595, 0.3661];

%------------------------------------------------------------
```

FIGURE K.32 Sections (for step 2) of file *motion_interior_point.m,* with sample data in bold type.

positions in groups 1 and 2, respectively (identical to the M and N values prescribed in the file *motion_prepare.m*). The user includes specific precision position values (defined by coupler points $\mathbf{p}$, $\mathbf{q}$, and $\mathbf{r}$, as described in Equation (9.7)). The number of *PM, QM, RM, PN, QN,* and *RN* variables and values should match the values of M and N. The user also specifies initial values for $\mathbf{a}_0$, $\mathbf{a}_1$, $\mathbf{a}_1^*$, $\mathbf{b}_0$, $\mathbf{b}_1$, $\mathbf{b}_1^*$, α_j, α_k, θ_j, and θ_k (in degrees). The number of *alphaM, alphaN, thetaM,* and *thetaN* variables and values should match the values of M and N. Figures K.32 and K.33 illustrate these user input sections in the file *motion_interior_point.m*, with sample data in bold type.

In step 3, the user specifies either clockwise or counterclockwise crank rotation constraints for the first and second group of precision positions for order defect elimination. Counterclockwise constraints are specified for both groups in the file *motion_interior_point.m.* To select clockwise constraints, the characters *%{* must be included as instructed in the file (to deactivate the counterclockwise constraints). Also, the characters *%{* must be removed as instructed in the file (to activate the clockwise constraints). The user has three options for selecting order constraints. These options are **clockwise (groups M and N)**, **counterclockwise (groups M and N)**, and **clockwise (group M)** and **counterclockwise (group N)**.

In step 4 in Figure K.31, the user includes the names for each dyad displacement angle (angles α and β) to be calculated. Figure K.34 illustrates this user input section in the file *motion_interior_point.m,* with sample data in bold type.

Lastly, the file *motion_interior_point.m* is run (also in step 4). The calculated components of $\mathbf{a}_0$, $\mathbf{a}_1$, $\mathbf{a}_1^*$, $\mathbf{b}_0$, $\mathbf{b}_1$, and $\mathbf{b}_1^*$ and dyad displacement angles α_j, α_k, θ_j, and θ_k are displayed in the MATLAB command window after this file has been run. Figure K.35 illustrates the sample output from the file *motion_interior_point.m.*

When calculating new solutions, repeating step 1 in Figure K.31 is only necessary if the numbers of precision positions (M and N) for the new solution are different from those of the previous solution. If M and N are the same for the new solution, the user can begin at step 2 in Figure K.31.

```
%---------------------------------------------------------------
% Here, initial values for the 4R Spherical variables are
% assigned. Additional rows can be included to accommodate
% additional alphaM, alphaN, thetaM and thetaN variables and
% values.

a0x = 1;
a0y = 1;
a0z = 1;

a1x = 1;
a1y = 1;
a1z = 1;

a1sx = 1;
a1sy = 1;
a1sz = 1;

b0x = 1;
b0y = 1;
b0z = 1;

b1x = 1;
b1y = 1;
b1z = 1;

b1sx = 1;
b1sy = 1;
b1sz = 1;

alpha2 = 20 *pi/180;
alpha3 = 40 *pi/180;
alpha4 = 20 *pi/180;
alpha5 = 40 *pi/180;
alpha6 = 60 *pi/180;

theta2 = 20 *pi/180;
theta3 = 40 *pi/180;
theta4 = 20 *pi/180;
theta5 = 40 *pi/180;
theta6 = 60 *pi/180;

%---------------------------------------------------------------
```

FIGURE K.33 Section (for step 2) of file *motion_interior_point.m,* with sample data in bold type.

```
%---------------------------------------------------------------
% Here, the names for each alphaM, alphaN, thetaM and thetaN
% displacement angle are included
% (e.g., alpha2; alpha3; alpha4; alpha5; alpha6; alpha7; .....;)
% (e.g., theta2; theta3; theta4; theta5; theta6; theta7;
.....];).

x0 = [a0x; a0y; a0z; a1x; a1y; a1z; a1sx; a1sy; a1sz;
      b0x; b0y; b0z; b1x; b1y; b1z; b1sx; b1sy; b1sz;
      alpha2; alpha3; alpha4; alpha5; alpha6;
      theta2; theta3; theta4; theta5; theta6];

%---------------------------------------------------------------
```

FIGURE K.34 Section (for step 4) of file *motion_interior_point.m,* with sample data in bold type.

```
%%%%%%%%%%%%%%%%%%%%%%%%%%%%%%%%%%%%%%%%%%%%%%%%%%%%%%%%%%%%%%%%%%%%%%
Solve for a0, a1, a1*, b0, b1, b1* and displacement angles
%%%%%%%%%%%%%%%%%%%%%%%%%%%%%%%%%%%%%%%%%%%%%%%%%%%%%%%%%%%%%%%%%%%%%%
Iterations: 57
Func Evals: 199
Algorithm: interior-point
exit flag = 2
error = ( 2.629132e-001 )
a0 = [0.385138,0.487050,0.783869]
a1 = [-0.481312,0.873661,0.071106]
a1s = [0.960993,0.182312,-0.207978]
b0 = [0.674625,-0.589894,0.443759]
b1 = [-0.663551,-0.535422,-0.522517]
b1s = [-0.404301,-0.446391,-0.798298]
alpha2 = ( -19.833558 )
alpha3 = ( -24.805259 )
alpha4 = ( 0.214224 )
alpha5 = ( 1.870566 )
alpha6 = ( 35.892424 )
theta2 = ( 45.476208 )
theta3 = ( 87.529388 )
theta4 = ( 37.480790 )
theta5 = ( 69.299028 )
theta6 = ( 95.392270 )
Elapsed time is 309.772208 seconds.
Solution calculation is complete.
```

FIGURE K.35 Sample output from the file *motion_interior_point.m.*

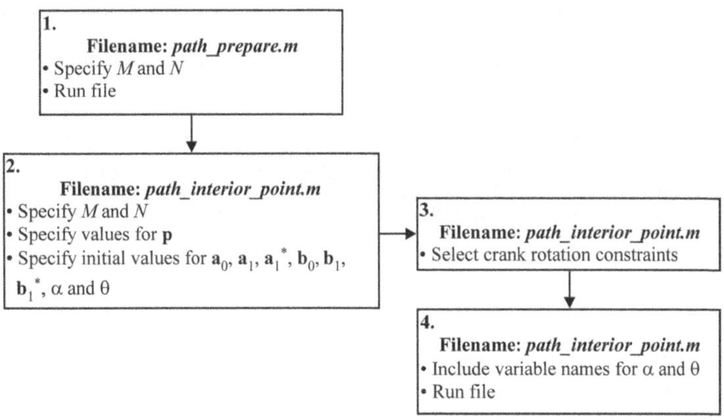

FIGURE K.36 Procedure for calculating $\mathbf{a}_0$, $\mathbf{a}_1$, $\mathbf{a}_1^*$, $\mathbf{b}_0$, $\mathbf{b}_1$, $\mathbf{b}_1^*$, α, and θ.

K.13 FIXED-LENGTH, MOVING PIVOT-ADJUSTABLE 4R SPHERICAL, PATH GENERATION, *M* AND *N* PRECISION POINTS WITH ORDER AND BRANCH CONSTRAINTS

The subfolder *Synthesis* within the folder *Appendix K.13* includes MATLAB files (filenames *path_prepare.m* and *path_interior_point.m*) for calculating the constant-length adjustable mechanism variables $\mathbf{a}_0$, $\mathbf{a}_1$, $\mathbf{a}_1^*$, $\mathbf{b}_0$, $\mathbf{b}_1$, $\mathbf{b}_1^*$, α, and θ for motion generation given *M* and *N* precision points. The folder *Common* (which includes supplementary MATLAB files) must be included alongside the folder *Synthesis* to properly run the MATLAB files. Figure K.36 includes the four-step procedure for calculating $\mathbf{a}_0$, $\mathbf{a}_1$, $\mathbf{a}_1^*$, $\mathbf{b}_0$, $\mathbf{b}_1$, $\mathbf{b}_1^*$, α, and θ.

Synthesizing the adjustable mechanism begins with the file *path_prepare.m* (step 1 in Figure K.36). In this file, the numbers of precision points in two groups are specified. After specifying the values of *M* and *N* (the numbers of precision points in groups 1 and 2, respectively) and running this file, the folder *pathfuncs* is produced. The folder *pathfuncs* includes functions required to run file *path_interior_point.m.*

After exiting the file *path_prepare.m* and opening the file *path_interior_point.m*, step 2 in Figure K.36 is started. In this file, the user specifies the values of M and N—the number of precision points in groups 1 and 2, respectively (identical to the M and N values prescribed in the file *path_prepare.m*). The user includes specific precision point values (defined by coupler point **p**, as described in Section 9.6). The number of *PM* and *PN* variables and values should match the values of M and N. The user also specifies initial values for $\mathbf{a}_0$, $\mathbf{a}_1$, $\mathbf{a}_1^*$, $\mathbf{b}_0$, $\mathbf{b}_1$, $\mathbf{b}_1^*$, α_j, α_k, θ_j, and θ_k (in degrees). The number of *alphaM, alphaN, thetaM,* and *thetaN* variables and values should match the values of M and N. Figures K.37 and K.38 illustrate these user input sections in the file *path_interior_point.m*, with sample data in bold type.

In step 3, the user specifies either clockwise or counterclockwise crank rotation constraints for the first and second group of precision positions for order defect elimination. Counterclockwise

```
%-----------------------------------------------------------------
% Here, values for precision point parameters PN are assigned.
% Group 1 includes points 1, 2...M and group 2 includes points 1,
% M+1...N. Additional rows can be included to accommodate
% additional PM and PN variables and values.

P1 = [1.5, 0, 0];
P2 = [1.299, 0.75, 0];
P3 = [0.75, 1.299, 0];
P4 = [1.299, 0.6797, -0.317];
P5 = [0.75, 1.1773, -0.549];

%-----------------------------------------------------------------
```

FIGURE K.37 Sections (for step 2) of file *path_interior_point.m,* with sample data in bold type.

```
%-----------------------------------------------------------------
% Here, initial values for the 4R Spherical variables are
% assigned. Additional rows can be included to accommodate
% additional alphaM, alphaN, thetaM and thetaN variables and
% values.

a0x = 1;
a0y = 1;
a0z = 1;

a1x = 1;
a1y = 1;
a1z = 1;

a1sx = 1;
a1sy = 1;
a1sz = 1;

b0x = 1;
b0y = 1;
b0z = 1;

b1x = 1;
b1y = 1;
b1z = 1;

b1sx = 1;
b1sy = 1;
b1sz = 1;

alpha2 = 20 *pi/180;
alpha3 = 20 *pi/180;
alpha4 = 20 *pi/180;
alpha5 = 20 *pi/180;

theta2 = 20 *pi/180;
theta3 = 20 *pi/180;
theta4 = 20 *pi/180;
theta5 = 20 *pi/180;

%-----------------------------------------------------------------
```

FIGURE K.38 Section (for step 2) of file *path_interior_point.m,* with sample data in bold type.

```
%------------------------------------------------------------
% Here, the names for each alphaM, alphaN, thetaM and thetaN
% displacement angle are included
% (e.g., alpha2; alpha3; alpha4; alpha5; alpha6; .....;)
% (e.g., theta2; theta3; theta4; theta5; theta6;  .....];).

x0 = [a0x; a0y; a0z; a1x; a1y; a1z; a1sx; a1sy; a1sz;
      b0x; b0y; b0z; b1x; b1y; b1z; b1sx; b1sy; b1sz;
      alpha2; alpha3; alpha4; alpha5;
      theta2; theta3; theta4; theta5];

%------------------------------------------------------------
```

FIGURE K.39 Section (for step 4) of file *path_interior_point.m,* with sample data in bold type.

```
%%%%%%%%%%%%%%%%%%%%%%%%%%%%%%%%%%%%%%%%%%%%%%%%%%%%%%%%%%%%%%%%%%%%%%%%%%%%
Solve for a0, a1, a1*, b0, b1, b1* and displacement angles
%%%%%%%%%%%%%%%%%%%%%%%%%%%%%%%%%%%%%%%%%%%%%%%%%%%%%%%%%%%%%%%%%%%%%%%%%%%%
Iterations: 34
Func Evals: 95
Algorithm: interior-point
exit flag = 2
error = ( 7.916637e-004 )
a0 = [-0.147365,-0.458442,0.876435]
a1 = [-0.771439,-0.477653,0.420411]
a1s = [-0.782066,-0.446142,0.435137]
b0 = [0.871137,0.135059,0.472134]
b1 = [0.036601,-0.185666,-0.981936]
b1s = [-0.533400,0.817500,-0.217216]
alpha2 = ( -35.079205 )
alpha3 = ( -51.449610 )
alpha4 = ( -59.858509 )
alpha5 = ( -93.112369 )
theta2 = ( 47.799308 )
theta3 = ( 87.290931 )
theta4 = ( 52.544630 )
theta5 = ( 94.011275 )
Elapsed time is 108.252144 seconds.
Solution calculation is complete.
```

FIGURE K.40 Sample output from the file *path_interior_point.m.*

constraints are specified for both groups in the file *path_interior_point.m.* To select clockwise constraints, the characters *%{* must be included as instructed in the file (to deactivate the counterclockwise constraints). Also, the characters *%{* must be removed as instructed in the file (to activate the clockwise constraints). The user has three options for selecting order constraints. These options are **clockwise (groups M and N)**, **counterclockwise (groups M and N)**, and **clockwise (group M)** and **counterclockwise (group N)**.

In step 4 in Figure K.36, the user includes the names for each dyad displacement angle (angles α and β) to be calculated. Figure K.39 illustrates this user input section in the file *path_interior_point.m,* with sample data in bold type.

Lastly, the file *path_interior_point.m* is run (also in step 4). The calculated components of $\mathbf{a}_0$, $\mathbf{a}_1$, $\mathbf{a}_1^*$, $\mathbf{b}_0$, $\mathbf{b}_1$ and $\mathbf{b}_1^*$ and dyad displacement angles α_j, α_k, θ_j, and θ_k are displayed in the MATLAB command window after this file has been run. Figure K.40 illustrates the sample output from the file *path_interior_point.m.*

When calculating new solutions, repeating step 1 in Figure K.36 is only necessary if the numbers of precision points (*M* and *N*) for the new solution are different from those of the previous solution. If *M* and *N* are the same for the new solution, the user can begin at step 2 in Figure K.36.

NOTE

1 All of the Appendix K MATLAB files can be downloaded from the publisher's web page or from the authors' server http://www.softalink.com/kruss/mech_des_vis_prog_app_2nd_ed/MATLAB_SIMMULT_FILES.zip.

Appendix L
User Instructions for Chapter 10 MATLAB® Kinematic Analysis Files

L.1 R-P-P ROBOTIC MANIPULATOR FORWARD KINEMATICS

The Appendix L.1 folder (which is available for download at https://www.routledge.com/9781032956800) includes two MATLAB files for R-P-P robotic manipulator forward kinematics.[1] These two files are described in Table L.1. To conduct an R-P-P forward kinematics analysis, the user specifies the R-P-P joint displacements in the file *RPP_Input.csv*. This file is compatible with Microsoft® Excel. Also, ${}^{3}\{\mathbf{p}_3\}$ is specified in the file *RPP_FK.m*. Figure L.1 illustrates the user input section of this file, with sample values in bold type.

After specifying the R-P-P joint displacements in *RPP_ Input.csv* and ${}^{3}\{\mathbf{p}_3\}$ in *RPP_FK.m*, the next step is to run *RPP_FK.m*. When running this file, 1 file (filename *RPP_p3.csv*) is written in a folder named *Results* (in a format compatible with Microsoft® Excel) that includes the ${}^{0}\{\mathbf{p}_3\}$ values calculated for the given ${}^{3}\{\mathbf{p}_3\}$ value and joint displacement values.

L.2 R-R-P ROBOTIC MANIPULATOR FORWARD KINEMATICS

The Appendix L.2 folder includes two MATLAB files for R-R-P robotic manipulator forward kinematics. These two files are described in Table L.2. To conduct an R-R-P forward kinematics analysis, the user specifies the R-R-P joint displacements in the file *RRP_Input.csv*. This file is compatible with Microsoft® Excel. Also, ${}^{3}\{\mathbf{p}_3\}$, l_1 and l_2 are specified in the file *RRP_FK.m*. Figure L.2 illustrates the user input section of this file, with sample values in bold type.

After specifying the R-R-P joint displacements in *RRP_Input.csv* and ${}^{3}\{\mathbf{p}_3\}$, l_1, and l_2 in *RRP_FK.m*, the next step is to run *RRP_FK.m*. When running this file, one file (filename *RRP_p3.csv*) is written in a folder named *Results* (in a format compatible with Microsoft® Excel) that includes the ${}^{0}\{\mathbf{p}_3\}$ values calculated for the given ${}^{3}\{\mathbf{p}_3\}$, length, and joint displacement values.

TABLE L.1
Section L.1 MATLAB Files

Filename	Use of File
RPP_Input.csv	To specify joint displacements
RPP_FK.m	To specify ${}^{3}\{\mathbf{p}_3\}$ and write the output (in a file compatible with Microsoft Excel)

```
%-----------------------------------------------------------
% Here, the coordinates for the end-effector point p3_3 are
% assigned.

p3_3 = [0, 0, -1];
%-----------------------------------------------------------
```

FIGURE L.1 Section of *RPP_FK.m*, with sample values in bold type.

TABLE L.2
Section L.2 MATLAB Files

Filename	Use of File
RRP_Input.csv	To specify joint displacements
RRP_FK.m	To specify ${}^{3}\{\mathbf{p}_3\}$, l_1, and l_2 and write the output (in a file compatible with Microsoft Excel)

```
%-------------------------------------------------------------
% Here, the coordinates for the end-effector point p3_3 and the
% lengths l1 and l2 are assigned.
l1 = 2;
l2 = 2;
p3_3 = [0, 0, 0];
%-------------------------------------------------------------
```

FIGURE L.2 Section of *RRP_FK.m,* with sample values in bold type.

TABLE L.3
Section L.3 MATLAB Files

Filename	Use of File
RRR_Input.csv	To specify joint displacements
RRR_FK.m	To specify ${}^{3}\{\mathbf{p}_3\}$, l_1, and l_2 and write the output (in a file compatible with Microsoft Excel)

```
%-------------------------------------------------------------
% Here, the coordinates for the end-effector point p3_3 and the
% lengths l1 and l2 are assigned.
l1 = 0.5;
l2 = 0.5;
p3_3 = [0, 1, 0];
%-------------------------------------------------------------
```

FIGURE L.3 Section of *RRR_FK.m,* with sample values in bold type.

L.3 R-R-R ROBOTIC MANIPULATOR FORWARD KINEMATICS

The Appendix L.3 folder includes two MATLAB files for R-R-R robotic manipulator forward kinematics. These two files are described in Table L.3. To conduct an R-R-R forward kinematics analysis, the user specifies the R-R-R joint displacements in the file *RRR_Input.csv.* This file is compatible with Microsoft® Excel. Also, ${}^{3}\{\mathbf{p}_3\}$, l_1, and l_2 are specified in the file *RRR_FK.m.* Figure L.3 illustrates the user input section of this file, with sample values in bold type.

After specifying the R-R-R joint displacements in *RRR_Input.csv* and ${}^{3}\{\mathbf{p}_3\}$, l_1, and l_2 in *RRR_FK.m,* the next step is to run *RRR_FK.m.* When running this file, one file (filename *RRR_p3.csv*) is written in a folder named *Results* (in a format compatible with Microsoft® Excel) that includes the ${}^{0}\{\mathbf{p}_3\}$ values calculated for the given ${}^{3}\{\mathbf{p}_3\}$, length, and joint displacement values.

L.4 R-R-C ROBOTIC MANIPULATOR FORWARD KINEMATICS

The Appendix L.4 folder includes two MATLAB files for R-R-C robotic manipulator forward kinematics. These two files are described in Table L.4. To conduct an R-R-C forward kinematics analysis, the user specifies the R-R-C joint displacements in the file *RRC_Input.csv*. This file is compatible with Microsoft® Excel. Also, ${}^{3}\{\mathbf{p}_3\}$, l_1, l_2, and l_3 are specified in the file *RRC_FK.m*. Figure L.4 illustrates the user input section of this file, with sample values in bold type.

After specifying the R-R-C joint displacements in *RRC_Input.csv* and ${}^{3}\{\mathbf{p}_3\}$, l_1, l_2, and l_3 in *RRC_FK.m*, the next step is to run *RRC_FK.m*. When running this file, one file (filename RRC_*p3.csv*) is written in a folder named *Results* (in a format compatible with Microsoft® Excel) that includes the ${}^{0}\{\mathbf{p}_3\}$ values calculated for the given ${}^{3}\{\mathbf{p}_3\}$, length, and joint displacement values.

L.5 R-P-P ROBOTIC MANIPULATOR INVERSE KINEMATICS

The Appendix L.5 folder includes two MATLAB files for R-P-P robotic manipulator inverse kinematics. These two files are described in Table L.5. To conduct an R-P-P inverse kinematics analysis, the user specifies the global R-P-P end-effector coordinates (${}^{0}\{\mathbf{p}_3\}$) in the file *RPP_Input.csv*. This file is compatible with Microsoft® Excel. Also, ${}^{3}\{\mathbf{p}_3\}$ is specified in the file *RPP_IK.m*. Figure L.5 illustrates the user input section of this file, with sample values in bold type.

TABLE L.4
Section L.4 MATLAB Files

Filename	Use of File
RRC_Input.csv	To specify joint displacements
RRC_FK.m	To specify ${}^{3}\{\mathbf{p}_3\}$, l_1, l_2, and l_3 and write the output (in a file compatible with Microsoft Excel)

```
%-----------------------------------------------------------
% Here, the coordinates for the end-effector point p3_3 and the
% lengths l1, l2 and l3 are assigned.

l1 = 1;
l2 = 0.5;
l3 = 0.5;
p3_3 = [1, 0, 0];
%-----------------------------------------------------------
```

FIGURE L.4 Section of *RRC_FK.m,* with sample values in bold type.

TABLE L.5
Section L.5 MATLAB Files

Filename	Use of File
RPP_Input.csv	To specify global end-effector coordinates
RPP_IK.m	To specify ${}^{3}\{\mathbf{p}_3\}$ and write the output (in a file compatible with Microsoft Excel)

```
%------------------------------------------------------------
% Here, the coordinates for the end-effector point p3_3 are
% assigned.

p3_3 = [0, 0, -1];
%------------------------------------------------------------
```

FIGURE L.5 Section of *RPP_IK.m,* with sample values in bold type.

TABLE L.6
Section L.6 MATLAB Files

Filename	Use of File
RRP_Input.csv	To specify global end-effector coordinates
RRP_IK.m	To specify ${}^{3}\{\mathbf{p}_3\}$, l_1, and l_2 and write the output (in a file compatible with Microsoft Excel)

```
%------------------------------------------------------------
% Here, the coordinates for the end-effector point p3_3 and the
% lengths l1 and l2 are assigned.

l1 = 0.5;
l2 = 0.5;
p3_3 = [0, 0, 0];
%------------------------------------------------------------
```

FIGURE L.6 Section of *RRP_IK.m,* with sample values in bold type.

After specifying the global R-P-P end-effector coordinates in *RPP_Input.csv* and ${}^{3}\{\mathbf{p}_3\}$ in *RPP_IK.m*, the next step is to run *RPP_IK.m*. When running this file, one file (filename *RPP_Joints.csv*) is written in a folder named *Results* (in a format compatible with Microsoft® Excel) that includes the joint displacement values calculated for the given ${}^{0}\{\mathbf{p}_3\}$ and ${}^{3}\{\mathbf{p}_3\}$ values.

L.6 R-R-P ROBOTIC MANIPULATOR INVERSE KINEMATICS

The Appendix L.6 folder includes two MATLAB files for R-R-P robotic manipulator inverse kinematics. These two files are described in Table L.6. To conduct an R-R-P inverse kinematics analysis, the user specifies the global R-R-P end-effector coordinates (${}^{0}\{\mathbf{p}_3\}$) in the file *RRP_Input.csv*. This file is compatible with Microsoft® Excel. Also, ${}^{3}\{\mathbf{p}_3\}$, l_1, and l_2 are specified in the file *RRP_IK.m*. Figure L.6 illustrates the user input section of this file, with sample values in bold type.

After specifying the global R-R-P end-effector coordinates in *RRP_Input.csv* and ${}^{3}\{\mathbf{p}_3\}$, l_1, and l_2 in *RRP_IK.m*, the next step is to run *RRP_IK.m*. When running this file, one file (filename *RRP_Joints.csv*) is written in a folder named *Results* (in a format compatible with Microsoft® Excel) that includes the joint displacement values calculated for the given ${}^{0}\{\mathbf{p}_3\}$, ${}^{3}\{\mathbf{p}_3\}$, l_1, and l_2 values.

L.7 R-R-R ROBOTIC MANIPULATOR INVERSE KINEMATICS

The Appendix L.7 folder includes two MATLAB files for R-R-R robotic manipulator inverse kinematics. These two files are described in Table L.7. To conduct an R-R-R inverse kinematics analysis, the user specifies the global R-R-R end-effector coordinates (${}^{0}\{\mathbf{p}_3\}$) in the file *RRR_Input.csv*.

TABLE L.7
Section L.7 MATLAB Files

Filename	Use of File
RRR_Input.csv	To specify global end-effector coordinates
RRR_IK.m	To specify ${}^{3}\{\mathbf{p}_3\}$, l_1, and l_2 and write the output (in a file compatible with Microsoft Excel)

```
%-------------------------------------------------------------
% Here, the coordinates for the end-effector point p3_3 and the
% lengths l1 and l2 are assigned.

l1 = 0.5;
l2 = 0.5;
p3_3 = [0, 1, 0];
%-------------------------------------------------------------
```

FIGURE L.7 Section of *RRR_IK.m,* with sample values in bold type.

TABLE L.8
Section L.8 MATLAB Files

Filename	Use of File
RRC_Input.csv	To specify global end-effector coordinates
RRC_IK.m	To specify ${}^{3}\{\mathbf{p}_3\}$, l_1, l_2, and l_3 and write the output (in a file compatible with Microsoft Excel)

```
%-------------------------------------------------------------
% Here, the coordinates for the end-effector point p3_3 and the
% lengths l1, l2 and l3 are assigned.

l1 = 1;
l2 = 0.5;
l3 = 0.5;
p3_3 = [0, 1, -1];
%-------------------------------------------------------------
```

FIGURE L.8 Section of *RRC_IK.m,* with sample values in bold type.

This file is compatible with Microsoft® Excel. Also, ${}^{3}\{\mathbf{p}_3\}$, l_1, and l_2 are specified in the file *RRR_IK.m*. Figure L.7 illustrates the user input section of this file, with sample values in bold type.

After specifying the global R-R-R end-effector coordinates in *RRR_Input.csv* and ${}^{3}\{\mathbf{p}_3\}$, l_1, and l_2 in *RRR_IK.m*, the next step is to run *RRR_IK.m*. When running this file, one file (filename *RRR_Joints.csv*) is written in a folder named *Results* (in a format compatible with Microsoft® Excel) that includes the joint displacement values calculated for the given ${}^{0}\{\mathbf{p}_3\}$, ${}^{3}\{\mathbf{p}_3\}$, l_1, and l_2 values.

L.8 R-R-C ROBOTIC MANIPULATOR INVERSE KINEMATICS

The Appendix L.8 folder includes two MATLAB files for R-R-C robotic manipulator inverse kinematics. These two files are described in Table L.8. To conduct an R-R-C inverse kinematics analysis, the user specifies the global R-R-C end-effector coordinates (${}^{0}\{\mathbf{p}_3\}$) in the file *RRC_Input.csv*. This file is compatible with Microsoft® Excel. Also, ${}^{3}\{\mathbf{p}_3\}$, l_1, l_2, and l_3 are specified in the file *RRC_IK.m*. Figure L.8 illustrates the user input section of this file, with sample values in bold type.

After specifying the global R-R-C end-effector coordinates in *RRC_Input.csv* and ${}^{3}\{\mathbf{p}_3\}$, l_1, l_2, and l_3 in *RRC_IK.m*, the next step is to run *RRC_IK.m*. When running this file, one file (filename *RRC_Joints.csv*) is written in a folder named *Results* (in a format compatible with Microsoft® Excel) that includes the joint displacement values calculated for the given ${}^{0}\{\mathbf{p}_3\}$, ${}^{3}\{\mathbf{p}_3\}$, l_1, l_2, and l_3 values.

NOTE

1 All of the Appendix L MATLAB files can be downloaded from the publisher's web page or from the authors' server http://www.softalink.com/kruss/mech_des_vis_prog_app_2nd_ed/MATLAB_SIMMULT_FILES.zip.

Appendix M

User Instructions for Chapter 10 MATLAB® and Simscape Multibody™ Kinematic Analysis Files

M.1 R-P-P ROBOTIC MANIPULATOR FORWARD KINEMATICS

The Appendix M.1 folder (which is available for download at https://www.routledge.com/9781032956800) includes four MATLAB and Simscape Multibody files for the forward kinematic analysis and simulation of R-P-P robotic manipulators. These files are described in Table M.1.[1] To conduct an R-P-P forward kinematics analysis, the user specifies the R-P-P joint displacements in the file *RPP_Input.csv*. This file is compatible with Microsoft® Excel. Also, ${}^{3}\{\mathbf{p}_3\}$ is specified in the file *RPP_FK_Simulate.m*. Figure M.1 illustrates the user input section of this file, with sample values in bold type.

After specifying the R-P-P joint displacements in *RPP_Input.csv* and ${}^{3}\{\mathbf{p}_3\}$ in *RPP_FK_Simulate.m*, the next step is to run *RPP_FK_Simulate.m*. When running this file, one file (filename *RPP_p3.csv*) is written in a folder named *Results* (in a format compatible with Microsoft® Excel) that includes the ${}^{0}\{\mathbf{p}_3\}$ values calculated for the given ${}^{3}\{\mathbf{p}_3\}$ value and joint displacement values.

When running *RPP_FK_Simulate.m* a graphical user interface appears where the motion of the R-P-P robotic manipulator is simulated over the joint displacements (see Appendix A.3). Also, the folder *slprq* is written. This folder includes functions that preclude recompiling the model in future simulations. This optional folder can be deleted however (e.g., if computer storage space is inadequate).

TABLE M.1
Appendix M.1 MATLAB and Simscape Multibody Files

Filename	Use of File
RPP_Input.csv	To specify joint displacements
RPP_FK_Simulate.m	To specify ${}^{3}\{\mathbf{p}_3\}$
RPP_FK_Model.slx	To calculate the manipulator output and simulate manipulator motion
Post_Simulation_Task.m	To write the manipulator output (in a file compatible with Microsoft Excel)

```
%-----------------------------------------------------------
% Here, the coordinates for the end-effector point p3_3 are
% assigned.

p3_3 = [0, 0, -1];
%-----------------------------------------------------------
```

FIGURE M.1 Section of *RPP_FK_Simulate.m*, with sample values in bold type.

TABLE M.2
Appendix M.2 MATLAB and Simscape Multibody Files

Filename	Use of File
RRP_Input.csv	To specify joint displacements
RRP_FK_Simulate.m	To specify ${}^{3}\{\mathbf{p}_3\}$, l_1, and l_2
RRP_FK_Model.slx	To calculate the manipulator output and simulate manipulator motion
Post_Simulation_Task.m	To write the manipulator output (in a file compatible with Microsoft Excel)

```
%-----------------------------------------------------------
% Here, the coordinates for the end-effector point p3_3 and the
% lengths l1 and l2 are assigned.

l1 = 0.5;
l2 = 0.5;
p3_3 = [0, 0, 0];
%-----------------------------------------------------------
```

FIGURE M.2 Section of *RRP_FK_Simulate.m*, with sample values in bold type.

M.2 R-R-P ROBOTIC MANIPULATOR FORWARD KINEMATICS

The Appendix M.2 folder includes four MATLAB and Simscape Multibody files for the forward kinematic analysis and simulation of R-R-P robotic manipulators. These files are described in Table M.2. To conduct an R-R-P forward kinematics analysis, the user specifies the R-R-P joint displacements in the file *RRP_Input.csv*. This file is compatible with Microsoft® Excel. Also, ${}^{3}\{\mathbf{p}_3\}$, l_1 , and l_2 are specified in the file *RRP_FK_Simulate.m*. Figure M.2 illustrates the user input section of this file, with sample values in bold type.

After specifying the R-R-P joint displacements in *RRP_Input.csv* and ${}^{3}\{\mathbf{p}_3\}$, l_1, and l_2 in *RRP_FK_Simulate.m*, the next step is to run *RRP_FK_Simulate.m*. When running this file, one file (filename *RRP_p3.csv*) is written in a folder named *Results* (in a format compatible with Microsoft® Excel) that includes the ${}^{0}\{\mathbf{p}_3\}$ values calculated for the given ${}^{3}\{\mathbf{p}_3\}$, length, and joint displacement values.

When running *RRP_FK_Simulate.m*, a graphical user interface appears where the motion of the R-R-P robotic manipulator is simulated over the joint displacements (see Appendix A.3). Also, the folder *slprq* is written. This folder includes functions that preclude recompiling the model in future simulations. This optional folder can be deleted however (e.g., if computer storage space is inadequate).

M.3 R-R-R ROBOTIC MANIPULATOR FORWARD KINEMATICS

The Appendix M.3 folder includes four MATLAB and Simscape Multibody files for R-R-R robotic manipulator forward kinematics. These files are described in Table M.3. To conduct an R-R-R forward kinematics analysis, the user specifies the R-R-R joint displacements in the file *RRR_Input.csv*. This file is compatible with Microsoft® Excel. Also, ${}^{3}\{\mathbf{p}_3\}$, l_1, and l_2 are specified in the file *RRR_FK_Simulate.m*. Figure M.3 illustrates the user input section of this file, with sample values in bold type.

After specifying the R-R-R joint displacements in *RRR_Input.csv* and ${}^{3}\{\mathbf{p}_3\}$, l_1, and l_2 in *RRR_FK_Simulate.m*, the next step is to run *RRR_FK_Simulate.m*. When running this file, one file (filename *RRR_p3.csv*) is written in a folder named *Results* (in a format compatible with

TABLE M.3
Appendix M.3 MATLAB and Simscape Multibody Files

Filename	Use of File
RRR_Input.csv	To specify joint displacements
RRR_FK_Simulate.m	To specify ${}^{3}\{\mathbf{p}_3\}$, l_1, and l_2
RRR_FK_Model.slx	To calculate the manipulator output and simulate manipulator motion
Post_Simulation_Task.m	To write the manipulator output (in a file compatible with Microsoft Excel)

```
%------------------------------------------------------------
% Here, the coordinates for the end-effector point p3_3 and the
% lengths l1 and l2 are assigned.

l1 = 0.5;
l2 = 0.5;
p3_3 = [0, 1, 0];
%------------------------------------------------------------
```

FIGURE M.3 Section of *RRR_FK_Simulate.m*, with sample values in bold type.

TABLE M.4
Appendix M.4 MATLAB and Simscape Multibody Files

Filename	Use of File
RRC_Input.csv	To specify joint displacements
RRC_FK_Simulate.m	To specify ${}^{3}\{\mathbf{p}_3\}$, l_1, l_2, and l_3
RRC_FK_Model.slx	To calculate the manipulator output and simulate manipulator motion
Post_Simulation_Task.m	To write the manipulator output (in a file compatible with Microsoft Excel)

Microsoft® Excel) that includes the ${}^{0}\{\mathbf{p}_3\}$ values calculated for the given ${}^{3}\{\mathbf{p}_3\}$, length, and joint displacement values.

When running *RRR_FK_Simulate.m*, a graphical user interface appears where the motion of the R-R-R robotic manipulator is simulated over the joint displacements (see Appendix A.3). Also, the folder *slprq* is written. This folder includes functions that preclude recompiling the model in future simulations. This optional folder can be deleted however (e.g., if computer storage space is inadequate).

M.4 R-R-C ROBOTIC MANIPULATOR FORWARD KINEMATICS

The Appendix M.4 folder includes four MATLAB and Simscape Multibody files for R-R-C robotic manipulator forward kinematics. These files are described in Table M.4. To conduct an R-R-C forward kinematics analysis, the user specifies the R-R-C joint displacements in the file *RRC_Input.csv*. This file is compatible with Microsoft® Excel. Also, ${}^{3}\{\mathbf{p}_3\}$, l_1, l_2, and l_3 are specified in the file *RRC_FK_Simulate.m*. Figure M.4 illustrates the user input section of this file, with sample values in bold type.

After specifying the R-R-C joint displacements in *RRC_Iinput.csv* and ${}^{3}\{\mathbf{p}_3\}$, l_1, l_2, and l_3 in *RRC_FK_Simulate.m*, the next step is to run *RRC_FK_Simulate.m*. When running this file, one file (filename *RRC_p3.csv*) is written in a folder named *Results* (in a format compatible zwith

```
%-------------------------------------------------------------
% Here, the coordinates for the end-effector point p3_3 and the
% lengths l1, l2 and l3 are assigned.

l1 = 1;
l2 = 0.5;
l3 = 0.5;
p3_3 = [1, 0, -3];
%-------------------------------------------------------------
```

FIGURE M.4 Section of *RRC_FK_Simulate.m*, with sample values in bold type.

Microsoft® Excel) that includes the ${}^{0}\{\mathbf{p}_3\}$ values calculated for the given ${}^{3}\{\mathbf{p}_3\}$, length, and joint displacement values.

Also, when running *RRC_FK_Simulate.m*, a graphical user interface appears where the motion of the R-R-C robotic manipulator is simulated over the joint displacements (see Appendix A.3). Also, the folder *slprq* is written. This folder includes functions that preclude recompiling the model in future simulations. This optional folder can be deleted however (e.g., if computer storage space is inadequate).

NOTE

1 All of the Appendix M Simscape Multibody files can be downloaded from the publisher's web page or from the authors' server http://www.softalink.com/kruss/mech_des_vis_prog_app_2nd_ed/MATLAB_SIMMULT_FILES.zip.

Index

E

F

G

H

I

J

K

L

M

N

O

P

Q

For Product Safety Concerns and Information please contact our EU representative GPSR@taylorandfrancis.com Taylor & Francis Verlag GmbH, Kaufingerstraße 24, 80331 München, Germany